Angiogenesis

Key Principles – Science – Technology – Medicine

Edited by Rudolf Steiner
Paul B. Weisz
Robert Langer

Birkhäuser Verlag
Basel · Boston · Berlin

Editor's addresses:

Dr. R. Steiner
Research Division
Dept. of Surgery
University Hospital Zurich
8091 Zurich
Switzerland

Dr. P. B. Weisz
University of Pennsylvania
Dept. of Chemical and Bioengineering
Room 312, Towne Building/D3
Philadelphia, PA 19104
USA

Dr. R. Langer
Dept. of Chemical Engineering and
Harvard – MIT
Division of Health, Science and Technology
Massachusetts Institute of Technology
Building E 25 Room 342
Cambridge, MA 02139
USA

Library of Congress Cataloging-in-Publication Data

Angiogenesis: key principles – science, technology, medicine/edited by R. Steiner, P. B. Weisz, R. Langer.
(Experientia. Supplementum; v. 61)
Includes bibliographical references and index.
ISBN 3-7643-2674-3—ISBN 0-8176-2674-3
1. Neovascularization. I. Steiner, R. (Rudolf), 1945– . II. Weisz, Paul B., 1921– . III. Langer, Robert S. IV. Series.
QP106.6.A53 1992
612.1′3—dc20

Deutsche Bibliothek Cataloging-in-Publication Data

Angiogenesis: key principles – science – technology – medicine/ed. by R. Steiner . . . – Basel; Boston; Berlin: Birkhäuser, 1992
(Experientia: Supplementum; Vol. 61)
ISBN 3-7643-2674-3 (Basel . . .)
ISBN 0-8176-2674-3 (Boston)
NE: Steiner, Rudolf [Hrsg.]; Experientia/Supplementum

Printed in Germany on acid-free paper

ISBN 3-7643-2674-3
ISBN 0-8176-2674-3

To
Judah Folkman
for his many services
to the experimental physiology and pathology
of angiogenesis

To
Paul Calabresi
in admiration
of his interdisciplinary spirit

Editorial Note

The book on Angiogenesis is a witness to history of the emergence of new scientific insight of a universally important basic phenomenon, and the concept of controlling it with respect to health and disease. It is an initial textbook of current and pertinent knowledge for readers of nearly all scientific branches providing material background for education and for research, much of which could easily be considered as obligatory reading in the study of medicine, molecular biology, chemistry and engineering. Thus the context of this book does not fit into the narrow niche of a single compartment. It shouldn't. On the contrary it is relevant to many.

The most challenging editorial concern was to effectively report major developments at the forefront of angiogenesis research and to reflect with short introductories the complex and inextricable interdependence of key principles in science, technology and medicine.

Rudolf Steiner
Paul B. Weisz
Robert Langer

Contents

Molecular Biology

Chemical Engineering

Cell Biology and Dynamics

Biomedical Technology

Pharmacology and Medicine

Notes

Angiogenesis: Key Principles - Science - Technology - Medicine
ed. by R. Steiner, P. B. Weisz & R. Langer

Angiogenesis – Historical perspective

R. Steiner

Research Division, Department of Surgery, University Hospital, CH-8091 Zürich, Switzerland

Introduction

The historical continuity of ideas in a relatively limited field of thought is nowhere more strikingly illustrated than in the development of knowledge on blood capillaries and their network formation in the living body [1].

With no means of seeing actually how the blood passes from the arteries into veins, William Harvey (1578–1657) in 1616 had inferred the existence of communicating 'pores', traditionally known as synanastomoses. René Descartes (1596–1650), in 'Le Discours de la Méthode', 1637, where he refers to Harvey's discovery, writes: "petits passages aux extrémités des artères par où le sang qu'elles reçoivent du coeur entre les petites branches des veines" [2].

Marcello Malpighi (1628–1694) discovered that the connections were in fact formed by microscopic tubelets, the capillaries. He first told of his discovery to Giovanni Borelli (1608–1679), the physiologist at Pisa in 1661, describing them in the frog's lung, and prefacing his letter with the quotation 'magnum certum opus oculis video' [3]. In 1672, in a letter to the Royal Society, London, he described the capillaries in the chorioallantoic membrane [CAM] of the chick embryo [4].

Antony van Leeuwenhoek (1632–1723) added knowledge of capillary structure in tissues other than the lungs [5]. Stephen Hales (1677–1761) observed that 'the small capillary vessels' may become constricted or dilated when exposed to certain chemical substances [6] long before the memorable lectures of August Krogh (1874–1949) were held at Yale University in autumn 1922 [7]. Krogh extended and confirmed the findings of Charles-Marie-Benjamin Rouget (1824–1904) who in 1873/79 communicated to the French Academy his discovery of a network of ramified, contractile cells attached to the capillaries of the dorsal fin of tadpoles – now called Rouget cells or pericytes [8].

In 1937 Benjamin William Zweifach (*1910) clarified the controversy over the contractibility of capillaries when he demonstrated the existence of a-v bridges, which were contractile, as distinct from the capillaries which were not [9].

Pericytes are now thought to be involved along with endothelial cells in the process of neovascularisation or angiogenesis – a term coined in 1935 by Arthur Tremain Hertig to describe the formation of blood vessels in the placenta [10].

The phenomenon of angiogenesis has long been familiar to many surgeons and pathologists dealing with inflammation, trauma and tumours. John Hunter (1728–1793) described it posthumously in 1794 in his 'treatise on the blood, inflammation and gun shot wounds', as Olga Hudlicka pointed out in her historical review on angiogenesis in 1984 [11]. In the 19th century, Johannes Müller (1801–1898), the founder of the great German school of physiology, deeply interested in diverse biological problems, inspired many of his pupils to investigate blood vessel formation in easily accessible tissues under normal and pathological conditions [12].

The 'how' of angiogenesis was soon explored by morphologists and physiologists who in the late 19th and in the first quarter of the 20th century started to describe this basic phenomenon in great detail. The study of tumour neovascularisation was initiated by E. Goldman with a series of papers in 1907. He summarised his findings as follows: "The normal blood vessels of the organs ⟨in which the tumour is developing⟩ are disturbed by chaotic growth – there is dilatation and spiralling of the affected vessels, marked capillary budding and new vessel formation, particularly at the advancing border" [13].

The development of the transparent ear chamber technique by E. R. Clark in the rabbit paved the way for a series of excellent radiographic studies on the developmental anatomy of tumor blood vessels which were made possible with improved contrast media [14].

Major advances to our knowledge in angiogenesis were made in 1939 by A. G. Ide who was first to suggest that there is a production of growth-stimulating factors by tumours [15]. For elegance of method and detail of observation these experimens are rivalled only by those of G. H. Algire's group in the years 1945 to 1956. These investigations revealed a sequential picture of the development of capillary neovascularisation [16].

The 'why' of angiogenesis in physiologic or pathophysiologic processes such as tumor-induced angiogenesis and wound healing only started to be discovered in the early seventies when J. Folkman entered the field [17]. It has taken almost twenty years and several well-characterised angiogenesis stimulating and inhibiting factors to convince the scientific community to take an active part in angiogenesis research [18].

C. S. Sherrington (1857–1952), studying the scientific background which eventually led to Harvey's discovery, reminds us that "to ask something which the time is not yet ripe to answer is of small avail. There must be means for reply and enough collateral knowledge to make the answer worthwhile" [19]. To our mind this has been achieved in the field of angiogenesis research in the last few years. However, the 'final common pathway' to influence angiogenesis is still hidden in a maze of new findings. For the time being we may find consolation in Jean Fernel's (1497–1558) view on scientific endeavour that "to journey is better than to arrive."

References

1 Fulton JF: Blood Capillaries. In: Selected Readings in the History of Physiology. Ed. by J. F. Fulton and L. G. Wilson, 2nd edn, Charles C Thomas, Springfield, 1966, pp. 66–111.

2 Sherrington CS: Obs. 14: ⟨On capillary blood-flow⟩. In: Mammalian Physiology. 2nd edn, Clarendon Press, London, 1929; 21.

3 Malpighi M: De Pulmonibus. Epist ii; Opera, tom ii, London, 1686.

4 idem. De Ovo Incubato. Opera, tom i, London, 1686.

5 Meyer AW: Leeuwenhoek as experimental biologist. Osiris 1938; 3: 103–122 (Leeuwenhoek's letter on capillary circulation is dated Sept. 7, 1688 and the original manuscript is at the Royal Society London).

6 Hales St: Statical essays: containing haemastaticks; or, an account of some hydraulick and hydrostatical experiments made on the blood and blood-vessels of animals. London, 1733, vol. 2, xxii, 361 pp.

7 Krogh SAS: The anatomy and physiology of capillaries. Silliman Memorial Lectures, New Haven, Yale University Press, 1922, xvii, 276 pp.

8 Rouget Ch-M-B: Mémoire sur le dévélopment de la tunique contractile des vaisseaux. C. R. Acad. Sci. Paris. 1873; 79: 559–562. idem. Sur la contractibilité des capillaires sanguins. C. R. Acad. Paris. 1879; 88: 916–918.

9 Zweifach BW: The structure and reactions of the small blood vessels in amphibia. Am. J. Anat. 1937; 60: 473–514; 478–479; 501–502.

10 Hertig AT: Angiogenesis in the early human chorion and in the primary placenta of the macaque monkey. Contrib. Embryol. 1935; 25(146): 39–81.

11 Hudlicka O: Growth of vessels – historical review. In: Prog. appl. Microcirc. vol. 4, Karger, Basel 1984, pp. 1–8.

12 Briefe von Theodor Billroth, 8. veränd. Aufl., Hannover/Leipzig, Hahnsche Buchhandlung, 1910, pp. 17–18.

13 Goldman E: The growth of malignant disease in man and the lower animals with special reference to the vascular system. Lancet 1907; 2: 1236.

14 Clark ER, Hitschler WJ, Kirby-Smith HT, Rex RO, Smith JH: General observations on the ingrowth of new blood vessels into standardized chambers in the rabbit ear, and the subsequent changes in the newly grown vessels over a period of months. Anat. Rec. 1931; 50: 129.

15 Ide AG, Baker NH, Warren SL: Vascularization of the Brown-Pierce rabbit epithelioma transplant as seen in the transplant ear chambers. Am. J. Roentgen. Radiother. 1939; 42: 891.

16 Algire GH, Chalkley HW, Earle WR: Vascular reactions of normal and malignant tissue in vivo. I. Vascular reactions of mice to wound and to normal and neoplastic transplants. J. Nat. Cancer Inst. 1945; 6: 73–85; and Merwin RW, Algire GH: The role of graft and host vessels in the vascularization of grafts of normal and neoplastic tissues. J. Nat. Cancer Inst. 1956; 17: 23–33.

17 Folkman J. Tumor angiogenesis: therapeutic implications. N. Engl. J. Med. 1971; 285: 1182–86.

18 International Symposium on Angiogenesis. Key Principles – Science-Technology-Medicine, St. Gallen, Switzerland, March 13–15, 1991. Abstract Book [printed privately]. Distr. by Atelier H. P. Kaeser, CH-9000 St. Gallen, Switzerland.

19 Sherrington, C.S. The Endeavour of Jean Fernel. Cambridge University Press, 1946, 223 pp.

Angiogenesis: Key Principles – Science – Technology – Medicine
ed. by R. Steiner, P. B. Weisz & R. Langer

Angiogenesis – Retrospect and outlook

Judah Folkman

Children's Hospital and Harvard Medical School, Boston, Massachusetts 02115, USA

Introduction

The process of angiogenesis, the growth of capillary blood vessels, is fundamental to reproduction, development and repair. Under these conditions, angiogenesis is highly regulated and of short duration. In many pathological states, this regulation is deranged so that the disease itself is driven by persistent, unabated neovascularization. Thus, tumor growth and metastasis are angiogenesis-dependent and a wide variety of non-neoplastic diseases are dominated by uncontrolled angiogenesis. These "angiogenic diseases" include arthritis, psoriasis, hemangiomas, and many types of ocular neovascularization, to name but a few.

The field of angiogenesis research which began almost three decades ago has now reached the stage of clinical trials. Therefore, it seems timely to re-visit its origins and to ask if there are central questions which may guide its future. A review of contributions by the many different investigators working in this field would be inappropriate here. Several up-to-date reviews are available [1–6].

It may be useful instead, to examine a few of the assumptions underlying earlier experiments. Progress was sometimes hindered by these assumptions. Comparing them to currently held views may suggest which of today's ideas about angiogenesis are likely to lead to fruitful experiments.

Early experiments and pitfalls

Angiogenesis research began in the early 1960s as an inquiry into the mechanism of tumor angiogenesis [7], when tumor cells implanted into isolated perfused organs failed to grow tumors beyond a diameter of 1–2 millimeters and also failed to become vascularized. We eventually realized the significance of this observation and proposed the hypothesis that tumor growth was dependent on tumor-induced neovascularization [8]. But the idea was not well accepted because of the prevailing assumption at that time that tumor hypervascularity was due to *dilated* host vessels and not to *new* vessels. To counter these arguments we showed that endothelial cells in tumor vessels incorporated H^3-thymidine [9]. However, before the advent of specific markers for endothelium, this early evidence of DNA synthesis in tumor blood vessels had to be supported by electron microscopic autoradiography to prove that the H^3-thymidine incorporation had been incorporated into vascular endothelium.

It was not until tumors were implanted in the rabbit cornea [10], and new vessels could be observed to grow across an avascular cornea, that it was finally accepted that tumors could induce the proliferation of *new* capillary blood vessels. However, critics then argued that perhaps any rapidly dividing cell could induce neovascularization. This assumption was put to rest when it was shown that irradiated, non-proliferating tumor cells in the cornea were just as angiogenic as proliferating cells [11].

The cornea experiments were instructive because intralamellar tumors continued to grow slowly in 2 dimensions *before* vascularization, but grew exponentially in 3 dimensions *after* vascularization. Only when 3-dimensional tumor spheroids were grown in the anterior chamber of the eye [12], did it become clear that even in vivo, tumor growth would be arrested at small volumes of less than 1 mm^3 in the absence of neovascularization (i.e., in the pre-vascular phase). This model was a closer approximation to human and animal cancer than the isolated perfused organ.

Purification of angiogenic factors

The cornea experiments also implied that angiogenesis was mediated by some soluble substance(s). Compelling evidence was provided only after it became possible to induce neovascularization of the cornea by an implanted sustained-release polymer pellet containing a tumor extract [13]. This led us to the arduous task of trying to purify an angiogenic factor from tumor cells [14].

In the early 1970s, purification of an angiogenesis factor was fraught with difficulties. Bioassays had to be developed pari passu with purification. Endothelial cells had never been grown in vitro before 1972, when the first human umbilical vein endothelial cells were cultured [15, 16]. The assumption that endothelial cells, like 3T3 fibroblasts, could be stimulated to proliferate from confluence, hampered purification. It is now clear that only sparse endothelial cells can proliferate [17]. Another assumption, i.e., that endothelial cells from large and small vessels were essentially the same, delayed the discovery that capillary endothelial cells were growth factor-dependent (whereas aortic endothelial cells were not) [18]. When endothelial proliferation was found to be a critical event in capillary formation [19], the idea arose that angiogenic factors should be endothelial mitogens. This idea was further strengthened because the first successful purification of an angiogenic factor, (bFGF), was guided by endothelial DNA synthesis in vitro [20]. However, the next angiogenic factor to be discovered, angiogenin [21], was not an endothelial mitogen, and we now know that not all angiogenic factors are endothelial mitogens [22]. The heparin-affinity chromatography employed to purify bFGF [20, 23], was at first thought to be only a technical maneuver for high resolution separation of proteins. However, it soon became clear that the heparin-affinity of FGFs had important biological ramifications [24, 25]. Thus, it was found that bFGF is stored in extracellular matrix bound to heparan sulfate [26]; that heparin protects bFGF from degradation [27]; that heparin potentiates the mitogenicity of aFGF [28]; and that heparin-like receptors on the cell surface prepare bFGF for binding to its high affinity receptor [29].

Regulation of angiogenesis

By the time pure angiogenic molecules became available, almost two decades of angiogenesis research had been conducted using crude tumor extracts. When pure molecules were finally produced in large quantity by recombinant technology, it was generally assumed that structural studies would immediately reveal how angiogenesis is activated and how it is controlled. But it was soon realized that angiogenesis is a complex cascade and that an understanding of its regulation would require a detailed knowledge of: (i) how an angiogenic molecule such as bFGF is exported out of cells; (ii) how bFGF is mobilized from extracellular matrix; (iii) how extracellular matrix itself governs the response of endothelial cells to growth factors [30, 31]; (iv) how non-endothelial cells, such as pericytes, mast cells, macrophages and keratinocytes may modify angiogenesis; (v) how natural inhibitors of endothelial proliferation normally maintain vascular endothlium in a quiescent state; (vi) how proteases such as plasminogen activator affect lumen size and branching [30]; and (vii) how other systems, such as the endocrine system, may regulate angiogenesis, e.g., in the ovarian follicle [32].

Inhibition of angiogenesis

The search for angiogenesis inhibitors was stimulated by the concept of "antiangiogenic therapy", which in its simplest terms stated that a putative inhibitor of blood vessel growth might be therapeutic because it could limit tumor growth, and non-toxic because angiogenesis is normally infrequent [8]. The first angiogenesis inhibitors to be identified were naturally occurring, or "endogenous" molecules, such as angiostatic steroids and platelet factor 4 [33]. It soon became clear that the efficacy of these inhibitors when employed at pharmacologic doses, depended upon their route of administration and upon the type of angiogenesis to which they were targeted. For example, angiostatic steroids are most effective when applied topically for corneal neovascularization [34], while systemic therapy is only effective against some tumors [35, 36]. In contrast, systemically administered alpha-interferon very effectively brings about regression of life-threatening hemangiomas in babies [37, 38], but it is ineffective most tumors. Protamine is toxic for reasons other than its biologic effect on endothelial cells [36]. An angiogenesis inhibitor derived from a fungus is a potent inhibitor of angiogenesis and of tumor growth when administered systemically [39].

Targets of angiogenesis inhibitors

Because the mechanisms by which tumors induce angiogenesis depend upon one or more pathways, an angiogenesis inhibitor could be directed against any of the following components of the angiogenic process (summarized in Fig. 1).

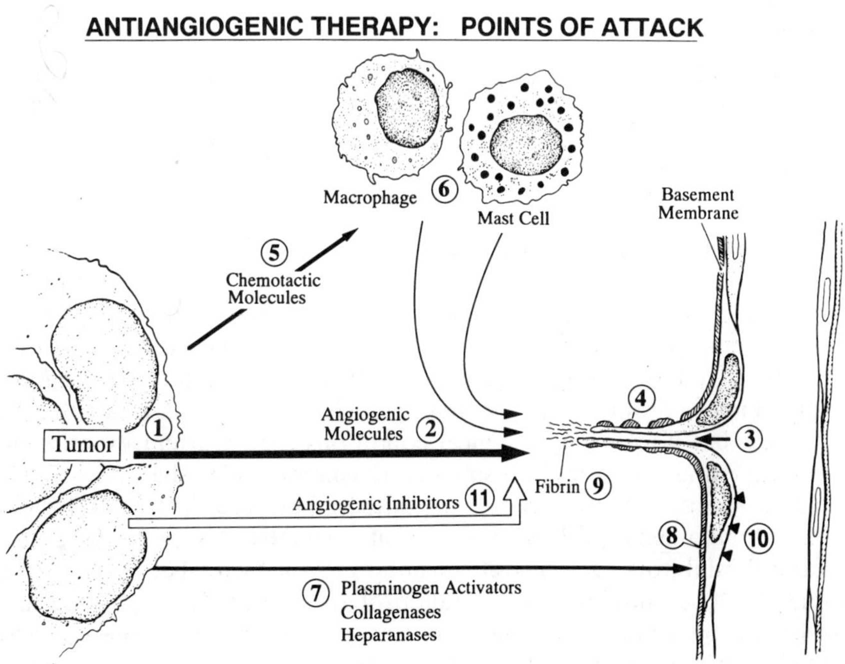

Figure 1. Diagram of different targets for antiangiogenic therapy.

1. Pathways for the export of angiogenic molecules from tumor cells.
2. Angiogenic molecules in transit.
3. Endothelial cell migration and/or proliferation.
4. Basement membrane production and degradation by endothelial cells of newly formed capillaries.
5. Tumor-derived factors that are chemotactic for macrophages or mast cells.
6. Angiogenic factors or angiogenic amplifiers secreted by macrophages and/or mast cells.
7. Tumor-derived (or endothelial-derived) plasminogen activators, collagenases, and heparanases [40–42].
8. bFGF stored in basement membrane.
9. Hyperpermeability of the neovascular bed (which leads to leakage of fibrin products into the extravascular space, among other effects) [43].
10. Endothelial markers by which to target tumor vasculature [44].
11. Endogenous inhibitors of endothelial cell growth secreted by the tumor [45], or normally present in the circulation, or in adjacent tissues. (Augmentation of such endogenous inhibitors would be the antiangiogenic approach here).

To develop antiangiogenic therapy for clinical trial, it is useful to categorize these components under two central mechanisms of tumor angiogenesis:

I. *Switch to the angiogenic phenotype during tumorigenesis* (1, 2, 5, 6, 7, 8, 11);
II. *Vascular endothelial cell response to stimulation by angiogenic factors* (3, 4, 9, 10).

Switch to the angiogenic phenotype

Other endogenous angiogenesis inhibitors, recently discovered (such as thrombospondin), while not yet available in sufficient quantities for testing as pharmacologic agents, have provided insight into the more fundamental problem of how angiogenesis is switched on, not only in tumors, but also in physiological conditions [45]. Current experiments in different laboratories [46] support an emerging idea that normal cells may constitutively produce polypeptides [45] or other factors [47] which inhibit endothelial proliferation and angiogenesis. Also, other endothelial cell inhibitors such as transforming growth factor-beta (TGF-beta) and the tissue inhibitors of metalloproteinases (TIMP-1) [46] are also released by tumor cells. During transformation to a tumor, these inhibitors are down-regulated but synthesis and release are not completely stopped. Thus, some tumor cells may produce both angiogenic factors and inhibitors [48]. It is quite possible, therefore, that a tumor does not become angiogenic until its production of angiogenic molecules exceeds its production of "endogenous" inhibitors.

Amplification of angiogenic activity

The excess production of angiogenic stimulators over inhibitors may also be amplified by other mechanisms which tumors employ to amplify or potentiate their own angiogenic molecules. For example: (i) Certain tumors attract macrophages and activate them to secrete angiogenic factors [49]. (ii) Other tumors release heparanases or collagenases which can mobilize bFGF from its storage sites in extracellular matrix [50]. (iii) Some tumor cells are chemotactic to mast cells [51] which may release heparin. Heparin itself can stabilize FGF and may also potentiate angiogenesis. (iv) The vascular endothelial cells of certain human tumors, such as colon carcinoma and neuroblastoma, contain an overabundance of bFGF compared to the endothelial cells in other tumors [52]. This finding suggests that these tumors release some unknown mediator which stimulates excessive production of bFGF by local endothelial cells. In summary, the switch to the angiogenic state may involve one or more mechanisms. In some tumors, angiogenesis may become an extremely redundant process.

Clinical applications of angiogenesis research

Clinical applications of angiogenesis research are just beginning to enter clinical trial in at least three areas of medicine.

Diagnosis and prognosis. The visualization of blood vessels on histological sections from biopsies can be greatly improved by staining the sections with antibody to Factor VIII-related antigen, a specific marker for endothelial cells [53]. This staining allows quantitation of vessels in a given microscopic field. In patients with invasive breast cancer, increasing vessel number is directly proportional to increasing risk of metastasis. By this simple method, metastatic risk can be predicted with high accuracy [53].

Stimulation of angiogenesis. Animal studies show that healing of chronic wounds can be accelerated by topical application of bFGF, which stimulates angiogenesis [54]. A clinical trial is underway in which bFGF spray is being used to treat patients with chronic wounds and burns (Robson, M. et al., unpublished).

Inhibition of angiogenesis. Life-threatening hemangiomas in babies have been successfully treated to complete disappearance with daily subcutaneous injections of alpha-interferon for up to a year, without apparent toxicity [55, Ezekowitz et al., submitted]. Systemically administered alpha-interferon also caused regression of iris neovascularization in monkey eyes [56]. It was recently reported that alpha-interferon was effective therapy for a small number of patients with retinal neovascularization secondary to macular degeneration [57]. Angiogenesis inhibitors such as "angioinhibin" [39] have not yet been used clinically, but from extensive animal studies it is possible to speculate that such angiogenesis inhibitors may be used in future antiangiogenic therapy of cancer as: (i) adjuvant therapy; (ii) prophylactic therapy to prevent tumor recurrence or (iii) anti-metastatic therapy. If new angiogenesis inhibitors are discovered, therapeutic indications may widen.

Questions for future research

From this brief review it may be apparent that the field of angiogenesis research developed not from a single discovery, but incrementally, from a series of discoveries and hypotheses, each opening an unexpected and usually unpredicted avenue for futher research. Many of these findings overturned previous assumptions.

On the basis of our current understanding of the angiogenic process, several new central questions may now be considered for future investigation: How are signal-less polypeptides such as bFGF exported from cells into the extracellular matrix? Are tumor cells heterogenic for angiogenic activity or for the production of specific angiogenic factors? What percent of tumor cells in a given tumor must switch to the angiogenic phenotype before the tumor can induce neovascularization? What does the term "established tumor vessels" mean at the cellular and biochemical level? Can TGF-beta operate as an angiogenic factor in the extravascular space, but as an endothelial cell inhibitor when it circulates intravascularly [58]? Why do new capillaries regress or involute when an angiogenic stimulus is removed? By what mechanisms does extracellular matrix govern growth of capillary blood vessels? Is the herterogeneity of different vascular beds such that angiogenesis inhibitors may also be selective? Will angiogenesis inhibitors employed as anti-tumor therapy circumvent the problem of acquired drug resistance [59]?

Beyond these questions lie considerations for the more distant future. Can the onset of angiogenic activity be detected in blood or other body fluids for use in diagnosis or prognosis? Can the process of angiogenesis itself be manipulated by genetic therapies, for example by administration of anti-sense DNA? These questions and others provide the basis for continuing optimism about the productive future of angiogenesis research.

References

1 Folkman J: Tumor Angiogenesis. In: Cancer Medicine. Ed. by J. F. Holland, 3rd edn, Lea & Febiger, Melbourne, Pennsylvania, USA, 1991, in press.

2 Folkman J: "What is the evidence that tumors are angiogenesis dependent?" J Natl Canc Inst 1990; 82 (No. 1): 4–6.

3 Hudlicka O, Tyler KR: Angiogenesis – The growth of the vascular system. Academic Press, Harcourt Brace Jovanovich, Publishers, London, New York, 1986, pp. 1–221.

4 Denekamp J: The current status of targeting tumour vasculature as a means of cancer therapy: an overview. Int J Radiat Biol 1991; 60: 401–408.

5 Moore JV, West DC: Vasculature as a target for anti-cancer therapy. In: Proceedings of the 16th L.H. Gray Conference 1991. Taylor & Francis, London, Washington D.C. (pp. 1–421). [A special tissue of the International Journal of Radiation Biology, Volume 60, Numbers 1/2, July/August 1991].

6 Klintworth GK: Corneal angiogenesis – A comprehensive critical review 1990. Springer-Verlag, New York, Berlin, London, pp. 1–135.

7 Folkman J, Long D, Becker F: Growth and metastasis of tumor in organ culture. 1963; 16: 453–467.

8 Folkman J: Tumor angiogenesis: therapeutic implications. New Eng J Med 1971; 285: 1182–1186.

9 Carvallo T, Sade R, Folkman J, Cotran RS: Tumor angiogenesis; rapid induction of endothelial mitosis demonstrated by autoradiography. J Cell Biology 1972; 54: 408–420.

10 Gimbrone MA Jr, Cotran R, Leapman S, Folkman J: Tumor growth neovascularization: An experimental model using rabbit cornea. J Natl Canc Inst 1974; 52: 413–427.

11 Auerbach R, Arensman R, Kubai L, Folkman J: Tumor-induced angiogenesis: lack of inhibition by irradiation. International Journal of Cancer 1975; 15: 241–245.

12 Gimbrone MA Jr, Leapman S, Cotran RS, Folkman J: Tumor dormancy in vivo by prevention of neovascularization. J Exp Med 1972; 136: 261–276.

13 Langer R, Folkman J: Polymers for the sustained release of proteins and other macromolecules. Nature 1976; 263: 797–800.

14 Folkman J, Merler E, Abernathy C, Williams G: Isolation of a tumor factor responsible for angiogenesis. J Exp Med 1971; 133: 275–288.

15 Jaffe EA, Nachman RL, Becker CG, Minick CR: Culture of human endothelial cells derived from umbilical veins: identification by morphologic and immunologic criteria. J Clin Invest 1972; 52: 2745–2756.

16 Gimbrone MA Jr, Cotran RS, Folkman J: Endothelial regeneration and turnover. Studies with human endothelial cell cultures. Ser Haematol 1973; 6: 453–455.

17 Haudenschild CC, Zahniser D, Folkman J, Klagsbrun M: Human endothelial cells in culture. Lack of response to serum growth factors. Exp Cell Res 1976; 98: 175–183.

18 Folkman J, Haudenschild CD, Zetter BR: Long-term culture of capillary endothelial cells. Proc Natl Acad Sci USA 1979; 76: 5217–5221.

19 Ausprunk DH, Folkman J: Migration and proliferation of endothelial cells in preformed and newly formed blood vessels during tumor angiogenesis. Microvasc Res 1977; 14: 53–65.

20 Shing Y, Folkman J, Sullivan R, Butterfield C, Murray J, Klagsbrun M: Heparin-affinity: purification of a tumor-derived capillary endothelial cell growth factor. Science 1984; 223: 1296–1299.

21 Fett JW, Strydom DJ, Lobb RF, Alderman WM, Bethune JL, Riordan JF, Vallee BL: Isolation and characterization of angiogenin, and angiogenic protein from human carcinoma cells. Biochemistry 1985; 24: 5480–5486.

22 Folkman J, Klagsbrun M: Angiogenic factors. Science 1987; 235: 442–447.

23 Maciag T, Mehlman T, Friesel R, Schreiber AB: Heparin binds endothelial cell growth factor, the principal cell mitogen in bovine brain. Science 1984; 225: 932–935.

24 Klagsbrun M: The affinity of fibroblast growth factors (FGFs) for heparin: FGF-heparan sulfate interactions in cells and extracellular matrix. Current Opinion in Cell Biology 1990; 2: 857–863.

25 Folkman J: Control of angiogenesis by heparin and other sulfated polysaccharides. In: Heparin and Related Polysaccharides. Ed. by U. Lindahl, D. Lane, 1991, in press.

26 Vlodavsky I, Folkman J, Sullivan R, Fridman R, Ishai-Michaeli R, Sasse J, Klagsbrun M: Endothelial cell-derived basic fibroblast growth factor: synthesis and deposition into subendothelial extracellular matrix. Proc Natl Acad Sci USA 1987; 84: 2292–2296.

27 Folkman J, Klagsbrun M, Sasse J, Wadzinski M, Ingber D, Vlodavsky I: A heparin-binding angiogenic protein – basic fibroblast growth factor – is stored within basement membrane. Am J Pathol 1988; 130: 393–400.

28 Thornton S, Mueller S, Levine E: Human endothelial cells: Use of heparin in cloning and long-term serial cultivation. Science 1983; 222: 623–625.

29 Yayon A, Klagsbrun M, Esko J, Leder P, Ornitz D: Cell surface heparin-like molecules are required for the binding of basic fibroblast growth factor to its high affinity receptor. Cell 1991; 64: 841–848.

30 Montesano R, Vassali JD, Baird A, Guillemin R, Orci L: Basic fibroblast growth factor induces angiogenesis in vitro. Proc Natl Acad Sci USA 1986; 83(19): 7297–7301.

31 Ingber DE, Folkman J: Mechanochemical switching between growth and differentiation during fibroblast growth factor-stimulated angiogenesis in vitro: role of extracellular matrix. J Cell Biol 1989; 109: 317–330.

32 Motro B, Itin A, Sachs L, Keshet, E: Pattern of interleukin 6 gene expression in vivo suggests a role for this cytokine in angiogenesis. Proc Natl Sci USA 1990; 87(8): 3092–3096.

33 Folkman J: Angiogenesis and its inhibitors. In: Important Advances in Oncology 1985. Ed. by V. T. DeVita Jr, S. Hellman, S. A. Rosenberg, J.B. Lippincott Co, Philadelphia, 1985, pp. 42–62.

34 Li WW, Casey R, Gonzales EM, Folkman J: Angiostatic steroids potentiated by sulfated cyclodextrins inhibit corneal neovascularization. Investigative Ophthalmology & Visual Science 1991; 32: 2898–2905.

35 Sakamoto N, Tanaka NG: Mechanism of synergistic effect of heparin and cortisone against angiogenesis and tumor growth. Cancer Journal 1988; 2(1): 9–13.

36 Folkman J, Ingber DE: Angiostatic steroids, In: Anti-inflammatory Steroid Action – Basic and Clinical Aspects. Ed. by R. P. Schleimer, H. N. Claman, A.L. Oronsky. Academic Press, 1989, pp. 330–350.

37 White CW, Sondheimer HM, Crouch EC, Wilson H, Fan LF: "Treatment of pulmonary hemangiomatosis with recombinant interferon alfa-2a. New Engl J Med 1989; 320: 1197–1200.

38 Orchard P, Smith C, Woods W, Dehner LP, Day DL, Shapiro RS: Treatment of hemangioendotheliomas with alpha-interferon. Lancet 1989; 2(8662): 565–567.

39 Ingber D, Fujita T, Kishimoto S, Sudo K, Kanamaru T, Brem H, Folkman J: Synthetic analogues of fumagillin that inhibit angiogenesis and suppress tumor growth. Nature 1990; 348: 555–557.

40 Nagy JA, Brown LF, Senger DR, Lanir N, Van de Water L, Dvorak AM, Dvorak HF: Pathogenesis of tumor stroma generation: a critical role for leaky blood vessels and fibrin deposition. Biochim Biophys Acta 1989; 948: 305–326.

41 Herron GS, Banda MJ, Clark EJ, Gavrilovic J, Werb Z: Secretion of metalloproteinases by stimulated capillary endothelial cells. II. Expression of collagenase and stromelysin activities is regulated by endogenous inhibitors. J Biol Chem 1986; 261: 2814–2818.

42 Liotta LA, Steeg PS, Stetler-Stevenson WG: Cancer metastasis and angiogenesis: an imbalance of positive and negative regulation. Cell 1991; 64: 327–336.

43 Dvorak HF, Nagy JA, Dvorak AM: Structure of solid tumors and their vasculature – implications for therapy with monoclonal antibodies. Cancer Cells 1991; 3(3): 77–85.

44 Denekamp J: Vasculature as a target of tumour therapy. In: Progress in Applied Microcirculation. Ed. by F. Hammersen, O. Hudlicka. Karger, Basel, 1984, pp. 28–38.

45 Bouck N: Tumor angiogenesis – the role of oncogenes and tumor suppressor genes. Cancer Cells 1990; 2(6): 179–185.

46 Stetler-Stevenson WG, Brown PD, Onisto M, Levy AT, Liotta LA: Tissue inhibitor of metalloproteinases-2 (TIMP-2) mRNA expression in tumor cell lines and human tumor tissues. J Biol Chem 1990; 265: 13933–13938.

47 Moses MA, Sudhalter J, Langer R: Identification of an inhibitor of neovascularization from cartilage. Science 1990; 248: 1408–1410.

48 Kandel J, Bossy-Wetzel E, Radvanyi F, Klagsbrun M, Folkman J, Hanahan D: Neovascularization is associated with a switch to the export of bFGF in the multistep development of fibrosarcoma. Cell 1991; 66: 1095–1104.

49 Polverini PJ, Leibovich SJ: Induction of neovascularization in vivo and endothelial cell proliferation in vitro by tumor-associated macrophages. Lab Invest 1984; 51: 635–642.
50 Vlodavsky I, Korner G, Ishai-Michaeli R, Baskin P, Bar-Shavit R, Fuks Z: Extracellular matrix-resident growth factors and enzymes: possible involvement in tumor metastasis and angiogenesis. Cancer Metastasis Rev 1990; 3: 203–226.
51 Dethlefesen SM, Matsuura N, Zetter BR: Tumor growth and angiogenesis in wild type and mast cell deficient mice. Fed Amer Soc Exp Biol 1990; 4(3): A623 (Abs #2070).
52 Schulze-Osthoff K, Risau W, Vollmer E, Sorg C: In situ detection of basic fibroblast growth factor by highly specific antibodies. Amer J Pathol 1990; 137: 85–92.
53 Weidner N, Semple JP, Welch WR, Folkman J: Tumor angiogenesis and metastasis – correlation in invasive breast carcinoma. New Engl J Med 1991; 324: 1–8.
54 Thomas KA: Fibroblast growth factors. FASEB J 1987; 1: 4343–4340.
55 Folkman J: Successful treatment of an angiogenic disease. New Engl J Med 1989; 320(8): 1211–1212.
56 Stinson WG, Miller JW, Puliafito CA, Folkman J: Alpha-interferon treatment of experimental iris neovascularization. Investigative Ophthalmology and Visual Science 1991; 32(4): 1046 (Abstract #1855).
57 Fung W: Interferon Alpha 2A for treatment of age-related macular degeneration. American Journal of Ophthalmology 1991; 112(3): 349–350.
58 Folkman J, Klagsbrun M: A family of angiogenic polypeptides. Nature 1987; 329: 671–672.
59 Kerbel RS: Inhibition of tumor angiogenesis as a strategy to circumvent acquired resistance to anti-cancer therapeutic agents. BioEssays 1991; 13(1): 31–36.

Angiogenesis: Key Principles - Science - Technology - Medicine
ed. by R. Steiner, P. B. Weisz & R. Langer

Angiogenesis – The interdisciplinary challenge

Paul B. Weisz

Department of Chemical Engineering, University of Pennsylvania, Philadelphia, Pennsylvania 19104, USA

A unique revelation

One day in the 1970's, in a conversation with James Wei, then Professor of Chemical Engineering at the University of Delaware (we were both immersed in fundamentals of chemical systems), my attention was drawn to revelations by Judah Folkman and associates [1]: A speck of tumor tissue implanted in a cornea would be dormant for 7 days, then initiate neovascularization from the limbus. Tumor growth would commence when the capillaries reached the implant. This, of course, was a key demonstration of the link of tumor growth to angiogenesis. Interestingly, the same speck implanted in the anterior chamber at the same distance from the limbus would leave the implant dormant.

We must transport molecules

A simple back-of-the-envelope calculation convinced me that the 7 days was just the time magnitude required by the laws of diffusion for a molecular "factor" to travel the distance from the tumor cells to the edge of the cornea. An examination of the elementary laws of diffusion for the two different physical geometries in Folkman's report showed that the effective concentration triggering endothelial sprouting in the corneal implant may indeed never be reached in the geometry of the anterior chamber [2]. This confronted us with the fact that the physical architecture rather than sophisticated biochemistry or molecular biology, can provide a reason for dormancy vs. malignancy!

Indeed, mass transport phenomena are an integral part of modulating or controlling *any* system of molecular processes. They have long dictated design parameters for effective chemical catalyst particles, processes and plants. They are even more dominating in molecular systems in the liquid phase: Biochemical systems such as cell, organ or body [3–5]. Diffusion allows or disallows success [3] over relatively small distances; convection may or may not aid [4]. The hydrodynamics of pumped flow (the cardio-vascular system must be relied upon over longer distances). These principles, in fact, are the basic ingredients to Folkman's key observation and proposition that the growth of a tumor mass beyond a certain small size is not possible without angiogenesis.

The dynamics of rate processes of chemical transformations and the role of mass transport parameters have largely been examined by the scientist in chemical engineering and catalytic chemistry. They are relevant and important to the life sciences.

Keeping our equilibria

The basic elements of physical chemical behavior of equilibria are also universally applicable to all of the molecular disciplines. Classical work of the 1970's and 1980's [6–8] has related the population of glycosaminoglycans like heparin and heparan sulfate on endothelial cells to their concentrations in the surrounding fluid, by quantitative observation of an equilibrium constant of a magnitude of $K_d \approx 10^{-6}$ to 10^{-7} M. We should be aware that reversibility is inherent in the ability to determine an equilibrium constant. It menas that there is a constant rate of molecular interchange of desorbing and adsorbing molecules on the endothelial surface.

The use of the term "desorption constant" for K_d is unfortunate, as it is actually the ratio k_d/k_a of the desorption rate constant and the adsorption rate constant. It is even more misleading to use the non-capitalized k_d, since k is reserved for *rate* constants, while K is used to describe an *equilibrium*.

The basic physico-chemical implications of equilibrium are important for interpreting and dealing with many phenomena of biological interest: Protein or saccharide species thus held on cell surfaces will have a finite concentration in the surrounding medium even if currently below the limit of detection. These molecules, while at *low concentration* at any moment, are nevertheless hopping on and off at *high rate*. It is *their* presence and action that allows observations of rapid net "release", replacement or "inhibition" by interaction of other agents with that very small but very lively species in solution. We discuss some aspects of this in a subsequent paper of this book [9].

It is a historical fact that we have often thought a factor to be absent (i.e. to have a zero concentration), to find later that we could accumulate a measurable amount, of FGF for example on a high affinity surface, by waiting until its favorable equilibrium with a large volume of sparsely populated fluid has been reached.

We could probably determine the actual although very low concentration in a medium by combining physical chemical principles of equilibrium and of diffusional transport by a method illustrated in Fig. 1: A large volume of the medium to be tested is contacted, with constant agitation with one side of a membrane. On the other side of the membrane is a small volume of constantly agitated fluid containing particles of a high affinity "trap" for the factor to be measured (e.g. heparin sepharose). If after time T, we determine the amount of factor accumulated on the particle "trap", we can determine the concentration C in the medium from

$$C = N \times X/T \times S \times D \times P,$$

where S is the cross-sectional surface, X the thickness and P the porosity fraction of the membrane, and D the diffusion coefficient (which can be taken as $10^{-6}\,cm^2/s$).

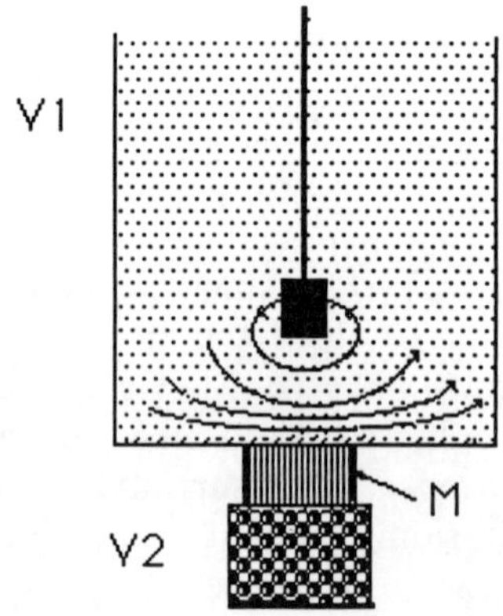

Figure 1. System for determination of very low concentration of a factor in volume V1. A large volume of fluid, rapidly stirred, in V1 is circulated past a membrane M. The membrane connects to a small volume V2, constantly stirred, containing particles of immobilized affinity receptor for the factor. See text.

Matters of equilibria of "sorption on surface sites, or of competitive sorption or poisoning of surface sites" are familiar grounds to the surface and catalytic science oriented chemist or chemical engineer. Translated for the bio-scientist, these are "binding on receptors, or of displacement or release from or inhibition of receptors". The major skill in the pursuit of interdisciplinary science is the carrying of language dictionaries!

Merging sophistication and simplicity

Chemical science began from the recognition of elements, and the curiosity about structure and interaction of simple atoms and small molecules and progressed up the ladder toward polymers and macromolecules. Molecular biology began from the point of greatest complexity, from DNA and RNA. It is now progressing by taking complexity down step by step.

To understand activities of heparin, the behavior of some detailed structural modifications and of some of the parts (fragments) of heparin have been examined. In a later paper of this symposium, we describe [9] the attack from the end of simplicity, that is the identification of "heparin behavior" starting from the simplest conceivable chemical model, and adding to same until we achieve that behavior.

The joining of the approaches of "from the complex" and "from the simple" end promises some of the most unifying, broad and useful insights. A recent example of this mixing is the examination of the binding of the HIV-1 Tat protein to an RNA structure [10]: The role of a suspected binding region of Tat protein having a sequence of RKKR-RQRRR was examined by synthesizing a series of peptides between KKKKKKKKK and RKKRRKKKR, and the affinity of these peptides themselves could be assessed successfully. The importance of the arginine entities could be specifically determined.

This is of considerable related relevance to the biology of angiogenesis, because here the interaction (complexing) of FGF with heparin like entities on or off cell surfaces appears to also center around arginine residues in FGF [11] creating electrostatic complexes with high density sulfate groups [9]. We should examine the question as to the simplest configuration of arginines (how many, how close) by a similar approach using synthetic small peptides. Does the FGF-heparin interaction also require an "arginine fork" [10]? The GAG-dye interaction experience, from electronic structure considerations in spectroscopy (metachromasia) and the observations of the histologist (sulfate-dependent staining of GAG's), may indeed link to that possibility.

Climbing the levels of molecular systems – to medicine

Angiogenesis is an outstanding example of a field of research in the life sciences that deals with *systems* of many interacting molecular entities. They have not only chemical (molecular) but physical parameters as we have seen above. Also there are systems at several levels of complexity: There are interacting simple ions, molecules, macromolecules, cellular components, cells, tissue components, organs and organ assemblies (the whole animal).

Goal-oriented basic research and its utilization must couple the entire spectrum of levels since the basics are molecular and the goal lies in the realm of life-preserving medicine.

Cell biology and the art and science of cell culture are the most promising and powerful stepping stones to the goal. There is much activity relating molecular chemistry to cell behavior now. Cell culture techniques promise to aid developments in pharmacology and medical therapy, in drug testing and development.

We can see such techniques develop as models and tools in diagnostic medicine. We have seen the rabbit cornea used as an in vivo "model" to test for the existence of angiogenic factors in mammary tissue samples, in relation to early detection of tumor potentials [12, 13].

Similarly, we have seen such techniques used to indicate the existence of angiogenic factors in synovial fluid relative to arthritis and bone destruction [14, 15]. It is an obvious future step to evolve suitable cell culture models for such diagnostic investigations and future developments for diagnostic practice.

Balancing the future

There is natural reluctance to turn to in vitro models of inherently lower level of complexity than is assumed to be operative in the in vivo environment. However, there is a current trend to seeing some unifying truths, such as many detailed variants of factors that might easily be lumped into one promoting category, and similar developments regarding GAG properties, etc. [16]. There are developing important insights that many factors can effect promotion and/or inhibition of angiogenic properties by virtue of the dynamics of the interactive equilibria (see above). All this may mean that a test for angiogenic *balance*, regardless of knowledge of all the participating partners, may constitute a huge step of progress for future developments in all of the related sciences and practices.

It is one thing to understand and control angiogenic balance. It is a further major challenge to learn how to apply such modulation of that balance to only those or that place of the system where it is needed. This becomes a major challenge for targeting technology.

Indeed angiogenesis is the result of the interaction of many parameters of a molecular *system*, and that system contains parts of knowledge and skills practiced in innumerable disciplines. We need to get acquainted across institutional and language boundaries to fully appreciate and exploit our capabilities for the benefit of society.

References

1 Gimbrone MA, Leapman SB, Cotran RS, Folkman J: Tumor dormancy in vivo by prevention of neovascularization. J Exp Med 1972; 136: 261–276. See also: Folkman J: Tumor angiogenesis, in: Cancer, Vol. 3, Becker FF Ed., pp. 355–388; Plenum Publ. Co., New York.

2 Weisz PB, Marshall JF: Tumors and chemical reaction engineering. Chemtech 1981; 11: 615.

3 Weisz PB: Enzymatic reaction sequences and cytological dimensions. Nature 1962; 195: 722.

4 Swabb EA, Wei J, Gullino PM: Diffusion and convection in normal and neoplastic tissues: Cancer Res 1974; 34: 2814–2822; see also Jain RK, Weissbrod JM and Wei J: Mass transport in tumors: Characterization and applications to chemotherapy. Adv Cancer Res 1980; 33: 251–310.

5 Weisz PB: Diffusion and chemical transformation – An interdisciplinary excursion. Science 1973; 197: 443–447.
6 Glimelius B, Busch C, Hook M: Binding of Heparin on the surface of human endothelial cells. Throm Res 1978; 12: 773–782.
7 Barzu T, Molho P, Tobelem G, Petitou M, Caen J: Binding and endocytosis of heparin by human endothelial cells in culture. Biochim biophys Acta 1985; 196–203.
8 Barzu T, Van Rijn JLML, Petitou M, Molho P, Tobelem G, Caen JP: Endothelial binding sites for heparin: Biochem J 1986; 238: 847–854.
9 Weisz PB, Herrmann HC, Joullié MM, Kumor K, Levine E, Macarak EJ, Weiner D: Angiogenesis and GAG-Mimics: *this book*.
10 Calnan BJ, Tidor B, Biancalana S, Hudson D, Frankel AD: Arginine-mediated RNA recognition: The ariginine fork. Science 1991; 252: 1167–1170.
11 Zhu X, Komiya H, Chirino A, Faham S, Fox GM, Arakawa T, Hsu BT, Rees DC: Three-dimensional structures of acidic and basic fibroblast growth factors. Science 1991; 251: 90–93.
12 Baum SS, Gullino PM, Medina D: Angiogenesis: A marker for neoplastic transformation of mammary papillary hyperplasia. Science 1977; 195: 880–882.
13 Jensen HM, Chen I, DeVault MR, Lewis AE: Angiogenesis induced in "normal" human breast tissue. A probable marker for precancer. Science 1981; 218: 293–295.
14 Brown RA et al: Angiogenic factor from synovial fluid resembling that from tumors. The Lancet 1980; March 29: 682–685.
15 Sano R, Forough et al: Detection of high levels of heparin binding growth factor-1 (aFGF) in inflammatory arthritic joints. J Cell Biol 1990; 110: 1417–1426.
16 Howard PS, Myers JC, Gorfien SF, Macarak EJ. Progressive modulation of endothelial phenotype during in vitro blood vessel formation. Dev Biol 1991; 146: 325–338.

Angiogenesis: Key Principles – Science – Technology – Medicine
ed. by R. Steiner, P. B. Weisz & R. Langer

Angiogenesis – Cytokines as part of a network

Leo Sachs

Department of Molecular Genetics and Virology, Weizmann Institute of Science, Rehovot 76100, Israel

The exciting developments in elucidating the control of angiogenesis which have been pioneered by J. Folkman [1] raise some interesting questions of the relationship to the controls that regulate the development of other cell types. Among these are cells of the hematopoietic system [2–5]. The following remarks may be of relevance to the molecular control of angiogenesis.

What one finds in nature are the consequences of biologic evolution. What is present is not only what the body uses today, but also what may be useful for further evolution. The production of different types of blood cells with a limited lifespan, both under normal conditions and in different emergency situations, such as infections, wound healing, and various diseases, requires a system with considerable plasticity. The development of several different interacting factors during evolution, therefore, is more useful for the functions required today and for adaptation to functions that may be required in the future, than would be the existence of only single factors with high specificity where a lack of function and a lack of flexibility would be lethal. The evolution of a family of factors, some of which have overlapping functions, is a useful safeguard; if one factor does not function effectively under certain conditions, another factor can take over.

The multigene family of proteins, including the colony-stimulating factors, interleukins, and erythropoietin, presumably originating from ancestral genes, which now exists to regulate the development of hematopoietic cells in health and disease, is an effective systems that contains such safeguards.

For example, with the four colony-stimulating factors, each of which is now coded for by a different gene, the production of macrophages can be initiated by macrophage colony-stimulating factor (M-CSF), granulocyte-macrophage colony stimulating factor (GM-CSF), or interleukin-3 (IL-3), and the production of granulocytes can be initiated by granulocyte colony-stimulating factor (G-CSF), GM-CSF, or IL-3. All four colony-stimulating factors can induce the production of interleukin-6 (IL-6), which does not induce the formation of colonies but can induce among myeloid cells their differentiation to macrophages, granulocytes, or megakaryocytes [2–6]. In a colony with differentiated cells, induction of growth by the colony-stimulating factors is thus followed by production of another factor, IL-6, which can induce differentiation of various cell lineages. This induction of a differentiation factor by a growth factor serves as an effective mechanism to couple growth and differentiation. Consequently, there are safeguards and economy in this system. IL-6 also may switch on other, so far unidentified factors that are required to determine the specificity of the final cell type.

A good way to obtain flexibility would be for different factors to function within a network of interactions, and this is what hematopoietic regulatory proteins actually do. A network of interactions allows considerable flexibility depending on which part of the network is activated. It also allows a ready amplification of response to a particular stimulus. This amplification can occur by autoregulation and by transregulation of genes for regulatory proteins [3]. There is also a transregulation by these proteins of receptors for other proteins [7].

In addition to the flexibility of this network, a network may also be necessary to stabilize the whole system. Hematopoietic regulatory proteins induce sustained levels of transcription factors during differentiation that can regulate and maintain gene expression in the differentiation program [8]. Interactions between the network for hematopoietic factors and transcription factors can thus ensure the production of specific cell types and stability of the differentiated state.

Production of specific cell types has to be induced when new cells are required and has to be stopped when sufficient cells have been produced. This requires an appropriate balance between inducers and inhibitors of development. The network of interactions between hematopoietic regulatory proteins, therefore, includes proteins that can function as inhibitors, such as tumor necrosis factor (TNF). Another inhibitory protein, transforming growth factor beta (TGF beta), which is also part of this network, can selectively inhibit the activity of some colony-stimulating factors and interleukins. It can also inhibit the production of some of these proteins [9].

Another important function of the network is that colony-stimulating factors and other factors, such as IL-1 and IL-6, which do not induce colonies, all function as viability factors. Cells have a limited life span and require cytokines that are part of the network for viability in the undifferentiated and differentiated state. The different proteins that induce cell viability have no apparent similarity. This and other evidence indicates that there are different pathways for inducing viability by preventing apoptosis (programmed cell death) [10, 11]. The use of alternative pathways to prevent apoptosis provides a useful safeguard to regulate the number of cells.

Parts of this network function not only within the hematopoietic cell system, but also for some nonhematopoietic cell types. In endothelial cells there is an induction of IL-6 when new blood vessels are being formed, and the production of IL-6 is switched off when angiogenesis has been completed [12]. The transient expression of IL-6 in the endothelial cells indicates a role for IL-6 in angiogenesis in addition to its role in regulating the development of myeloid and lymphoid hematopoietic cells.

The pleiotropic effects of a protein such as IL-6 raise the question whether these effects on different cell types are direct or indirect by IL-6 switching on production of other regulators that vary in the different cell types. Interpretation of experimental data on the effect of each regulator, therefore, has to take into account that the regulator functions in a network of interactions to avoid an incorrect assignment of a specific effect to a direct action of a particular protein.

The concept of the way cells communicate will provide a new dimension for the clinical use of these factors. What can be therapeutically useful may be due to the direct action of an injected factor or to an indirect effect due to other factors that are switched on in vivo. The application of these principles of 'cytokines as part of a network' should help to devise strategies for an innovative and economic clinical development of regulatory molecules including angiogenesis factors to improve existing therapies.

References

1 For reference, see Folkman J., in this book.
2 Sachs L: The molecular control of blood cell development. Science 1987; 238: 1374–1379.
3 Sachs L: The control of growth and differentiation in normal and leukemic blood cells. The 1989 Alfred P. Sloan Prize of the General Motors Cancer Research Foundation. Cancer 1990; 65: 2196–2206.
4 Lotem J, Shabo Y, Sachs L: The network of hematopoietic regulatory proteins in myeloid cell differentiation. Cell Growth Differ 1991; 2: 421–427.
5 Sachs L: Keynote Address for Symposium on: A Visionary Assessment of the Scientific, Clinical and Economic Implications of Hematopoietic Growth Factors. Cancer (May 15 Supplement), 1991; 67: 2681–2683.
6 Lotem J, Shabo Y, Sachs L: Regulation of megakaryocyte development by interleukin 6. Blood 1989; 74: 1545–1551.
7 Lotem J, Sachs L: Induction of dependence on hematopoietic proteins for viability and receptor upregulation in differentiating myeloid leukemic cells. Blood 1989; 74: 579–585.
8 Shabo Y, Lotem J, Sachs L: Induction of genes for transcription factors by normal hematopoetic regulatory proteins in the differentiation of myeloid leukemic cells. Leukemia 1990; 4: 797–801.
9 Lotem J, Sachs L: Selective regulation of the activity of different hematopoietic regulatory proteins by transforming growth factor beta-1 in normal and leukemic myeloid cells. Blood 1990; 76: 1315–1322.
10 Lotem J, Cragoe EJ, Sachs L: Rescue from programmed cell death in leukemic and normal myeloid cells. Blood 1991; 78: 953–960.
11 Yonish-Rouach E, Resnitzky D, Lotem J, Sachs L, Kimchi A, Oren M: Wild type p53 induces apoptosis of myeloid leukaemic cells that is inhibited by interleukin 6. Nature 1991; 352: 345–347.
12 Motro B, Itin A, Sachs L, Keshet E: Pattern of interleukin 6 gene expression in vivo suggests a role for this cytokine in angiogenesis. Proc Natl Acad Sci USA 1990; 87: 3092–3096.

Angiogenesis: Key Principles - Science - Technology - Medicine
ed. by R. Steiner, P. B. Weisz & R. Langer

Angiogenesis – Biomedical technology

Robert Langer

MIT E25-342, Dept of Chemical Engineering, 77 Mass Ave, Cambridge, MA 02139, USA

Introduction

One of the major consequences of angiogenesis research has been that it has led to new technologies that can aid many areas of science and medicine. In addition, a variety of biomedical technologies that are being developed can be used to aid in the understanding and quantification of the angiogenic response in vivo. There are a number of chapters in this book which deal with the exciting advances of biomedical technology as it relates to angiogenesis. These advances generally fall into three major areas: 1) drug delivery systems, 2) biomaterials and artificial organs, and 3) new imaging methods. Each of these areas is discussed below briefly as a prelude to the chapters in this book.

Drug delivery

Angiogenesis research has been responsible for spawning a number of novel drug delivery systems. These delivery systems were developed principally because of the great difficulty investigators had in studying angiogenesis stimulators and inhibitors. These molecules generally have very short half-lives, and are available in small quantities. In addition, it has been very hard to localize them specifically to desired body sites (e.g. cornea) where the vascular response can be studied. Finally, it can take days of even weeks depending on the in vivo site being studied for an angiogenic response to be fully apparent. To address these problems, small polymeric systems have been developed that can release proteins such as angiogenesis inhibitors for weeks or even years. These systems have been used in rabbit cornea as well as the chick chorioallantoic membrane (CAM) to study the angiogenic response. They provided some of the earliest methods by which angiogenesis stimulators and inhibitors could be studied in bioassay systems. It is of note, however, that the first biocompatible systems to deliver peptides or proteins were developed because of angiogenesis research [1, 2]. Today, in part as a consequence of these studies, many controlled drug delivery systems for peptides and proteins are being researched, and some are already being used clinically. Examples of situations where controlled release of peptides and proteins are important include the use of luteinizing hormone releasing hormone analogues for the treatment of prostate cancer and endometriosis, release of interferon for cancer treatment, release of growth hormones for growth control, release of growth factors for wound healing, and the release of erythropoietin for red blood cell production.

Another area where angiogenesis research is important in drug delivery involves understanding tumor vascular system. This will prove valuable for the important area of drug targeting. For example, one of the major challenges in cancer chemotherapy is to design a drug delivery system that will enable the drug to reach its target cells or target tissue. It has been difficult to do this in the case of solid tumors. However, by having a better understanding of tumor blood flow and angiogenesis, targeting strategies may be designed and in many cases facilitated.

Biomaterials and artificial organs

Angiogenesis also plays a major role in biomaterials. In some cases it is essential to know the vascular response to implanted biomaterials. For example, this is important in vascular grafts. Another example of where the vascular response to biomaterials can be significant involves research where new organs can be created from transplants of mammalian cells on appropriately designed polymeric scaffolds. In this case, polymeric scaffolds are designed so that a desired cell (e.g. hepatocytes) will attach to them. However, for such cellular transplants to be successful, it is essential that an angiogenic response be induced to create a new vascular supply. The ability to induce angiogenesis in this way could be critical in the design of these new cellular transplantation devices.

New imaging methods

Understanding tumor blood flow is extremely important both for diagnosis as well as from the standpoint of designing appropriate kinds of therapies. A number of techniques are being explored. One of the exciting applications in this area is NMR imaging using perfluorocarbon emulsions to examine tumor blood flow. In addition, methods previously used to detect the presence of tumors such as CT scanning, are now being investigated to estimate blood flow. In fact, some studies are underway which examine dynamic CT scanning, dynamic NMR, and thermoclearance methods. Finally laser Doppler flowmetry as well as osseointegrated probes are being used to understand blood perfusion.

Summary

All of these studies show that angiogenesis research can benefit from new biomedical technology tools currently being developed, as well as contribute by providing new technologies that can be used in other areas of medicine. It is hoped that the chapters in this book in this area will provide the reader with an up-to-date appreciation of some of the exciting research that is currently being pursued.

References

1 Langer R, Folkman J: Polymers for the sustained release of proteins and other macromolecules. Nature 1976; 263: 654–656.
2 Langer R, Brem H, Falterman K, Klien M, Folkman J: Isolation of a cartilage factor that inhibits tumor neovascularization. Science 1976; 193: 70–72.

Angiogenesis: Key Principles - Science - Technology - Medicine
ed. by R. Steiner, P. B. Weisz & R. Langer

Angiogenesis – Time to review progress

K. Hellmann

Radiotherapy Department, Westminster Hospital, London SW1P 2AP, England

Most biomedical research has two inter-related strategic objectives; that of gaining better understanding of fundamental processes and that of utilizing this knowledge to arrive at improvements in therapeutics. Although these solemn goals are often reiterated, particularly for grant giving bodies, there is no doubt that both objectives are greatly helped by serendipity. It is useful therefore to pause from time to time to review progress. The diversity of approach, the analytical resources and the ingenuity now brought to bear on the subject of tumour blood vessels is also witness to the enormous complexity of the problem of tumour growth. At the same time one must hope that the fundamentals of tumour pathology are not forgotten.

The importance of these fundamentals lie in the fact that while experimental results presented by individual groups undoubtedly reflect the truth of their experiments they cannot necessarily be extrapolated into broad generalizations. One fundamental point which was addressed was that although tumours consist of masses of heterogeneous cancer cells (most of which are dead, dying or doomed) the neovasculature which supplies them consists of cells which are essentially normal even if the blood vessels as a whole are not. This implies that the tumour neovasculature has malignant characteristics of its own which are extremely difficult to reproduce in isolation.

Although the development of tumour blood vessels has been studied for more than a century by many of the great medical investigations it was the papers of Folkman that stimulated much of the recent research on tumour angiogenesis [1–3]. His idea of specifically blocking tumor-induced angiogenesis, attractive as it is, has yet to demonstrate that it can be successfully applied to tumour therapy in man and translate into solid therapeutic benefit.

Despite a great deal of research over the last 20 years which aimed to prevent the development of tumour blood vessels as a means of controlling tumour growth and in particular on the factors which may be responsible, the key targets of the process of neovascularization have proved extremely elusive. Nevertheless it has had the effect of greatly enlarging our understanding of the control of blood vessel growth in health and disease. Recently two encouraging reports of a beneficial anti-angiogenic therapy with interferon alpha in children with pulmonary haemangiomatosis, an 'angiogenic disease' [4], have been reported [5, 6].

While Folkman's concept in March 1972 for therapy of solid tumours was aiming for anti-angiogenesis [3], the observations of Le Serve and Hellmann [7], also published in March 1972, revealed that more complex approaches to tumour control through its vasculature were possible. These authors showed they could normalize the *developing* neovasculature (not that which had already developed) with razoxane, an antitumour agent structurally related to EDTA [8], with important consequences for the growth of the tumour and its effects on the host. These results were amplified by the group of Maragoudakis, who found that this drug could, at therapeutically attainable concentrations, inhibit type IV collagen and basement membrane degradation in growing tumour-induced capillaries by an enzyme isolated from malignant tumours [9]. This may explain the normalization of the developing neovasculature which in turn could prevent tumour cells entering the circulation and thus deleting one of the most malignant characteristics, that of metastasis formation. In contrast to the anti-angiogenic idea it therefore seems that it is possible to permit tumour-induced angiogenesis to continue but to change its characteristics.

Other consequences followed from these observations, one of them being the degree of oxygenation which could reach the tumour if vessels and blood flow to the tumour were normalized. Since oxygen is one of the most powerful if not the most powerful radiosensitizer there is a theoretical enhancement of radiation sensitivity bonus from the normalization of the tumour vasculature. In subsequent papers our group showed that the normalization effect could only be expected to produce a reduction of metastases in those tumours in which the neovascular growth was abnormal. Rhomberg recently reported in patients with recurrent rectal carcinoma treated with radiotherapy alone or combined with razoxane a rise of the median survival from 13 month to 24 month in a small pilot trial which needs to be confirmed. Interestingly no new metastases occured after discontinuation of the short-course drug treatment [10].

Furthermore, in an adjuvant clinical trial lasting more than 7 years and involving 272 colorectal cancer patients the experimental findings were confirmed in that in the Duke's C patients only half as many developed liver secondaries who were treated with razoxane as compared with those in the control group, and those who developed liver metastases took twice as long to do so [11].

In psoriasis, a non-malignant condition, where the neovasculature is very similar to that of malignant tumours, razoxane was shown to normalize the pathologic vasculature and to be one of the most effective agents available for this disease [12].

Clearly, however interesting and exciting the search for knowledge may be in the biomedical field the ultimate test is whether the results can also apply to man. This volume is adequate testimony to the tremendous effort which is being made at the moment to increase fundamental understanding of angiogenesis so as to apply it effectively to the prevention and treatment of tumours and other 'angiogenic diseases' [13].

References

1 Folkman J, Merler, F, Abernathy C, Williams G: Isolation of a humoral factor responsible for angiogenesis. J Exp Med 1971; 133: 275–288.
2 Folkman J: Tumor angiogenesis: Therapeutic implications. New Engl J Med 1971; 285: 1182–1186.
3 Folkman J: Anti-angiogenesis: A new concept for therapy of solid tumors. Ann Surg 1972; 175: 409–416.
4 Folkman J: Successful treatment of an angiogenic disease. Editorial. New Engl J Med 1989; 320: 1211–1212.
5 White CW, Sondheimer HM, Crouch EC, Wilson H, Fan LL: Treatment of pulmonary haemangiomatosis with recombinant interferon alfa-2a. New Engl Med J 1989; 320: 1197–1200.
6 Orchard PJ, Smith CM, Woods WG, Day DL, Dehner LP, Shapiro R: Treatment of hemangioendotheliomas with alpha interferon. Lancet 1989; 2: 565–567.
7 Le Serve AW, Hellmann K: Metastases and the normalization of tumor blood vessels by ICRF-159: A new type of drug action. Brit Med J 1972; 1: 597–601.
8 Bakowski MT ICRF-159, 1,2-di (3,5-dioxopiperazin-1-yl) propane; NSC 129, 943; Razoxane. Cancer Treat Rev 1976; 3: 95–107.
9 Karakiulakis G, Missirlis E, Maragoudakis ME: Mode of action of razoxane: inhibition of basement membrane collagen-degradation by a malignant tumor enzyme. Methods Find Exp Clin Pharmacol 1989; 11: 255–261.
10 Rhomberg W, Eiter H: Radio-oncological treatment possibilities in local recurrence of rectal carcinoma. An interdisciplinary comparison. Wiener Med Wschr 1990; 140: 206–211.
11 Hellman K, Gilbert J, Evans M, Cassell P, Taylor R, Stoodley B, Ellis H, Wastell C: Effect of razoxane on metastases from colorectal cancer. Clin Exp Metastasis 1987; 5: 3–8.
12 Horton JJ, Wells RS: Razoxane: a review of 6 years' therapy in psoriasis. Brit J Dermatol 1983; 109: 669–673.
13 Folkman J, Szabo S, Stovroff M, McNeil P, Li W, Shing Y: Duodenal ulcer. Discovery of a new machanism and development of angiogenic therapy that accelerates healing. Ann Surg 1991; 214: 414–427.

Angiogenesis: Key Principles – Science – Technology – Medicine
ed. by R. Steiner, P. B. Weisz & R. Langer

Angiogenesis – A new fascinating enigma for surgeons?!

Georg K. Uhlschmid

Research Division, Department of Surgery, University Hospital, CH-8091 Zürich, Switzerland

Introduction

It was my first impression of this topic, that the various popes of this new field of biology are domiciled in practically all faculties except the theological, so far.

Angiogenesis is a two-edged sword and therefore very "easy" to explain. In oncology you may want to suppress it and in almost all types of healing you try to enforce it. This is the key issue for future surgical ideas and imaginations and subsequent testing in experimental setups. For the time being, however, the surgeon is bewildered and blurred and he might turn his attention to the year 1856 when a certain Dr. Billroth from Berlin published his "Investigations on the development of blood vessels". This just to reconsider the fact, that human and scientific hybris is out of place concerning our mental and imaginative capacities independent from modern technologies.

Angiogenesis today is the family name of a daily increasing number of specified factors, which makes the field more and more complex, as mirrored by their obfuscating initials.

The different surgical specialities and subspecialities will be faced in the near future with their particular "factors" in order to find out more about their practical application. This implies close collaboration between clinical and experimental research. Angiogenesis is one of the many fields where the surgeon inevitably will be confronted in a varying degree with molecular biology, biophysics and technology. He must find his scientific position between the: "Goal oriented basic research, the basics being molecular and the realm of life preserving medicine" (P. B. Weisz). Diagnostic and therapeutic tools will emergy from this "new science". It is up to the surgeons to not just wait for "products" but actively participate with ideas in research and development.

At a first glance, the concept of angiogenesis evokes hopes in the surgical mind concerning the improvement of all types of wound healing [1]. Up to now angiogenesis in wound healing has been dealt with

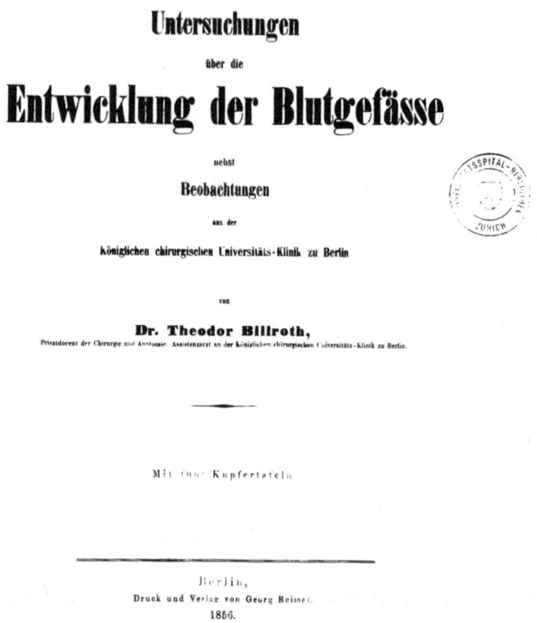
Untersuchungen

über die

Entwicklung der Blutgefässe

nebst

Beobachtungen

aus der

königlichen chirurgischen Universitäts-Klinik zu Berlin

von

Dr. Theodor Billroth,

Privatdocent der Chirurgie und Anatomie, Assistenzarzt an der königlichen chirurgischen Universitäts-Klinik zu Berlin.

Mit fünf Kupfertafeln

Berlin,

Druck und Verlag von Georg Reimer.

1856.

Plate 1. The title-page of Billroth's book on the development of blood vessels, 1856 [8].

mainly in exploratory research. Dramatic improvement of wound healing has been shown in chronic ulcers, even in diabetic patients [2]. Most of these results have been obtained by topical application of angiogenesis factors and led to a significant change in wound behaviour.

Once the "factors" are freely available, a vast number of applications with adequate biodegradable carriers is imaginable. The "loading" of biodegradable suture materials to improve healing of intestinal, bronchial and other suturelines will be realized in the near future. Conditioning of free and pedicular flaps or autologous skin transplants will soon be a matter of practical consideration [4]. The growing of "skin" cultures in combination with improved attachment will keep burn specialists busy [3]. Sheeting of nerve transplants [5] or nerve stures may enhance both healing and sprouting, if correct direction of growth is provided.

New dimensions appear with the rapid development of adequate biodegradable polymers which will act as drug carriers for angiogenesis factors. Vascular stents or grafts designed in such a way could be used for all types of vascular surgery. Adhesives loaded with growth factors might be useful for a wide range of applications (e.g. pleurodesis, flaps, nerves, skin and hair transplants, etc.). The forthcoming development of polymers as cell or organoid carriers in combination with angiogenesis factors [6, 7] opens up a vast field for imaginative ideas in research. The same is true for the pathophysiological processes occurring in traumatology be it in soft tissue or bone, where angiogenesis may be a functional tool in combination with biodegradable osteosynthesis.

The rôle of angiogenesis factors in the reestablishment of circulation in all types of muscles – myocardial or other – either in combination with new laser techniques or with electrical and magnetic transportation is another challenging prospect.

Some of the "factors" show bidirectional potencies and act as promotors or inhibitors respective to their environment.

Angiogenesis suppression as a therapeutic concept is another complex issue for the surgeon of today. Topical, regional or systemic application of factors at the time of surgical intervention might reduce the number of metastases due to operative manipulation. Smooth muscle cell proliferation with subsequent, narrowing hyperplasia following vascular dilatation, endarterectomy or laser treatment might be reduced or suppressed by adequate inhibitors, administered systemically or attached to appropriate stents. Even the old problem of keloid formation could be approached in a new way.

The arising possibilities of applications at the horizon of surgical oncology are so manifold that the contemplative surgeon would rather like to turn your attention to the following pages for further enlightenment.

References

1 Van Brunt J, Klausner A: Growth factors speed wound healing. Biotechnology 1988; 6: 25–30.
2 Barnhill RL, Wolf JE: Angiogenesis and the skin. J Am Acad Dermat 1987; 16(6): 1226–42.
3 Ronfard V, Broly H, Mitchell V, Galizia JP, Hochart D et al.: Use of human keratinocytes cultured on fibrin glue in the treatment of burn wounds. Burns 1991; 17(3): 181–84.
4 Höckel M, Burke JF: Angiotropin treatment prevents flap necrosis and enhances dermal regeneration in rabbits. Arch Surg 1989; 124: 693–98.
5 Duprez K, Bour C, Merle M, Duprez A: Human nerve xenografting in the nude mouse: experimental study of graft revascularization. Microsurgery 1991; 12(1): 1–8.
6 Vanholder R, Ringoir S: Bioartificial pancreas and liver. Internat J Artif Organs 1991; 14(7): 398–402.
7 For reference, see Stagner JI, Samols E: Induction of angiogenesis by growth factors: Relevance to pancreatic islet transplantation. *In this book.*
8 Billroth Th: Untersuchungen über die Entwicklung der Blutgefässe. Georg Reimer, Berlin, 1856, 81 pp. + V plates.

Angiogenesis: Key Principles - Science - Technology - Medicine
ed. by R. Steiner, P. B. Weisz & R. Langer

Intussusceptive microvascular growth, a new mechanism of capillary network formation

Peter H. Burri

Institute of Anatomy, University of Berne, CH-3000 Berne, Switzerland

Abstract. Growth by *intussusception* is defined as growth by deposition of new particles or pieces of formative material among those already embodied in a tissue or structure. In the context of capillary growth the term stands for the extension of the capillary system by the insertion of new capillary meshes within the existing network. New meshes arise as slender transcapillary tissue pillars. These are formed initially by a circumscribed fusion of opposite endothelial leaflets. Following reorganization of the junctional complexes, the pillar is invaded by interstitial tissue. This means that for the formation of new capillaries no sprouting of endothelial cords or tubules is required. The mechanism is described in the growing lung, but may occur in other tissues too. There may be a chance for in vivo observation of the process in the capillaries of the chorio-allantoic membrane of the chicken.

Introduction

The lungs of rats and humans, both increase in volume more than 23 times between birth and adult age. At the same time, the capillary surface area increases 20, and the capillary volume more than 35 times. Despite this expansion the capillary network is denser in adult lungs than shortly after birth. This means that further capillary units or segments have to be added as long as growth proceeds. So far, it has tacitly been assumed that the growth of vascular networks in general, but also in the lung, is achieved by capillary sprouting, a phenomenon well known and described in wound healing [1], in tumor growth [reviews by 2–4] and in many other in vivo and in vitro conditions.

Investigating Mercox casts of the pulmonary vasculature of growing rats by scanning electron microscopy (SEM), we were struck by the presence of extremely small holes (diameters around 1 μm) in sheet-like areas of the capillary bed in the interalveolar walls [5]. Dismissing the idea that these holes were mere casting artefacts, we developed the concept that they could represent newly formed slender tissue pillars across the capillary lumina. [We recently noticed, that in a study dealing primarily with the postnatal fate of the alveolar epithelium, Short [6] had already proposed that "growth takes place by multiplication of intercapillary spaces . . .".] Such pillars would give rise to new capillary meshes (Fig. 1). As opposed to capillary growth by sprouting, we

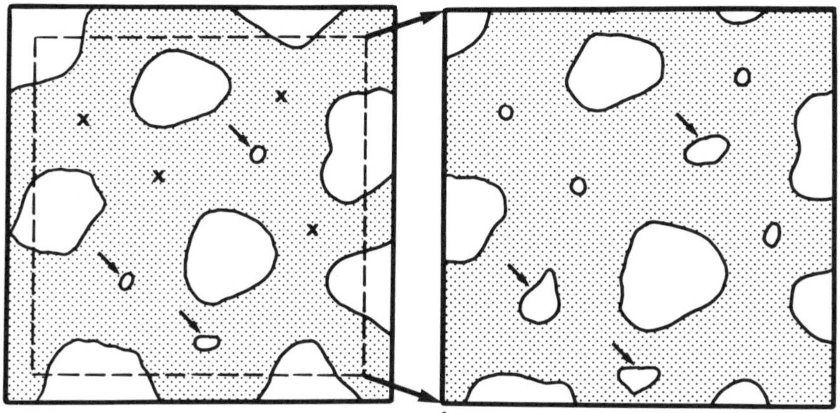

Figure 1. Schematic illustration of intussusceptive growth principle for capillaries. Capillary network (dotted area) is viewed from above; intercapillary meshes (clear areas) are of variable size. *a* Within wide capillary segments some tiny meshes are present (arrows); they may just have been formed and are going to grow (see Fig. b, arrows). Crosses indicate places where new meshes (transcapillary tissue pillars) are eventually being formed. *b* Capillary network has enlarged by growth of small meshes and by addition of new ones. (Growth of area delineated by broken line amounts to 38%.)

termed this process capillary growth by intussusception (or intussusceptional or intussusceptive capillary growth). This hypothesis needed confirmation by further investigations, both by SEM and transmission electron microscopy (TEM).

Materials and methods

Preparation of casts of pulmonary vasculature

Rats aged between birth and 139 days were deeply anesthetized and their lungs and trachea exposed. The collapsed lungs were expanded by intratracheal instillation of Ringer's solution. After cannulation of the pulmonary artery and an incision of the right auricle, the pulmonary vascular bed was flushed with Ringer's solution containing 1% Liquemin^{R} and 1% procain. Mercox [Mercox is a trade name for a methylmethacrylate of the Japan Vilene Company, Tokyo] (0.1 ml accelerator in 5 ml resin [7]) was then perfused until flow stopped due to polymerisation (~5 min). After hardening of the resin, the lungs were removed from the chest and immersed in 15% KOH for 1–3 weeks to dissolve all tissues. The casts were then prepared for SEM [5].

Preparation of lung specimens for TEM

Rat lungs were collapsed by transabdominal perforation of the diaphragm and fixed by standardized intratracheal instillation of a buffered 2% glutaraldehyde solution. Following fixation they were removed from the chest and sliced. Tissue blocks were postfixed in buffered 1% OsO_4 and further processed for TEM analysis. Several batches of 80–100 perpendicular serial sections of interalveolar walls were cut and screened for tissue posts crossing the capillary lumina [8].

Results and discussion

Searching the Mercox casts for the typical tiny holes, it became apparent that the latter were present at all ages investigated, but were more frequent during phases of rapid growth.

As has already been observed by Miller [9], the pulmonary capillary bed shows two different morphological patterns: a dense and a coarse network (Fig. 2a). The dense network is located in the interalveolar walls of the parenchyma and the coarse network is found in the alveolar walls abutting with their back side to the peribronchial and perivascular sheaths and to the pleural fibrous envelope. The holes were present in both, the dense and the coarse networks. They were especially prominent in the subpleural capillaries (Fig. 2b). These observations all fit to the idea that, if these holes are involved in capillary growth, they would occur more frequently whenever lung is growing fast, and in those locations with the highest rates of surface expansion, as e.g. in the subpleural region.

In all the casts some filling defects were manifested (Figs. 2a and b). They could easily be interpreted as tubular sprouts burgeoning from capillary lumina. Very often such "sprouts" were observed by pairs, pointing towards each other. This latter observation and the increased frequency of these "sprouts" in poor casts made us assume that such pictures were due to filling defects. This assumption was corroborated by the TEM study, which allowed us to confirm the presence of tissue pillars in the right locations and with the critical dimensions of the holes in the casts, but not of blind ending sprouts. The analysis of the ultrathin serial sections revealed 4 different types of tissue pillars, all with diameters between 1 and 2.5 μm.

A condition for a transcapillary tissue bridge to be recognized as a true pillar was that, screening through the serial sections, a capillary segment had to appear open, then traversed by a tissue bridge, and then open again. The 4 pillar types can be described as follows:

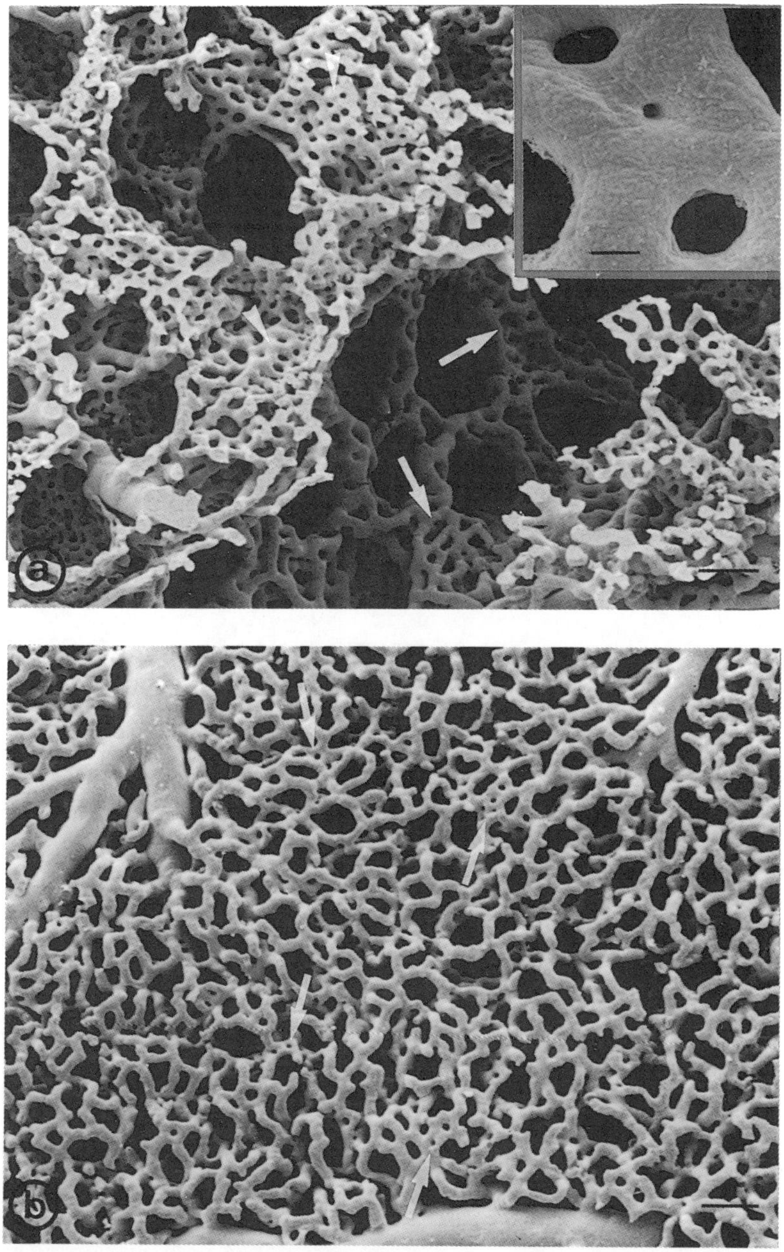

Figure 2. *a* Mercox cast of rat pulmonary vasculature. Notice two different patterns in the capillary system: a coarse network ensheathing an airway (arrows) and a dense network in the interalveolar walls (arrow heads). Tiny holes (=small capillary meshes) are present in both types of networks. Bar = 20 μm. Inset: detail of case at higher magnification. Bar = 2 μm. *b* Mercox cast of subpleural alveolar capillaries. They are of the coarse network type. Signs for intussusceptive growth are prominent (arrows). Bar = 20 μm.

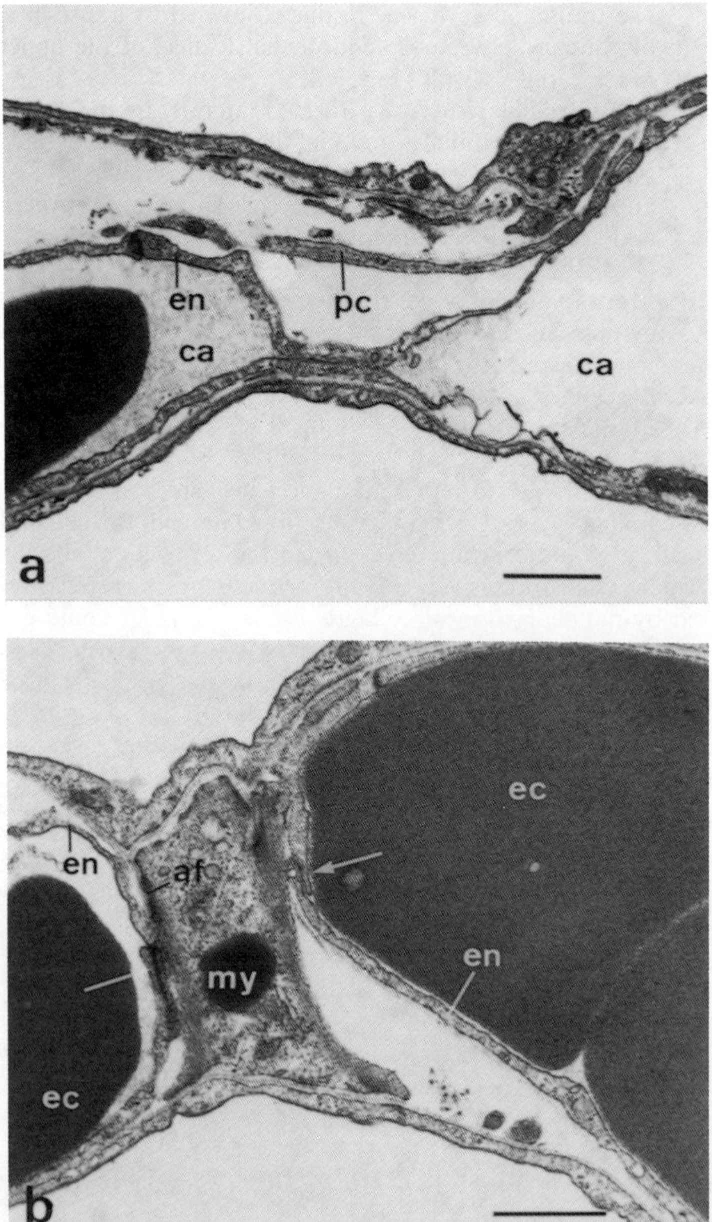

Figure 3. *a* Type I transcapillary pillar. Opposite endothelial walls form a disc-like zone of contact, which was 14 sections deep (~ 1 μm); maximal width was 1.1 μm. Note pericytic cell process (pc) roofing over the endothelial fold. ca = capillary lumen; en = endothelium. Bar = 1 μm. *b* Type II pillar. Dimensions of this cylindrical transcapillary tissue bridge were 1.1 μm in depth and 1.5 μm in width. The core of the pillar is formed by a myofibroblast (my) with intracytoplasmic actin filaments (af). Note the lateral interendothelial junctions, which circle the pillar (arrows). en = endothelial cells; ec = erythrocytes. Bar = 1 μm.

Type I: The transcapillary tissue bridge is formed by a disc-like zone of contact between the endothelial leaflets of the upper and lower capillary wall (Fig. 3a).

Type II: The pillar-like structure is formed laterally by the endothelial cells of the capillary and, in the core, by the cytoplasmic process of a myofibroblast. It has been shown by Kapanci and coworkers [10] that these cells contain actin filaments and are actually contractile elements of the interalveolar wall (Fig. 3b).

Type III: Morphology of type II pillar complemented, however, by cytoplasmic extensions of pericytes enwrapping parts of the lateral aspects of the capillary walls.

Type IV: Resembles type III pillar but additionally contains collagen fibrils.

We propose the 4 types to represent 4 successive steps in the formation of a new capillary mesh: Once a first disc-like interendothelial contact is established, it would centrally give way to the elements of the interstitium, which then successively invade the core of the pillar. When stabilized by the deposition of collagen fibrils, the pillar could grow in diameter and become a capillary mesh.

So far, there are only speculations regarding the mechanics involved in pillar formation. For the first step, the initiation of the transcapillary interendothelial contact, it has been hypothetized that intracapillary flow dynamics or pericytic contractility (Fig. 3a) could be involved.

For an interendothelial contact to proceed from a type I to a type II pillar, the disc-like area has obviously to become perforated. To this effect junctional complexes have first to be created between the two endothelial leaflets all around the disc, in order to seal the capillary lumen and to prevent leakage. Because the transition from a pillar I to a pillar II structure may last only minutes, the chances to detect this process in TEM are minimal. There is, however, strong structural evidence for such a transition: The constant presence of interendothelial junctions at the mid-lateral circumference of all the pillars of types II to IV (Fig. 3b). These junctional complexes can be considered to represent the ultrastructural markers of the endothelial restructuring.

It is evident that only the in vivo observation of the actual formation of such pillars can provide the final proof for the intussusceptive growth principle. Following a suggestion made by C. Haudenschild, we therefore started to investigate the capillaries of the chorio-allantoic membrane (CAM) of the chicken. Using the shell-free culture technique of chicken embryos in a petri dish [11], we performed live video microscopy of the CAM at various ages. Figure 4 illustrates that, indeed, extremely slender tissue pillars are also

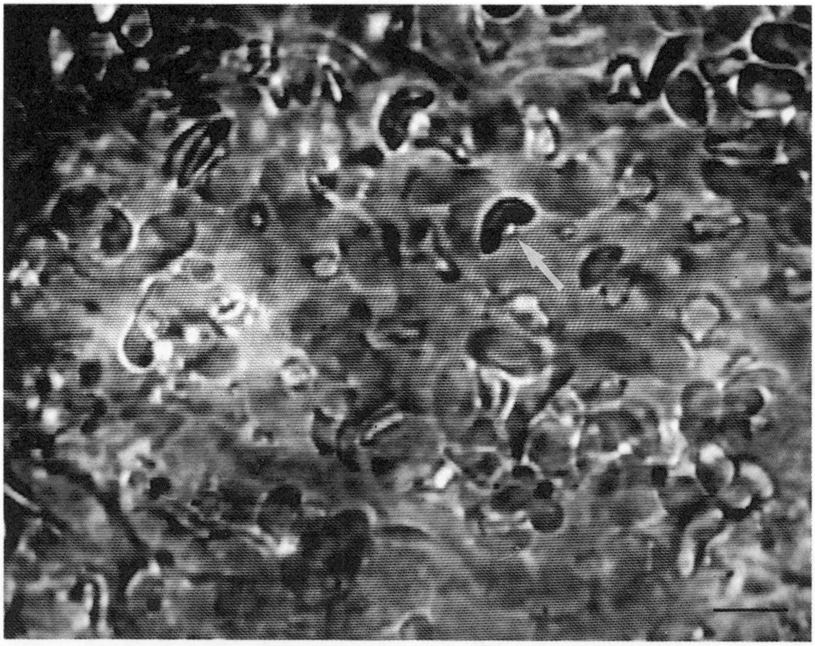

Figure 4. Top view of chorio-allantoic capillary network (screen photograph of frozen video image). Note intercapillary meshes of varying sizes and tiny transcapillary tissue pillars (arrow) with erythrocyte stuck to it. Bar = 5 μm.

present in the CAM circulation, so that the CAM could be an adequate model for the in vivo study of the intussusceptive capillary growth [12].

In summary, the electron microscopic evidence of 4 different types of slender tissue pillars crossing the capillary lumina strongly supports the concept of intussusceptive microvascular growth in the lung. The in vivo observations in the CAM hint at the possibility that the concept may also apply to other capillary networks and to other species.

Acknowledgements. This work was supported by SNSF grants No. 3.492-0.86 and 3100-2775.89.

References

1 Burger PC, Chandler DB, Drysdale DB, Tano Y, Crapo JD, Freeman BA, Klintworth GK: Scanning electron microscopy of vascular casts in experimental ocular vasoproliferation. Scan Electron Microsc 1984; 4: 1893–1898.
2 Folkman J: Angiogenesis: Initiation and control. Ann NY Acad Sci 1982; 401: 212–227.
3 Folkman J, Klagsbrun, M: Angiogenic factors. Science 1987; 235: 442–447.
4 Caduff JH, Fischer LC, Burri PH: Scanning Electron Microscope study of the developing microvasculature in the postnatal rat lung. Anat Rec 1986; 216: 154–164.
5 Furcht LT: Critical factors controlling angiogenesis: Cell products, cell matrix, and growth factors. Editorial. Lab Invest 1986; 55: 505–509.

6 Short RHD: Alveolar epithelium in relation to growth of the lung. Philos Trans Soc Biol 1950; 235: 35–87.
7 Tano Y, Chandler DB, Machemer, R: Vascular casts of experimental retinal neovascularization. Am J Ophthalmol 1981; 92: 110–120.
8 Burri PH, Tarek MR: A novel mechanism of capillary growth in the rat pulmonary microcirculation. Anat Rec 1990; 228: 35–45.
9 Miller WS: The Lung, 2nd Edn: Springfield, IL, Thomas, CC, 1947.
10 Kapanci Y, Assimacopoulos A, Irle C, Zwahlen A, Gabbiani G: "Contractile interstitial cells" in pulmonary alveolar septa: A possible regulator of ventilation/perfusion ratio? J Cell Biol 1974; 60: 375–392.
11 Auerbach R, Kubai L, Knighton D, Folkman J: A simple procedure for the long-term cultivation of chicken embryos. Dev Biol 1974; 41: 391–394.
12 Patan S, Haudenschild C, Burri PH: Capillary growth in the chicken chorio-allantoic membrane: A model for the intussusceptive microvascular growth? Experientia 1990; 46: 564.

Angiogenesis: Key Principles - Science - Technology - Medicine
ed. by R. Steiner, P. B. Weisz & R. Langer

Combined morphological approaches in the study of network formation in tumor angiogenesis

Moritz A. Konerding[a], Claudia van Ackern[a], Fritz Steinberg[b] and Christian Streffer[b]

[a]*Institut für Anatomie*, [b]*Institut für Medizinische Strahlenbiologie, Universitätsklinikum Essen, D-4300 Essen 1*

Introduction

The vascular system of malignant tumors differs from that of normal tissues in many aspects. These include, among others, lack of hierarchy, the capillary-sinusoidal vessel wall, absence of differentiated arteries and veins as well as the low structural stability [1–3]. Detailed investigations on the tumor vessel system are required to assess the particular clinical relevance of the tumor vessel system in terms of growth and metastasation as well as for diagnosis and therapy. In particular, this applies to tumor angiogenesis: Concepts such as "Endothelium as a vulnerable element in tumor therapy" [4] or the "angiogenic attack" [5] highlight the therapeutic significance and implications.

The central significance of the vascular system and the impetus given by the discovery of tumor-inherent angiogenic factors [6, 7] have led to numerous in vitro and in vivo studies in different areas, that have contributed to a better understanding of the events occurring in secondary angiogenesis. However, since the initial extensive studies on capillary sprouting by Thoma in the last century [8], the prevailing view is that all morphologically tangible events of primary and secondary angiogenesis are similar even in such differing processes as normal growth, wound healing, and chronic inflammations as well as tumor angiogenesis [7].

In spite of this, however, there are comparatively few papers dealing with the morphology of angiogenesis: Over the period from 1983 to 1991, roughly 900 papers on angiogenesis were to be found in the Medline data bank, but of those, only 8% or 70 reports were mainly or at least partially concerned with morphological aspects (Fig. 1a). Of these, only 21 publications were concerned with the morphology of tumor angiogenesis (Fig. 1b).

In this respect, it is interesting that numerous papers on angiogenesis were carried out on in vitro and not on suitable in vivo tumor models.

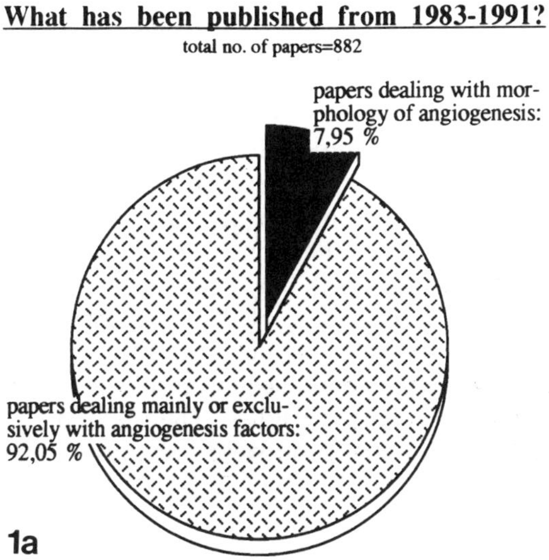

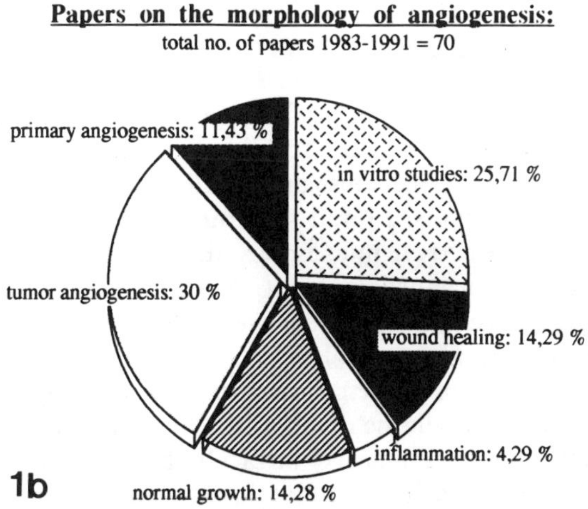

Figure 1a and b. Literature on angiogenesis: Medline Research 1983–1991 (as of March, 1991). b: Areas of interest of papers dealing with the morphology of secondary angiogenesis.

Studies on primary tumors, as has been frequently suggested in the past [9], are also extensively missing.

Against this background, the aim of the present study is to investigate the characteristics of tumor angiogenesis and the (re-) structuring of the vessel system resulting therefrom in a suitable model using complementary morphological and morphometrical methods. Particularly emphasis is placed on the type of sprout and lumen formation, sprout fusion as well as the tumor-specific network formation.

Methodology

Two human tumor lines, an undifferentiated squamous carcinoma 4197 and an amelanotic melanoma MeWo were xenotransplanted on each of 48 thymus-plastic immune deficient nude mice as described in detail previously [2, 10]. The melanoma was in the 2nd, while the squamous carcinoma was in the 8th passage. After the occurrence of macroscopic tumor growth, the animals were sequentially perfused and the tumor material removed. The melanomas were investigated on day 8, 12, 18, 25 and 79 and those of the squamous carcinoma on day 4, 7, 10, 15 and 23. In order to determine the vascular density, in addition, at least two equatorial slices were cut from each tumor and embedded for light microscopic studies. The methods used for animal maintenance and tissue preparation were previously described in detail [2, 11, 12].

Results

Sprout and lumen formation

Structurally stabilized as well as destroyed and newly formed vessels are found closely side by side at all times in the case of tumor angiogenesis. Sprout formations and early forms can be definitely distinguished from established and structurally stabilized vessels above all by transmission electron microscopy due to their distinctly higher endothelium and cell organelle content. As described elsewhere (Fig. 2), the latter, however, do not show such a high degree of differentiation [11, 12], as from normal terminal vessels.

During the process of lumen formation, in principle, at least two different types can generally be distinguished as a general principle. The first is characterized by an initial intracellular formation of vacuoles giving rise to a primitive intracellular lumen with consecutive connection to already established lumina (Fig. 3). In these cases, the high endothelium contains a relatively large amount of endoplasmic reticulum and numerous polysomes as well as mitochondria. The vacuole, which forms the lumen later, stems from cisterna-like dilated sections of the endoplasmic reticulum or the Golgi-apparatus. However, it should be noted that not all intracellular vacuoles are indicative of an initial sprout formation, on the contrary, serial sections show that these vacuoles frequently only serve as lumen extensions (Fig. 4). This intracellular type of lumen formation is mainly found in early tumor stages; later, this is only seen in rare cases.

The second type of lumen formation involves a migration of two or more endothelial cells. Here, more plump pseudopodia or the eccentrically situated perikaryon of the involved endothelial cells veer out from the endothelial structure leading to the formation of a slit-like lumen (Fig. 5). An intercellular lumen is thus already formed at the start of bud formation and not, as usually postulated, after formation of a solid endothelial cell cord. Initially, no continuous connection to the blood vessel can be seen (Fig. 5). This can also be explained by the breadth of the lumina, often only 0.1 μm in size, which are frequently still attached to opposite cell membranes through point-like or cell-attached oppositions and thus not yet traversable for cellular blood constituents. The extent to which the basal lamina is digested prior to this type of lumen formation cannot be answered by us with certainty, since this is frequently only seen as an incomplete, small, electron-sparse zone and apparently only encompasses the endothelium completely in exceptional cases.

A more rare form of lumen formation is in the protrusion of one or two cell ramifications with reflexive contact formation between these branches (Fig. 3).

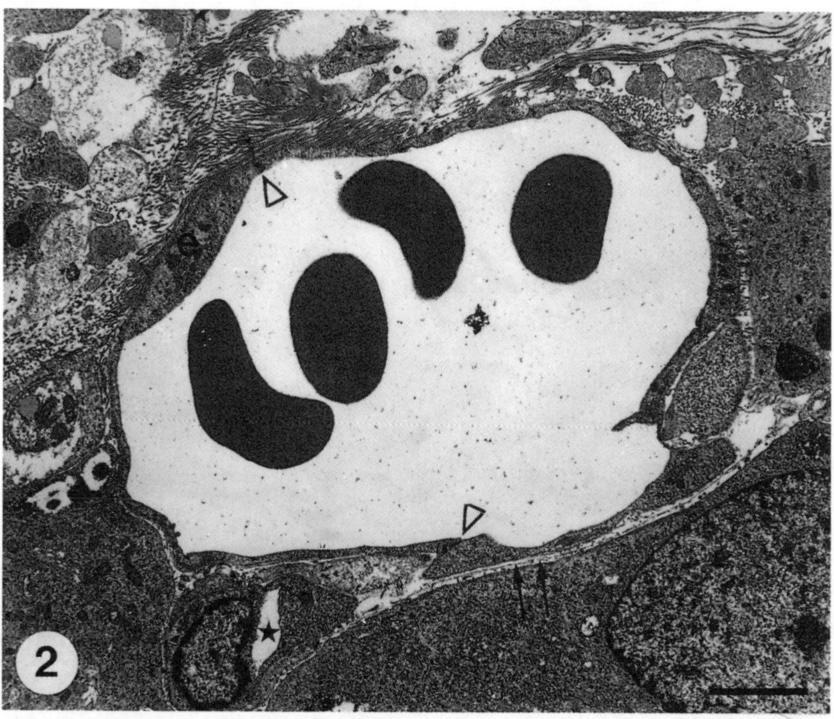

Figure 2. Established tumor vessel with sinusoidal wall structure. Note the in places thinned endothelium (e) and the simply structured cell contacts (△) as well as the poorly developed basement lamina (arrows). Compare to the early form in the lower left (*). Bar = 3 μm.

Pericytes already occur early on, although a distinct relation to the genesis of these cells is frequently not possible. A differentiation from endothelial cells is also rendered more difficult because the endothelia from the capillaries and venoles of the host animal form no endothelial-specific organelles such as Weibel-Palade-bodies.

The structural similarity to endothelial cells and fibroblasts as well as bundles of collagen fibers enclosed in pericytic cells shows that in tumor angiogenesis both the angioblastic as well as the mesenchymal route of pericyte formation and endotheliogenesis are involved (Fig. 6).

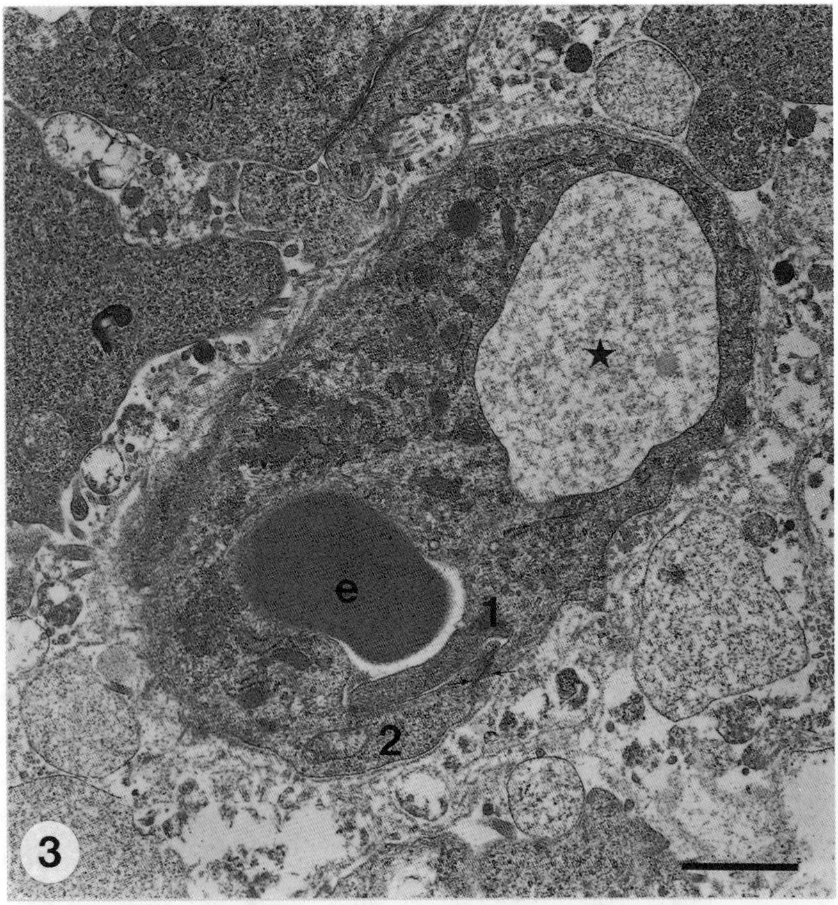

Figure 3. Vascular sprout with intracellular lumen formation (*). Note the high content of cell organelles and the height of the endothelium. The second lumen, which is occluded by an erythrocyte (e) was formed by an extensive reflexive overlapping of two ramifications (1, 2) from the same cell that have developed contacts between them (small arrows). Bar = 5 μm.

The cell contacts of endothelial cells involved in sprout wall formation are seen as simple, sporadically incomplete appositions. As in the case of more mature types, principally only poorly differentiated contact structures are observed, which appear as point-like or elongated diffuse compressions (Fig. 3, 5, 6). Notable overlapping or interdigitations are exceptionally seen in early stages, giving the impression of increased fragility. Notwithstanding, the continuity of the endothelial formation remains intact; no structural gaps or discontinuities can be observed, which frequently occur in more mature vessels of these tumors and favor the formation of extravasates depending on their size and reduced

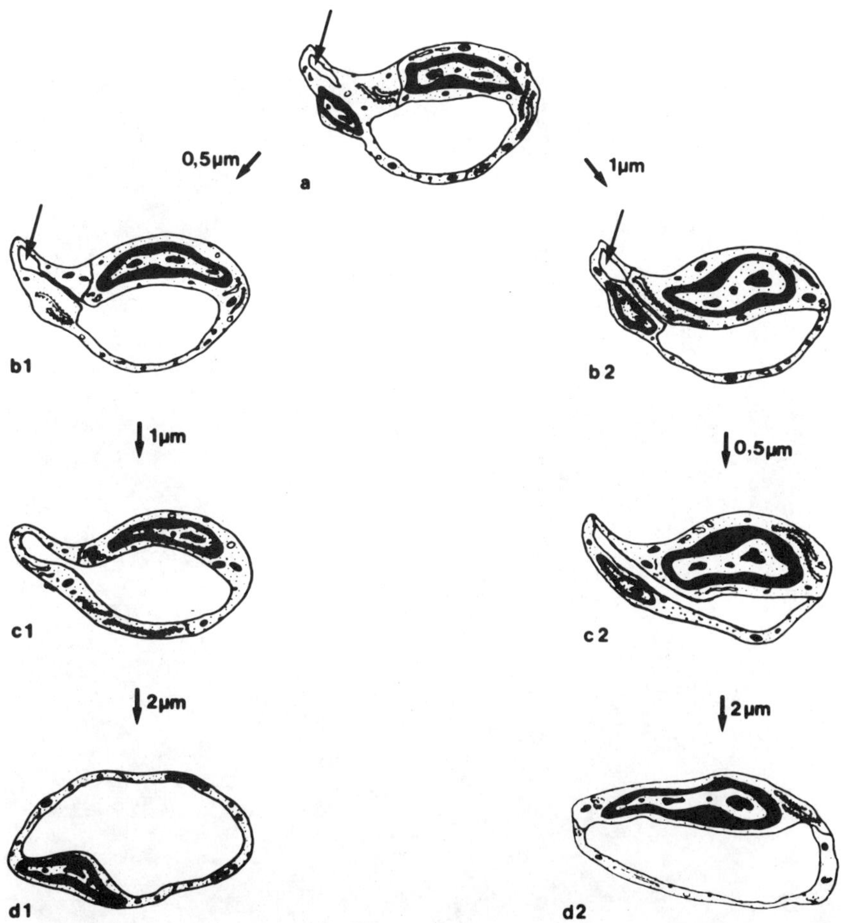

Figure 4. Intracellular vacuole formation as a mechanism for lumen extension. The vacuole in a (arrow) does not form another lumen but, instead, gets into contact with the lumen (b1, b2) in order to enlarge it (c–d). Drawn after serial sections of an amelanotic melanoma (MeWo) 8 days after transplantation. The numbers between the single drawings indicate the distances between the section planes.

resistance. Thus, the semiquantitative evaluation of light microscopic sections also shows a far lower number of extravasates in earlier than later stages (Fig. 7).

Sprout fusion

The newly formed vascular sprouts after elongation either become attached to an established vessel or form an arcade with the parent-vessel or happen in a further sprout. From our studies, the latter possibility, which leads to the formation of new vascular stretches, quantitatively occurs the most often. The variety of sprouts that

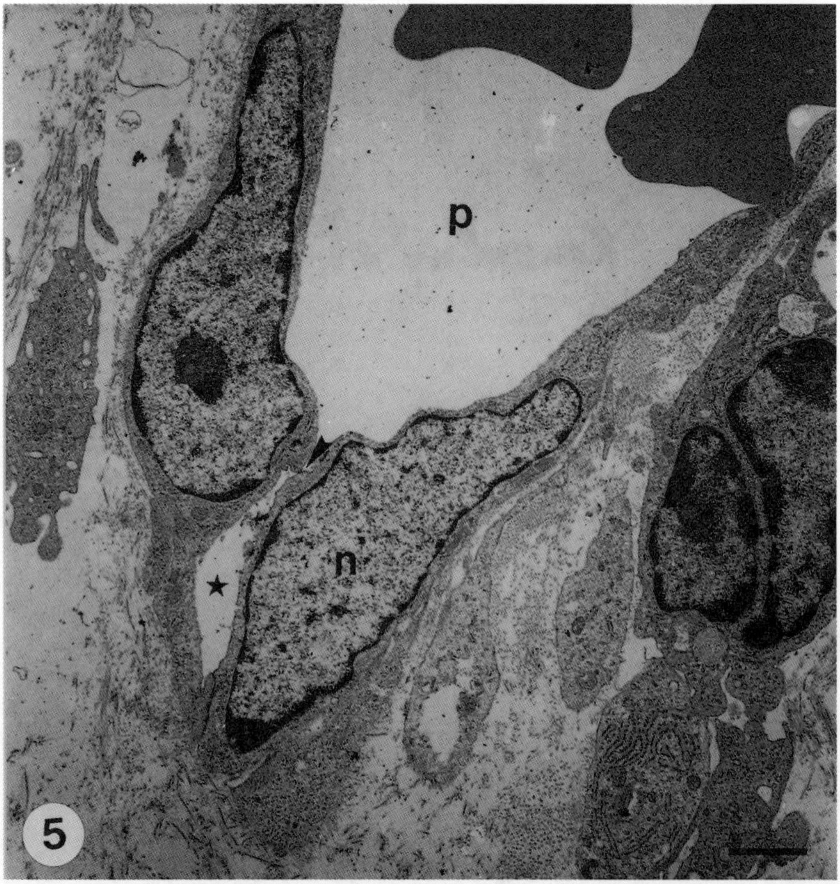

Figure 5. Intercellular sprout development through endothelial cell migration: The eccentrically situated perikaryon and nuclei (n) leading to the formation of a fine slit (*) veer out from the endothelial structure. Due to a punctiform contact (arrowhead) the newly formed lumen (*) is not yet connected to the parent vessel (p).

degenerated or are destructed prior to fusion and structural stabilization, however, indicates that only a small amount of vascular sprouts is finally effectively attached and perfused.

Reconstructive serial sections have shown that, on average, sprout fusion takes place over a distance of 10–15 μm (Fig. 8). This is, however, not to be confused with the length of newly formed stretches of vessels, which are considerably longer. However, since the endothelia involved in sprout formation differentiate very rapidly, only the cells at the tip of the sprout reveal their characteristic features: the endothelium becomes higher, and the slit-like lumina become more tighter. Depending on the stage of fusion, more or less distinctly pronounced cell contacts can be shown amongst the endothelia involved.

Network formation

The angiogenic activity mentioned results in a characteristic vascular architecture [2, 11], as studied by corrosion cast specimens. It thus

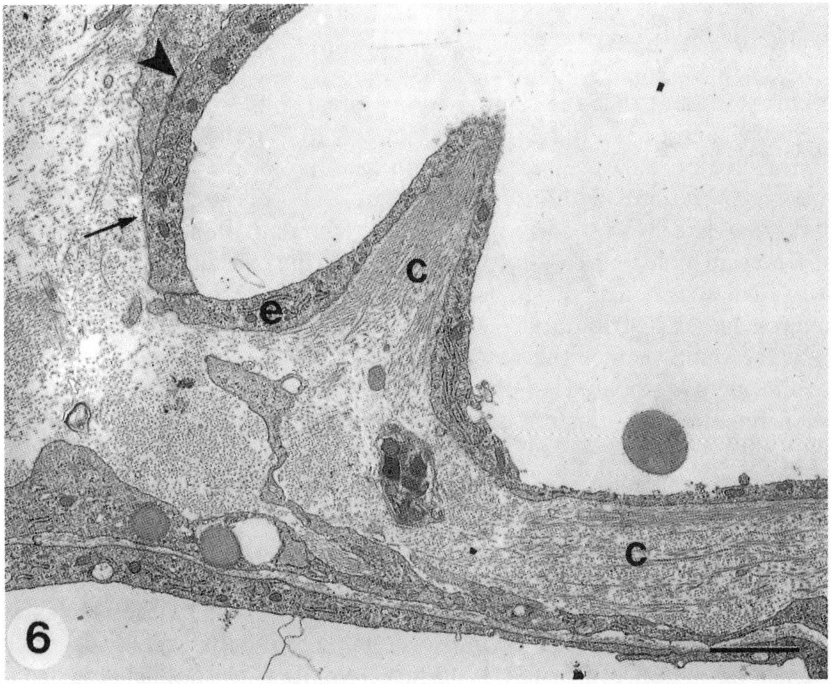

Figure 6. Fine structure of vessels of an amelanotic melanoma. Note the comparatively high content of endoplasmic reticulum in the endothelial cells (e) and the lack of micropinocytotic vesicles. The basal lamina is poorly developed (arrows). Note the slightly flattened endothelium and the simply structured contact zones (arrow heads). c = collagenous fiber bundles. Bar = 3 μm.

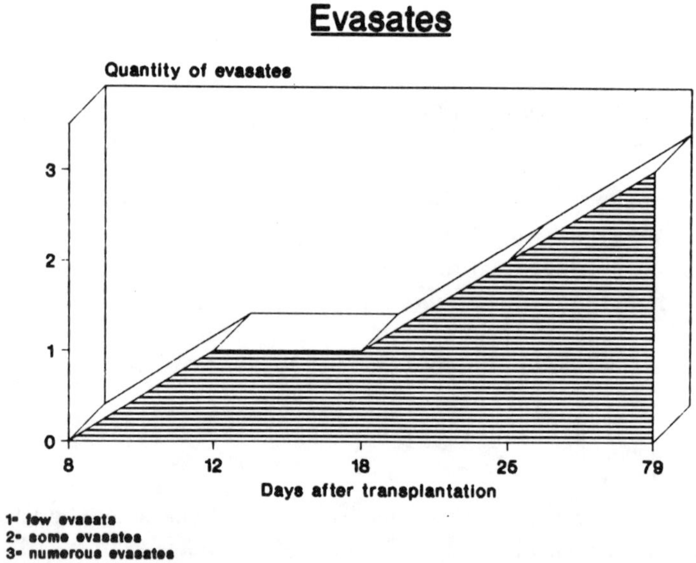

Figure 7. Semiquantitative determination of the amount of extravasates. For all stages, at least 10 sections of five tumors were evaluated by light microscopy.

becomes evident that the tumor vessel system shows a certain degree of hierarchy at best in the capsule (Fig. 9a, b). In the subcapsular and central region new vessel formation results in the formation of a sinusoidal system without any recognizable hierarchy. Frequently, corkscrew-like vessels (Fig. 10a), and redundant capillary loops are observed, which lead to the formation of further vascular sprouts (Fig. 10b). These can also be seen in later stages of growth, with a very heterogeneous distribution.

In the tumor center the vascular system shows a much more chaotic arrangement compared to the main branches of the periphery. Very often regions with clusters of vessels could be seen just beneath areas almost free of vessels (Fig. 11). In this case, no relationship to the topographical localization could be established. The size and extent of these avascular regions varied considerably in the individual xenotransplants.

A major characteristic of the network formation is the formation of a sinusoidal system with numerous blind ends without clearly discernible endothelial cell impressions. Frequently sinusoids with diameters of between 5 and 100 μm originating or confluencing from large caliber vessels could be shown (Fig. 12). This is of considerable consequence for the nutritive blood flow insofar as that the blood in these vessels can no longer show any appreciable oxygen fusion.

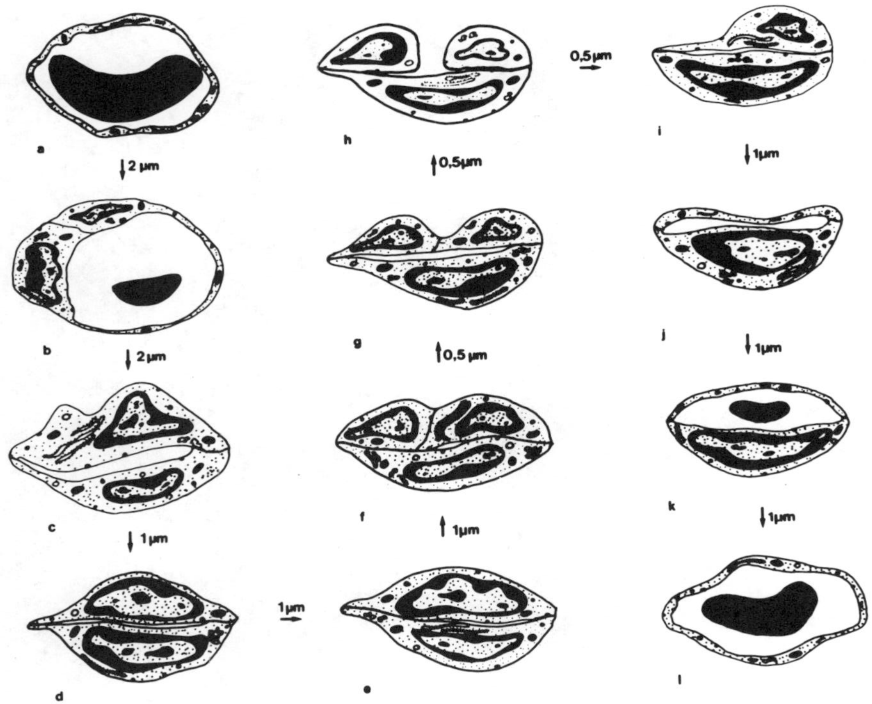

8A

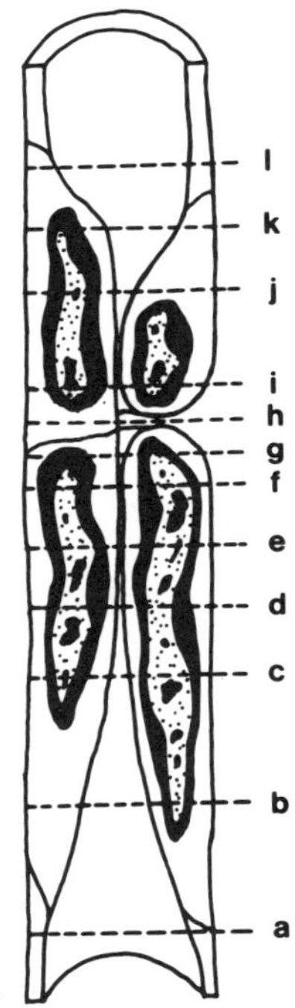

B

Figure 8A and B. Fusion of two sprouts with intercellular lumina. The already differentiated parts of the sprouts (a, b and k, l) show an increased height of the endothelium when reaching the tip of the sprout (c–g and i–j), whereby the presumptive lumen is still a narrow slit. In h, the section plane of the fusion, this lumen is only partially confined by endothelial cells since the distance between the tips is not yet overbridged. Compare the single sections (A) with the reconstruction (B). Drawings of serial sections of a melanoma (MeWo) 8 days after transplantation. Numbers indicate the distance between the single section planes.

Morphometric data

The angiogenic activity should be manifest as a change in vascular density. Since this parameter can be of direct and indirect decisive consequence for the viability of a tissue, the percentage of necrotic vs viable regions was determined in light microscopic sections apart from the number of perfused and functionally compressed vessels.

Figure 13a shows the number of vessel per mm^2 of the squamous carcinoma as a function of time. Despite large ranges of deviation, there was a general tendency for a decrease from an initially high to a low value and which is also reflected by the number of perfusable vessels. Analogously, this also applies to the number of occluded vessels which thus no longer contribute to nutritive blood flow.

On comparing the percentage of necroses during the course of time with the number of vessels (Fig. 13b), particularly perfused ones, no direct correlation can be observed. An analysis of individual specimens has shown that even at high vascular densities many necroses can occur,

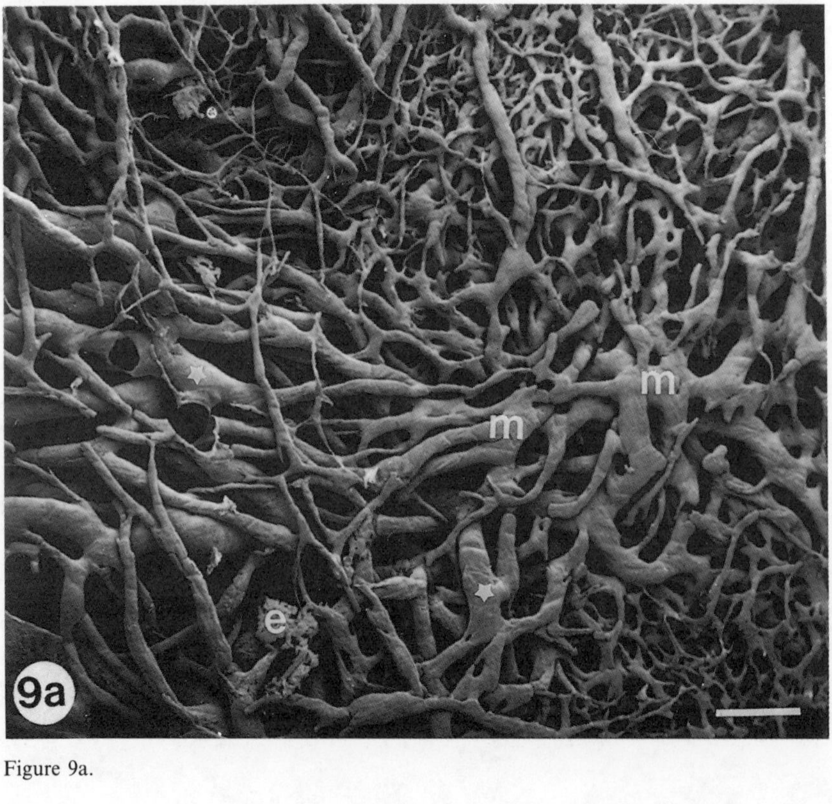

Figure 9a.

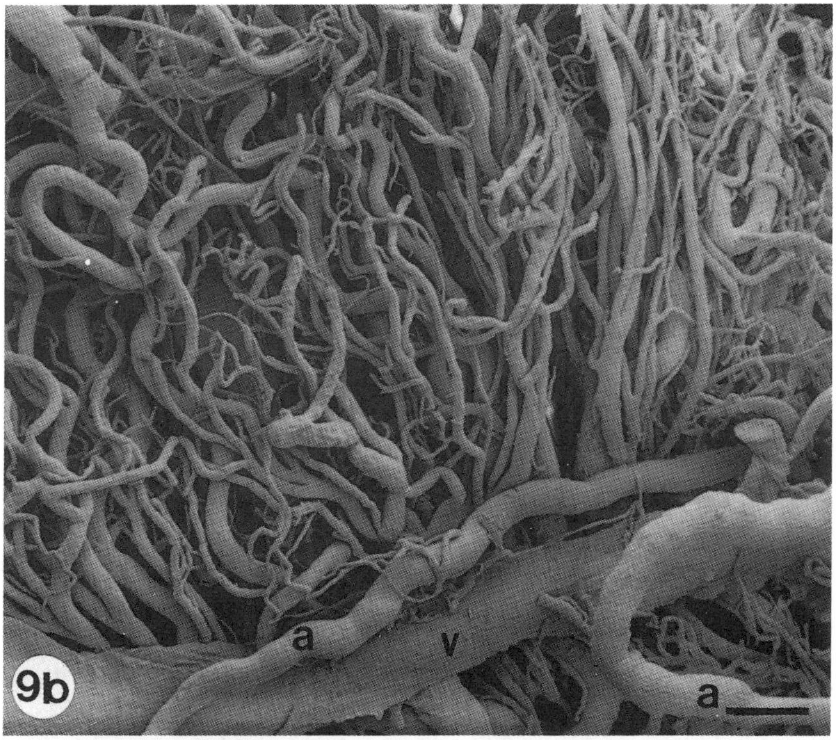

Figure 9a and b. Vascular corrosion cast of the tumor vascular envelope of a xenografted carcinoma 4 days (a) and 15 days after transplantation. Note the low degree of hierarchy and the lack of differentiation as well as the absence of typical branching patterns of the main vessels (m). e = extravasates, a = artery, b = vein, * = flattenings. Bars = 300 μm.

and vice versa. This is a further indication that the number of vessels per unit area alone is not the only determining factor for tissue viability. The existence of vessels is thus only the pre-requisite, but, not the guarantee for sufficient nutritive blood flow.

Discussion

Neoangiogenesis is a phenomenon neither restricted to tumors nor is it surely representative of a pathological condition, but, instead is rather an integral part of numerous normal processes. It is agreed that tumor growth can only occur to any significant extent after establishing a sufficient microcirculation [6, 7]. The significance of angiogenesis for the transition of solid tumors into the growth phase was so overwhelming that the first concept of anti-angiogenesis as a therapeutic measure was already formulated before inhibitors of angiogenesis were even discovered [13].

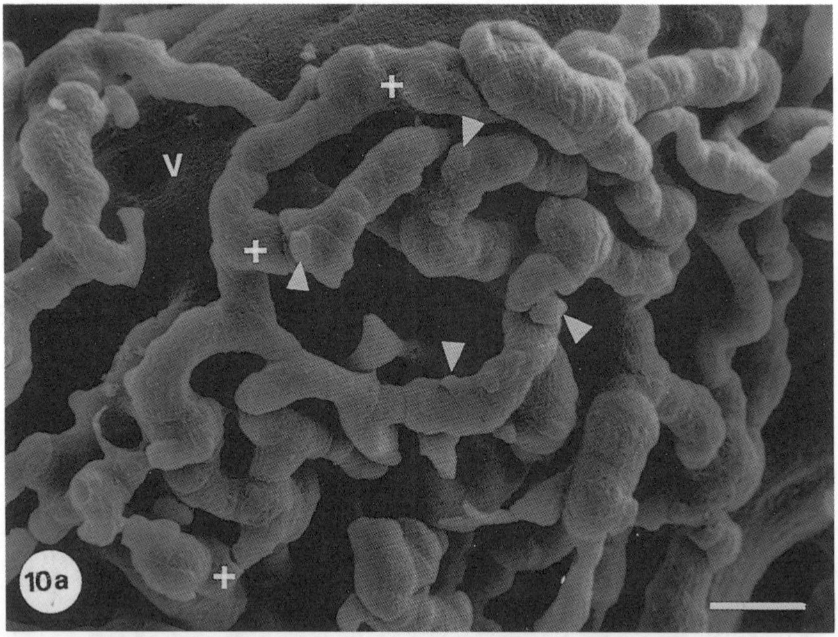

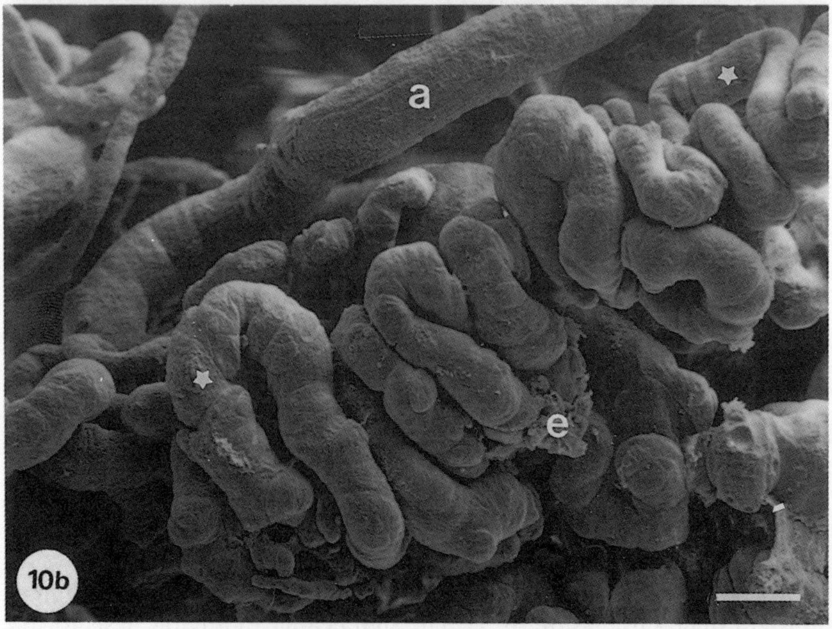

Figure 10a and b. Loop formation in the periphery of a carcinoma. Vascular corrosion casts. Note the newly formed vascular stretches with changes in diameter (+) in front of a large drainage vein (v) in a. Arrows indicate new sprouts. b: corkscrew-like, tortuous vessel in the periphery with circular constriction profiles (*) which are indicative for further sprouting. Bars in a = 20 μm, in b = 30 μm.

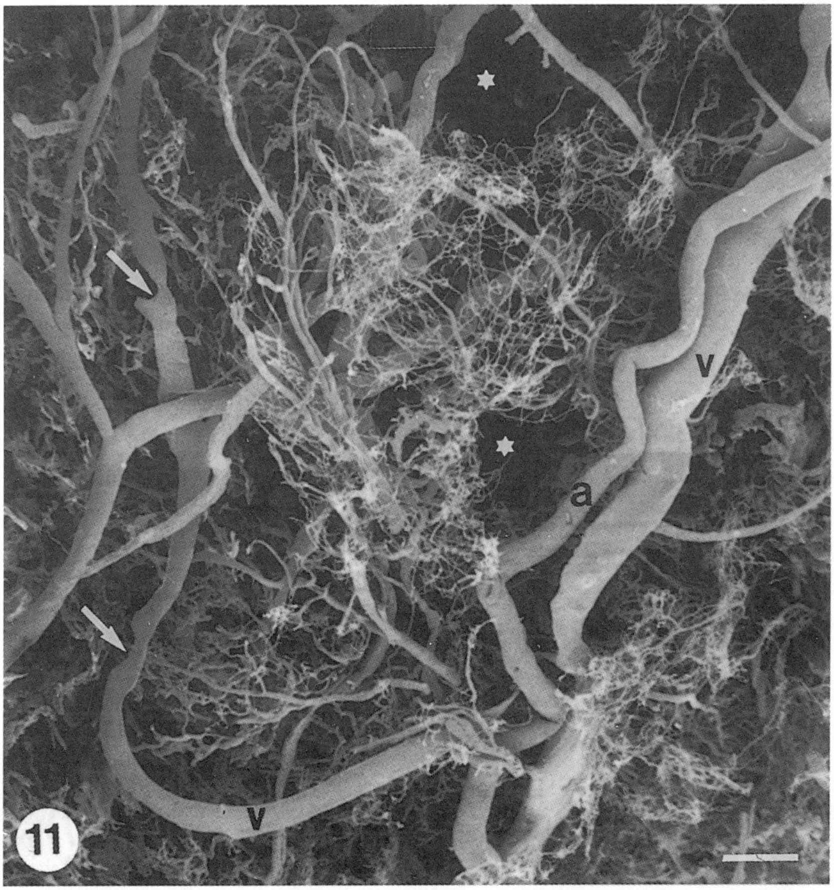

Figure 11. Section from the periphery of a melanoma. Large avascular areas (*) adjacent to a disordered vessel convolute in middle of the figure. a = arterial, v = venous main vessel. Note the compressions and variations in caliber (arrows) of the vein indicated, which shows no influxes over large stretches. Bar = 300 μm.

In spite of this, however there are comparatively few papers dealing with the morphology of tumor angiogenesis [e.g., 1, 14–16]. Most have been concerned with the "why" of angiogenesis and paid less attention to the morphological "how".

This may be grounded on the fact that morphological studies on angiogenesis can be carried out on primary tumors only to a limited extent [17], since the vasculature in the material available is predominantly established as a matter of course. On the other hand, the vascular system of clinical tumors also shows such high turnover rates that descriptive, morphological studies on the angiogenesis in primary tumors are equally possible.

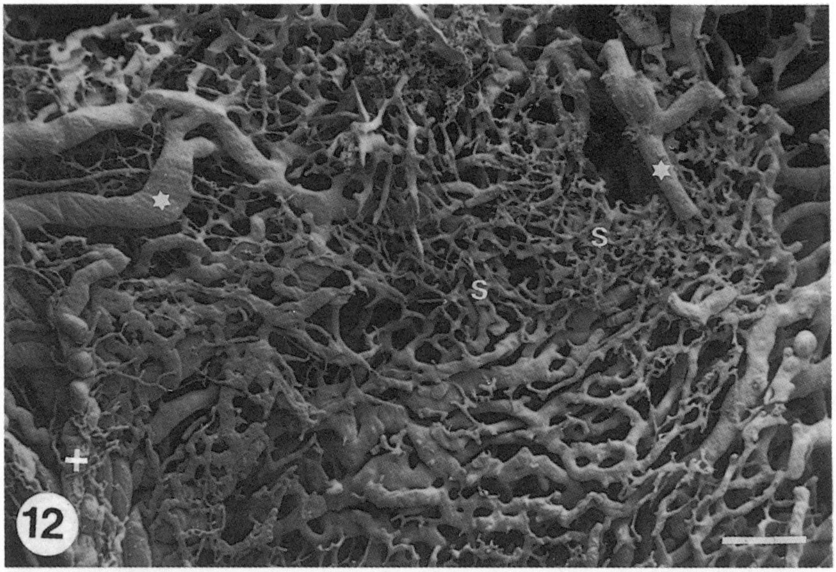

Figure 12. View over the central sinusoidal system (s) of a sarcoma. Note the flattened (*) and partially tortuous veins (+), that drain or supply the system. Bar = 300 μm.

Based on the tumor-biological and experimental-therapeutic implications, we have sequentially studied the morphology of angiogenesis in xenotransplanted human tumors. Up to now, there are no detailed studies available although several authors have concluded that the nude mouse tumor model offers many morphological parallelisms to the clinical situation [18].

Our studies have shown that there are different mechanisms of lumen formation as well as various modes of endotheliogenesis and pericyte formation during angiogenesis. The commonly held view [e.g. 7], that migrated endothelial cells form a solid cord initially which later becomes channalized [8] does not hold for tumor angiogenesis in its present form. Apart from the fact that lumen formation in vascular sprouts is not dependent on the existence of two or more endothelial cells, according to our results at least a partial association with the primary lumen can certainly exist in the case of multicellular sprouts right from the beginning. Intracellular lumen formation in sprouts derived from a single endothelial cell were frequently found at least in the early stages in the xenotransplants studied. In standard text-books on normal and pathological secondary neovascularization no allusions to intracellular lumen formation are yet to be found. This phenomenon, however, was already reported in the first in vitro study on angiogenesis in endothelial cell cultures by Folkman and Haudenschild [19]: They found intracellular vacuoles both in endothelia from tumors as well as normal tissues,

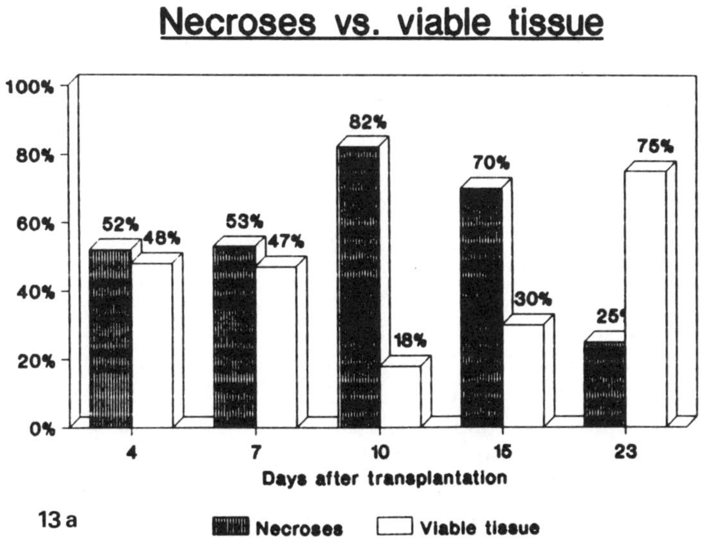

13 a

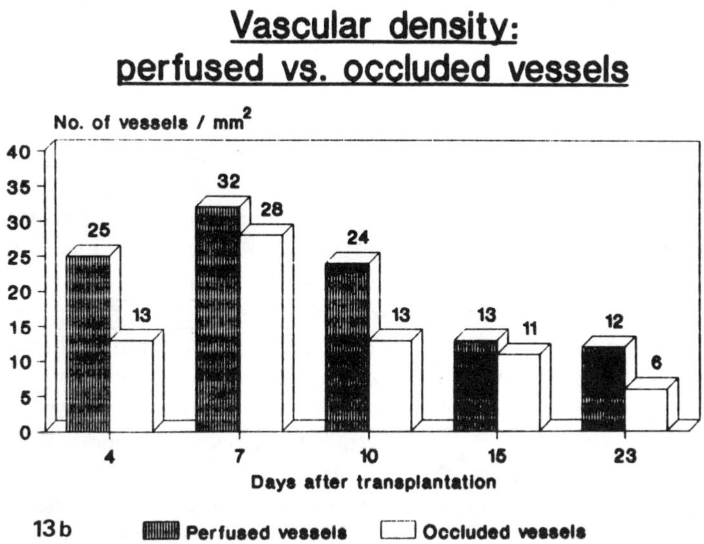

13 b

Figure 13a and b. Quantification of necroses and viable tissue as well as the vascular densities as seen in equatorial light microscopic sections.

forming regular tubular systems. By the way, the formation of such lumina is not specific to endothelia but can also be observed in tumor cells [7, 12]. We consider the occurrence of the different forms of lumen formation as an answer to the enhanced metabolic requirements and the poor nutritive situation.

However, in no case does the result show a balanced relation then despite the particularly high initial angiogenic activity: in the stage with pronounced hypervascularization in effect there already exists a reduced perfusion and tissue hypoxia [20, 21]. On the other hand, this effect is further supported by the overhasty neovascularization due to further missing stabilization of vessels.

The activation of both routes of endotheliogenesis and pericyte formation, which was already postulated by Hammersen et al. [22], is also plausible in terms of function since tumor growth is, among other things, critically dependent on the establishment of the vascular system. A more rapid neovascularization may be achieved through an enhanced recruitment and transformation of fibroblasts, which indeed partially form the stroma in the model used by us [23].

An involvement of tumor cells in the process of angiogenesis, as described by Hammersen 1985 for an experimental tumor [22], could not be shown with certainty in the case of the xenografted human melanoma and carcinoma, although we have previously found indications of tumor cell-lined sinusoids in xenotransplanted soft tissue sarcomas [2, 12]. This important finding requires further comparative studies.

According to the studies by Warren [3, 16] the initiation of the blood flow in the newly formed capillary sprouts should manifest itself in typical, pin-head-like microhemorrhages, which confluence to macroscopic hemorrhages at high capillary sprout density. From this he deduced an increased degree of fragility. Grunt et al. [15] also found an enhanced amount of extravasates around the sprouts in corrosion casts. Our observations do not confirm those of Warren [16]. We noticed extravasations and leaks in the vessel wall above all in ectatic, dilated late types and far less frequently in capillary sprouts in the early stages. These, in turn, are characterized by a high endothelium, which would provide some protection from microhemorrhages purely for mechanical reasons. The investigations by Hammersen et al. [1] also contradict this. The early formation of cell contacts and the lack of endothelial gaps which can be characteristic of late types provide a further means of protection from extravasates. We assume that the extravasates seen by Grunt et al. [15] more likely stemmed from blind-ending sprouts: these are affected by the introduction of the casting medium anyway as well as by the displacement of plasma possibly present resulting in an unphysiological high pressure.

Our morphometric data have shown that there is no simple correlation between the vascular density and necroses frequency [24]. Indeed, the existence of vessels is a major requirement for tissue viability although on no account do vessels guarantee this. This becomes particularly apparent against the background of the structured characteristics of the tumor vascular system, above all, the lack of differentiated vessels and the formation of veno-venous sinusoidal systems with correspond and the formation of veno-venous sinusoidal systems with corresponding low pO_2 [21].

In summary, it can be stated along with Hammersen et al. [1] that the results discussed here on the morphology of tumor angiogenesis show that "contrary to the current universal conceptions, tumor angiogenesis differs only to a minor extent from other neovascularization processes . . . , considerable structural differences . . . occur".

References

1 Hammersen F, Osterkamp-Baust U, Endrich B: Ein Beitrag zum Feinbau terminaler Strombahnen und ihrer Entstehung in bösartigen Tumoren, in Vaupel P, Hammersen F (eds): Mikrozirkulation in Forschung und Klinik. vol. 2. Basel-München-Paris-London-New York-Tokyo-Sydney, Karger, 1983, pp 15–51.

2 Konerding MA, Steinberg F, Budach V: The vascular system of xenotransplanted tumors – scanning and light microscopic studies. Scanning Microsc 1989; 3: 327–336.

3 Warren BA: Tumor angiogenesis, Peterson HI (ed.): Tumor Blood Circulation: Angiogenesis, Vascular Morphology and Blood Flow of Experimental and Human Tumors. Boca Raton (FL), CRC-Press, 1979, pp 49–75.

4 Denekamp J: Vascular endothelium as the vulnerable element in tumors. Acta Radiol Oncol 1984; 23: 217–225.

5 Denekamp J: Endothelial cell proliferation as a novel approach to targeting tumour therapy. Br J Cancer 1982; 45: 136–139.

6 Folkman J: Tumor angiogenesis. Adv Cancer Res 1985; 43: 175–203.

7 Paweletz N, Knierim M: Tumor-related angiogenesis. Crit Rev Oncol Hematol 1989; 9: 197–242.

8 Thoma R: Untersuchungen über die Histogenese und Histomechanik des Gefäß-Systems. Stuttgart, Enke, 1893.

9 Jain RK: Determinants of tumor blood flow: a review. Cancer Res 1988; 48: 2641–2658.

10 Budach V, Bamberg M, Donhuijsen K, Schmidt U, Beuningen van D, Stuschke M: Serial xenotransplantation of a human embryonal carcinoma in experimental urology. J Urol 1986; 136: 1143–1147.

11 Konerding MA, Steinberg F, Streffer C: The vasculature of xenotransplanted human melanomas and sarcomas. 1. Vascular corrosion casting studies. Acta Anat 1989; 136: 21–26.

12 Konerding MA, Steinberg F, Streffer C: The vasculature of xenotransplanted melanomas and sarcomas. 2. Scanning and transmission electron microscopic studies. Acta Anat 1989; 136: 27–33.

13 Folkman J: Anti-angiogenesis: new concept for therapy of solid tumors. Ann Surg 1972; 175: 409–416.

14 Eddy HA, Casarett GW: Development of the vascular system in the hamster malignant neurilemmoma. Microvasc Res 1973; 6: 63–82.

15 Grunt TW, Lametschwandtner A, Karrer K: The angioarchitecture of the Lewis lung carcinoma in laboratory mice. Scanning Microsc 1986; II: 557–573.

16 Warren BA: The ultrastructure of the microcirculation at the advancing edge of Walker 256 carcinoma. Microvasc Res 1970; 2: 443–453.

17 Weller RO, Foy M, Cox S: The development and ultrastructure of the microvasculature in malignant gliomas. Neuropathol appl Neurobiol 1977; 3: 307–322.

18 Donhuijsen K, van Beuningen D, Budach V, Schmidt U: Instability of xenotransplanted soft tissue sarcomas: morphologic and flow cytometric results. Cancer 1988; 61: 68–75.

19 Folkman J, Haudenschild C: Angiogenesis in vitro. Nature 1980; 288: 551–556.

20 Endrich, B: Hyperthermie und Tumormikrozirkulation – eine kritische Wertung experimenteller und klinischer Befunde, Eckhardt S, Holzner JH, Nagel GA (eds.): Beiträge zur Onkologie, vol. 31, Basel-München-Paris-London-New York-New Delhi-Singapore-Tokyo-Sydney, Karger, 1988.

21 Vaupel P, Fortmeyer HP, Runkel S: Blood flow, oxygen consumption, and tissue oxygenation of human breast cancer xenografts in nude rats. Cancer Res 1987; 47: 3496–3503.
22 Hammersen F, Endrich B, Messmer K: The fine structure of tumor blood vessels. I. Participation of non-endothelial cells in tumor angiogenesis. Int J Microcirc Clin Exp 1985; 4: 31–43.
23 Giovanella BC, Fogh J: Present and future trends in investigations with the nude mouse as a recipient of human tumor transplants, Fogh J, Giovanella BC (eds.): The Nude Mouse in Experimental and Clinical Research. New York, Academic Press, 1978, pp 281–297.
24 Steinberg F, Konerding MA, Streffer C: The vascular architecture of human xenotranplanted tumors: histological, morphometrical and ultrastructural studies. Int J Cancer Res 1991; 116: 517–524.

Angiogenesis: Key Principles - Science - Technology - Medicine
ed. by R. Steiner, P. B. Weisz & R. Langer

Maturation of newly-formed subretinal vessels

Tatsuro Ishibashi[a] and Stephen J. Ryan[b]

[a]*Department of Ophthalmology, Faculty of Medicine, Kyushu University, Fukuoka, Japan*; [b]*Doheny Eye Institute, Department of Ophthalmology, University of Southern California, School of Medicine, Los Angeles, California, USA*

Abstract. The maturation process of newly-formed subretinal vessels was studied ultrastructurally. Primitive new vessels have thick endothelial cells with cytoplasm that contains many ribosomes and rough endoplasmic reticulum. Some immature endothelial cells show budding, and many of the intercellular junctions have small gaps. As the newly-formed vessels mature, the endothelial cells become attenuated and fenestrated, and their junctions become tight. Mature subretinal new vessels closely resemble those of the normal choriocapillaris.

Introduction

Subretinal neovascularization (SRN) is a pathologic process common to many eye diseases [1]. The newly-formed vessels proliferate from the choroid and extend into the subretinal space, where they are diagnosed by fluorescein leakage during angiography [2].

We have developed a primate model in which SRN is induced by high-intensity laser photocoagulation [3, 4], this experimentally induced neovascularization closely correlates with the SRN in humans. In a previous study of this primate model, we showed that the early stages of SRN were characterized by endothelial cell migration and proliferation from existing choroidal vessels [5]. In the study described herein, we used transmission electron microscopy to study progression of the maturation progress of the newly-formed subretinal vessels in our primate model.

Materials and methods

SRN was induced in cynomolgus monkeys by intense laser photocoagulation. High-intensity laser burns were applied at and around the macula of one eye each of four monkeys, as previously described [3, 4]. The eyes were enucleated at 3, 8, 14, and 90 days after photocoagulation, opened and fixed overnight by immersion in 2% paraformaldehyde and 2.5% glutaraldehyde in 0.1 M phosphate buffer (pH 7.4). The laser lesions were then excised, post-fixed in 2% osmium tetroxide in 0.1 M phosphate buffer, pH 7.4, for two hours, dehydrated in a graded series of alcohol followed by propylene oxide, and embedded in epoxy resin. Thin sections were cut on an ultramicrotome, stained with uranyl acetate and lead citrate, and viewed with an electron microscope.

Results and discussion

Three days after photocoagulation, many newly-formed vessels were found in the subretinal space and choroid. These vessels had lumens of various sizes and the endothelial cells typically had bulky cytoplasm that did not contain fenestrations (Fig. 1). A major portion of the cytoplasm was occupied by ribosomes and rough endoplasmic reticulum. Mitochondria, Golgi apparatuses, microfilament bundles, and Weibel-Palade bodies were also present. The endothelial cells had a discontinuous, thin basement membrane, and their nuclei contained large amounts of chromatin. Endothelial cell cytoplasmic processes of various shapes and lengths were abundant in the newly-formed vessels; the tips of these cytoplasmic processes were not covered by basement membrane. These phenomena represent budding of endothelial cells (Fig. 2). Although small numbers of ribosomes and some rough endoplasmic reticulum are found commonly in mature endothelial cells, these are particularly abundant in developing cells [6]. The presence of numerous free ribosomes suggests a capacity for synthesis of substantial amounts of protein; the primary function of rough endoplasmic reticulum is the production of secretory or export protein [7]. These findings suggest that the endothelial cells of these primitive new vessels are metabolically very active.

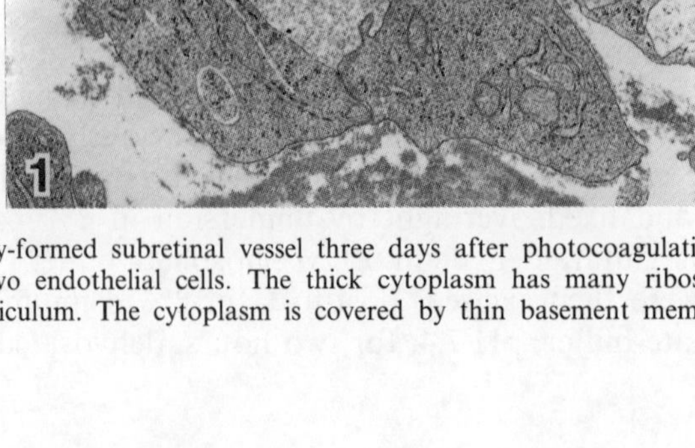

Figure 1. Newly-formed subretinal vessel three days after photocoagulation. The vessel is composed of two endothelial cells. The thick cytoplasm has many ribosomes and rough endoplasmic reticulum. The cytoplasm is covered by thin basement membrane. L; lumen. (×14,800)

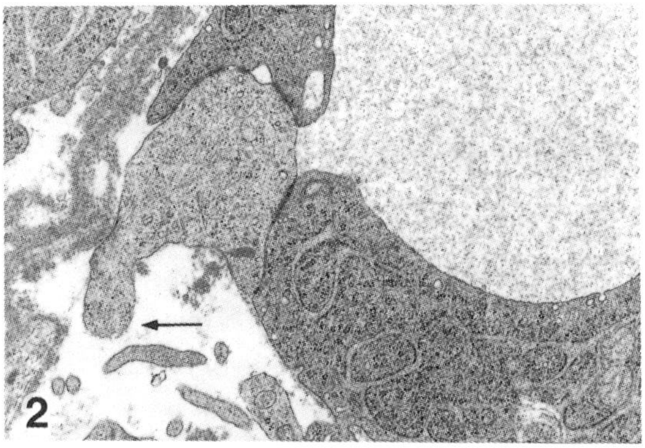

Figure 2. Budding endothelial cell three days after photocoagulation. The cytoplasmic process of the endothelial cell (arrow) extends into the perivascular space. The tip of the process has no basement membrane. (× 17,000)

Intercellular junctions of various lengths were observed between endothelial cells in the newly-formed vessels; some junctions were long and tortuous, while others were very short. The junctions were open, with short adherens regions in which the junctional membranes showed increased electron density (Fig. 3). These junctions were classified as intermediate junctions [8], although definition was often poor. Some junctions showed the formation of gaps (Fig. 4), and the intermediate junctions with gaps between endothelial cells would afford access of free plasma into the extravascular space.

Pericytes were sometimes observed around the endothelial cells (Fig. 5). The cytoplasm of these pericytes was also rich in ribosomes and rough endoplasmic reticulum.

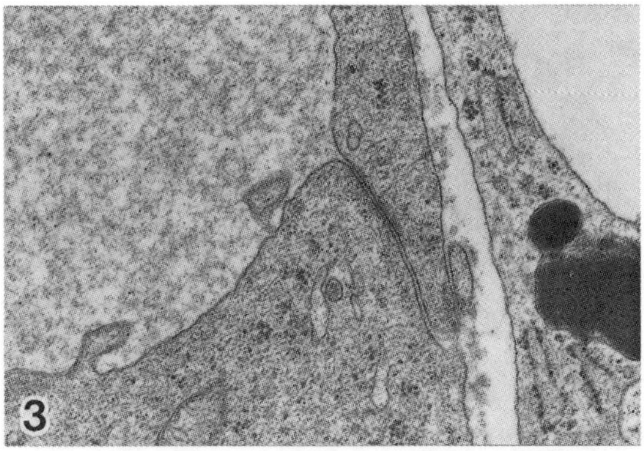

Figure 3. Higher magnification of the intercellular junction between endothelial cells shown in Fig. 1. The junction is open with short adherens regions in which the junctional membranes show increased electron density. (× 44,200)

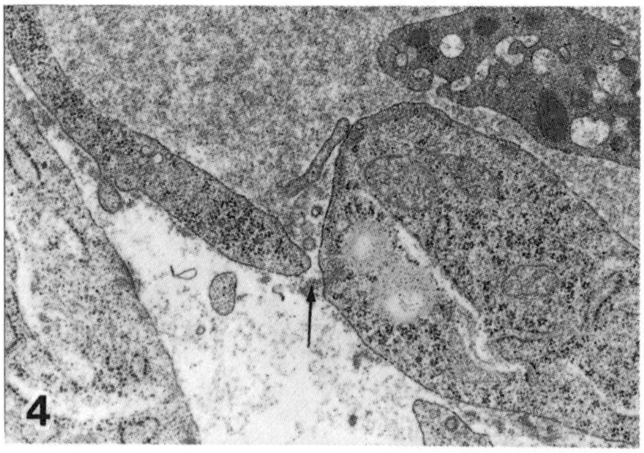

Figure 4. Intercellular junction of newly-formed subretinal vessel three days after photocoagulation. The intercellular junction between endothelial cells is open and has a small gap (arrow). L; lumen. (× 23,700)

By eight days after photocoagulation, some of the newly-formed vessels appeared quite mature. These vessels had attenuated cytoplasm that demonstrated characteristic fenestrations, which were 50 to 60 nm in diameter and which were closed by diaphragms. There were fewer ribosomes and less rough endoplasmic reticulum, and the endothelial cells were covered now by continuous basement membrane, in some areas with associated pericytes.

By 14 days after photocoagulation, the mature vessels had increased in number, and by 90 days after photocoagulation, all of the new vessels had a mature appearance. The endothelial cell cytoplasm of these mature new vessels had many fenestrations with diaphragms (Fig. 6), and the mature subretinal vessels resembled closely those of the normal

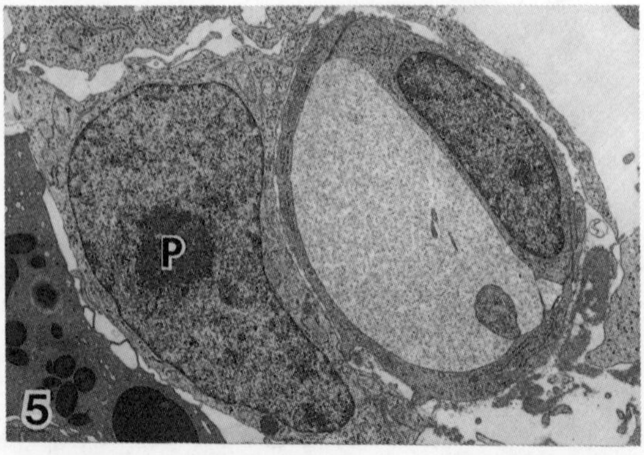

Figure 5. Newly-formed subretinal vessel three days after photocoagulation. The vessel has a pericyte (P). The cytoplasm of the pericyte has many ribosomes and rough endoplasmic reticulum. (× 7,500)

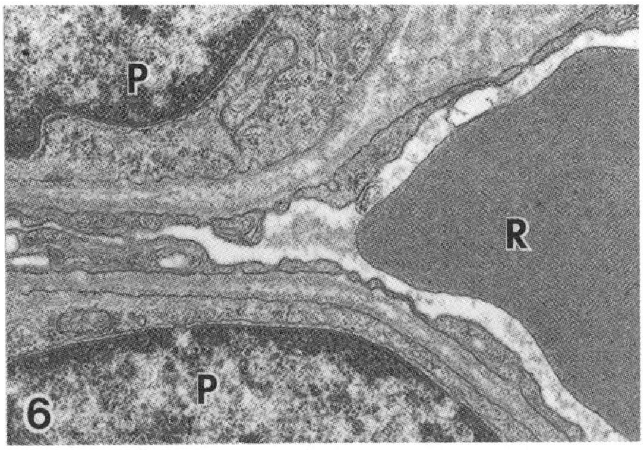

Figure 6. Newly-formed subretinal vessel 90 days after photocoagulation. The thin cytoplasm of endothelial cells has many fenestrations with diaphragms. R; Red blood cell, P; Pericyte. (× 29,000)

choriocapillaris, which is a layer of fenestrated capillaries in the choroid of the eye [9]. As the subretinal new vessels originated from extant choroidal vessels [5], it is not surprising that their appearance was characteristic of choriocapillaris. Some of the junctions between endothelial cells had focal fusion points between opposing plasma membranes, and, thus, represented tight junctions. As the intercellular junctions of the new vessels became mature, tight junctions were established.

References

1 Ryan SJ, Mittl RN, Maumenee AE: The disciform response: a historical perspective. Graefes Arch Clin Exp Ophthalmol 1980; 215: 1–20.
2 Gass JDM: Pathogenesis of disciform detachment of the neuroepithelium. IV. Fluorescein angiographic study of senile disciform macular degeneration. Am J Ophthalmol 1967; 63: 645–659.
3 Ryan SJ: The development of an experimental model of subretinal neovascularization in disciform macular degeneration. Trans Am Ophthalmol Soc 1979; 77: 707–745.
4 Ryan SJ: Subretinal neovascularization: natural history of an experimental model. Arch Ophthalmol 1982; 100: 1804–1809.
5 Ishibashi T, Miller H, Orr G, Sorgente N, Ryan SJ: Morphologic observations on experimental subretinal neovascularization in the monkey. Invest Ophthalmol Vis Sci 1987; 28: 1116–1130.
6 Ausprunk DH, Knighton DR, Folkman J: Differentiation of vascular endothelium in the chick chorioallantois: a structural and autoradiographic study. Dev Biol 1974; 38: 237–248.
7 Palade GE, Porter KR: Studies on the endoplasmic reticulum. I. Its identification in cells in situ. J Exp Med 1954; 100: 641–656.
8 Townes-Anderson E, Raviola G: The formation and distribution of intercellular junctions in the rhesus monkey optic cup: the early development of the cilio-iridic and sensory retinas. Dev Biol 1981; 85: 209–232.
9 Hogan MJ, Alvarado JA, Weddell JE: Histology of the Human Eye: An Atlas and Textbook. Philadelphia, WB Saunders, 1971, pp. 363–371.

Angiogenesis: Key Principles - Science - Technology - Medicine
ed. by R. Steiner, P. B. Weisz & R. Langer

Embryonic development of the CNS microvasculature in the mouse: new insights into the structural mechanisms of early angiogenesis

H. C. Bauer[a], M. Steiner[b] and H. Bauer[c]

[a]*Institut für Molekularbiologie, ÖAW, Billrothstr. 11, A-5020 Salzburg, Austria,*
[b]*Institut für Histologie und Embryologie, Universität Wien, A-1090 Wien, Austria, and*
[c]*Institut für Zoologie, Universität Salzburg, A-5020 Salzburg, Austria*

Abstract. Vascularization and expression of blood-brain barrier (bbb) associated morphological characteristics were studied by ultrastructural analysis during the embryonic development of the mouse CNS. At day 9 (E9) capillaries were only found in the perineural mesenchymal tissue exhibiting fenestrations and loosely attached pericytes. At E10 endothelial cells (EC) together with pericytes and other mesenchymal cells invaded the intraneural domain. The immigrating EC showed numerous protrusions and lost their fenestrations as soon as the new intraneural capillaries were formed. Intraneural EC showed prominent nuclei, many pinocytotic vesicles and junctional complexes which appeared tight in some places. These results indicate that from the first day of intraneural vascularization onwards, the morphological properties of the bbb are present in the early embryonic mouse cerebral cortex.

Introduction

The gate-keeping role of the brain endothelium, which is induced by specific neural environment [1–6] is basically maintained by a series of special endothelial transport properties: tight junctions, lack of fenestrations and the scarcity of pinocytotic vesicles are considered to be the morphological equivalent of the blood-brain barrier (bbb) [7, 8]. Apart from extensive investigations on the fetal and neonatal human brain [9] most of the studies have been done with embryonic and newborn rat [10, 11], hamster [12], rabbit [13] and chick [14]. Very little information exists about the early events which may be crucial during the formation of the cerebral microvasculature in the mouse. Therefore our attention has been focused on murine cerebral ontogenesis, beginning with the ninth day of development.

In this paper we demonstrate for the first time, that EC of the developing mouse cerebral cortex already display the morphological characteristics for a subsequent bbb function as early as day 10 of gestation.

Material and methods

Balb/c mice were used as experimental animals. Pregnant mice were anesthetized with diethyl ether, the fetus was removed from the uterus and killed in ice water. Either whole fetuses (younger than E14) or brains from fetuses were fixed, osmicated, dehydrated and embedded in Epon 812 resin. Semithin sections of the prosencephalic region were stained with Azur II-methylene blue and used for overview pictures. Ultrathin sections were contrasted with 1% uranyl acetate and lead citrate and examined with an electron microscope (Zeiss EM 10).

Results

The overview picture in Fig. 1a shows that the prosencephalon consists of two distinct domains, the perineural and the intraneural domain. The perineural mesenchymal tissue is already vascularized. Pericytes (PC) were attached to these capillary EC which had numerous junctional complexes (JC) and fenestrations (Fig. 1b, Tab. 1). One day later, the neural groove was closed and vascularization of the intraneural part took place. Fig. 1c shows that not EC alone, but also other mesenchymal cells, f.i. PC, invaded the intraneural section. Immigration took place at performed "tents" of the separating lamina. The immigrating EC were elongated and exhibited numerous protrusions (Fig. 1d). Intraneural EC did not show any fenestrations and in perineural EC less fenestrations were observed compared to E9 (Tab. 1). Between E10 and E14 the intraneural domain greatly expanded, capillary lumina became more rounded (Fig. 1e). At day 17 the capillaries were tightly embedded in the neopallic tissue (Fig. 1f). PC were closely attached to the EC and astroglial footprocesses were in intimate contact with the capillary wall.

Table 1. Frequency of endothelial fenestrations (f) and junctional complexes (j.c.) in cEC of E9 to E17

	Perineural capillaries		Intraneural capillaries	
	f	j.c.	f	j.c.
E9	0.9	3.2	no capillaries present	
E10	0.13	2.9	n.d.	1.7
E12	n.d.	3.25	n.d.	2.58
E14	n.d.	2.43	n.d.	2.29
E17	n.d.	2.75	n.d.	2.25

Values = $\frac{\text{number of f, j.c.}}{\text{numbers of vessels examined}}$; n.d.: not detected; vessels examined: n = 7–17, means are given.

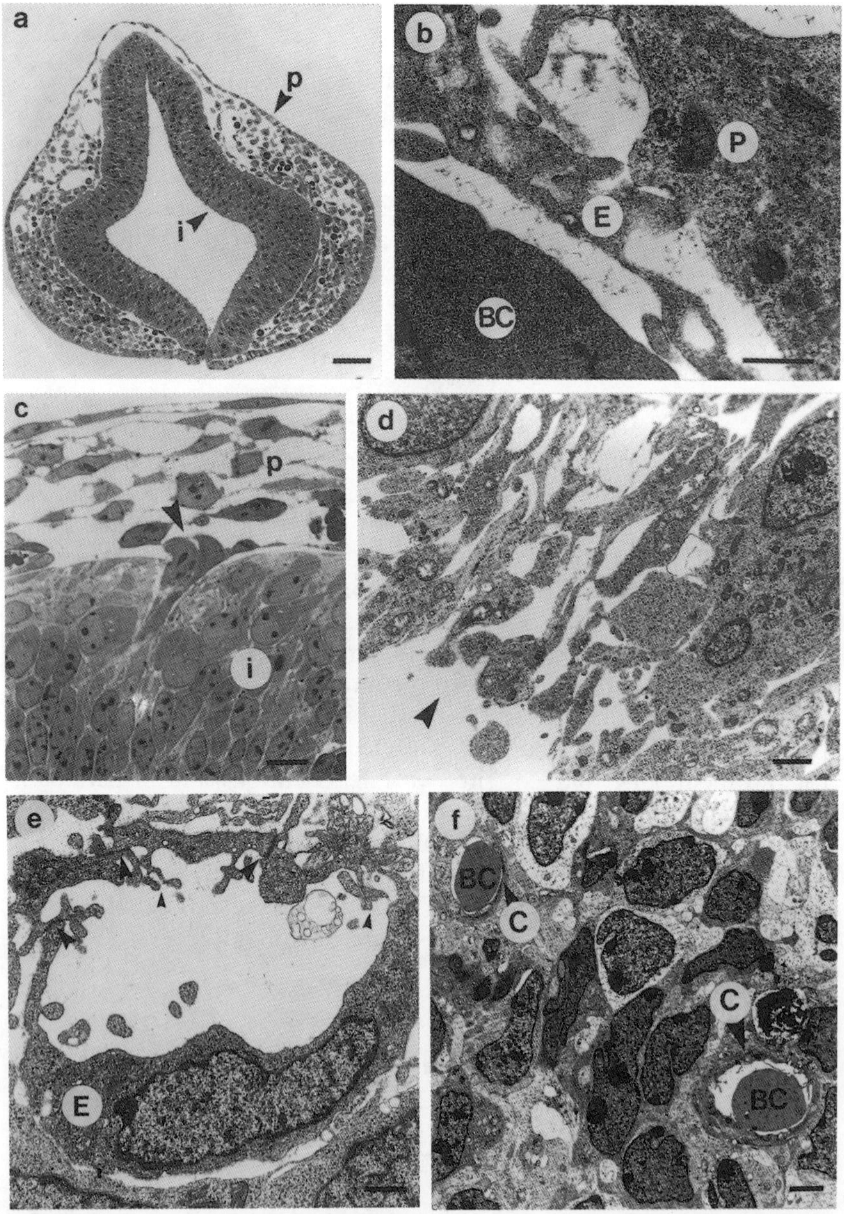

Figure 1. *a,c*: Semithin sections; *b,d,e,f*: ultrathin sections. *a*) Overview of an E9 prosencephalon. p: perineural domain, i: intraneural domain. bar: 20 μm. *b*) Transsection of a perineural capillary. E: endothelial cell, P: pericyte, BC: blood cell. bar: 1 μm. *c*) Immigrating mesenchymal cells ▲ at E10. p: perineural domain, i: intraneural domain. bar: 2.5 μm. *d*) Protrusions of an immigrating EC ▲ bar: 2 μm. *e*) Endothelial cell (E) at E14; ▲ jc, ▲ pinocytotic activity. bar: 1 μm. *f*) Intraneural domain at E17. C: capillaries, BC: blood cells. bar: 2 μm.

Discussion

The onset of intraneural vascularization took place at E9, which is in good agreement with data reported by Herken et al. [15]. Compared to the embryonic rat and hamster, there are considerable timing differences in the beginning of cortex vascularization. In both rodent species, immigration of capillaries into the intraneural domain occur later compared to the mouse [10, 12]. We have demonstrated, that not only EC cells but also other mesenchymal cells invade the intraneural tissue at E10. From the shape of nuclei and the texture of these cells we conclude that probably PC and microglia accompany the immigrating EC. This was also confirmed by our observations that PC were already loosely attached to perineural EC before the onset of intraneural vascularization. In perineural vessels fenestrations existed throughout embryonic development, even though they were markedly decreased after E10. Thus perineural EC lost their fenestrations as soon as they invaded the intraneural tissue. This is in contrast to early rat brain capillaries, where fenestrations did not disappear intraneurally before E17 [10]. We have further shown that the number of endothelial JC in intraneural capillaries, appearing "tight" in some places, were frequent and remained constant throughout embryonic development.

The "maturity" of the fetal bbb is still being discussed. Numerous experiments with intravenously administered tracers have been carried out, suggesting, that horse radish peroxidase is not excluded from the mouse brain before E16 [16]. On the other hand, there are findings indicating that the bbb to proteins is acquired much earlier in fetal development [9]. In summary, our data suggest that morphological bbb characteristics appear simultaneously with intraneural capillary formation, thus supporting the hypothesis of an early bbb function.

Acknowledgement. This work was supported by a grant from the Austrian FWF, project number 7654.

References

1 Bradbury MWB: The concept of a blood-brain barrier. Wiley & Sons, New York 1979.
2 Goldstein GW, Betz AL: The blood-brain barrier. Sc Am 1986; 255: 70–79.
3 Stewart PA, Wiley MJ: Developing nervous tissue induces formation of blood-brain barrier characteristics in invading endothelial cells: A study using quail-chick transplantation chimera. Dev Biol 1981; 84: 183–192.
4 DeBault LE, Cancilla PA: Gamma-glutamyl transpeptidase in isolated brain endothelial cells. Induction by glial cells in vitro. Science 1980; 207: 653–655.
5 Tao-Cheng J, Nagy Z, Brightman MW: Tight junctions of brain endothelium in vitro are enhanced by astroglia. J Neurosci 1987; 7: 3293–3299.
6 Tontsch U, Bauer HC: Glial cells and neurons induce blood-brain barrier related enzymes in cultured cerebral endothelial cells. Brain Res 1991; 539(2): 247–253.

7 Reese TA, Karnovsky MJ: Fine structural localization of a blood-brain barrier to exogenous peroxidase. J Cell Biol 1967; 34: 207–217.
8 Brightman MW, Reese TA: Junctions between intimately apposed cell membranes in the vertebrate brain. J Cell Biol 1969; 40: 648–677.
9 Mollgard K, Dziegielewska KM, Saunders NR, Zakut H, Soreq H: Synthesis and localization of the plasma proteins in the developing human brain. Dev Biol 1988; 128: 207–221.
10 Yoshida Y, Yamada M, Wakabayashi K, Ijuta F: Endothelial fenestrae in the rat fetal cerebrum. Dev Brain Res 1988; 44: 211–219.
11 Bär T., Wolff JR: The formation of capillary basement membranes during internal vascularization of the rat's cerebral cortex. Z Zellforsch 1972; 133: 231–248.
12 Marin-Padilla M: Early vascularization of the embryonic cerebral cortex: golgi and electron microscopic studies. J Comp Neurol 1985; 241: 237–249.
13 Braun LD, Cornford EM, Oldendorf WH: Newborn rabbit blood-brain barrier is selectively permeable and differs substantially from the adult. J Neurochem 1980; 34: 147–152.
14 Wakai S, Hirokawa N: Development of the blood-brain barrier in the chick embryo. Cell Tiss Res 1978; 195: 195–203.
15 Herken R, Goetz W, Wattjes KH: Initial development of capillaries in the neuroendothelium of the mouse. J Anat 1989; 164: 85–92.
16 Risau W: Differentiation of blood-brain barrier endothelium. TINS 1990; 4: 151–153.

Angiogenesis: Key Principles - Science - Technology - Medicine
ed. by R. Steiner, P. B. Weisz & R. Langer

Vascular supply in embryonic neural transplants

Daniela Virgintino, Domenico Ribatti, Beatrice Nico,
Mirella Bertossi and Luisa Roncali

Institute of Human Anatomy, Histology and Embryology, University of Bari Medical School, Bari, Italy

Introduction

Neural transplantations in the mammalian central nervous system have been used to investigate the problems of brain plasticity, blood-brain barrier formation, and vascular astroglial relations [1–7].

The source of the blood allowing the reperfusion and survival of neural tissues grafted into the central nervous system has been evidenced in detailed microscopic and ultrastructural studies demonstrating that the vessels of adult recipients are stimulated to grow towards the graft and to anastomose with its pre-existing vasculature [8, 9].

In the present study the plasticity of the brain vascular components and the mechanisms controlling the transplant vascularization are investigated during early embryonic development, after an implant of highly differentiated and vascularized neural tissues of quail embryos into the cerebral ventricles of very young chicken embryos, at the start of intraneural vascularization [10–11].

Preliminary reports have been published [12, 13].

Materials and methods

Small fragments of the dorsal wall of the mesencephalic vesicle (optic tectum) of quail embryos (donors) at the 12th incubation day were inserted into the mesencephalon ventricles of fifty 6-incubation day chick embryos (hosts). Twenty-four hours after the implant the chick embryos were sacrificed and their mesencephalic vesicles were fixed in Carnoy's solution, or in 0.1 M phosphate-buffered 3% glutaraldehyde, and embedded in paraffin, or in Epon 812. The observations were made on thick and semi-thin sections, stained with Toluidine blue or Schiff reagent in the Feulgen reaction. The latter staining allows the quail nuclei (with a strongly Feulgen-positive, large, central mass of chromatin) to be distinguished from the chick nuclei supplied with evenly distributed chromatin [14].

Results

24 hours after surgery, the graft was seen either suspended from the host ventricle or adhering to its neuroepithelium.

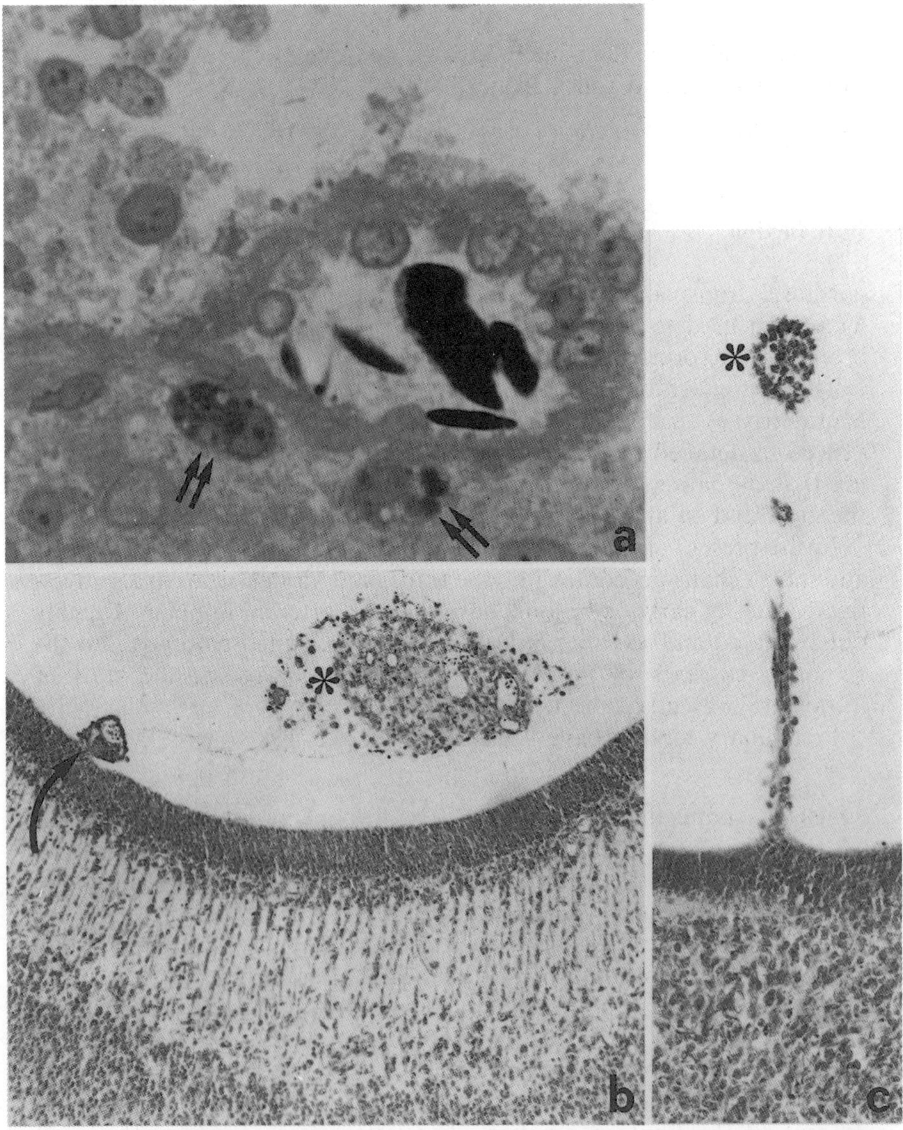

Figure 1. *a* Semi-thin section of a microvessel at the ventricular surface of the host tectum; two perivascular phagocytic cells are recognizable (arrows). *b* On the left, a vascular sprout (arrow) juts out into the host ventricle containing the graft (asterisk). *c* A vascular twig arising from the host tectum is directed towards a small fragment of transplanted tissue (asterisk).

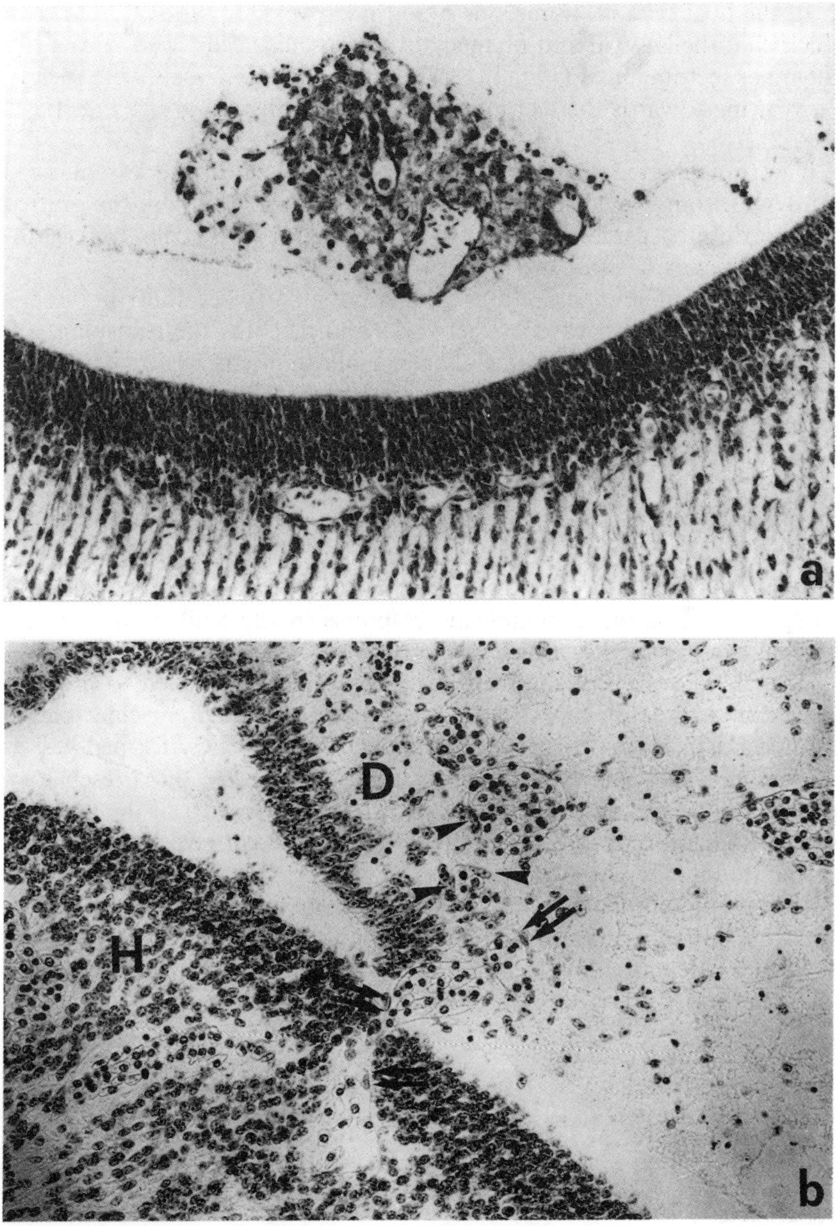

Figure 2. *a* The grafted neural tissue, suspended from the wall of the recipient tectum, is vascularized by numerous and dilated blood vessels. *b* The host(H)-donor(D) vascular connection shows chick endothelial cells (arrows) replaced by quail ones (head arrow), at the transplant periphery.

In the host tectum, numerous blood microvessels, characterized by a thick endothelial wall and phagocytic perivascular cells, were located in the inner tectum laters (Fig. 1a). A number of these vessels were seen to be growing towards the tectum cavity and entering the graft (Fig. 1b, c; Fig. 2).

The transplanted tissue was normal in appearance and vascularized by patent, thin-walled, blood microvessels (Fig. 2a). Where the grafted tissue exhibited some signs of degeneration, surviving cells were gathered in clusters around the implant vessels.

By the Schiff-stained sections it was possible to see that the vessels arising from the host vascular networks and reaching the transplant are lined by a continuous endothelial layer built up of chick (host) endothelial cells as far as the graft periphery, where they are replaced by quail (donor) cells (Fig. 2b).

Discussion

The results indicate that the more mature neural tissues of the donor embryos elicit a rapid angiogenic response in the wall of the mesencephalic vesicle of the younger host embryos.

The blood vessels of the recipient mesencephalon elicited to proliferate and to invade the graft, belong to the periventricular vascular plexus located beneath the neuroepithelial layer [10]. The newly-formed vessels migrate towards the transplant, possibly helped by the presence of phagocytic perivascular cells, may also cross a large area of the ventricle cavity, and end at the graft periphery attaching themselves to the vessels of the donor tissue networks.

It is very likely that the host vascular sprouting and the formation of the host-donor connection tracts is stimulated by the grafted tissues since the chicken embryo brain produces an angiogenic factor whose specific activity increases remarkably from the 12th to the 14th incubation day [15, 16].

Acknowledgements. This work was supported in part by the 'Ministero dell'Università e della Ricerca Scientifica e Tecnologica', Rome, Italy (40%, 1991).

References

1 Bjorklund A, Stenevi U: Reconstruction of the nigrostriatal dopamine pathway by intracerebral nigral transplants. Brain Res 1979; 177: 555–560.
2 Lund RD, Harvey AR: Transplantation of the tectal tissue in rats. I. Organization of host afferens within them. J Comp Neurol 1981; 201: 191–209.
3 Broadwell RD, Charlton HM, Ganong WF, Salcman M, Sofroniew M: Allografts of CNS tissue possess a blood-brain barrier. I. Grafts of medial preoptic area in hypogonadal mice. Exp Neurol 1989; 105: 135–151.

4 Rosenstein JM: Neocortical transplants in the mammalian brain lack a blood-brain barrier to macromolecules. Science 1987; 235: 772–774.
5 Swenson RS, Shaw P, Alones V, Kozlowski G, Zimmer J, Castro AJ: Neocortical transplants grafted into the newborn rat brain demonstrate a blood-brain barrier to macromolecules. Neurosci Lett 1989; 100: 35–39.
6 Krum JM, Rosenstein JM: The fine structure of vascular-astroglial relations in transplanted fetal neocortex. Exp Neurol 1989; 103: 203–212.
7 Anagnostopoulos J, Knoth R, Duffner T, Klein PJ, Volk B: Vascularization of fetal cerebellar tissue transplanted into the striatum of adult rats. In Cervòs–Navarro J, Ferszt R (eds): Stroke and microcirculation, New York, Raven Press, 1987, pp. 171–176.
8 Krum JM, Rosenstein JM: Patterns of angiogenesis in neural transplant model: I. Autonomic tissue transplants. J Comp Neurol 1987; 258: 420–434.
9 Krum JM, Rosenstein JM: Patterns of angiogenesis in neural transplant model: II. Fetal neocortical transplants: J Comp Neurol 1988; 271: 331–345.
10 Roncali L, Ribatti D, Ambrosi G: Vasculogenesis in the chick embryo optic tectum. Acta anat 1985; 122: 229–234.
11 Roncali L, Ribatti D, Ambrosi G: Ultrastructural basis of the vessel wall differentiation in the chick embryo optic tectum. J Submicrosc Cytol 1985; 17: 83–88.
12 Roncali L, Bertossi M, Marzullo A, Quondametteo F, Virgintino D: Survival and effects of neural omografts in the CNS cavities. Eur J Neurosci 1989; 2S: 138.
13 Bertossi M, Virgintino D, Nico B, Ribatti D, Roncali L: Glial cell plasticity after neural transplants in the developing central nervous system. Neurosci Lett 1990; 39S: S26.
14 Le Douarin NM: Particularités du noyau interphasiqué chez la caille japonaise (*Coturnix coturnix japonica*). Utilisation de ces particularités comme marquer biologique dans les recherches sur les interactions tissulaires et les migrations cellulaires au cours de l'ontogenèse. Bull Biol Fr Belg 1969; 103: 435–452.
15 Risau W: Developing brain produces an angiogenesis factor. Proc Natl Acad Sci USA 1986; 83: 3855–3859.
16 Risau W, Gautschi-Sova P, Bohlen P: Endothelial cell growth factors in embryonic and adult chick brain are related to human acidic fibroblast growth factor. EMBO J 1988; 7: 959–962.

Angiogenesis: Key Principles - Science - Technology - Medicine
ed. by R. Steiner, P. B. Weisz & R. Langer

Microvascular network architecture in a mammary carcinoma

Joanne R. Less, Thomas C. Skalak, Eva M. Sevick[a], and Rakesh K. Jain[a]

Department of Biomedical Engineering, University of Virginia, Charlottesville, VA 22908, and [a]Department of Chemical Engineering, Carnegie Mellon University, Pittsburgh, PA 15213, USA

Introduction

The distribution of blood flow in a given tumor regulates the exchange and uptake of relevant molecules in chemo-, immuno-, and radiation therapy, thereby determining the efficacy of present day cancer treatments. Because it plays a significant role in heat transfer, the distribution of blood flow is also important in both thermographic detection and hyperthermia treatment [1, 2]. But what determines the distribution of blood flow in a microvascular network? In both normal and pathologic tissues, blood flow through a vascular bed is determined by several factors including the vascular network topology and the dimensions of the blood vessels. Previous workers have made qualitative comparisons of host and tumor vasculature including both animal tumors (for review see [3]) as well as human tumor xenotransplants [4]. To date, however, there exists no quantitative data describing the microvascular network architecture of any tumor type. If such data can be obtained, then a network model capable of predicting the blood pressure and flow distribution in a given tumor can be constructed, thus allowing insight into growth rate variability among tumors and the mechanisms governing current cancer therapies. Below is a brief presentation of the results of such a quantitative analysis of the vascular network topology in a mammary adenocarcinoma from Less et al. [5].

Materials and methods

Experimental protocol

Ovarian "tissue-isolated" tumors were prepared according to the methods of Gullino and Grantham [6] by surgically implanting mammary adenocarcinoma R3230AC in the left ovarian fat pad of adult female Fischer 344 rats. Following tumor excision and perfusion with a saline solution, modified Batson's # 17 polymer was infused. Upon complete polymerization, the tissue surrounding the Batson's cast was macerated with KOH (384 g/l distilled water).

Vessel classification

Microvascular network architecture or topology is the pattern of interconnections of the blood vessels. Topology can be quantified if a classification scheme is employed which uniquely classifies every vessel in a tissue according to its location in the network without regard to its dimensional or histological characteristics. Several classification schemes have been devised for normal tissues. For example, Engelson et al. [7] described the arteriolar architecture formed by the planar arcades and transverse vessels in skeletal muscle. Ellsworth et al. [8] used branch angles to classify the daughter vessels at bifurcations in skeletal muscle. Tumor vasculature, however, has specialized features (Figs. 1 and 2) and so could not be classified by any preexisting branching scheme. Because tumor vascular morphology not only varies from region to region within a tumor, but also among tumors, the classification rules need to be independent of the physical dimensions of the vessel as well

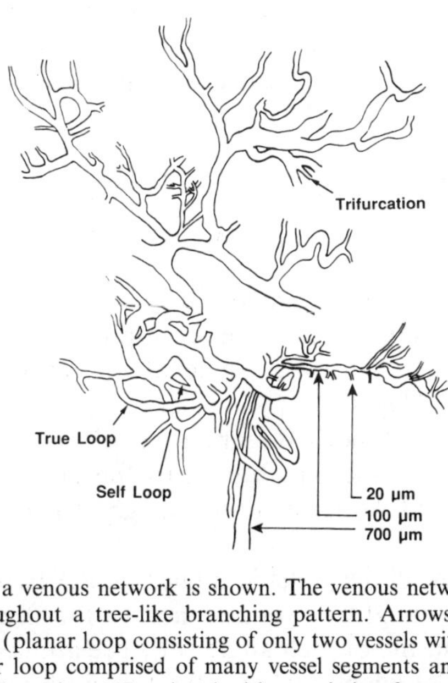

Figure 1. A tracing of a venous network is shown. The venous network topology consists of loops distributed throughout a tree-like branching pattern. Arrows indicate examples of a trifurcation, a self loop (planar loop consisting of only two vessels without any side branches), a true loop (nonplanar loop comprised of many vessel segments and having numerous side branches), and vessel dimensions. (Reprinted with permission from *Cancer Research*, Less et al., ibid [5], 1991).

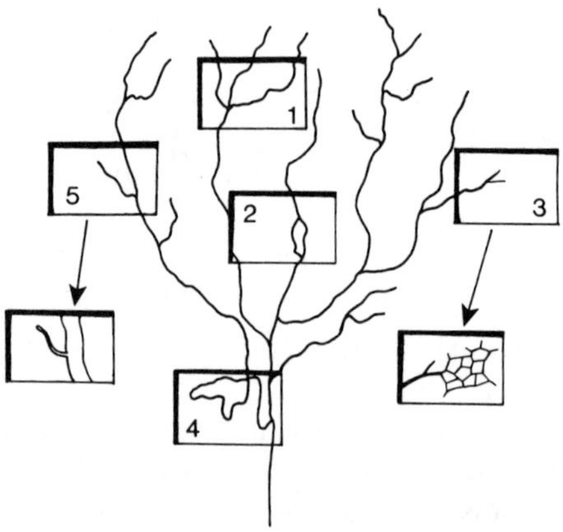

Figure 2. Schematic diagram depicting several characteristic features of the tumor microvascular network topology, including trifurcations (1), self loops (2), the characteristic polygonal structure of the capillary meshworks (3), venous convolutions in proximity to the feeding/draining vessels (4), and small (20–40 μm diameter) vessels branching off of large (200 μm) vessels (5). (Reprinted with permission from Cancer Research, Less et al., 1991).

as the vessel's spatial position within the tumor. In addition, since blood vessels in tumors periodically change their direction of flow and their function in the circulation, the rules must not depend on either of these parameters. Therefore, a new set of rules, based upon a centrifugal ordering scheme and the branch angle rules proposed by Ellsworth et al. [8], was devised and is given below:

1. The main feeding/draining vessel is assigned to order 1.

Outside a loop, the following branch angle rules are used:

2. At a bifurcation, if the vessel continues past the branch point at an outside angle of 160–180°, the vessel retains its original order. If the vessel diverges at an outside angle of less than 160°, the vessel's order is increased by 1.
3. At a bifurcation, if both daughter vessels diverge to an outside angle of 160–180°, the daughter with the larger angle retains the order of the parent while the other daughter is assigned to the next higher order.
4. At a bifurcation, if the outside angles of the two daughter vessels are equal, both daughters are assigned to the next higher order.
5. At a trifurcation, the branch angle rules are applied to the two pairs of vessels, each of which includes the vessel of largest branch angle (denoted as the middle vessel).

Inside a true loop, the following rules are used:

6. All vessels in a loop are assigned to the same order.
7. To enter a loop, the above branch angle rules are used. If this results in two segments of the loop having different orders, the lower order is chosen.
8. To exit a loop, the branch angle rules are applied to both pairs of vessels and the lowest order is assigned to the exiting vessel.

In a self loop, the following rules are used:

9. Both daughters retain the same order as the parent no matter what the outside angles are.
10. When the daughters converge, the same order is retained.

Results

Gross tumor structure

Qualitative examination of the tumor vasculature revealed a concentric arrangement of the arteriolar, capillary, and venular networks. The feeding arteriole penetrated to the center of the tumor and branched peripherally into the smaller arteriolar network that supplies the capillaries. The capillaries then drained into the small venules, which finally drained into the large venules. In a given vascularized region, the arteriolar supply vessel was typically located toward the center of the tumor and the venular drainage was more peripherally located, although the terminal endings of arterioles and venules may have interdigitated within the capillary meshwork itself.

Vascular topology

The overall arterial and venous branching patterns consisted of loops distributed throughout a tree-like framework. These vascular loops were of two types, which we denoted as self loops and true loops. Self loops are planar loops consisting of only two vessels without any side branches (Figs. 1 and 2), while true loops are usually nonplanar, comprised of many vessel segments, and have numerous branches (Fig. 1). Several other features such as trifurcations, venous convolutions, and small (20–40 μm diameter) vessels branching off of large (200 μm) vessels were observed in the vascular topology of tumors and are depicted schematically in Figure 2.

Vessel dimensions

Arterioles. The arteriolar networks exhibited two distinct branching patterns which occurred in different regions of the same arteriolar tree as well as in different trees. One arteriolar network exhibited a pattern seen in many normal tissues [8] in which vessel diameter and length decrease as branching order increases, while the second network displayed an atypical structure in which vessel length increases with progression toward the capillaries (Fig. 3).

Capillaries. The capillaries formed nonplanar meshworks of interconnected vessels with an undetermined number of arteriolar and venular feeder vessels. The meshes had irregular, polygonal shapes and were comprised of 4–6 capillary segments (see Fig. 2). In vascularized regions, the meshes were homogeneous networks where the mean capillary diameter, length, and intercapillary distance were 10 μm, 67 μm,

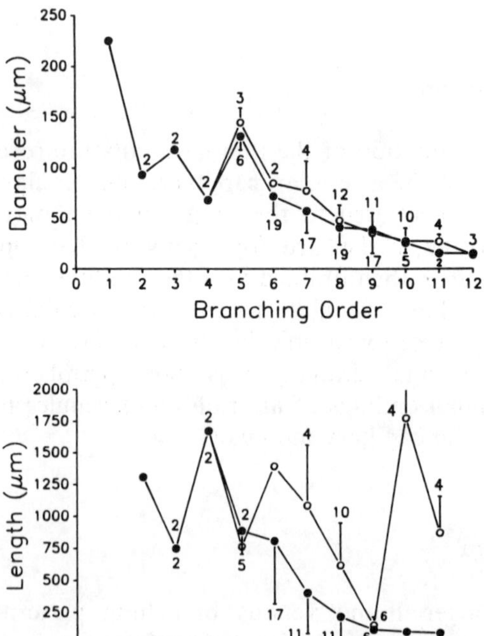

Figure 3. Arteriolar diameter (top) and arteriolar length (bottom) are plotted as functions of branching order. ● indicate mean values from two arterial trees, while ○ indicate mean values from a third arterial tree. Numbers above and below the SD error bars indicate the number of vessels. Diameters decreased monotonically for all trees above order five. Lengths, however, exhibited some variation among the three trees. ● show a monotonic decrease in length above generation four, whereas ○ display large fluctuations in length throughout the eleven generations. (Reprinted with permission from Cancer Research, Less et al., 1991).

and 49 μm, respectively. Capillary pathlengths, measured from the arteriolar entrance of the capillary mesh to the venular exit, ranged from 0.5–1.5 mm.

Venules. Vessel diameters and lengths were measured starting with the draining vein and progressing towards the capillaries. Venular diameter decreased with increasing branch order so that at the ninth generation, the highest order examined, the diameters were approximately 20 μm. Vessel lengths also exhibited a monotonic decrease with increasing branch order until the fourth generation, after which vessel length became fairly uniform at 0.25 mm.

Discussion

A tumor vessel classification scheme is presented that is capable of accounting for all vessels anatomically present in a vascular network. Because this classification scheme is not specific to a particular tumor type, a quantitative comparison of the vascular network architectures of various animal tumor models is possible. As discussed earlier, it also provides a useful tool for describing the heterogeneity present in tumors, which can then be used to accurately model tumor blood flow and drug delivery.

Currently, the proposed classification scheme is being used to study the vascular network topology in human tumors. This work, along with an examination of the geometric resistance to blood flow [9, 10] and the interstitial fluid pressure in human neoplasms, may help to elucidate the mechanisms governing the delivery of anticancer agents.

Acknowledgements. Supported by American Cancer Society Grants PF3551 (J.R.L.) and IN149B (T.C.S.) and National Cancer Institute Grant CA37239 (R.K.J.).

References

1 Jain RK: Delivery of novel therapeutic agents in tumors: Physiological barriers and strategies. J Natl Cancer Inst 1989; 81: 570–576.
2 Vaupel P, Kallinowski F, Okunieff P: Blood flow, oxygen, and nutrient supply, and metabolic microenvironment of human tumors: A review. Cancer Res 1989; 49: 6449–6459.
3 Jain RK: Determinants of tumor blood flow. A review. Cancer Res 1988; 48: 2641–2658.
4 Lauk S, Zietman A, Skates S, Fabian R, Suit HD: Comparative morphometric study of tumor vasculature in human squamous cell carcinomas and their xenotransplants in athymic nude mice. Cancer Res 1989; 49: 4557–4561.
5 Less JR, Skalak TC, Sevick EM, Jain RK: Microvascular architecture in a mammary carcinoma: branching patterns and vessel dimensions. Cancer Res 1991; 51: 265–273.
6 Gullino PM, Grantham FH: Studies on the exchange of fluids between host and tumor. I. A method for growing "tissue-isolated" tumors in laboratory animals. J Natl Cancer Inst 1961; 27: 679–693.

7 Engelson ET, Skalak TC, Schmid-Schönbein GW: The microvasculature in skeletal muscle. I. Arteriolar network in rat spinotrapezius muscle. Microvasc Res 1985; 30: 29–44.
8 Ellsworth ML, Liu A, Dawant B, Popel AS, Pittman RN: Analysis of vascular pattern and dimension in arteriolar networks of the retractor muscle in young hamsters. Microvasc Res 1987; 34: 168–183.
9 Sevick EM, Jain RK: Geometric resistance to blood flow in solid tumors perfused ex-vivo: Effects of tumor size and perfusion pressure. Cancer Res 1989; 49: 3506–3512.
10 Posner MC, Less JR, Wolmark N, Jain RK: Geometric resistance to blood flow in human colon carcinoma perfused ex-vivo. Proc Amer Assoc for Cancer Res 1991; 32: 60.

Angiogenesis: Key Principles - Science - Technology - Medicine
ed. by R. Steiner, P. B. Weisz & R. Langer

Tumor angiogenesis: evidence of new blood channels from plasma infiltrations

I. Freitas, G. F. Baronzio[a], S. Barni, V. Bertone, P. Griffini, P. Accossato and P. Pontiggia[b]

Department of Animal Biology, University of Pavia, [a]APMCTP, Turbigo, and [b]Hyperthermia Service, Clinica Città di Pavia, Pavia, Italy

Abstract. Stroma formation in Ehrlich carcinoma, studied with histochemical and TEM techniques, is similar to wound healing. In this tumour mast cells, macrophages, adipocytes, platelets and fibrin seem to co-operate locally with malignant cells in regulating stroma formation. The gaps opened in the tumor parenchyma by plasma outpouring from local blood vessels seem to offer easy routes for endothelial cell migration towards ill-nourished areas, and may explain the irregular aspect of tumor microvascularity.

A tumor is an organoid entity, whose growth is conditioned by local interactions between the stroma and the parenchyma [1]. The angiogenesis phenomenon is, in particular, enclosed in the global stroma induction process [2–3]. Histochemical and ultrastructure analyses performed in Ehrlich carcinoma at different growth phases allowed us to visualize the outcome of hypoxia and the ensuing process of stroma induction [4–5]. The latter was indeed similar to wound healing [2]. In its initial phase, hyperpermeability of local blood vessels caused the infiltration of plasma forcing irregular openings in the tumor parenchyma [5]. Lastly, an irregular vascular network developed. This sequence of events was described elsewhere for demonstrating that the edema leads to the reoxygenation of hypoxic tumor cells [5]. Here the same experiments will be reconsidered specifically under the angiogenesis light.

Materials and methods

A carcinoma was obtained after i.p. inoculation of Ehrlich ascites in the hind muscle of Swiss mice. Tumor morphology was visualized under fluorescence microscopy with Acridine Orange and neutral lipid with Nile Reed. The ultrastructure was studied after fixation with glutaraldehyde/osmium tetroxide for preserving lipid droplets [4, 5].

Figure 1. Electron micrograph of an edema channel (e. ch.; arrows) lined by a cell with lipid droplets (star). ×10,800.

The former relevant observations

Cells with hypoxia-associated LDH patterns and neutral membrane lipids were identified in the perinecrotic region and in 'tentacules' between the necrosis and blood vessels [4, 5]. High numbers of mast cells were seen at the tumor/host interface. Venules were dilated and paved by leukocytes and edema infiltrated the tumor reaching the necrosis. Malignant cells with lipid droplets were seen lining edema channels (Fig. 1). "Tentacules" of lipid vacuoles were later seen associated with edema channels (Fig. 2). In the same areas, macrophages, mast cells and adipocytes were closely associated (Fig. 3). Mast cells were also present along local arterioles and in the perinecrotic region. Platelet aggregates filled some channels. Endothelioid cells were later observed in patterns suggesting migration and proliferation along the channels. Lumina cavitation processes and lipid extrusion into the channels were seen. Eventually, a plexus of microvessels developed; no vacuoles were associated with them, only mast cells.

Discussion

The association and distribution of different cell types and edema in the areas where the stroma builds up suggest non-random intercellular

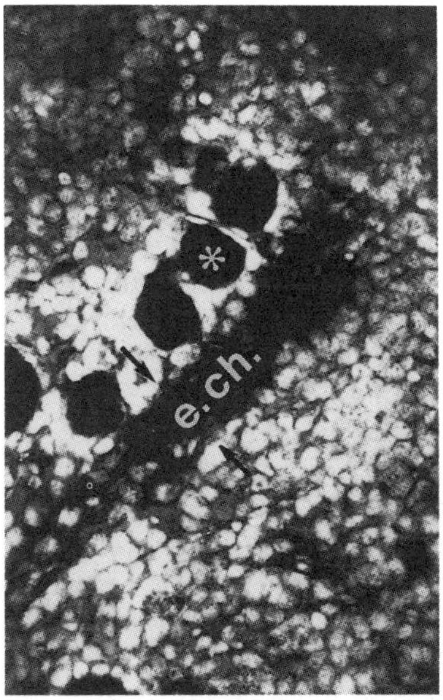

Figure 2. Fluorescence microscopy of an Acridine Orange-stained section. Association of a "tentacule" of lipod vacuoles (*) with an edema chennel (e. ch.; arrows). ×310.

Figure 3. Electron micrograph of mesenchymal cells association close to an edema channel. ×4100.

interaction and paracrine control of the process [1]. We speculated that lipid-laden cells may be a source of energetical and regulatory substances evoking angiogenesis [5]. Lipid-storing cells produce the basal lamina-specific type IV collagen in some inflammatory diseases [6]. Locally activated macrophages and mast cells are well-known modulators of angiogenesis [3]. The complex nature of mast cells secretory products and their distribution areas suggest that these cells promote: a) the invasion of normal tissue and that of the tumor by endothelial cells; b) endothelial cell proliferation; c) activation of lipases for lipid utilization; d) vascular permeability and activation of the kinin and clotting cascade. Fibrin clots may induce endothelial tube-like structures [2, 7]. Ultrastructural organization of the clots provide mechanical guiding tracks for oriented migration (and plexus formation) of nerve fibers and fibroblasts [8]; it may also orient the migration of endothelial cells. Platelets may deliver growth factors to sites of rapid cell turnover [3]. The capacity of neoplasms to activate the hemostatic system was proven to contribute to their invasive and metastitic behavior [9]. The gaps forced by the edema seem to facilitate endothelial cell migration towards the necrosis: this would account for the irregular morphology of the new vasculature. It is therefore likely that experiments with models for assessing individually the roles in angiogenesis of different cell types and matrix, that is, aside from the true in vivo mutual influence, may lead to misleading conclusions.

References

1 Zipori D: Stromal cells in tumor growth and regression. Cancer J 1990; 3: 164–169.
2 Nagy JA, Brown LF, Senger DR, Lanir N, Van De Water L, Dvorak AM, Dvorak HF: Pathogenesis of tumor stroma generation: a critical role for leaky blood vessels and fibrin deposition. Biochim Biophys Acta 1988; 948: 305–326.
3 Blood CH, Zetter BR: Tumor interactions with the vasculature: angiogenesis and tumor metastasis. Biochim Biophys Acta 1990; 1032: 89–118.
4 Freitas I, Pontiggia P, Barni S, Bertone V, Parente M, Novarina A, Roveta G, Gerzeli G, Stoward PJ: Histochemical probes for the detection of hypoxic tumour cells. Anticancer Res 1990; 10: 613–622.
5 Freitas I, Baronzio GF, Bertone V, Griffini P, Gerzeli G, Pontiggia P, Stoward PJ: Stroma formation in Ehrlich carcinoma: I. Oedema phase. A mitosis burst as an index of physiological reoxygenation? Anticancer Res, in press.
6 Rojkind M: From regeneration to scar: the collagen way. Lab Invest 1991; 64: 131–134.
7 Montesano R, Pepper MS, Mohle-Steinlein U, Risau W, Wagner EF, Orci L: Increased proteolytic activity is responsible for the aberrant morphogenetic behaviour of endothelial cells expressing the middle T oncogene. Cell 1990; 62: 435–445.
8 Patterson WB (ed.): Wound Healing and Tissue Repair. University of Chicago Press, 1956, pp. 1–24.
9 Gasic GJ, Tuszynski G, Gorelik E: Interaction of the hemostatic and immune systems in the metastatic spread of tumor cells. Int Rev Exp Pathol 1986; 29: 173–212.

Angiogenesis: Key Principles - Science - Technology - Medicine
ed. by R. Steiner, P. B. Weisz & R. Langer

Receptors for platelet-derived growth factor on microvascular endothelial cells

Julie G. Beitz, In-Soon Kim, Paul Calabresi and A. R. Frackelton, Jr.

Roger Williams Cancer Center and Brown University, Providence, RI, USA (J.G.B., P.C., A.R.F., Jr.) and Catholic University Medical College, Seoul, Korea (I.-S.K.)

Abstract. Endothelial cells are widely thought to be unresponsive to platelet-derived growth factor (PDGF) and devoid of PDGF receptors. However, in examining the growth factor responses of microvascular endothelial cells isolated from human omental adipose tissue, we detected PDGF-induced tyrosine phosphorylation of an 180-kD glycoprotein, subsequently identified as the cellular receptor for PDGF by specific immunoprecipitation. Scatchard analysis of ^{125}I-PDGF binding to human microvascular endothelial cells revealed 30,000 PDGF receptors/cell with a kD of 0.14 nM. PDGF stimulated tyrosine phosphorylation of PDGF and other cellular proteins in a dose- and time-dependent manner, with half-maximal receptor phosphorylation occuring at 0.3 nM recombinant human PDGF-BB within 1 min of PDGF exposure. Normal cellular consequences of receptor activation were also observed, including tyrosine phosphorylation of a 42-kD protein and serine phosphorylation of ribosomal protein S6. Furthermore, PDGF was mitogenic for these cells. The finding of functional PDGF receptors on human microvascular endothelial cells suggests an important direct role for PDGF in neovascularization.

Introduction

Neovascularization plays an important role in many physiologic and pathologic conditions, including wound healing, ovulation, diabetic retinopathy, and tumor angiogenesis [1]. New capillary growth appears to be controlled by an interplay of several polypeptide growth factors, primarily fibroblast growth factor, and transforming growth factors α and β [1]. However, endothelial cells have been widely believed to be unresponsive to platelet-derived growth factor (PDGF), a growth factor released primarily at sites of vascular injury [2], and to lack receptors for PDGF [3–6]. In this report, we describe the surprising finding of functional PDGF receptors on microvascular cells of human origin.

Materials and methods

Cell culture

Human microvascular endothelial cells (HMVECs) were obtained from omental adipose tissues (Clonetics, Inc., San Diego, CA) and maintained in MCDB-131 medium supplemented with 2% (vol/vol) FCS, 1.7 nM epidermal growth factor (EGF), and 2.5 μM hydrocortisone, and were used between passages 1 and 5. Human foreskin fibroblasts (HFFs), obtained from C. Eil (Roger Williams Hospital), were maintained in RPMI 1640 medium supplemented with 10% FCS, and used between passages 19 and 23. Recombinant human PDGF-BB was obtained from AMGEN Biologicals.

Immunoaffinity purification of phosphotyrosyl proteins

Cells were initially seeded in 6-well tissue culture plates at 25% confluency. After 4 days of culture, cells were rinsed with phosphate-free minimal essential medium, then incubated in this medium with ^{32}P-orthophosphate (ICN; 1 mCi/ml/well) for 3 h at 37°C, exposed to growth factors for the final 5–10 minutes of the labeling period, and extracted at 0°C with 1% Triton X-100 buffer containing proteinase, kinase and phosphatase inhibitors as described [7]. Tyrosine-phosphorylated proteins were isolated from the extracts by microbatch immunoaffinity chromatography on the 1G2 monoclonal antibody to phosphotyrosine as described [7]. In some cases, tyrosine-phosphorylated proteins were further purified by wheat germ agglutinin (WGA) affinity chromatography. PDGF receptors were immunoprecipitated from the WGA eluate using a 1:50 dilution of rabbit antiserum (antiserum 88) specific for PDGF receptors and coprecipitated using protein-A Sepharose [8–11]. Purified proteins were resolved by SDS/PAGE on 7.5% gels, and visualized by autoradiography using Kodak XAR-5 pre-exposed film with Lightning Plus intensifying screens at −70°C.

Results

PDGF receptors on HMVECs

Surprisingly, exposure of HMVECs to PDGF-BB resulted in the appearance of a prominent phosphoprotein of 180-kD that co-migrated with the 180-kD PDGF receptor from PDGF-stimulated HFFs (Fig. 1). To determine whether the 180-kD protein in HMVECs might be the glycosylated PDGF receptor, we tested for its ability to bind to the lectin WGA. When phosphotyrosyl proteins from PDGF-stimulated HMVECs were incubated with WGA-Sepharose, the 180-kD protein bound to WGA and was specifically eluted by N-acetylgucosamine. Elution was identical to that of the 180-kD PDGF receptor from HFFs (Fig. 1). As a final test of the identity of the 180-kD protein, the WGA eluates were subjected to immunoprecipitation with a rabbit antibody

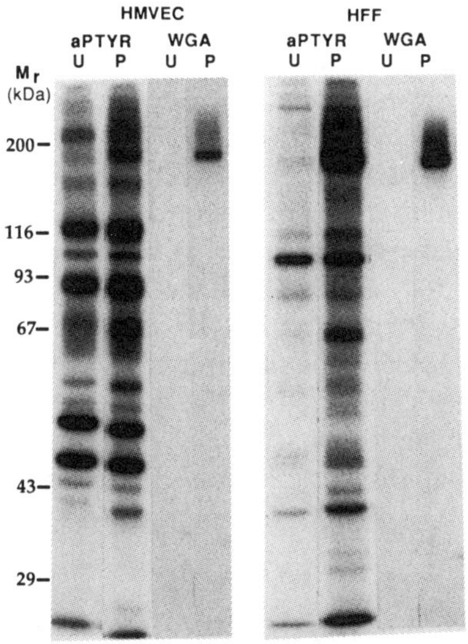

Figure 1. PDGF-stimulated tyrosine phosphorylations in HMVECs and HFFs. Representative autoradiograms of ^{32}P-labeled phosphotyrosine-containing proteins, resolved by SDS/PAGE are shown. aPTYR, proteins purified from cell extracts by immunoaffinity chromatography on the 1G2 monoclonal antibody to phosphotyrosine; WGA, proteins first purified by antiphosphotyrosine immunosorbant, then secondarily purified by WGA affinity chromatography; U, unstimulated cells; P, cells exposed to 1 nM PDGF for 5 min at 37°C.

specific for the PDGF receptor [8–11]. This antiserum immunoprecipitated the 180-kD PDGF receptor candidate proteins from HMVECs as well as known PDGF receptors from HFFs. Maximal phosphorylation of the HMVEC PDGF receptor was extremely rapid, occuring within 1 min of PDGF addition, then diminishing over the course of an hour. Similar kinetics of PDGF receptor phosphorylation were observed with HFFs. Half-maximal PDGF receptor phosphorylation in HMVECs occured with 0.3 nM PDGF-BB.

PDGF receptors on HMVECs were characterized further by ligand binding using a radioreceptor assay [5]. Scatchard analysis of ^{125}I-labeled PDGF binding to HMVECs revealed 30,000 receptors per cell, with a single population of high-affinity receptors having a kD of 0.14 nM.

PDGF stimulates biochemical events in HMVECs

Exposure of HMVECs and HFFs to PDGF resulted in prominent tyrosine phosphorylation of a 42-kD protein (Fig. 1), an expected finding characteristic of several mitogens [12]. Phosphorylation of the 42-kD protein is believed to result from a cascade of events that includes activation of protein kinase C as well as activation of an as yet unidentified cellular tyrosine kinase [13].

The binding of mitogens to cells causes serine phosphorylation of ribosomal protein S6, believed to be involved in protein synthesis [13]. HMVECs were incubated with $^{32}P_i$ for 3 h, and exposed to PDGF-BB for the last 30 min of the labeling period. Ribosomes were extracted with detergent, isolated by ultracentrifugation, and phosphorylated S6 proteins were resolved by SDS/PAGE on 10% gels, and visualized by autoradiography. As expected, PDGF markedly stimulated serine phosphorylation in HMVECs, with maximal S6 phosphorylation observed after a 30-min exposure to 10^{-10} M PDGF-BB. This finding further demonstrated that intracellular events were evolving within HMVECs following the interaction of PDGF with its receptor on these cells.

PDGF-BB mitogenicity for HMVECs

We examined the ability of PDGF to cause HMVECs to proliferate in tissue culture under low serum conditions (0.7% FCS). When HMVECs were cultured with 1 nM PDGF-BB for 72 h, we observed a significant increase in HMVEC cell numbers, $p < 0.001$ by Student's t-test (Fig. 2).

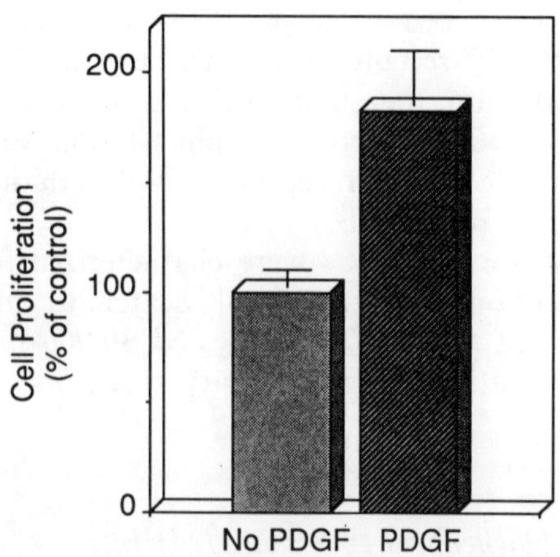

Figure 2. PDGF-stimulated proliferation of HMVECs. HMVECs were seeded into 48-well tissue culture plates at 7×10^3 cells per well in medium containing 0.7% FCS. Where indicated, cells were incubated with 1 nM PDGF-BB for 72 h, then trypsinized and counted. Results are expressed as the number of PDGF-treated cells per well, normalized to media controls (n = 10 wells, bars indicate standard errors). For PDGF-treated HMVECs, $p < 0.001$ by Student's t-test (Fig. 2).

Discussion

We report here that microvascular endothelial cells derived from human adipose tissues possess high-affinity PDGF receptors, and that these cells are capable of responding to PDGF biochemcially and mitogenically. In contrast to the microvascular cells studied here, the many previous studies reporting endothelial cells to lack PDGF receptors and responses dealt primarily with macrovascular endothelial cells [3–6]. Because PDGF is the major growth factor released by platelets at sites of vascular injury [2], and because microvascular endothelial cells are uniquely involved in neovascular events such as tumor angiogenesis, these findings suggest a direct role for PDGF in these events. In fact, several tumors are known to produce PDGF [5, 14–16]. Further supporting the role of PDGF in tumor neovascularization is the report of Hermansson et al. [17] describing PDGF receptor mRNA in situ in hyperproliferating microvascular endothelial cells growing adjacent to PDGF-secreting malignant glioma cells.

Obviously, it will be crucial to examine human microvascular endothelial cells isolated from a variety of tissues for the presence of cell surface PDGF receptors and for PDGF responses. When human microvascular cells become more readily available, such studies may further our understanding of the role of PDGF in vivo.

References

1 Folkman J, Klagsbrun M: Angiogenic factors. Science 1987; 235: 442–447.

2 Ross R, Raines EW, Bowen-Pope, DF: The biology of platelet-derived growth factor. Cell 1986; 46: 155–169.

3 Haudenschild CC, Zahniser D, Folkman J, Klagsbrun M: Human vascular endothelial cells in culture: Lack of response to serum growth factors. Exp Cell Res 1976; 98: 175–181.

4 Heldin C-H, Westermark B, Wasterson A: Specific receptors for platelet-derived growth factor on cells derived from connective tissue and glia. Proc Natl Acad Sci (USA) 1981; 78: 3664–3668.

5 Bowen-Pope DF, Ross R: Platelet-derived growth factor. II. Specific binding to cultured cells. J Biol Chem 1982; 257: 5161–5171.

6 Kazlauskas A, DiCorleto PE: Cultured endothelial cells do not respond to a platelet-derived growth factor-like protein in an autocrine manner. Biochem Biophys Acta 1985; 846: 405–412.

7 Frackelton AR Jr, Tremble PM, Williams LT: Evidence for the platelet-derived growth factor-stimulated tyrosine phosphorylation of the platelet-derived growth factor receptor in vivo: Immunopurification using a monoclonal antibody to phosphotyrosine. J Biol Chem 1984; 259: 7909–7915.

8 Keating MT, Williams LT: Processing of the platelet-derived growth factor. Biosynthetic and degradation studies using anti-receptor antibodies. J Biol Chem 1987; 262: 7932–7937.

9 Keating MT, Harryman CC, Williams LT: Platelet-derived growth factor receptor inducibility is acquired immediately after translation and does not require glycosylation. J Biol Chem 1989; 264: 9129–9132.

10 Escobedo JA, Keating MT, Ives HE, Williams LT: Platelet-derived growth factor receptors expressed by cDNA transfection couple to a diverse group of cellular responses associated with cell proliferation. J Biol Chem 1988; 263: 1482–1487.

11 Escobedo JA, Navankasattusas S, Cousens LS, Coughlin SR, Bell GI, Williams LT: A common PDGF receptor is activated by homodimeric A and B forms of PDGF. Science 1988; 240: 1532–1534.
12 Cooper JA, Hunter T: Major substrate for growth factor-activated protein-tyrosine kinases is a low-abundance protein. Mol Cell Biol 1985; 5: 3304–3309.
13 Kazlauskas A, Cooper JA: Protein kinase C mediates platelet-derived growth factor-induced tyrosine phosphorylation of p42. J Cell Biol 1988; 106: 1395–1402.
14 Montesano R, Orci L: Phorbol esters induce angiogenesis in vitro from large-vessel endothelial cells. J Cell Physiol 1987; 130: 284–291.
15 Nister M, Heldin C-H, Wasteson A, Westermark B: A platelet-derived growth factor analog produced by a human clonal glioma cell line. Proc Natl Acad Sci (USA) 1984; 81: 926–930.
16 Sariban E, Sitaras NM, Antoniades HN, Kufe DW, Pantazis P: Expression of platelet-derived growth factor (PDGF)-related transcripts and synthesis of biologically active PDGF-like proteins by human malignant epithelial cells. J Clin Invest 1988; 82: 1157–1164.
17 Hermansson M, Nister M, Betsholtz C, Heldin C-H, Westermark B, Funa K: Endothelial cell hyperplasia in human glioblastoma: Coexpression of mRNA for platelet-derived growth factor (PDGF) B chain and PDGF receptor suggests autocrine growth stimulation. Proc Natl Acad Sci (USA) 1988; 85: 7748–7752.

Angiogenesis: Key Principles - Science - Technology - Medicine
ed. by R. Steiner, P. B. Weisz & R. Langer

Five FGF receptors with distinct expression patterns

Jaana Korhonen, Juha Partanen, Elina Eerola, Satu Vainikka, Riitta Alitalo, Tomi P. Mäkelä, Minna Sandberg[a], Harri Hirvonen[a] and Kari Alitalo

Cancer Biology Laboratory, Departments of Virology and Pathology, University of Helsinki, 00290 Helsinki, and Department of Medical Biochemistry[a], University of Turku, 20520 Turku, Finland

Introduction

The fibroblast growth factors (FGF) constitute a polypeptide family whose members can interact with cell surface receptors to modulate gene expression for cell proliferation and differentiation. Seven structurally related heparin binding FGFs are known. The first discovered members of the family were aFGF and bFGF, also called HBGF-1/ECGF (heparin binding growth factor/endothelial cell growth factor) and HBGF-2, respectively. The tertiary fold of bFGF is similar to that of interleukin-1 [1–3]. Amino acid sequence analysis demonstrates 55% identity between aFGF and bFGF. Sequence homology and similar biological responses suggest that they are derived from a common ancestral gene. Both aFGF and bFGF also lack an N-terminal signal peptide sequence (Fig. 1) and thus their mechanism of secretion is unknown.

FGFs stimulate the growth and survival of hippocampal and cerebral neurons [4, 5] and PC12 pheochromocytoma cells [6]. In vivo they support the survival of lesioned cholinergic neurons [7]. Both aFGF and bFGF are capable of inhibiting the differentiation of skeletal muscle myoblasts [8, 9], but they induce the differentiation of adipocytes [10]. During embryogenesis, bFGF can act as an inducer of mesoderm of Xenopus embryos [11, 12]. Both aFGF and bFGF also exhibit angiogenic activities. They have chemotactic and mitogenic effects on endothelial cells and may be involved in the repair mechanisms of tissue injury, for example in wound healing [13].

The other known members of the FGF family are *int*-2 [14], K-FGF/*hst*-1 [15, 16], FGF-5 [17], FGF-6/*hst*-2 [18] and KGF (keratinocyte growth factor; [19]). They share 30–55% amino acid identity (Fig. 2),

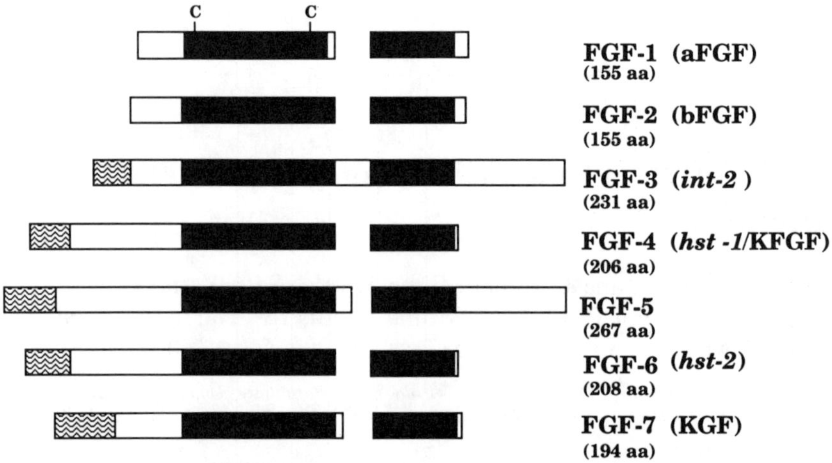

Figure 1. Schematic structures of FGFs. Conserved cysteine residues are marked with C. Signal sequences and homologous regions are indicated with wavy and black boxes, respectively. The lengths of the primary translation products are also indicated.

similar gene structures, and may also play crucial roles in tumorigenesis. Despite their current nomenclature (Tab. 1), some FGFs, such as KGF in fact lack a fibroblast stimulating activity [19]. There is evidence that constitutive expression of FGFs can induce cellular transformation of transfected cells [20, 21]. In fact, *int*-2 and K-FGF/*hst*-1 were identified as possible oncogenes; the *int*-2 gene was found at the proviral integration site in virally induced mammary carcinomas [22] and K-FGF/*hst*-1 as oncogenes in hyman stomach cancer and Kaposis sarcoma [16, 23]. However, despite their amplification in breast carcinoma and other squamous cell carcinomas, no expression of either transcription unit has been detected, and it now seems that the recently cloned PRAD-1 gene, which belongs to the superfamily of cyclin-like genes is the driver for this amplicon [24]. In general, aFGF and bFGF are also expressed in adult tissues, while the other FGFs have predominantly been described in the developing embryo.

FGF-receptors

Fibroblast growth factors induce their effects in responding cells by binding to cell surface receptor tyrosine kinases of subclass IV. All fibroblast growth factor receptors (FGFRs) share common structural features; they have three immunoglobulin-like domains in their extracellular part and their cytoplasmic tyrosine kinase domains are interrupted by short hydrophilic insertion sequences, the so-called kinase inserts (Fig. 2).

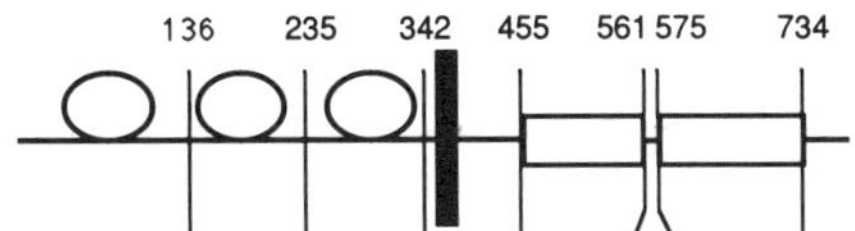

	I	II	III	TM + JM	TK1	KI	TK2	CT	overall identity
FGFR-1 vs. FGFR-2	**38**	**81**	**79**	**99**	**88**	**50**	**92**	**60**	**71**
FGFR-1 vs. FGFR3	**21**	**66**	**82**	**41**	**83**	**43**	**91**	**44**	**61**
FGFR-2 vs. FGFR-3	**27**	**72**	**81**	**45**	**87**	**50**	**92**	**56**	**65**
FGFR-4 vs. FGFR-1	**22**	**63**	**74**	**32**	**75**	**14**	**86**	**38**	**56**
FGFR-4 vs. FGFR-2	**21**	**70**	**72**	**31**	**78**	**29**	**84**	**48**	**57**
FGFR-4 vs. FGFR-3	**26**	**78**	**74**	**34**	**78**	**36**	**86**	**46**	**60**
domains	**I**	**II**	**III**	**TM + JM**	**TK1**	**KI**	**TK2**	**CT**	

Figure 2. Schematic structure of FGFRs, with regional amino acid identities between different receptors indicated. I–III refer to the immunoglobulin-like loops I–III, TM and JM transmembrane and juxtamembrane domains, TK1 + TK2 tyrosine kinase domain, KI kinase insert, CT carboxyterminal domain. Numbers on the top of the figure show the positions of the lines (in FGFR-4 amino acids) used to demarcate the various domains.

Table 1. Different fibroblast growth factor receptors and their localization in human chromosomes. (*Michael Jaye, personal communication; **Michael Hayman, personal communication)

Human	Chromosome	Ref.	Mouse	Ref.	Chick	Ref.
FGFR-1 (*flg*)	8p12	(25)	m*flg*	[37]	*Cek*-1	[49]
FGFR-2 (*bek*)	10q25.3–26	*	bek (KGFR)	[30]	*Cek*-3	[51]
FGFR-3	4	**	—		*Cek*-2	[51]
FGFR-4	5q33-ter	(52)	—		—	
FGFR-5 (*flg*-2)	?	(35)	—	—		

The *flg* gene (*fms*-like-gene) was the first identified FGFR (FGFR-1). Originally it was isolated as a 2 kb insert from a human endothelial cell cDNA library [25], later the full length cDNA was cloned from human placenta [26], human endothelial cells [27] and Tera-2 teratocarcinoma cells [28]. FGFR-1 expression is increased 2–5-fold during the differentiation of endothelial cells from proliferative to tube-like structures [29]. A partial clone of the FGFR-2 (*bek*) was obtained by screening a mouse liver cDNA library with antiphosphotyrosine antibodies [30]. A rearranged human homologue of FGFR-2 was identified as an amplified DNA fragment in a gastric cancer cell line KATO III and it was named K-*sam* (KATO II-*am*plified in *s*tomach cancer [31]). The K-*sam* clones apparently contained tumor-specific rearrangements; the normal human FGFR-2 cDNA was isolated from a brainstem library [27].

The full length cDNA of FGFR-3 was clones from the human leukemia cell line K562 [32, 33]. FGFR-3 encodes a protein of 125 kD. Treatment with the inhibitor of N-linked glycolysation, tunicamycin leads to the biosynthesis of a 97 kD FGFR-3 protein. FGFR-4 was also found from a human K562 library [32, 34]. According to the deduced amino acid sequence the length of the FGFR-4 polypeptide is 802 amino acids (aa). Fig. 2 shows the homology of FGFR-4 with the other FGFRs. Immunoprecipitation of FGFR-4 with antibodies against its carboxy terminus revealed 110 kD glycoprotien with a protein core of 88 kD. Table 1 shows a summary of the different FGFRs cloned from humans, mice and chickens.

The most recently cloned FGFR is *flg*-2 (FGFR-5) which was isolated from a human keratinocyte cDNA library [35]. The predicted *flg*-2 amino acid sequence shows 68% and 64% identity with *bek* and *flg*, respectively. It is expressed predominantly in skin, brain and lung. Keratinocytes, breast and colon tumor cell lines express *flg*-2 but no expression has been observed in hematopoietic cell lines. The ligand specificity is not known but it is possible that *flg*-2 may encode the KGF (FGF-7) receptor.

The overall identity of different human FGFRs is about 56–71% on the basis of amino acid sequence analysis. The regional distribution of similarities for FGFR1-4 is shown in Fig. 2. The most homologous portions of these receptors are their immunoglobulin-like domains II and III (63–82% identity) and the intracellular tyrosine kinase domain (75–92%). The first Ig-like domain, the transmembrane region and the kinase insert have lesser degrees of homology between the FGFRs. FGFR-1 and FGFR-2 are more closely related than the other FGFRs. The different FGFRs are distributed in different chromosomes (Tab. 1).

Functions of FGFRs

All FGFRs bind both acidic and basic FGF. For example, FGFR-1 and FGFR-2 have dissociation constants of $(2–15) \times 10^{-11}$ M [27, 28]. aFGF and bFGF binding to FGFR-3 has been studied by expressing FGFR-3 in *Xenopus* oocytes and measuring the ability of the receptor to mobilize intracellular calcium stores upon stimulation with the ligand [33]. A rapid and large efflux of calcium was observed when oocytes were stimulated with either 0.5 nM bFGF or aFGF. The KGF receptor (a form of *bek*) and FGFR-4 bind aFGF with a higher affinity than bFGF [32, 36]. FGFR-1 and FGFR-2 also are also able to bind K-FGF

[36, 37]. Besides these receptor tyrosine kinases an additional high-affinity receptor, which is not a tyrosine kinase has been recently identified from chicken embryo [38].

The FGFs also bind to low affinity receptors (K_d about 2×10^{-9}) formed by cell-associated heparan sulfate molecules. Low affinity binding sites for aFGF and bFGF are present on the surface of many types of cells. Heparin potentiates the biological activity of aFGF, apparently by prolonging its half-life. It can induce an antigenic epitope to aFGF, increase the affinity of aFGF for its receptor and restore the biological activity of aFGF. Heparin protects both aFGF and bFGF from heat and acid inactivation and prevents proteolysis of aFGF caused by trypsin, plasmin and thrombin (reviewed by Burgess & Maciag, [39]). Recent evidence indicates that interaction with a low-affinity receptor and an apparently associated conformational change is a prerequisite for FGF binding to the high-affinity tyrosine kinase receptor [40] (reviewed by Ruoslahti, 1991 [41]).

The complexity of the FGF receptor system is further increased by differential splicing, which may produce alternative polypeptide forms. The coding capacity of the different FGFR-1 and mRNA forms so far identified are schematically illustrated in Fig. 3 [42, 43]. The various possible combinations of these changes could increase the heterogeneity of FGFRs tremendously, if their corresponding protein products are indeed made in cells and tissues. One of the splicing variants of FGFR-1 predicts a protein with no signal sequence, suggesting that the corresponding protein, if it exists, may encode an intracellular growth factor receptor. As already mentioned, aFGF and bFGF also lack signal sequences and they have been detected within the cytoplasm. A nuclear

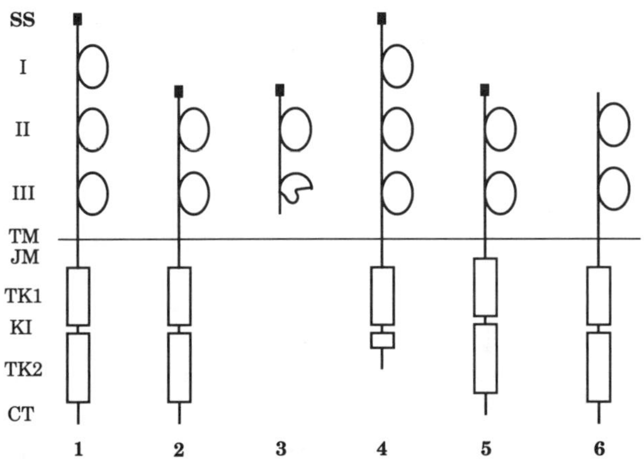

Figure 3. Schematic view of the deduced polypeptide structures of six different splicing variants of FGFR-1 presented in the literature [42, 43]. The reading frame of the fifth alternative has a two amino acid deletion in the juxtamembrane region of the receptor.

localization of bFGF has also been reported for *int*-2 and bFGF initiated from upstream CUG codons (personal communications from C. Dickson and R. Florkiewicz, respectively). Besides these splicing variants, one has to account for the likely possibility that the different high-affintiy tyrosine kinase receptor polypeptides undergo hetero-dimerisation upon ligand binding as has been described for the platelet-derived growth factor receptors [44]. These combinatorial possibilities are also great in cells such as the Tera-2 human teratocarcinoma cells where at least four different FGFRs are expressed [45].

According to our preliminary data, FGFR-1 is the major endothelial cell FGFR transcript. Tab. 2 shows the distribution of the different FGFR mRNAs in human fetal tissues [32]. As can be seen from the table, there are several striking examples of differential regulation of FGFR mRNAs. For example, FGFR-4 mRNA is abundant in fetal lung tissue, and the transcripts localize mainly to epithelial cells of developing alveoli (Fig. 4). We have also detected FGFR1, FGFR-3 and FGFR-4 mRNA in several leukemia cell lines [46]. At least aFGF can be crosslinked to a cell surface receptor on the K562 leukemia cells, indicating that FGFRs are expressed also in some hematopoietic cell lines or that there is aberrant activation of FGFR expression in leukemia cells. The role of FGFRs in oncogenesis is under investigation as Adnane et al. [47] have shown that FGFR-1 and FGFR-2 are each amplified in about 10% of breast carcinomas. Even if mutations of FGFR genes do not prove important, the receptors are certainly

Table 2. Expression of FGFR mRNAs in different human fetal tissues. The hybridization signal intensities have been classified into four groups (−, +, ++, +++) after normalization to the signals of two "household" genes, β-actin and glyceraldehyde-3-phosphate-dehydrogenase

Tissue	FGFR-1	FGFR-2	FGFR-3	FGFR-4
Brain	++	+	+	−
– Cerebellum	+	+	++	+
– Choroid Plexus	+	+++	+	+
– Temporal lobe	+	+	++	−
– Meninges	++	+	+	+
Lung	+	+	++	+++
Thymus	+	−	(+)	(+)
Liver	−	+	+	+
Spleen	+	−	+	+
Kidney	+	+	+	+
Adrenal glands	+	+	−	+++
Heart	+	−	(+)	+
Pancreas	+	−	+	+
Intestine	+	+	++	+
Striated muscle	+	−	(+)	+
Skin	++	+++	+++	++
Calvarial bond	++	+	+++	−
Growth plate (bone)	+++	+	+++	+

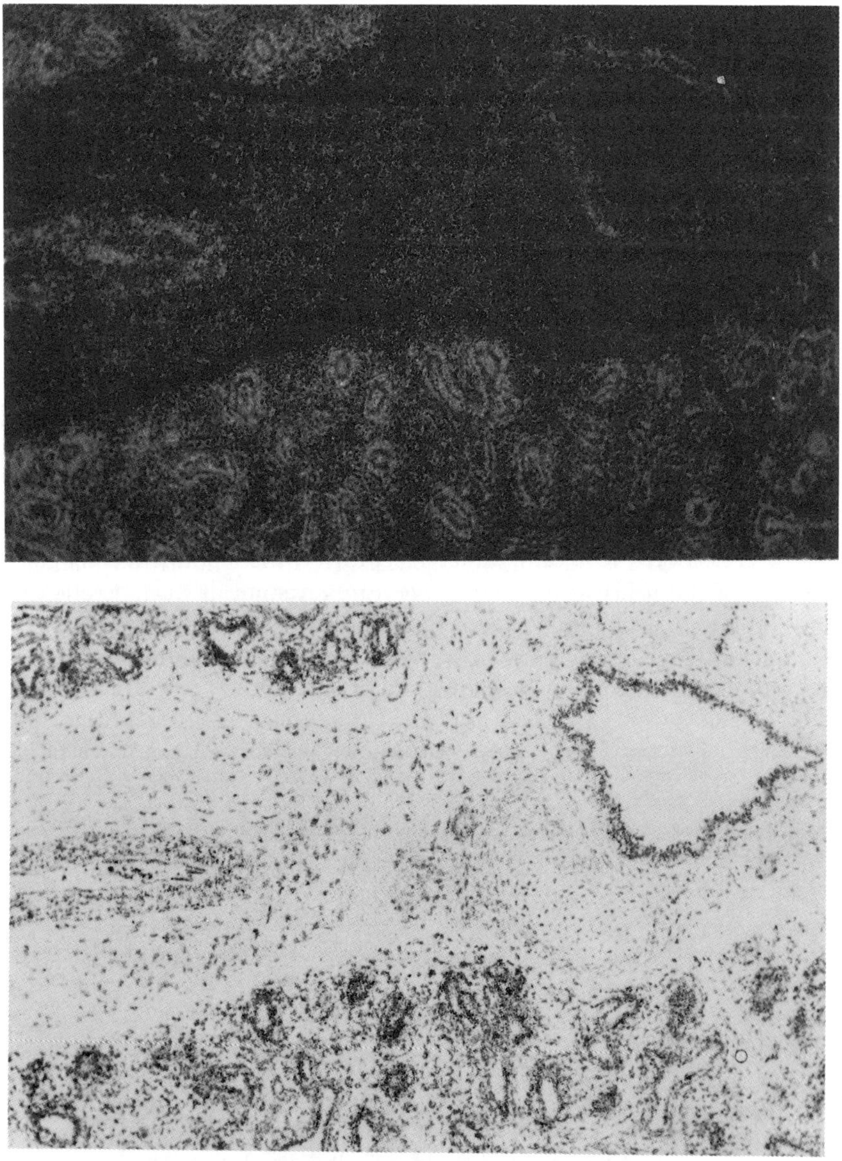

Figure 4. Location of FGFR-4 in fetal lung detected by in situ hybridization. The samples were obtained from 15–19-week old human fetuses from therapeutic abortions (with permission of the joint ethical commitee of the University Central Hospital and University of Turku). The signal of the FGFR-4 cDNA probe is strongest in the epithelial cells of developing alveoli.

involved in growth factor signal transduction and angiogenesis in tumors which contain FGFs.

Future perspectives

As is evident from the above discussion, one of the most acute research problems concerning FGFs and their receptors is the unexpected complexity of this growth factor-receptor system. Among major questions are the importance and function of intracytoplasmic FGFs and their putative receptors and the mode of release of FGFs from cells and extracellular matrices. The specificity of expression of the tyrosine kinase FGFRs and their developmental regulation are of obvious interest. They are being studied by in situ hybridization and immunohistochemical methods using specific antibodies. With cloned receptors we can begin to reconstruct and analyze this multicomponent system and approach its functions also in transgenic animals. Clinical interest in the activities of FGFs is high in angiogenesis, repair of endothelial damage, diseases of the heart, eyes, central nervous system and wound repair. It may be possible to inhibit some of the activities of FGFs in vivo, for example angiogenesis, by use of suitable recombinant molecules, such as the receptor extracellular domains.

Acknowledgements. Our studies have been supported by the Finnish Cancer Research Fund, the Finnish Academy, the Sigrid Juselius Foundation and by the University of Helsinki.

Abbreviations. FGF, fibroblast growth factor; aFGF, acidic FGF; bFGF, basic FGF; FGFR, FGF receptor; PCR, polymerase chain reaction.

References

1 Zhu X, Komiya H, Chirino A, Faham S, Fox GM, Arakawa T, Hsu BT: Three-dimensional structure of acidic and basic fibroblast growth factors. Science 1991; 251: 90–93.
2 Eriksson AE, Cousens LS, Weaver LH, Matthews BW: Three-dimensional structure of human basic fibroblast growth factor. Proc Natl Acad Sci USA 1991; 88: 3441–3445.
3 Zhang J, Cousens LS, Barr PM, Sprang SR: Three-dimensional structure of human basic fibroblast growth factor, a structural homolog of interleukin 1. Proc Natl Acad Sci USA 1991; 88: 3446–3450.
4 Walicke P, Cowan WM, Ueno N, Baird A, Guillemin R: Fibroblast growth factor promotes survival of dissociated hippocampal neurons and enhances neurite extension. Proc Natl Acad Sic USA 1986; 83: 3012–3016.
5 Morrison RS, Sharma A, De Vellis J, Bradshaw RA: Basic fibroblast growth factor supports the survival of cerebral cortical neurons in primary culture. Proc Natl Acad Sci USA 1986; 83: 7537–7541.
6 Togari A, Dickens G, Kuzuya H, Guroff G: The effect of fibroblast growth factor on PC12 cells. J Neurosci 1985; 5: 307–316.
7 Anderson KJ, Dam D, Lee S, Cotman CW: Basic fibroblast growth factor prevents death of lesioned cholinergic neurons in vivo. Nature 1988; 332: 360–361.

8 Lathrop B, Olson E, Glaser L: Control by fibroblast growth factor of differentiation in the BC3H1 muscle cell line. J Cell Biol 1985; 100: 1540–1547.

9 Linkhart TA, Clegg CH, Hauschka SD: Myogenic differentiation in permanent clonal mouse myoblast cell lines: regulation by macromolecular growth factors in the culture medium. Dev Biol 1981; 86: 19–30.

10 Broad TE, Ham RG: Growth and adipose differentiation of sheep preadipocyte fibroblasts in serum-free medium. Eur J Biochem 1983; 135: 33–39.

11 Kimelman D, Kirschner M: Synergistic induction of mesoderm by FGF and TGF-beta and the identification of an mRNA coding for FGF in the early Xenopus embryo. Cell 1987; 51: 869–877.

12 Slack JM, Darlington BG, Heath JK, Godsave SF: Mesoderm induction in early Xenopus embryos by heparin-binding growth factors. Nature 1987; 326: 197–200.

13 Folkman J, Klagsbrun M: Angiogenic factors. Science 1987; 235: 442–447.

14 Dickson C, Peters G: Potential oncogene product related to growth factors. Nature 1987; 326: 833.

15 Delli Bovi P, Curatola AM, Kern FG, Greco A, Ittmann M, Basilico C: An oncogene isolated by transfection of Kaposi's sarcoma DNA encodes a growth factor that is a member of the FGF family. Cell 1987; 50: 729–737.

16 Yoshida T, Miyagawa K, Odagiri H, Sakamoto H, Little PF, Terada M, Sugimura T: Genomic sequence of *hst*, a transforming gene encoding a protein homologous to fibroblast growth factors and the *int*-2-encoded protein. Proc Natl Acad Sci USA 1987; 84: 7305–7309.

17 Zhan X, Bates B, Hu X, Goldfarb M: The human FGF-5 oncogene encodes a novel protein related to fibroblast growth factors. Mol Cell Biol 1988; 8: 3487–3495.

18 Marics I, Adelaide J, Raybaud F, Mattei M-G, Coulier F, Planche J, de Lapeyriere O, Birnbaum D: Characterization of the *HST*-related *FGF.6* gene, a new member of the fibroblast growth factor gene family. Oncogene 1989; 4: 335–340.

19 Finch PW, Rubin JS, Miki T, Ron D, Aaronson SA: Human KGF is FGF-related with properties of a paracrine effector of epithelial cell growth. Science 1989; 245: 752–755.

20 Jaye M, Lyall RM, Mudd R, Schlessinger J, Sarver N: Expression of acidic fibroblast growth factor cDNA confers growth advantage and tumorigenesis to Swiss 3T3 cells. EMBO J 1988; 7: 963–969.

21 Rogelj S, Weinberg RA, Fanning P, Klagsbrun M: Basic fibroblast growth factor fused to a signal peptide transforms cells. Nature 1988; 331: 173–175.

22 Peters G, Brookes S, Smith R, Dickson JC: Tumorigenesis by mouse mammary tumor virus: evidence for a common region for provirus integration in mammary tumors. Cell 1983; 33: 369–377.

23 Barr PJ, Cousens LS, Lee-Ng CT, Medina-Selby A, Masiarz FR, Hallewell RA, Chamberlain S, Bradley J, Lee D, Steimer KS, Poulter L, Burlingame AL, Esch F, Baird A: Expression and processing of biologically active fibroblast growth factors in the yeast *Saccharomyces cerevisiae*. J Biol Chem 1988; 263: 16471–16478.

24 Motokura T, Bloom T, Kim HG, Jüppner H, Ruderman JV, Kronenberg HM, Arnold A: A novel cyclin encoded by a *bcl1*-linked candidate oncogene. Nature 1991; 350: 512–515.

25 Ruta M, Howk R, Ricca G, Drohan W, Zabelshansky M, Laureys G, Barton DE, Francke U, Schlessinger J, Givol D: A novel protein tyrosine kinase gene whose expression is modulated during endothelial cell differentiation. Oncogene 1988; 3: 9–15.

26 Isacchi A, Bergonzoni L, Sarmientos P: Complete sequence of a human receptor for acidic and basic fibroblast growth factors. Nuc Acids Res 1990; 18: 1906.

27 Dionne CA, Crumley G, Bellot F, Kaplow JM, Searfoss G, Ruta M, Burgess WH, Jaye M, Schlessinger J: Cloning and expression of two distinct high-affinity receptors cross-reacting with acidic and basic fibroblast growth factors. EMBO J 1990; 9: 2685–2692.

28 Wennström S, Sandström C, Claesson-Welsh L: cDNA cloning and expression of a human FGF receptor which binds acidic and basic FGF. Growth Factors 1991; 4: 197–208.

29 Ruta M, Burgess W, Givol D, Epstein J, Neiger N, Kaplow J, Crumley G, Dionne C, Jaye M, Schlessinger J: Receptor for acidic fibroblast growth factor is related to the tyrosine kinase encoded by the *fms*-like gene (FLG). Proc Natl Acad Sci USA 1989; 86: 8722–8726.

30 Kornbluth S, Paulson KE, Hanafusa H: Novel tyrosine kinase identified by phosphotyrosine antibody screening of cDNA libraries. Mol Cell Biol 1988; 8: 5541–5544.

31 Hattori Y, Odagiri H, Nakatani H, Miyagawa K, Naito K, Sakamoto H, Katoh O, Yoshida T, Sugimura T, Terada M: K-*sam*, an amplified gene in stomach cancer, is a member of the heparin-binding growth factor receptor genes. Proc Natl Acad Sci USA 1990; 87: 5983–5987.
32 Partanen J, Mäkelä TP, Eerola E, Korhonen J, Hirvonen H, Claesson-Welsh L, Alitalo K: FGRFR-4, a novel acidic fibroblast growth factor receptor with a distinct expression pattern. EMBO J 1991; 10: 1347–1354.
33 Keegan K, Johnson DE, Williams LT, Hayman MJ: Isolation of an additional member of the fibroblast growth factor receptor family, FGFR-3. Proc Natl Acad Sci USA 1991; 88: 1095–1099.
34 Partanen J, Mäkelä TP, Alitalo R, Lehväslaiho H, Alitalo K: Putative tyrosine kinases expressed in K-562 human leukemia cells. Proc Natl Acad Sci USA 1990; 87: 8913–8917.
35 Avivi A, Zimmer Y, Yayon A, Yarden Y, Givol D: Flg-2, a new member of the family of fibroblast growth factor receptors. Oncogene 1991; 6: 1089–1092.
36 Miki T, Fleming TP, Bottaro DP, Rubin JS, Ron D, Aaronson SA: Expression cDNA cloning of the KGF receptor by creation of a transforming autocrine loop. Science 1991; 251: 72–75.
37 Mansukhani A, Moscatelli D, Talarico D, Levytska V, Basilico C: A murine fibroblast growth factor (FGF) receptor expressed in CHO cells is activated by basic FGF and Kaposi FGF. Proc Natl Acad Sci USA 1990; 87: 4378–4382.
38 Burrus LW, Olwin BB: Isolation of a receptor for acidic and basic fibroblast growth factor from embryonic chick. J Biol Chem 1989; 264: 18647–18653.
39 Burgess WH, Maciag T: The heparin-binding (fibroblast) growth factor family of proteins. Annu Rev Biochem 1989; 58: 575–606.
40 Yayon A, Klagsbrun M, Esko JD, Leder P, Ornitz DM: Cell surface, heparin-like molecules are required for binding of basic fibroblast growth factor to its high affinity receptor. Cell 1991; 64: 841–848.
41 Ruoslahti E, Yamaguchi Y: Proteolgycans as modulators of growth factor activities. Cell 1991; 64: 867–869.
42 Johnson DE, Lee PL, Lu J, Williams LT: Diverse forms of a receptor for acidic and basic fibroblast growth factors. Mol Cell Biol 1990; 10: 4728–4736.
43 Hou J, Kan M, McKeehan K, McBride G, Adams P, McKeehan WL: Fibroblast growth factor receptors from liver vary in three structural domains. Science 1991; 251: 665–668.
44 Heldin CH, Westermark B: Platelet-derived growth factor: mechanism of action and possible in vivo function. Cell Regulation 1990; 1: 555–566.
45 Pertovaara L, Tienari J, Partanen J, Vartio T, Lehtonen E, Alitalo K: Modulation of fibroblast growth factor receptor expression and signalling during retinoic acid-induced differentiation of Tera-2 teratocarcinoma cells. 1991; manuscript in preparation.
46 Eerola E, Partanen J, Pertovaara L, Korhonen J, Alitalo R: Expression of functional fibroblast growth factor receptors in several human leukemia cell lines. 1991; in press.
47 Adnane J, Gaudray P, Dionne CA, Crumley G, Jaye M, Schlessinger J, Jeanteur P, Birnbaum D, Theillet C: *BEK* and *FLG*, two receptors to members of the FGF family, are amplified in subsets of human breast cancers. Oncogene 1991; 6: 659–663.
48 Safran A, Avivi A, Orr-Urtereger A, Neufeld G, Lonai P, Givol D, Yarden Y: The murine *flg* gene encodes a receptor for fibroblast growth factor. Oncogene 1990; 5: 635–643.
49 Pasquale EB, Singer SJ: Identification of a developmentally regulated protein-tyrosine kinase by using anti-phosphotyrosine antibodies to screen cDNA expression library. Proc Natl Acad Sci USA 1989; 86: 5449–5453.
50 Lee PL, Johnson DE, Cousens LS, Fried VA, Williams LT: Purification and complementary DNA cloning of a receptor for basic fibroblast growth factor. Science 1989; 245: 57–60.
51 Pasquale EB: A distinctive family of embryonic protein-tyrosine kinase receptors. Proc Natl Acad Sci USA 1990; 87: 5812–5816.
52 Eerola E, Partanen J, Cannizzaro L, Huebner K, Alitalo K: Localization of fibroblast growth factor receptor-4 gene to chromosome 5q33-qter. 1991; in press.

Angiogenesis: Key Principles - Science - Technology - Medicine
ed. by R. Steiner, P. B. Weisz & R. Langer

Activation of the tyrosine kinase receptor is not sufficient for the full biological activity of bFGF

A. Isacchi, L. Bergonzoni, M. Statuto[a], R. Chiesa[a], M. Rusnati[a], M. Presta[a], G. Ragnotti[a] and P. Sarmientos

Department of Biotechnology, Farmitalia Carlo Erba, Viale Bezzi 24, 20146 Milano, Italy and [a]Department of Biomedical Sciences and Biotechnologies, School of Medicine, Università di Brescia, 25124 Brescia, Italy

Introduction

Basic Fibroblast Growth Factor (bFGF) is an 18,000-D protein produced by various mesodermal and neuroectodermal tissues and especially by tissues with high angiogenic activity. It belongs to a well-characterized family of endothelial cell mitogens also known as heparin-binding growth factors for their high affinity to heparin [1, 2].

The responses of endothelial cells to bFGF in vitro include changes in cell morphology, stimulation of cell proliferation and migration and increased synthesis and release of lytic enzymes such as plasminogen activator and collagenase. In vivo, bFGF was shown to induce neo-angiogenesis and to promote wound healing [2].

Several high affinity receptors for basic FGF have already been identified through both cellular and molecular biology approaches [3, 4]. While the function of each of these receptors in the angiogenic process remains to be established, all of them seem to share the same structural characteristics of typical tyrosine-kinase receptors.

Low affinity receptors have also been identified [5]. However, the correlation between receptor binding and activation and the subsequent biological activities displayed by this type of cell mitogen is poorly understood.

Using synthetic peptides derived from the 155-amino acid sequence of bFGF, Presta and coworkers have identified specific sequences comprised between residues 38–61 and 82–101 which behave as partial agonists and antagonists of bFGF activity in mitogenic assays on endothelial cells [6]. However, interestingly, these peptides do not affect plasminogen activator (PA) production in bFGF-treated and untreated endothelial cells, suggesting that the PA-inducing activity of bFGF may take place through mechanisms different from those mediated by high affinity receptor binding and may be independent from mitogenesis.

Through a different approach based on the investigation of the behavior of mutants of bFGF, we have studied the biochemical and biological properties of a bFGF derivative lacking the amino acid residues 27 through 32, named M1-bFGF [7]. This mutant, obtained by expression in recombinant *E. coli*, interacts with cell membrane receptors and activates the corresponding tyrosine-kinase activity in a manner that is undistinguishable from the wild type bFGF. However, while stimulation of proliferation of endothelial cells and fibroblasts does not seem to be affected, the ability to induce PA production is considerably diminished. The observed differential behavior of M1-bFGF suggests the possibility that the PA induction by bFGF follows a pathway independent from the tyrosine-kinase receptor activation.

Receptor-interactions of M1-bFGF

We have used the technique of site-directed mutagenesis to obtain the deletion of the amino acid sequence from residue 27 to 32 of the bFGF polypeptidic chain. A schematic representation of M1-bFGF is shown in Fig. 1. The deleted sequence was selected for the high degree of homology existing among the different growth factors of the FGF family within that region. The recombinant protein was efficiently expressed in *E. coli* and purified from the soluble fraction of cell extracts by Heparin-Sepharose chromatography. Indeed, M1-bFGF retains the same heparin binding affinity as wild type bFGF (data not shown).

We first compared the ability of M1-bFGF to bind to the high affinity receptors present on endothelial cells. Accordingly, GM7373 cells were incubated for 2 h at 4°C with a constant amount of ^{125}I-bFGF in the presence of increasing concentrations of cold M1-bFGF or wild type bFGF. As shown in Fig. 2A, M1-bFGF was as effective as wild type bFGF in competing with labeled bFGF for its high affinity binding sites. In addition, chemical cross-linking experiments demonstrated that both molecules bind to the same 140,000 plasma membrane receptor (data not shown).

When bFGF binds to the high affinity receptors, a down-regulation of the number of the binding sites is observed [8]. We have therefore investigated whether binding of M1-bFGF is also able to cause the same disappearance of binding sites. As shown in Fig. 2B, both bFGF and M1-bFGF elicit the same dose-dependent disappearance of ^{125}I-bFGF binding sites on GM7373 cells following a 2-h incubation time. Moreover, the kinetics of down-regulation of ^{125}I-bFGF high affinity binding sites after bFGF or M1-bFGF treatment were found to be identical (data not shown).

These data clearly indicate that deletion of the amino acid residues from 27 through 32 does not alter the properties of interaction of bFGF

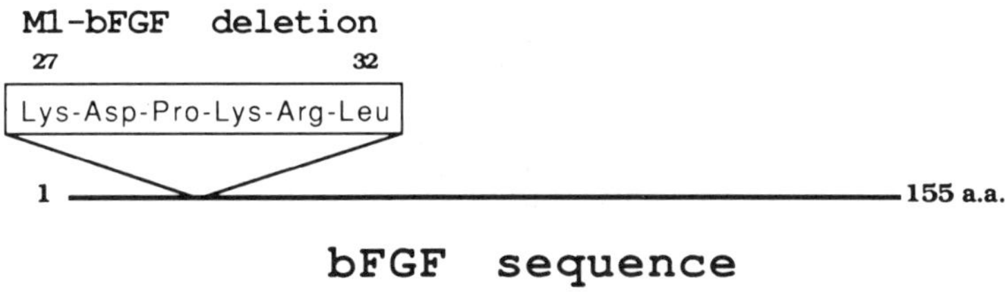

Figure 1. Schematic representation of the M1-bFGF mutant. Amino acid residues 27–32, which were deleted in the M1-bFGF mutant, are indicated in the box.

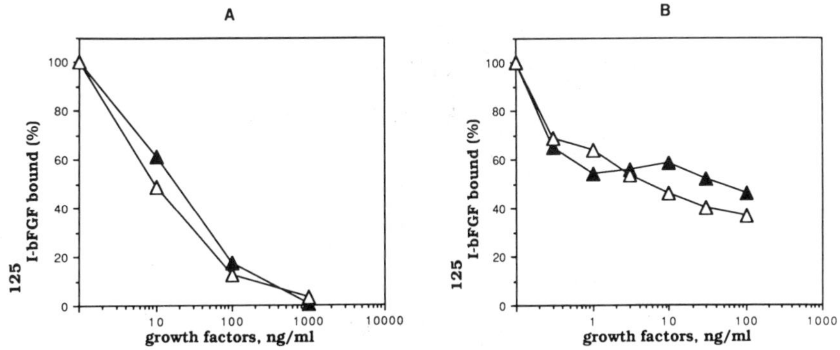

Figure 2. *A* Competition for binding of ^{125}I-bFGF to GM7373 cells by bFGF and M1-bFGF. GM 7373 cells were incubated for 2 h at 4°C with ^{125}I-bFGF (8 ng/ml) and increasing concentrations of unlabelled bFGF (open triangles), or M1-bFGF (closed triangles). The incubation was performed in the presence of 10 μg/ml heparin. Then, binding of ^{125}I-bFGF to plasma membrane receptors was quantitated and expressed as percent of the binding in the absence of unlabeled molecules. *B* Down-regulation of bFGF plasma membrane receptor. GM 7373 cells were incubated at 37°C with increasing concentrations of bFGF (open triangles) or of M1-bFGF (closed triangles). After 2 h, cells were washed with 2 M NaCl buffered at pH 4.0, to remove bound bFGF high affinity binding sites.

with the high affinity receptors. We looked then at the ability of M1-bFGF to activate the tyrosine-kinase domain of the receptor and to promote phosphorylation of cytoplasmic proteins as compared to wild type bFGF. As shown in Fig. 3, using anti-phosphotyrosine antibodies, the same pattern of tyrosine phosphorylation is observed when bFGF and M1-bFGF are added for 15 min at 37°C to NIH 3T3 cells under serum-free conditions. Tyrosine-phosphorylated proteins include the typical Mr 90,000 substratum and a Mr 140,000 molecule which probably corresponds to the activated bFGF receptor.

Biological activities of M1-bFGF

Basic FGF stimulates cell proliferation and PA production in different endothelial cell types. The mitogenic activity of M1-bFGF has been

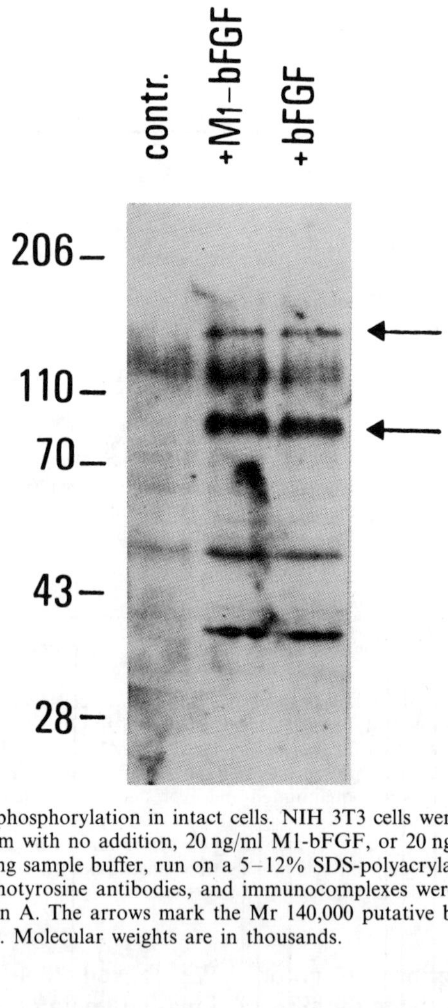

Figure 3. Tyrosine phosphorylation in intact cells. NIH 3T3 cells were incubated for 15 min in serum-free medium with no addition, 20 ng/ml M1-bFGF, or 20 ng/ml bFGF. Then, cells were lysed in reducing sample buffer, run on a 5–12% SDS-polyacrylamide gel, immunoblotted with anti-phosphotyrosine antibodies, and immunocomplexes were visualized by incubation with ^{125}I-Protein A. The arrows mark the Mr 140,000 putative bFGF receptor and the Mr 90,000 substrate. Molecular weights are in thousands.

evaluated as the capacity to support proliferation of ABAE cells as well as to stimulate DNA synthesis on 3T3 fibroblasts. Fig. 4 shows the dose-response curves obtained for M1-bFGF and bFGF in these two assays which clearly indicate that M1-bFGF has the same mitogenic activity as wild type bFGF.

However, M1-bFGF is at least 100 times less potent than bFGF in stimulating PA production in all endothelial cell types tested. As shown in Fig. 5, we have assayed PA production in GM7373, AG7680 and ABAE cells. The inability of M1-bFGF to stimulate PA production has also been confirmed at the mRNA level. Indeed, a 5–10-fold increase of the levels of urokinase-type plasminogen activator (uPA) mRNA was observed by Northern blot analysis of bFGF-treated GM7373 cells but not of M1-bFGF-treated cells (data not shown).

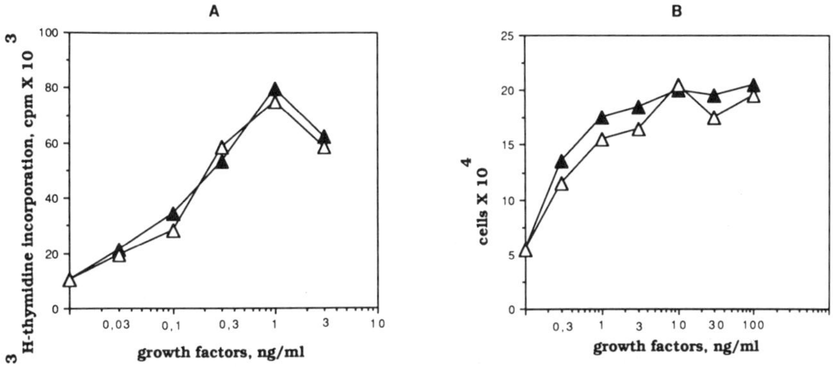

Figure 4. *A* Proliferation activity on ABAE cells. ABAE cells were seeded in 35-mm dishes 2000 cells/dish. After 16 h, cells were treated with bFGF (open triangles) or M1-bFGF (closed triangles) in the presence of 10% FCS. Then, the medium was changed every other day. Cells were trypsinized and counted after 6 days. *B* DNA synthesis in NIH 3T3 cells. Quiescent cultures of NIH 3T3 cells were incubated in fresh medium containing 3H-thymidine and increasing concentrations of bFGF (open triangles) or M1-bFGF (closed triangles). After 24 h, the amount of radioactivity incorporated into the trichloroacetic acid-precipitable was measured.

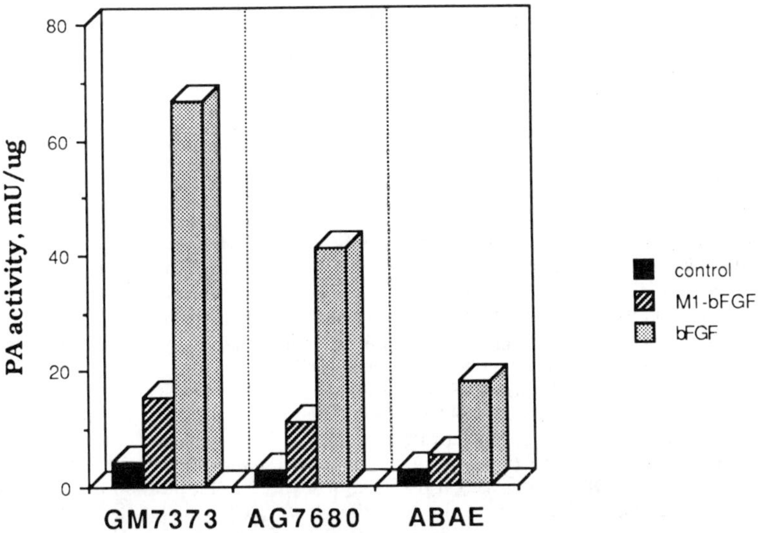

Figure 5. PA-inducing activity in different endothelial cell lines. FBAE GM 7373 cells, FBAE AG 7680 cells and ABAE cells were incubated in the presence of 30 ng/ml of bFGF or of M1-bFGF. After 24 h, cells were extracted and assayed for cell-associated PA activity as described in [2].

Discussion

We have prepared by recombinant DNA methods a mutant of bFGF which has lost the amino acid sequence from residue 27 through 32 of the 155-amino acid bFGF form. This region was chosen since it corresponds to a highly conserved sequence among the different members of the FGF family. Such a new molecule, M1-bFGF, has conserved many of the biochemical and biological characteristics of the wild type form. In particular, it has retained the full ability to recognize and activate the high affinity bFGF receptors leading to the phosphorylation of specific cytoplasmic proteins in a manner undistinguishable from the wild type molecule. Interestingly, receptor activation and the subsequent phosphorylation do not seem to be sufficient to transduce the intracellular signals for the stimulation of PA production. Instead, the ability to stimulate cell proliferation seems to be kept by M1-bFGF. However, presently, we do not know whether the intracellular signals that lead to cell proliferation depend on the activation of the tyrosine-kinase receptor or they may be transduced by separate mechanisms.

Following interaction with endothelial cell receptors, bFGF is internalized and translocated to the cell nucleus where additional biological functions may take place. It is thus possible that bFGF has to remain active in the intracellular environment to elicit its functions. We do not know whether the observed loss of PA production stimulated by M1-bFGF is due to a reduced intracellular stability or to the lack of the specific amino acid sequence between residues 27 and 32. Experiments to verify these hypotheses are currently going on in our laboratory.

Finally, the availability of bFGF mutants such as M1-bFGF allows us to study the importance of particular biological functions in angiogenesis assays.

References

1 Moscatelli D, Presta M, Rifkin DB: Purification of a factor from human placenta that stimulates capillary endothelial cell protease production, DNA synthesis and migration. Proc Natl Acad Sci USA 1986; 83: 2091–2095.

2 Presta M, Moscatelli D, Joseph-Silverstein J, Rifkin DB: Purification from a human hepatoma cell line of a basic fibroblast growth factor-like molecule that stimulates capillary endothelial cell plasminogen activator production, DNA synthesis and migration. Mol Cell Biol 1986; 6: 4060–4066.

3 Lee PL, Johnson DE, Cousens LS, Fried VA, Williams LT: Purification and complementary DNA cloning of a receptor for basic fibroblast growth factor. Science 1989; 245: 57–60.

4 Isacchi A, Bergonzoni L, Sarmientos P: Complete sequence of a human receptor for acidic and basic fibroblast growth factors. Nucl Acid Res, 1990; 18: 1906.

5 Saunders S, Jalkanen M, O'Farrel S, Bernfield M: Molecular cloning of syndecan, an integral membrane proteoglycan. J Cell Biol 1989; 108: 1547–1556.

6 Presta M, Rusnati M, Urbinati C, Statuto M, Ragnotti G: Functional domains of basic fibroblast growth factor (bFGF): possible role of the Asp-Gly-Arg sequences in the mitogenic activity of bFGF. Annls NY Acad Sci, 1991, In press.

7 Isacchi A, Statuto M, Chiesa R, Bergonzoni L, Rusnati M, Sarmientos P, Ragnotti G, Presta M: A 6-amino acids deletion in basic fibroblast growth factor dissociates its mitogenic activity from its plasminogen activator-inducing capacity. Proc Natl Acad Sci USA 1991; 88: 2628–2636.

8 Moscatelli D: Metabolism of receptor-bound and matrix-bound basic fibroblast growth factor by bovine capillary endothelial cells. J Cell Biol 1988; 107: 753–759.

Angiogenesis: Key Principles - Science - Technology - Medicine
ed. by R. Steiner, P. B. Weisz & R. Langer

Angiogenesis and heparin mimics

P. B. Weisz[a,b], H. C. Hermann[c], M. M. Joullié[d], K. Kumor[a], E. M. Levine[e], E. J. Macarak[f] and D. B. Weiner[c,e]

Departments of [a]Chemical Engineering, [b]Bio-Engineering, [c]Medicine, [d]Chemistry, University of Pennsylvania; [e]Wistar Institute and [f]Connective Tissue Research Institute, Philadelphia, Pennsylvania 19104, USA

Following the naming of the first "growth factor", we are threatened by a growing plethora of "name factors". They promote or inhibit some cellular function – and, confusingly, they sometimes appear to do both. Fortunately, the number of basic phenomena is usually much smaller than the inital complexity of names and manifestations. The goal of science is unification by recognition of the fewest key elements that determine macroscopic behavior.

The variety of growth factors has begun to shrink. Maciag and co-workers [1] demonstrated the virtual identity of ECGF and EDGF2 with aFGF, and presented reasons why RDGF, α-HGF, and acidic HDGF may also belong to the same family. Int-2 hst, FGF-5, k-FGF and KGF followed this trend [2], which has been continuing. Also, for a given function, the "business" portion of a proteinic factor appears to be much smaller than the total molecule.

The glycosaminoglycans (GAG's) are another set of elaborate macromolecules bearing many names. Each describes a class of polysaccharides. They exist as non-uniform mixtures in terms of size (molecular weight) as well as structural detail. Heparin is perhaps the most prominent one.

Some aspects of their behavior, such as their ability to modulate angiogenesis can vary greatly among "heparins" of different origin or production lots, as demonstrated by Folkman et al. [3].

Heparin, fragments and mimics

The ability to modulate angiogenesis (with cortisone) extends from whole heparin to fragments as small as a hexasaccharide [3]. Similarly, Castellot et al. [4] found the anti-proliferative role of heparin for smooth muscle cells to persist down to a fragment size of some twelve saccharide units.

Thus the challenge arises: what is the simplest effective structure ("model") that is responsible for some of the biological functions of heparin, in terms of size and chemical detail? We consider the cyclodextrins [5] to be suitable starting materials to explore the creation of simpler heparin mimics.

The most common one, β-cyclodextrin (β-CD), consists of seven glucose units forming a toroidal shaped molecule. Its rigidity adds structural simplicity by eliminating the conformational flexibility of a saccharide chain. Its 21 hydroxyl groups can be substituted by a variety of chemical groups.

Angiogenesis

With J. Folkman we demonstrated [6] that β-cyclodextrin when bearing about 14 sulfates (β-CD-14S) becomes a heparin mimic for inhibiting angiogenesis (with cortisone) more effectively than heparin. The same CD molecule, bare or with only seven sulfates, or with methyl or propyl substituents was ineffective.

Complexation with cationic dyes: Metachromasia

Complexation of heparin with cationic dyes leads to marked spectral shifts, called metachromasia. Langer and co-workers [7] demonstrated that the degree of metachromasia induced in the dye Azure A for various heparin fragments remains nearly constant from the value of whole heparin down to a molecular weight of about 3000. Fig. 1

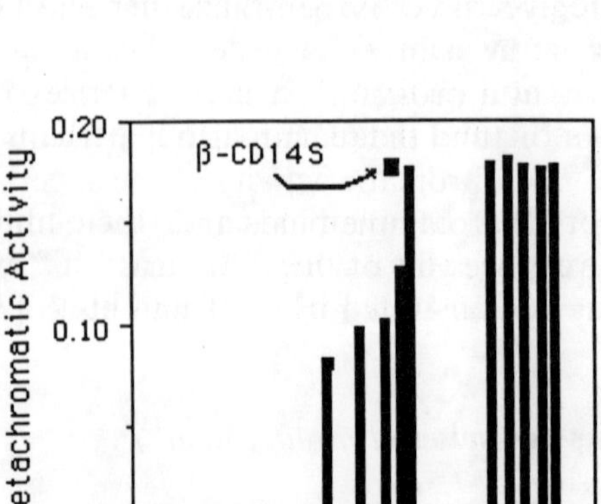

Figure 1. Metachromatic activity of heparin and heparin fragments as found by Grant et al. [7]; and for beta-cyclodextrin tetradecasulfate (β-CD-14S).

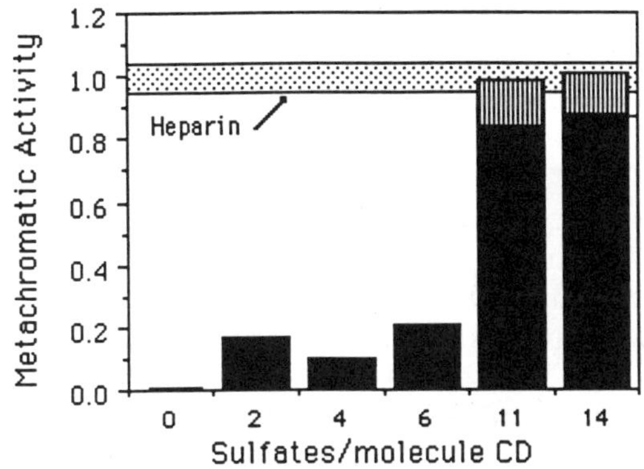

Figure 2. Metachromatic activity of beta-cyclodextrin sulfates of various degrees of sulfate substitution, compared to heparins.

reproduces their findings. Using the same technique [7] we find that β-CD-14S (M.W. 2563) shows metachromasia as strong as the heparins (Fig. 1). When we examine this property against the number of sulfate substitutions (Fig. 2), we find the maximal metachromatic activity down to near 10 sulfate groups per molecule, below which there is a sharp drop.

Interaction with protamine

Protamine administration is used clinically when needed to control, limit or optimize the effective heparin concentration. Typically [8], 1.3 mg of protamine is used to neutralize the effect of about 0.6 to 0.7 mg (100 units) of heparin.

We have found that a protamine/heparin complex is readily precipitated by centrifugation, and the remaining heparin concentration can be determined by the metachromatic activity of the supernatant. Using this technique we find that protamine binds and thus eliminates about 70% of an initially equal weight of heparin; and that β-CD-14S reacts equally with protamine to 80% for equal weights combined.

Inhibition of smooth muscle cell proliferation

Injury of endothelial cells can lead to the proliferation of smooth muscle cells (SMC). This phenomenon is relevant to atherosclerosis. In the clinical practice of angioplasty it causes restenosis. Heparin has been shown to inhibit SMC proliferation [9], while other classes of GAGs having less sulfate contents do not exhibit such effect [10].

We find that β-CD-14S also inhibits SMC proliferation. It causes a marked reduction of intimal hyperplasia following endothelial cell injury in a rabbit atherosclerosis model [11].

Potentiation of FGF for endothelial cell proliferation

Heparin has been shown [12–14] to potentiate the effectiveness of acidic fibroblast growth factor (aFGF) in endothelial and other cultured cells. In Fig. 3 we show results of relative cell density achieved with adult human endothelial cells, using 30 ng of aFGF in the presence of heparin (10 or 100 μg/ml) and in the presence of various sulfated cyclodextrins (10 and 100 μg/ml). β-CD-14S is equally or more effective at lower concentrations than heparin. We again find a critical requirement of sulfate density, requiring some 10 or more sulfates per CD molecule.

Cell protection against viral infection

Inhibition of viral infection has been observed for heparin and also for linear dextran sulfate [15]. We have used the cyclodextrins to examine the structural requirement for anti-viral (HIV-1 and HIV-2) activity.

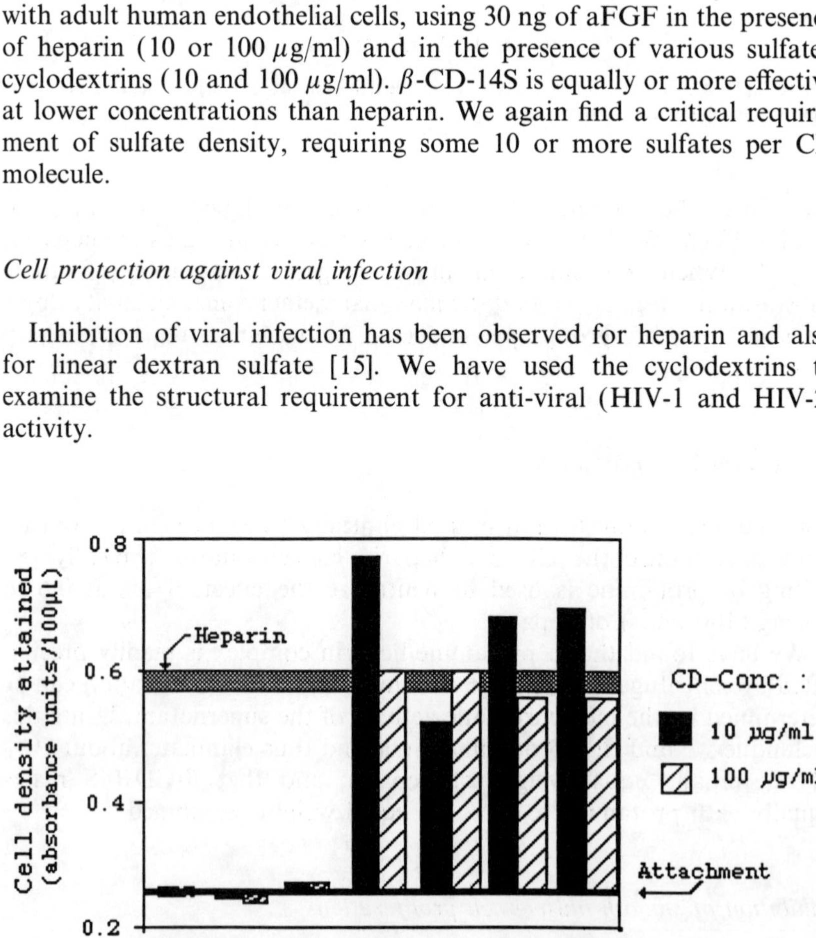

Figure 3. Promotion of endothelial cell growth achieved by cyclodextrins of various degrees of sulfate substitution, and with heparin. The 12-sulfated CD is alpha-CD, the 16-sulfated is gamma-CD. All others are beta-CD's. The ordinate is optical absorbance in units/100 μl.

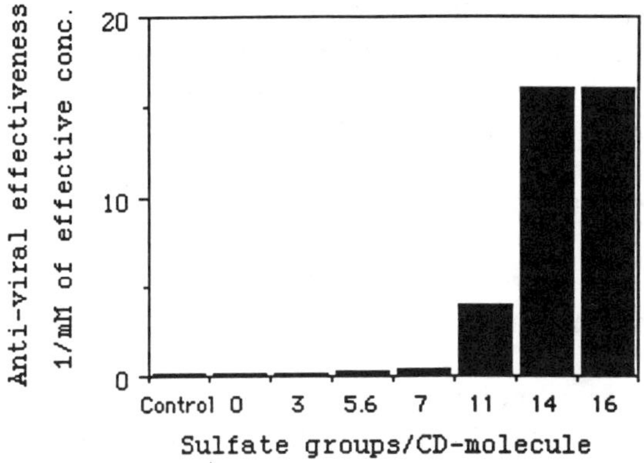

Figure 4. Inhibition of Human Sup T1 cell infection by HIV-1 virus, in the presence of beta-cyclodextrins of varying degrees of sulfate substitution.

We have found β-CD-14S to be very effective in protecting human Sup T1 cells against HIV-1 virus infection [16]. Both syncytia formation as well as reverse transcriptase production are prevented at doses below 80 μM of the sulfated CD. As Fig. 4 shows, we again find that more than 10 sulfate entities per CD molecule must be present to produce measurable activity. HTLV-IIIb cell free virus stocks were inoculated onto target cells in the presence and absence of the CD-agents. After incubation, wells were harvested and reverse transcriptase activity scored. Fig. 4. displays the lowest concentration of compound that completely inhibits infection.

Hemolytic activity

Many cyclodextrins have been reported to have various degrees of hemolytic activity [17, 18]. It is therefore of some interest to inquire into

Table I. Concentration for onset of hemolysis of several β-cyclodextrins

Substituents in place of —OH		Concentration
Type	Number	mg/ml
—	—	4
$—O—CH_3$	14	1
$—O—C_2H_5$	4	10
$—O—C_3H_7$	4	30
$—O—SO_3^+$	14	none
Heparin	—	none

the hemolytic activity of the sulfated cyclodextrin. We have re-examined various cyclodextrins and the β-CD-14S [19] using a standard methodology for hemolytic testing [18]. The critical concentrations above which hemolysis sets in for various cyclodextrins are shown in Table I. In contrast the sulfated β-CD-14S shows no hemolytic activity at the highest concentrations; nor does heparin.

Common denominators

These experiences point to key common denominators for heparin and the heparin mimic. In the phenomena discussed above, the highly sulfated cyclodextrin has performed similarly or better than heparin. We will call such heparin type materials HTMs:

Chemistry and cell biology

A high density of sulfate groups on a molecule, corresponding to some 10 sulfate groups within a hexasaccharide – or about 10 sulfate anions per 100 A^2 of molecular area – appears to be a controlling factor to the exclusion of all other structural detail.

Structural detail beyond the sulfate anion requirement may be necessary for heparin's anti-coagulant and other activities, but the sulfate density may in that case still be a necessary but not sufficient requirement.

The multiple sulfate anions possess a unique capability to strongly complex with molecules containing multiple cationic centers. This includes cationic dyes (e.g. in metachromasia, histology of staining GAGs), and peptides or proteins with neighboring basic (cationic) amino acid residues like arginine and lysine (e.g. protamine or FGF, and probably many others). The HTM entity will complex with such proteins whether it is mobile in solution or a portion of an immobilized structure.

Hence we have "heparin binding growth factors" and "heparin affinity chromatography"; we have already shown how immobilized β-CD-14S displays similar protein affinity for chromatography [20].

In angiogenesis, and in endothelial growth promotion we deal with HTM complexing to both FGF protein and to cell and extra-cellular matrix; in protamine antagonism we deal with simple complex formation with heparin; in smooth muscle cell (SMC) growth inhibition, heparin complexing with FGF must surely be involved; we suggest that the complex with FGF may be non-mitogenic to SMC, in contrast to the case of endothelial cell mitogenicity. In inhibiting viral cell entry, we

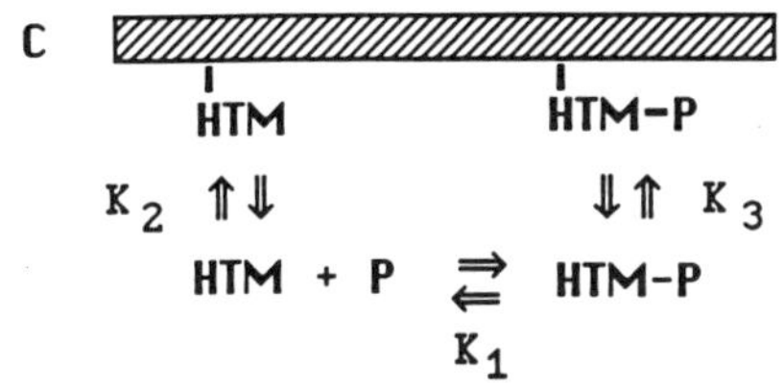

Figure 5. The interacting major equilibria involving solution species of protein (growth factor) P, heparin-like entity HTM, and the surface S (cell surface protein or basement membrane protein entities). K_1, K_2 and K_3 complexing equilibrium constants are all large (towards complexed product). The direct P to C binding is considered to have a small equilibrium constant compared to those pictured. Introduction of another protein species (Q) or a different surface (S_1) would alter the pictured equilibrium concentrations.

believe that surface bound HTM inhibits the mobility (diffusion) of virus on the cell surface to locate CD4 receptors. The dye responses (metachromasia) are a result of the physico-chemically analogous complexing phenomena.

Fig. 5 demonstrates the basic physical chemistry of a (simplest) model of the complexing interactions between protein (P), such as FGF, HTM molecules such as heparin and a surface (cell, basement membrane) in cell biology.

The complexing (attachment) constant of heparin-like GAGs and fragments with endothelial cells (K_2 and K_3) has been shown to be large, and reversible in classical work with endothelial cells, and desorption or re-equilibration rates were measured on a time scale of the order of an hour [21–23].

The amount of biologically effective entities on the cell or matrix surfaces will depend on the relative size of the equilibrium constants for the complex formation between *all* interacting (complexing) species (K_1, K_2, K_3) and *all* their respective concentrations.

This simple scheme assumes that the direct interaction (complexing, attachment, etc.) of protein to the cell is negligibly small compared to the dominance of the protein-HTM as the protein delivery species. The evidence for this for the case of FGF-mediated cell responses is indeed mounting [23].

If we add to a given system another protein, say Q, or an altered or different surface, say S′, the concentration of the species previously covering S, and other species concentrations, will be altered by virtue of the necessary re-equilibration between *all* interaction species in *all* of the surrounding system.

There is another observational indicator for the ruling mechanism of binding between a strongly cationic protein site and the high density anionic localization represented by the sulfates located on some 5 to 7 linked glucose groups (coiled in heparin or cyclic in our cyclodextrins): The repeated experience of achieving similar biological effect with similar weight rather than molar quantities of the HTM (heparin fragment vs. heparin and now for CD-derivative vs. heparin).

The number of bound protein will depend on how many moles of high density anionic *centers* are presented, not on how many moles of molecules are involved. In other words, a heparin chain of 12,000 MV bearing 5 such anionic configurations along its chain, after fragmentation into five fragments of 2400 MW would still be able to present 5 binding configurations, although now located on a five times larger molar quantity.

Thus it is the weight quantity which remains comparable in effectiveness, as was observed in previous work [3, 4, 7, 14] and in this work comparing the CD and heparin entities. Indeed we note the observation by Van Rijn et al. [24] that, in competition experiments of heparin fragments of different sizes binding (complexing) to endothelial cells, "I_{50} was inversely correlated with molecular weight". This would follow from above since I_{50} is expressed in molar (!) and not in weight concentration. The common denominator in many of the above described biological phenomena is the binding of the HTM on the endothelial cell.

Cell biology and biomedicine

Re-equilibration provides an apparent mechanism of mass transport. For example, if the cell surface is not fully heparin (HTM) saturated, additional heparin (HTM), due to its ready complexing of FGF (P) will "bring" a higher FGF concentration to the cell surface.

Re-equilibrium can cause unexpected "opposite" effects. For example, addition of large amounts of a heparin, at a given amount of proteinic factor may redistribute a portion of surface-heparin-complexed growth factor into complexes with excess heparin in solution, thus diminishing the growth factor effectiveness on the cell.

Re-equilibration can produce an apparent mechanism for "releasing" factors. Factors complexed to a surface-complexed HTM, for example a heparin sulfate will be replaced through reequilibration by exposure to added amounts of equal or higher sulfate density HTM such as heparin (see e.g. Bashkin et al. [25]).

It is notable that the "release" is not due to an action by the cell.

"Release" can also be achieved due to a metabolic action at the cell: Even a mild desulfation of the surface located HTM would lower the lower equilibrium constant sufficiently to result in desorption. An example is the heparitinase addition to extracellular matrix examined by Baskin et al. [25].

In general, we must note that any agent of sufficiently polyionic character added to an existing system can *act as an apparent promoter or inhibitor, by shifting the equilibrium* of the true promoter/inhibitor balance.

An example is the finding by Taylor and Folkman [26]: In a system where the balance of FGF and HTM is causing angiogenesis, protamine, capable of equally strong complexing with surface or solution HTM, will shift the available concentration of surface FGF to turn off its mitogenic effectiveness. A similar effect of protamine on growth factor/heparin mitogenicity for human endothelial cells was observed by Thornton et al. [27].

We have demonstrated reasonable evidence to believe that the *complexation chemistry that involves the high-sulfate density molecular partners is basic to a large variety of cell-biological and thus biomedical phenomena.*

The molecular sulfate density achieved, with its resulting capability for electrostatic (polyionic) complexation, appears to be the principal if not sole key to protein and cell attachment. This is an important guidepost to pharmacological chemistry. Indications are that pharmacological design of HTM agents for specific purposes involving cell attachment of protein delivery may be sought in a manner decoupled from other activities of heparin. It also may not require heparin's complex chemistry.

This also has the advantage of being able to concentrate on a specific target objective in therapy. The nature of the remainder of the molecule must, of course, be compatible with the biological environment, be non-toxic and not prone to interfere with other than the target organs or functions.

A "simple" saccharide may have inherent advantages in such compatibility. That the proper choice of the "remaining" structure, even within the saccharides is not trivial is seen, for example, by our observation in anti-viral activity [16] that the cyclic sugar (cyclodextrin polysulfate) has a longer active life than its straight chain counterpart (dextran sulfate, [28]), presumably because chemically and enzymatically, hydrolytic attack is known to progress easiest from the terminus of a saccharide; the cyclodextrins have none.

Acknowledgements. We wish to acknowledge efforts in chemical synthesis by E. J. Schweiger, and thank Dr. J. Bello of Roswell Park Memorial Institute for making available some cyclodextrin sulfates and American Maize-Products Company for supplying us with larger quantities of the β-CD-14S.

References

1 Schreiber AB, Kenney J, Kowalski J, Thomas KA, Gimenez-Gallego G, Rios-Candelore M, Di Salvo J, Barritault D, Courty J, Courtois Y, Moenner M, Loret C, Burgess WH, Mehlman T, Friesel R, Johnson R, Maciag T: A unique family of endothelial cell polypeptide mitogens: The antigenic and receptor cross-reactivity of bovine endothelial cell growth factor, brain-derived acidic fibroblast growth factor, and eye-derived growth factor-II. Cell Biol 1985; 101: 1623–1626.

2 Burgess WH, Maciag T: The heparin-binding (fibroblast) growth factor family of proteins: Annu Rev Biochem. 1989; 58: 575–606. – Rifkin DB, Moscatelli D: Recent developments in the cell biology of basic fibroblast growth factor. J Cell Biol 1989: 109: 1–6. – Finch PW, Rubin JS, Miki T, Ron D, Aaranson SA: Human KGF is FGF

related with properties of a paracrine effector of epithelial cell growth. Science 1989; 245: 752.
3 Folkman J, Langer R, Lindhardt RJ, Haudenschild C, Taylor S: Angiogenesis inhibition and tumor regression caused by heparin or a heparin fragment in the presence of cortisone. Science 1983; 221: 719–725.
4 Castellot JJ, Beeler DL, Rosenberg RD, Karnovsky MJ: Structural determinants of the capacity of heparin to inhibit the proliferation of vascular smooth muscle cells. J Cell Physiol 1984; 120: 315–320.
5 Bender ML, Komiyama M: Cyclodextrin Chemistry, 1978, Springer-Verlag, Berlin.
6 Folkman J, Weisz PB, Joullié MM, Li WW, Ewing WR: Control of angiogenesis with synthetic heparin substitutes. Science 1989; 243: 1490–1493.
7 Grant AC, Linhardt RJ, Fitzgerlad GL, Park JJ, Langer R: Metachromatic activity of heparin and heparin fragments. Analyt Biochem 1984; 137: 25–32.
8 Bull BS, Huse WM, Brauer FS, Korpman RA: Heparin therapy during extracorporal circulation II. The use of a dose response curve to individualize heparin and protamine dosage. J Thorac Cardiovasc Surg 1975; 69: 685–689.
9 Clowes AW, Karnovsky MJ: Suppression by heparin of smooth muscle cell proliferation in injured arteries. Nature 1977; 265: 625–626.
10 Castellot J, Cochran D, Karnovsky M: Effect of heparin on vascular smooth muscle cells. J Cell Physiol 1981; 124: 21–28.
11 Herrmann HC, LeVeen RF, Tomaszewski JE, Weisz PB, Barnathan ES: Inhibition of experimental angioplasty restenosis by oral administration of the heparin analogue beta-cyclodextrin tetradecasulfate. Circulation 1991: submitted.
12 Damon DH, Lobb RR, D'Amore PA, Wagner JA: Heparin potentiates the action of acidic fibroblast growth factor by prolonging its biological half-life. J Cell Physiol 1989; 138: 221–226.
13 Mueller SN, Thomas KA, DiSalvo J, Levine EM: Stabilization by heparin of acidic fibroblast growth factor mitogenicity for human endothelial cells in vitro. J Cell Physiol 1989; 140: 439–448.
14 Sudhalter J, Folkman J, Svahn CM, Bergendal K, D'Amore PA: Importance of size, sulfation, and anticoagulant activity in the potentiation of acidic fibroblast growth factor by heparin. J Biol Chem 1989; 264: 6892–6897.
15 Ito M, Baba M, Sato A, Pauwels R, DeClercq E, Shigeta S: Inhibitory effect of dextran sulfate and heparin on the replication of human immunodeficiency virus (HIV) in vitro. Antiviral Res 1987; 7: 361–367.
16 Weiner DB, Williams WV, Weisz PB, Greene MI: Synthetic cyclodextrin derivatives inhibit HIV infection in vitro. Pathobiology 1991, in press.
17 Irie T, Otagiri M, Sunada M, Uekama K, Ohtani Y, Yamada Y, Sugiyama Y: Cyclodextrin induced hemolysis and shape changes of human erythrocytes in vitro. J Pharm Dyn 1982; 5: 741–744.
18 Jodál I, Nánási P, Szejtli J: Investigation of the hemolytic effect of the cyclodextrin derivatives; in Proceedings of the Fourth International Symposium on Cyclodextrins, O. Huber and J. Szejtli (eds.), 1988: pp. 421–425; Kluwer Academic Publishers, Boston.
19 Macarak EJ, Kumor K, Weisz PB: Sulfation removes hemolytic activity of cyclodextrin: Biochem Pharmac 1991, in press.
20 Shing Y, Folkman J, Weisz PB, Joullié MM, Ewing WR: Affinity of fibroblast growth factors for β-cyclodextrin tetradescasulfate. Analyt Biochem 1990; 185: 108–111.
21 Barzu T, Molho P, Tobelem G, Petitou M, Caen J: Binding and endocytosis of heparin by human endothelial cells in culture. Biochim Biophys Acta 1985; 196–203.
22 Barzu T, Van Rijn JLML, Petitou M, Molho P, Tobelem G, Caen JP: Endothelial binding sites for heparin. Biochem J 1986; 238: 847–854.
23 Rapraeger AC, Krufka A, Olwin BB: Requirement of heparan sulfate for bFGF-mediated fibroblast growth and myoblast differentiation. Science 1991; 252: 1705–1708.
24 Van Rijn JLML, Trillou M, Mardiguian J, Tobelem G, Caen J: Selective binding of heparins to human endothelial cells; implications for pharmacokinetics. Thromb Res 1987; 45: 211–222.
25 Bashkin P, Doctrow S, Klagsbrun M, Svahn CM, Folkman J, Vlodavsky I: Basic fibroblast growth factor binds to subendothelial extracellular matrix and is released by heparatinase and heparin-like molecules. Biochemistry 1989; 28: 1737–1743.

26 Taylor S, Folkman J: Protamine is an inhibitor of angiogenesis. Nature 1982; 297: 307–312.
27 Thornton SC, Mueller SN, Levine EM: Human endothelial cells: Use of heparin in closing and long-term serial cultivation. Science 1983; 222: 623–625.
28 Mitsuya H, Looney DJ, Kuno S, Ueno R, Wong-Stall F, Broder S: Dextran sulfate suppression of viruses in the HIV family: Inhibition of virion binding to $CD4^+$ cells. Science 1988; 240: 646.

Angiogenesis: Key Principles - Science - Technology - Medicine
ed. by R. Steiner, P. B. Weisz & R. Langer

Endothelial cell migration and chemotaxis in angiogenesis

Cynthia L. Stokes

Department of Chemical Engineering, University of Houston, Houston, TX, USA

Abstract. Our goal has been to provide a quantitative analysis of the random motility and chemotaxis of microvessel endothelial cells (MEC) in order to understand the role of these functions in the development of new capillaries. A major difference of our work from previous investigations has been our use of mathematical analysis to interpret experimental results, and to relate in vitro migration measurements to in vivo angiogenesis observations. In this paper we present some of our methods and their rationale, with recent results from both experiment and mathematical models.

Introduction

A number of observations have demonstrated that the locomotion of MEC is an integral part of angiogenesis, along with cell proliferation, protease production and basement membrane production. A combination of quantitative experiments and mathematical models has been used to study how the migration of MEC determines the growth rate and microvessel network structure of developing of capillary beds. The under-agarose assay with a linear geometry was used to measure the random migration and chemotaxis of MEC in reponse to several chemical angiogenesis modulators. Additional observations were done to characterize the cell paths of individually migrating cells for comparison to the paths of growing blood vessels. Finally, using data from these experiments, we have constructed a mathematical model of angiogenesis to test hypotheses regarding MEC proliferation, random migration and chemotaxis, and the coordination of these functions, in angiogenesis.

Methods

Under-agarose assay

MEC from human subcutaneous adipose tissue were used in the linear under-agarose assay as described [7]. Briefly, 3×10^5 cells/cm^2 were plated in a rectangular well (3×19 mm) formed in the agarose on a gelatin-coated polystyrene microscope slide. In random motility experiments, the chemical factor of interest was mixted uniformly with both the agarose and cell suspension medium [assay medium: M199, 0.1% bovine serum albumin (BSA)]. For chemotaxis experiments, the factor was added only to a second well parallel to the cell well to set up a chemical gradient [7]. After 24, 48 or 72 h, the cell density was measured at increasing distances from the edge of the original cell well, resulting in a "cell density profile".

From these data we determined values of the random motility coefficient (in uniform concentrations) and the chemotaxis coefficient (in chemical gradients). These parameters are defined in a mathematical model that describes random motility as a persistent random walk and chemotaxis as a biased persistent random walk [1, 5]. For random motility, the model is a diffusion equation for cell density containing the random motility coefficient (μ) in place of a diffusion coefficient. To obtain a value of μ for each experiment, the diffusion equation was solved, the experimental cell density profile was inserted, and regression analysis used to obtain the value of μ that best described the data. For chemotaxis, the model states that the rate of change of the cell density profile equals the contributions of random migration, chemokinesis (stimulated random migration) and chemotaxis. A value of χ was obtained for each experiment in a gradient by comparing model predictions of cell density profile for various values of χ to the experimental cell density profile. The previously obtained values for μ for various uniform chemical concentrations were used in these calculations to account for random motility and chemokinesis in the gradient, so that χ depicts *only* the migration caused by true directional migration, not chemokinesis. See refs. [1, 5, 7] for details.

Single cell experiments

Cell paths were obtained by videotaping sparsely plated cells on the microscope. From these paths the mean square displacement, $\langle D^2 \rangle$, for different time increments, T, was calculated [9]. These paths can be described as a random walk using a simple stochastic differential equation (the Ornstein-Uhlenbeck process; 11) from which the relationship of mean square displacement versus time increment is derived:

$$\langle D^2 \rangle = S^2 P_v^2 \left(\frac{T}{P_v} - 1 + e^{-T/P_v} \right)$$

The two parameters, speed, S, and persistence time, P_v, characterize the rate of movement and the time a cell migrates in one direction, respectively. S and P_v were obtained for each cell path using nonlinear regression of the $\langle D^2 \rangle$ versus T data in this equation.

Results

Population migration experiments

As explained above, we have interpreted the results of the under-agarose assay in terms of μ and χ The random motility coefficient, μ, is analogous to a diffusion coefficient for Brownian motion and measures the rate of "dispersion" of a cell population. The chemotaxis coefficient, χ, is similar but characterizes only the rate of "drift" of the population (the directed response) caused by biased movement in a concentration gradient. These parameters quantitatively reflect only the *intrinsic* cell migration response, regardless of extraneous system variables (experiment duration, assay geometry, attractant diffusivity, etc.). In addition, this analysis accounts for the stimulated random motility (chemokinesis) in a chemical gradient, so that the true directed, or chemotactic component can be measured unambiguously, a major reason for selecting this method of analysis [1, 5, 7].

From the cell population experiments we found that human MEC exhibit a baseline level of motility and that they are stimulated by both uniform concentrations (random motility) and gradients (chemotaxis) of acidic fibroblast growth factor (aFGF). For the control (no aFGF), the random motility coefficient μ has a value of $(4.9 \pm 0.8) \times 10^{-9}$ cm^2/s (mean $\pm$ SEM). (For comparison, this is about 10^{-3} the size of the diffusion coefficient for a 15 kD globular protein in water). μ increased between 10^{-11} M and 10^{-8} M aFGF when 1 μg/ml heparin (not stimulatory alone) was added, with a maximum value of $(9.9 \pm 0.3) \times 10^{-9}$ cm^2/s. Stimulation only occurred in 10^{-8} M aFGF ($\mu = (8.3 \pm 0.4) \times 10^{-9}$ cm^2/s) in the absence of heparin [7]. Gradients of aFGF (with uniform heparin concentration of 1 μg/ml present) elicited a chemotactic response with 10^{-10} M and 3×10^{-10} M initially in the attractant well, with a peak of $\chi = 2600 \pm 750$ cm^2/s-M at 10^{-10} M. No chemotaxis was observed above (to 10^{-8} M) and below (to 10^{-12} m) these concentrations [7]. The chemotatic concentration is near the dissociation constant for aFGF binding to cell surface receptors, supporting the idea that chemotaxis is a receptor-mediated phenomenon.

The random motility of MEC in β-cyclodextrin tetradecasulfate and hydrocortisone was also measured [8]. 400 μg/ml of either chemical alone inhibited migration up to 50% in the absence or presence of 10^{-8} M aFGF. The combination completely abolished the migratory response (without causing cell death) in the absence of aFGF and reduced it by 85% in 10^{-8} M aFGF ($\mu = (1.5 \pm 0.2) \times 10^{-9}$ cm^2/s). These results suggest that inhibition of MEC migration is at least part of the mechanism by which angiogenesis is inhibited by these chemicals in vivo [4].

Single cell experiments

We observed individual migrating cells to study the characteristics of cell paths. In the assay medium, we measured $S = 25 \pm 3$ μm/h and $P_v = 5.5 \pm 1.0$ h, whereas inculture medium (M199, 20 μg/ml unpurified aFGF, 10% serum, 0.1% BSA, 90 μg/ml heparin), $S = 42 \pm 5$ μm/h and $P_v = 2.9 \pm 1.1$ h. Thus, stimulatory factors caused the cells to move more quickly but turn more often, so that their mean free path length ($S \times P_v$) remained approximately constant. If the turn rate determines the likelihood of sprout collision and anastomosis, then a constant mean free path length should provide constant loop size. The rate of network formation would increase, however, due to increased speed. These cell speeds are similar to the rate at which sprouts elongate in vivo [3, 12], indicating that cell speed is likely an important determinant in the rate of vessel formation.

The persistence time result has implications for experimental measurements of endothelial cell migration. Because the persistence time is several hours long, a given cell will travel in one direction for that time. Since changes in direction of MEC are gradual, the random walk character of the migration will be evident only after several persistence times. This implies that any experiment of shorter duration in a gradient will not discern between directional persistence of a random walk and true directional bias (i.e. chemotaxis) and should be interpreted carefully.

Angiogenesis model

To further investigate the quantitative relationship between cell migration and vessel growth, we constructed a probabilistic mathematical model in which vessel networks are simulated by computer. For simplicity, the model is confined to two dimensions. This should be a good approximation of the vessel growth in certain experimental situations such as thin (<50 μm) rabbit ear chambers [2, 12].

The model works as follows: We model the existing vasculature as a single, straight, parent vessel located at $y = 0$ (see Fig. 1). New buds grow off the parent vessel at random positions at time $t = 0$. The path of each capillary sprout is determined by the trajectory of an actively migrating cell located at its tip. This cell migrates with a velocity generated by a stochastic ordinary differential equation that describes a random walk (used above to characterize the single cell experiments). All other cells in the sprout follow directly behind the tip cell. The cells proliferate to provide additional cells. In addition, cells in one sprout can move into an attached sprout, causing redistribution of cells among vessels. There is a minimum cell density in the sprout, ρ_{min}, which is

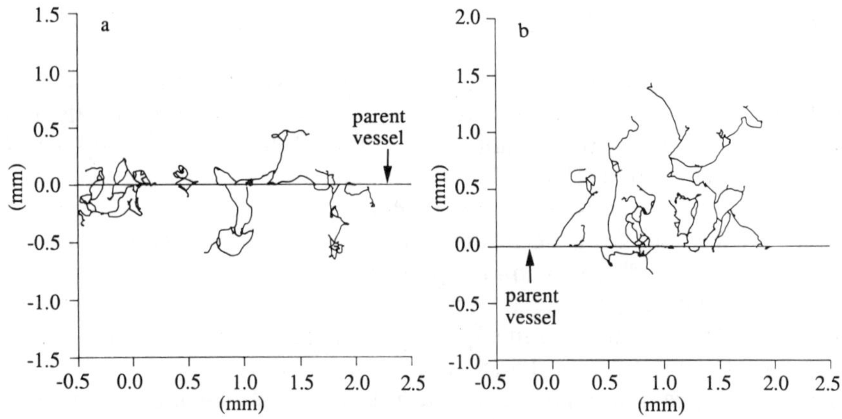

Figure 1. Simulations of vessel networks from mathematical model. Results of growth after four days for (a) random motility only and (b) random motility and a moderate chemotactic response. The attractant source is centered at (1 mm, 2 mm) and a chemical gradient is formed by diffusion. See text for details.

required to maintain a contiguous vessel, and a maximum density, ρ_{max}, which reflects the confluent density at which MEC stop growing. If a sprout's density drops to ρ_{min}, then the sprout cannot further elongate and new buds cannot start growing off it unitl ρ increases. Thus, if the proliferation and cell redistribution rates are not great enough in comparison to the migration rate, the vessel will not be able to elongate, because additional cells will not be available to maintain a contiguous vessel. Branches bud off existing sprouts and loops with a probability based on data from [6]. Anastomosis occurs whenever a sprout tip runs into another sprout or loop. The equations used to described this model are presented elsewhere [10].

Many simulations of the model have been done to investigate the effects of different levels of random motility, chemotaxis and proliferation rate on the network growth. Here we present only two examples. Figure 1a illustrates a simulation for random migration only, while Figure 1b includes a chemotactic response similar to that measured in the aFGF gradient above ($S = 40\ \mu m/h$, $P_v = 3$ h, cell doubling time was 18 h, and initial direction of growth of all sprouts was random for both). The first figure illustrates that random migration alone does not result in networks similar to those in vivo [2, 12], especially because the direction of growth is haphazard. We obtained this result even when P_v was increased to 18 h and initial sprout direction was towards the attractant source. Thus, persistent random migration is not sufficient to account for the ongoing directed growth of new vessels. In Figure 1b, most of the growth is in the general direction of the attractant source, similar to angiogenesis in vivo. While the vessel density is low because of computational limitations, comparison of the average vessel length and network expansion rate to experimental measurements of angiogenesis in vivo [2, 12] indicate that these simualted networks are also quantitatively similar to real networks. These simulations predict that the measured level of chemotaxis in aFGF is sufficient to account for directed vessel growth in vivo, and that the rate of migration is a strong determinant of network growth rate.

Discussion

Our experiments and models have been designed to determine the correspondence between MEC migration characteristics and the capillary growth rate and network structure during angiogenesis. The experiments indicate that the chemical factors tested have essentially the same effects on MEC migration in vitro as they have on angiogenesis in vivo. This suggests that MEC motility and chemotaxis may be a useful in vitro screen for potential angiogenic mediators. Quantitatively, the rate at which MEC migrate is similar to that of vessel elongation, suggesting that migration strongly influences the rate of vessel growth. The measurements of MEC chemotaxis in response to aFGF are the first unambiguous demonstration of chemotaxis by MEC, because our methods of quantitation are the first that can quantitatively account for chemokinesis in a gradient. Our computer simulations predict that cell proliferation is not a rate determining factor (though it can be limiting if no proliferation occurs), and that directional bias of movement is required for directed network growth. In summary, this work indicates that the rate of vessel outgrowth is primarily determined by migration rate and the moderate chemotactic response we measured in vitro is sufficient to account for the directed growth of vessels in vivo.

Acknowledgements. I gratefully acknowledge the contributions of Maria A. Rupnick, Stuart K. Williams, Douglas A. Lauffenburger, and Paul B. Weisz to this work and the ideas contained here.

References

1 Alt W: Biased random walk models for chemotaxis and related diffusion approximations. J Math Biol 1980; 9: 147–177.
2 Clark ER, Clark EL: Microscopic observations on the growth of blood capillaries in the living mammal. Am J Anat 1939; 64: 251–301.
3 Eddy HA, Casarett GW: Development of the vascular system in the hamster malignant neurilemmoma. Microvasc Res 1973; 6: 63–82,
4 Folkman J, Weisz PB, Joullie MM, Li WW, Ewing WR: Control of angiogenesis with synthetic heparin substitutes. Science 1989; 243: 1490–1493.
5 Lauffenburger DA; Measurement of phenomenological parameters for leukocyte motility and chemotaxis. Agents Actions Suppl 1983; 12: 34–53.

6 Sholley MM, Ferguson GP, Seibel HR, Montour JL, Wilson JD; Mechanisms of neovascularization: Vascular sprouting can occur without proliferation of endothelial cells. Lab Invest 1984; 51: 624–634.
7 Stokes CL, Rupnick MA, Williams SK, Lauffenburger DA: Chemotaxis of human microvessel endothelial cells in response to acidic fibroblast growth factor. Lab Invest 1990; 63: 657–668.
8 Stokes CL, Weisz PB, Williams SK, Lauffenburger DA: Inhibition of microvascular endothlial cell migration by β-cyclodextrin tetradecasulfate and hydrocortisone. Microvasc Res 1990; 40: 279–284.
9 Stokes CL, Lauffenburger DA, Williams SK: Migration of individual microvessel endothelial cells: stochastic model and parameter measurement. J Cell Sci 1991 (in press).
10 Stokes CL, Lauffenburger DA: Analysis of the roles of microvessel endothelial cell random motility and chemotaxis in angiogenesis. J theor Biol 1991; 152: 377–403.
11 Uhlenbeck GE, Ornstein LS: On the theory of Brownian motion. Phys Rev 1930; 36: 823–841.
12 Zawicki DF, Jain RK, Schmid-Schoenbein GW, Chien S: Dynamics of neovascularization in normal tissue. Microvasc Res 1981; 21: 27–47.

Angiogenesis: Key Principles - Science - Technology - Medicine
ed. by R. Steiner, P. B. Weisz & R. Langer

Microenvironment and angiogenic response

Pietro M. Gullino

Department of Biomedical Sciences, University of Turin, 10126 Turin, Italy

Angiogenic capacity, i.e. the property of inducing neovascularization is the rule for most neoplastic cells but the exception for normal tissues. When neoplastic cells from a V2 carcinoma were implanted in the vitreous body of a rabbit's eye and their location was more than 2 mm from the retina, a nodule developed that reached a diameter of less than 1 mm and remained practically unchanged for weeks. However, if this nodule drifted towards the retina, within a few days it was colonized by microvessels and an explosive growth followed with formation of a solid tumor that destroyed the eye. Similar effects were observed when a mouse glioma was implanted in the rabbit vitreous [1].

One interpretation of these observations is that the neoplastic cells produced molecules having a paracrine-like effect on microvessels and that the range of this effect was extended to about 2 mm in the avascular vitreous body. Indirect support of this interpretation was supplied by the isolation from the vitreous of a fraction with an inhibitory effect on angiogenesis [2].

Over the last decade a number of molecules have been described as capable of inducing new formation of vessels in the rabbit cornea or other suitable models [3]. The list of these molecules shows a peculiar heterogeneity for both chemical characteristics and biological properties. Indeed, common properties among these molecules on which to ground an interpretation of their angiogenic capacity are difficult to find.

It should be stressed that any treatment applied in vivo to induce neovascularization alters the biologic milieu of the tissue. This alteration may be extensive, for instance when angiogenesis is one component of the granulation tissue, or may be histologically inexistent when the mobilization of inflammatory-type cells does not occur. We feel that any attempt to clarify the mechanism of the angiogenic response should consider this last condition as an indispensable postulate to minimize the influence of interfering factors.

One can hypothesize that the heterogeneous group of angiogenic molecules already described might have in common this property of changing the microenvironment of the tissue to be invaded by the capillaries. Angiogenesis occurs prominently in embryonal and growing tissues but, under physiological conditions, it is kept to almost zero level in the adult organism (uterus and ovaries are obvious exceptions). The down regulation of angiogenesis in the adult organs probably depends on a balance among a set of molecules with either stimulatory or inhibitory action on endothelium and related cell populations. If this is the case, one would expect that, at least in part, down regulation could be obtained via paracrine-like modalities able to modify total concentration and relative proportions of molecules normally present in the microenvironment. Although molecules of different nature added to the organ microenvironment could interfere with the down regulation of angiogenesis, the relevant molecules should be the ones modulating the angiogenic response under physiological conditions.

As an approach to find these relevant molecules we assumed that the rabbit cornea could be a useful model for three reasons: (1) the organ is avascular, therefore blood components that may alter tissue composition depending simply on hypo- or hyperemia bound to sampling manipulations may be excluded, (2) the organ is transparent, therefore any inappropriate treatment during angiogenic stimulation is revealed by an opacity consequent to an inflammatory response, and (3) the organ is bilateral, therefore one cornea can be treated and the contralateral cornea of the same animal can be utilized as control.

In our experience, 250 ng of PGE1 incorporated into a "slow release pellet" of ethylene-vinyl acetate copolymer according to Langer and Folkman [4] induced a large number of microvessels within the cornea in about 70% of implants without mobilization of inflammatory-type cells. Buds of endothelial cells appear on the limbal vessels by 70 to 90 h after pellet implantation; by day 5–6 the newly-formed vessels have penetrated about 1 mm within the cornea and by day 8–10 the vessels surround the pellet located about 2 mm from the limbus.

In the same rabbit, one cornea received a pellet with 250 ng PGE1, and the contralateral cornea was implanted with either a pellet without an angiogenesis trigger, or a pellet with PGI2, not angiogenic, or a pellet with PGE2, angiogenic only at doses higher than those of PGE1. At the end of day 3 or the beginning of day 4 after grafting when endothelial buds started to appear on the limbal vessels, the cornea was removed and the difference in composition was evaluated between the fragment ready to be colonized by capillaries (about 1/3 of the cornea) and the similar contralateral fragment taken as control. So far two relevant observations have been obtained: the cornea under angiogenic stimulation becomes richer in copper ions and gangliosides. This occurs before the newly-formed microvessels have invaded the tissue [5–8].

Accumulation of copper ions

The content of most tissue ions was determined for the rabbit cornea stimulated by PGE1 and copper was the only one found to be consistently increased by about one third as compared to the contralateral cornea of the same animal. When the rabbits receiving the angiogenesis stimulation were kept on a copper deficient diet for a time sufficient to reduce serum copper to about one half the physiological level, the angiogenic response did not occur [5, 6]. Ceruloplasmin, the normal carrier of copper in blood was angiogenic, but removal of copper from the molecule nullified the angiogenic capacity. Molecules not angiogenic, like heparin or the tripeptide Glycyl-L-Histidyl-L-Lysine became angiogenic when bound to copper. We have been unable to induce angiogenesis with copper salts, in the sense that small doses were ineffective and higher doses produced keratitis [6]. Indirect evidence that copper ions are involved in tumor growth is given by the increment of serum ceruloplasmin in tumor-bearing subjects followed by rapid decrease upon tumor removal [9–11] and by the finding of ceruloplasmin-m-RNA in the neoplastic tissue [12]. Presently the evidence for copper involvement in the angiogenic response is rather convincing, the role of copper ions in the sequence of biological events conducing to microvascularization is still obscure.

Accumulation of gangliosides

Under angiogenic stimulation the ganglioside content of the cornea doubled and their relative proportions changed. The increment occurred before penetration of capillaries within the tissue and regardless of the angiogenic stimulus utilized [7]. As a first approach to evaluate the role of gangliosides in angiogenesis, we observed that addition of gangliosides to the culture medium improved motility and growth of cloned microvascular endothelium [13]. The conditions observed in vitro were reproduced in vivo by implanting a copolymer pellet bearing 50 ng of GM1 or GT1b in the cornea. The ganglioside released by the pellet enhanced the local concentration, and this was sufficient to promote formation of new vessels when the dose of the angiogenesis trigger was too small to be angiogenic by itself. Evidently the in vivo increment of total gangliosides observed in the corneal tissue under angiogenic stimulation favored neovascularization. Gangliosides were not angiogenic by themselves [7].

As an approach to assess the biological significance of changes in the relative proportions of the three major corneal gangliosides (GM3, GM2, GD3), microvascular endothelium was challenged in vitro with two mixtures, one constituted by GM3 + GM2 + GD3 with proportions found in the normal cornea and the second containing the three gangliosides with proportions found in the angiogenesis-stimulated cornea. This last mixture improved growth and motility of microvascular endothelium, the former did not. The major difference between the two mixtures was a change in the GM3/GD3 ratios. We observed that when this ratio decreased, motility and growth of microvascular endothelium increased. Moreover, increment in vivo of local GM3 concentration produced by a pellet implanted in the cornea, reduced the number of angiogenic responses to PGE1, i.e. the in vitro cytostatic effect of GM3 on microvascular endothelium corresponded to an angiostatic effect in vivo (manuscript in preparation).

In conclusion, I wish to stress three points: (1) under angiogenic stimulation the corneal tissue changes its composition. Increased content of copper ions and gangliosides are the two events observed so far; (2) changes in tissue composition occur before neovascularization is visible, and within the limits of our observation, aims at improving motility and growth rate of microvascular endothelium, two indispensable events during angiogenesis and (3) modification of tissue composition during angiogenic stimulation is probably a rewarding field to explore.

References

1 Brem S: The role of vascular proliferation in the growth of brain tumors. Clinical Neurosurgery 1976; 23: 440–453.

2 Preis I, Langer R, Brem H, Folkman J: Inhibition of neovascularization by an extract derived from vitreous. Am J Ophthalmol 1971; 84: 323–328.

3 Gullino PM: Angiogenesis factor(s), in Baserga R (ed): Handbook Exp Pharmacol. New York, Springer-Verlag, 1981, vol 57, pp 427–449.

4 Langer R, Folkman J: Polymers for the sustained release of proteins and other macromolecules. Nature 1976; 263: 797–800.

5 Ziche M, Jones J, Gullino PM: Role of prostaglandin E. and copper in angiogenesis. J Natl Cancer Inst 1982; 69: 475–482.

6 Raju KS, Alessandri G, Ziche M, Gullino PM: Ceruloplasmin, copper ions and angiogenesis. J Natl Cancer Inst 1982; 69: 1183–1188.

7 Ziche M, Alessandri G, Gullino PM: Gangliosides promote the angiogenic response. Lab Invest 1989; 61: 629–634.

8 Gullino PM, Ziche M, Alessandri G: Gangliosides, copper ions and angiogenic capacity of adult tissues. Cancer and Metastasis Reviews 1990; 9: 239–251.

9 Ungar-Waron H, Gluckman A, Spira E, Waron M, Trainin Z: Ceruloplasmin as a marker of neoplastic activity in rabbits bearing VX-2 carcinoma. Cancer Res 1978; 38: 1296–1299.

10 Shifrine M, Fisher GL: Ceruloplasmin levels in sera from human patients with osteosarcoma. Cancer 1976; 38: 244–248.

11 Turecky L, Kalina P, Uhlikova E, Namerova S, Krizko J: Serum ceruloplasmin and copper levels in patients with primary brain tumors. Klin Wochenschr 1984; 62: 187–189.

12 Kunapuli SP, Singh H, Singh P, Kumar A: Ceruloplasmin gene expression in human cancer cells. Life Science 1987; 40: 2225–2228.

13 De Cristan G, Morbidelli L, Alessandri G, Ziche M, Cappa APM, Gullino PM: Synergism between gangliosides and basic fibroblastic growth factor in favouring survival, growth and motility of capillary endothelium. J Cell Physiol 1990; 144: 505–510.

Angiogenesis: Key Principles – Science – Technology – Medicine
ed. by R. Steiner, P. B. Weisz & R. Langer

Modulation of angiogenesis in vitro

R. Montesano, M. S. Pepper, J.-D. Vassalli and L. Orci

Department of Morphology, University Medical Center, CH-1211 Geneva 4, Switzerland

Angiogenesis plays a critical role in a number of physiological and pathological settings, including wound healing, diabetic retinopathy, and tumor growth [1]. Until recently however, angiogenesis could only be studied in animal models, with the attendant drawbacks of not being able to precisely define the molecular components of this process in vivo. In an attempt to better understand the regulatory mechanisms of angiogenesis, we set out to develop novel experimental systems, in which essential components of the neovascularization process can be recapitulated and subsequently modulated and analyzed under well-defined in vitro conditions. In order to approximate as closely as possible the in vivo situation, we have designed three-dimensional culture systems allowing the re-establishment of the normal interactions between microvascular endothelial cells and the surrounding extracellular matrix.

In vivo, microvascular endothelial cells experience a different extracellular matrix environment, depending on whether they are in a resting state or are undergoing sprouting and migration during angiogenesis. In the normal, quiescent state, the endothelial cells rest on a specialized extracellular matrix, the basement membrane, which contains predominantly type IV collagen and laminin. During angiogenesis, however, they focally degrade their investing basement membrane, and subsequently migrate into the interstitial matrix of the surrounding connective tissue, which consists mainly of type I collagen [2]. In an attempt to understand the role of the extracellular matrix in the process of angiogenesis, we have studied the interactions of microvascular endothelial cells with three-dimensional gels of reconstituted type I collagen fibrils. When a monolayer of capillary endothelial cells on the surface of a collagen gel is covered with a second layer of collagen, it reorganizes into a network of branching and anastomosing tubules. Electron microscopy reveals the presence of basement membrane material at the interface between the abluminal surface of the endothelial cells and the collagen matrix, i.e. in the same location as observed in vivo. These experiments demonstrate that an appropriate interaction with collagen fibrils plays an important role in the organization of endothelial cells into correctly polarized capillary-like tubules [3].

The induction of the invasive phenotype

In the studies described above [3], formation of capillary-like tubules was experimentally induced by embedding endothelial cells within a three-dimensional collagen matrix. During angiogenesis in vivo, however, the morphogenetic events that culminate in the formation of new blood capillaries are intimately associated with the activation of an invasive process, i.e. the local breakdown of the microvascular basement membrane and the penetration of endothelial sprouts into the interstitial extracellular matrix [1, 2]. Central to the understanding of the mechanisms of angiogenesis is therefore the question of how normally quiescent endothelial cells acquire invasive properties that endow them with the ability to migrate through the extracellular matrix. Can in vitro systems help us to understand the regulatory mechanisms and factors that control endothelial cell invasiveness?

Our earlier studies [3] had shown that endothelial cells grown on a collagen gel form a monolayer on the surface of the gel and do not invade the underlying matrix. We then asked whether it might be possible to induce the endothelial cells to penetrate into the collagen matrix, as they do during angiogenesis in vivo. Cell invasiveness in angiogenesis and in other biological processes is believed to require the elaboration of matrix-degrading proteolytic enzymes, which include metalloproteases such as collagenases, and serine proteases such as plasminogen activators (PA's) [4]. To investigate the possible role of proteases in endothelial cell invasiveness in vitro, confluent monolayers of microvascular endothelial cells on collagen gels were treated with phorbol myristate acetate (PMA), a tumor promoter that markedly stimulates their production of collagenase and PA's [4]. Whereas control endothelial cells were confined to the surface of the gels, PMA-treated endothelial cells invaded the underlying collagen matrix, within which they formed capillary-like tubular structures [5]. The examination of serial sections showed that the tubules were in continuity with the monolayer of endothelial cells on the surface of the gel (Fig. 1a–c), and that they could branch like blood vessels in vivo (Fig. 1d–f). This phenomenon was associated with the degradation of collagen fibrils and was prevented by an inhibitor of metalloproteases, which suggests the involvement of a collagenase in the invasive process [5].

The experiments described above [4] indicated that endothelial cells, even after repeated passage in culture, retain the potential to express a latent "angiogenic program", that may be "switched on" by appropriate signals. To establish whether physiological messengers can elicit an

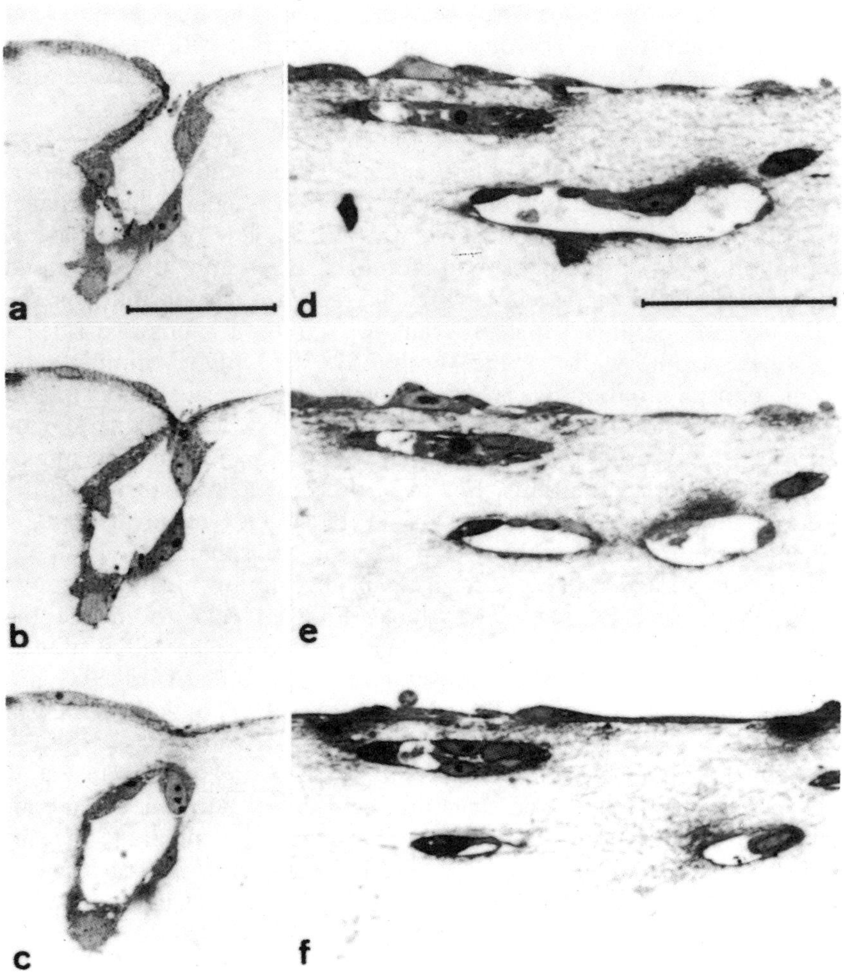

Figure 1. Invasion of collagen gels and formation of vessel-like structures by PMA-treated capillary endothelial cells. (*a–c*) Consecutive serial sections showing the continuity between the endothelial cells forming the surface monolayer and those delimiting a tube-like structure inside the collagen matrix. (*d–f*) Serial sections showing the division of a vessel-like structure into two smaller tubes that progressively separate from one another. Bars = 50 μm [reproduced from Cell, 1985, vol. 42, pp. 469-477, with copyright permission from Cell Press].

angiogenic response similar to that induced by PMA, microvascular endothelial cells grown on collagen gels were treated with basic fibroblast growth factor (bFGF), one of the best characterized angiogenic polypeptides [1]. These experiments showed that bFGF induces the endothelial cells to invade the underlying collagen matrix and to form capillary-like tubules as previously observed in response to PMA. Concomitantly, bFGF stimulated the endothelial cells to produce PA's [6].

These results provide clear evidence that bFGF is an angiogenic factor which acts directly on endothelial cells, in contrast with other factors that may stimulate angiogenesis indirectly, as for example during inflammation.

The intracellular mechanisms by which PMA and bFGF induce an angiogenic phenotype in capillary endothelial cells are not yet understood. However, the effects of both agents are thought to be mediated by two major signal transduction pathways, namely protein kinase C activation and tyrosine phosphorylation. In an attempt to understand the nature of the intracellular signals that are responsible for the induction of the angiogenic program, we explored a possible role of tyrosine phosphorylation. Since both PMA and bFGF stimulate tyrosine phosphorylation, we wondered whether pharmacological inhibition of phosphotyrosine phosphatases (which switch off the signals generated by tyrosine phosphorylation) could mimic the in vitro angiogenic activity of these agents. For this purpose, we incubated endothelial cell monolayers on collagen gels with sodium orthovanadate, a potent inhibitor of phosphotyrosine-specific phosphatases, which we have demonstrated induces tyrosine phosphorylation in our cells. We found that vanadate stimulates the endothelial cells to invade the underlying collagen matrix and to organize into capillary-like tubules resembling those observed previously in response to PMA or bFGF [7]. In addition, vanadate markedly increased PA activity in endothelial cell cultures, as well as the levels of mRNA for both urokinase-type PA (u-PA) and tissue-type PA (t-PA) [7]. The observation that PMA, bFGF and vanadate, three structurally-unrelated substances that increase tyrosine-specific protein phosphorylation by different mechanisms, are all capable of inducing a similar angiogenic response in vitro, suggests a role for tyrosine phosphorylation in the regulation of the angiogenic phenotype of capillary endothelial cells.

Protease inhibitors play an important permissive role in angiogenesis

The coordinate modulation of invasive behavior and PA production by PMA, bFGF and vanadate (see above), together with the demonstration of an increase in u-PA activity in endothelial cells migrating from the edge of an experimental wound in vitro [8; see also the paper by Pepper et al. in this book], support the proposed role for the PA-plasmin system [4] in angiogenesis. The expression of proteolytic activity by endothelial cells must however be tightly controlled in order to prevent inappropriate matrix degradation. That a precise regulation of extracellular proteolysis is essential for angiogenesis was demonstrated by experiments with endothelial cells grown on fibrin gels (formation of new capillary blood vessels during wound healing and other physiopathological settings takes place within a fibrin-rich matrix). When grown to confluence on the surface of a three-dimensional fibrin gel and treated with PMA [9] or bFGF [10], endothelial cells progressively lyse their underlying matrix, in the absence of which they are unable to form tube-like structures. However, when a plasmin inhibitor such as Trasylol is added to the cultures at the same time as the inducing agent, the structural integrity of the underlying matrix is preserved, at least in part, thereby providing a three-dimensional matrix scaffold into which endothelial cells can now migrate to form tube-like structures. These results suggest that neutralization of excess proteolytic activity by protease inhibitors plays an important permissive role in angiogenesis by preventing uncontrolled matrix degradation [9].

Unbalanced proteolysis results in aberrant vascular morphogenesis

A complementary approach for investigating the molecular mechanisms of capillary blood vessel formation might be the identification of genes whose inappropriate expression results in the subversion of the normal morphogenetic program of endothelial cells. The discovery that the polyoma virus middle T oncogene (mT) induces the formation of cyst-like endothelial tumors (cavernous hemangiomas) in mice [11, 12] prompted us to investigate the in vitro behavior of endothelial cells expressing this oncogene. We found that the morphogenetic properties of the mT-expressing hemangioma cells are dramatically different from those of normal endothelial cells. Whereas normal endothelial cells form a network of tubules when suspended within fibrin gels, mT-expressing endothelial cells form large cysts (Fig. 2a, b), which are reminiscent of the cavernous hemangiomas from which the cells were isolated [13].

The formation of rapidly expanding spherical cavities within fibrin gels suggested the possibility that mT-expressing endothelial cells produce unusually high levels of fibrinolytic activity, and that this might be responsible for their peculiar morphogenetic properties. To explore this possibility, we compared the levels of PA produced by mT-expressing endothelial cells with that of normal endothelial cells. We found that mT-expressing endothelial cells synthesize increased amounts of u-PA but virtually no PA inhibitor type 1 (PAI-1) (see chapter by Pepper et al. in this book).

We then asked what would happen if we attempted to correct the proteolytic imbalance of the hemangioma cells by adding protease inhibitors to the culture system. We found that when plasmin inhibitors such as Trasylol or ε-aminocaproic acid were added to the cultures, the mT-expressing endothelial cells formed capillary-like branching tubules (Fig. 2c, d) instead of the large hemangioma-like cysts observed in the absence of these inhibitors (cf. Fig. 2a, b). In other words, these results

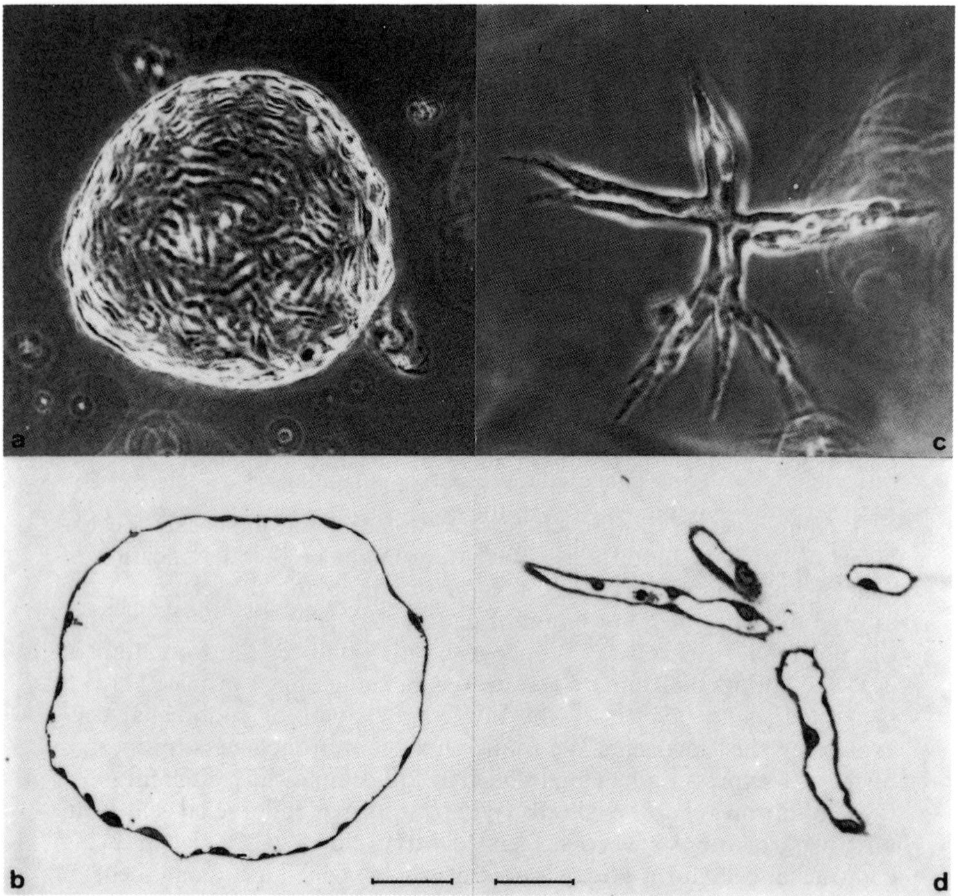

Figure 2. Morphogenetic behavior of mT-expressing endothelial cells grown within fibrin gels under control conditions or in the presence of exogenously added serine protease inhibitors. (*a*) spherical cyst formed by mT-expressing endothelial cells grown within a fibrin gel without the addition of protease inhibitors (phase-contrast microscopy). In this picture, the focus is approximately on the equatorial plane of the cyst. The endothelial cells lining the floor and the roof of the cavity appear as a blurred image in the center of the cyst. (*b*) Semithin section of an endothelial cyst. The large, hemangioma-like cavity is lined by a continuous monolayer of flattened endothelial cells. (*c*, *d*) Branching, capillary-like tubules formed by mT-expressing endothelial cells grown within a fibrin gel in the presence of the serine protease inhibitors ε-aminocaproic acid (*c*) or Trasylol (*d*). *a*, *b*: bar = 100 μm; *c*, *d*: bar = 50 μm [(*a*–*c*) are reproduced from Cell, 1990, Vol. 62, pp. 435-445, by copyright permission of Cell Press].

show that the aberrant morphogenetic behavior of mT-expressing endothelial cells can be corrected by neutralizing the excessive proteolytic activity of these cells. The experiments with the hemangioma cells therefore add significant support to our earlier proposal [9] (see above) that protease inhibitors play a critical regulatory role in the control of the angiogenic process. They also suggest that a disturbance of the endothelial proteolytic balance may represent one of the factors responsible for hemangioma formation [13].

Concluding remarks

The study of the intracellular and extracellular mechanisms that govern the angiogenic process represents the first necessary steps towards the elaboration of pharmacological strategies for controlling the growth of blood vessels. The in vitro models we have developed provide a unique opportunity for undertaking a systematic experimental dissection of the molecular mechanisms of angiogenesis. Using these models, we have contributed to the present understanding of the roles of endothelial cell-matrix interactions [14] and controlled extracellular proteolysis [15] in the process of neovascularization. In addition, in vitro models of angiogenesis represent a particularly attractive alternative to in vivo bioassays for the purpose of identifying additional angiogenesis factors as well as anti-angiogenic substances. Information gained through experimental analysis of the neovascularization process may ultimately provide clinicians with novel therapeutic approaches for the management of a variety of devastating angiogenesis-dependent diseases.

Acknowledgements. This work was supported by the Swiss National Science Foundation, the Juvenile Diabetes Foundation International, and the Sir Jules Thorn Charitable Overseas Trust.

References

1 Folkman J, Klagsbrun M: Angiogenic factors. Science 1986; 235: 442–447.
2 Madri JA, Pratt BM: Angiogenesis, in: Clark RAF, Henson PM (eds): The Molecular and Cellular Biology of Wound Repair. New York, Plenum Press, 1988, pp 337–358.
3 Montesano R, Orci L, Vassalli P: In vitro rapid organization of endothelial cells into capillary-like networks is promoted by collagen matrices. J Cell Biol 1983; 97: 1648–1652.
4 Gross JL, Moscatelli D, Jaffe EA, Rifkin DB: Plasminogen activator and collagenase production by cultured capillary endothelial cells. J Cell Biol 1982; 95: 974–981.
5 Montesano R, Orci L: Tumor-promoting phorbol esters induce angiogenesis in vitro. Cell 1985; 42: 469–477.
6 Montesano R, Vassalli J-D, Baird A, Guillemin R, Orci L: Basic fibroblast growth factor induces angiogenesis in vitro. Proc Natl Acad Sci USA 1986; 83: 7297–7301.
7 Montesano R, Pepper MS, Belin D, Vassalli J-D, Orci L: Induction of angiogenesis in vitro by vanadate, an inhibitor of phosphotyrosine phosphatases. J Cell Physiol 1988; 134: 460–466.
8 Pepper MS, Vassalli J-D, Montesano R, Orci L: Urokinase-type plasminogen activator is induced in migrating capillary endothelial cells. J Cell Biol 1987; 105: 2535–2541.
9 Montesano R, Pepper MS, Vassalli J-D, Orci L: Phorbol ester induces cultured endothelial cells to invade a fibrin matrix in the presence of fibrinolytic inhibitors. J Cell Physiol 1987; 132: 509–516.
10 Pepper MS, Belin D, Montesano R, Orci L, Vassalli J-D: Transforming growth factor-beta 1 modulates basic fibroblast growth factor-induced proteolytic and angiogenic properties of endothelial cells in vitro. J Cell Biol 1990; 111: 743–755.

11 Bautch VL, Toda S, Hassell JA, Hanahan D: Endothelial cell tumors develop in transgenic mice carrying polyoma virus middle T oncogene. Cell 1987; 51: 529–538.
12 Williams RL, Courtneidge SA, Wagner EF: Embryonic lethalities and endothelial tumors in chimeric mice expressing polyoma virus middle T oncogene. Cell 1988; 52: 121–131.
13 Montesano R, Pepper MS, Möhle-Steinlein U, Risau W, Wagner EF, Orci L: Increased proteolytic activity is responsible for the aberrant morphogenetic behavior of endothelial cells expressing the middle T oncogene. Cell 1990; 62: 435–445.
14 Montesano R, Pepper MS, Orci L: Angiogenesis in vitro: morphogenetic and invasive properties of endothelial cells. News Physiol Sci 1990; 5: 75–79.
15 Pepper MS, Montesano R: Proteolytic balance and capillary morphogenesis. Cell Differ Devel 1990; 32: 319–328.

Angiogenesis: Key Principles - Science - Technology - Medicine
ed. by R. Steiner, P. B. Weisz & R. Langer

Proteolytic balance and capillary morphogenesis in vitro

M. S. Pepper, J.-D. Vassalli, L. Orci and R. Montesano

Department of Morphology, University Medical Center, Geneva, Switzerland

The angiogenic process can be summarized as follows: in response to angiogenic factors, endothelial cells are activated, their adjacent basement membrane is degraded and thin cytoplasmic processes are extended in the direction of the angiogenic stimulus. The endothelial cells then migrate into the surrounding extracellular matrix within which they form a capillary sprout. Sprout maturation is completed by the formation of a patent capillary lumen and reconstitution of the basement membrane [1].

In order to form capillary sprout, endothelial cells must overcome the restraining mechanical barriers imposed by their basement membrane and surrounding matrix. How does this occur? It has been proposed that proteolytic enzymes localized to the cell surface facilitate this process by degrading the matrix components which they come into contact with. One of the enzyme systems which has been most extensively studied in this regard is the plasminogen activator-plasmin system. Plasmin, a tryptic protease capable of degrading many extracellular matrix components and also of activating metalloprotease zymogens, is generated from its inactive precursor plasminogen, which is widely distributed in most tissues of the organism, by the activity of two plasminogen activators (PAs), urokinase and tissue type-PA (t-PA). The activity of urokinase and t-PA is subject to inhibition by specific physiological PA inhibitors (PAIs), PAI-1 and PAI-2 [2].

Extracellular proteolysis may be involved in at least three phases of the angiogenic process: degradation of the investing basement membrane, cell migration through the surrounding matrix, and lumen formation, which by definition requires the formation of an extracellular space within a three-dimensional matrix.

To explore these possibilities, we wished to know whether migrating capillary endothelial cells produce plasminogen activators. To answer this question, migration was induced by scraping away cells from the central region of a confluent monolayer of microvascular endothelial cells, leaving the monolayer intact on either side. Without further

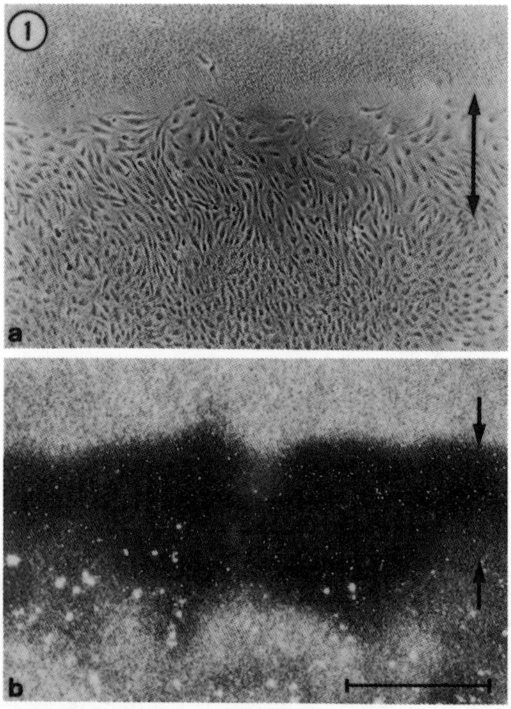

Figure 1. Wound edge caseinolytic assay. 24 h after wounding, bovine microvascular endothelial cell monolayers were overlaid with a thin layer of agar containing casein and plasminogen. Under phase contrast microscopy (*a*), cells can be seen to have migrated into the original wound (cell-free region in the upper third of the photograph) from the remaining intact monolayer (lower third of the photograph). The region of cell migration is indicated by the double-headed arrow (middle third of the photograph). Under dark field illumination (*b*), a zone of plasmin-dependent caseinolysis appears as a dark band delimited by white bands of unlysed casein above and below, thereby revealing the production of plasminogen activator by the migrating cells. The region of cell migration indicated by the two arrows corresponds to the same region of cell migration indicated in (*a*). Bar = 500 μm. (From Pepper et al., J Cell Biol 105: 2535–2541 (1987), with copyright permission from Rockefeller University Press.)

stimulus, the cells migrate in to cover the wound. The wounded monolayers were left for 24 h to allow migration to become well established, and were then overlaid with a thin layer of agar containing plasminogen and casein (casein is opaque under dark field illumination, and when degraded by plasmin, its opacity is lost). Macroscopically, we observed the appearance of sharp bands of plasmin-mediated casein lysis, which under phase contrast microscopy corresponded to regions in which endothelial cells were in the process of migrating in to cover the wound (Fig. 1). By co-incorporating specific inhibitors of either urokinase or t-PA into the overlay, we found that the PA being produced by the migrating cells was urokinase [3].

What about the other side of the proteolytic coin, that is to say what happens to production of the physiological PA inhibitor PAI-1 during wound-induced migration? Using the same wounded monolayer model, we assessed the levels of PAI-1 mRNA in migrating and non-migrating endothelial cells by in situ hybridization. We found that PAI-1 mRNA was increased in migrating endothelial cells when compared to cells in the intact monolayer (M. S. Pepper, A.-P. Sappino, R. Montasano, L. Orci and J.-D. Vassalli, in preparation). Thus, both urokinase and its physiological inhibitor PAI-1 are induced during microvascular endothelial cell migration.

One of our current objectives is to identify potential physiological inhibitors of angiogenesis. Cartilage is one of the few avascular tissues in the body, and therefore represents a potential source of anti-angiogenic factors. We therefore wished to determine whether chondrocytes might produce a factor which inhibits endothelial sprout formation in vitro.

We first devised an in vitro model of spontaneous endothelial sprout formation. Microvascular endothelial cells were seeded onto a non-adhesive agarose substrate; this results in the formation of solid cellular aggregates floating in the culture medium. These aggregates were then embedded into three-dimensional collagen or fibrin gels. After three days, endothelial sprouts had spontaneously developed, radiating out from the original aggregate into the surrounding gel (Fig. 2). Serial semi-thin sections revealed that the sprouts were in the form of hollow endothelial tubes, connected to the aggregate which had now become a hollow sphere (Fig. 2). We next co-incorporated chondrocytes into the gel, or added them to the culture medium above the gel, at the same time as the endothelial cell aggregates were embedded. We found that sprout formation normally seen in controls was markedly inhibited in the presence of chondrocytes [4].

The inhibition observed in cultures in which chondrocytes were added to the medium above the gel suggested that a diffusible factor might be involved. Amongst the physiological factors that inhibit endothelial cell functions which are relevant to angiogenesis, such as cell division, cell migration and extracellular proteolysis, is transforming growth factor-beta (TGF-β) [1]. To assess whether this factor might be responsible for the inhibitory effect of chondrocytes on sprout formation, anti-TGF-β antibodies were added to the co-cultures. We found that the antibodies significantly reduced the inhibitory activity of the chondrocytes, suggesting that TGF-β is at least one of the factors responsible for the inhibition of sprout formation in our system. Mean sprout length was also reduced by exogenously added human platelet-derived TFG-β1 in a dose-dependent manner [4].

Since inhibition of proteolytic activity might conceivably inhibit sprout formation, we wished to determine whether chondrocyte-conditioned

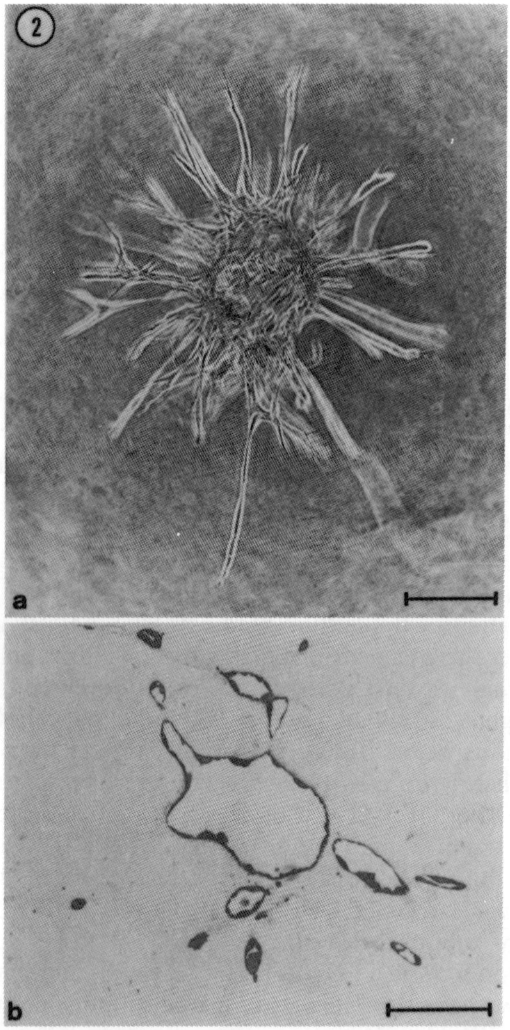

Figure 2. Spontaneous formation of endothelial sprouts from bovine microvascular endothelial cells cultured within collagen gels for three days. Phase contrast view (*a*) and semi-thin section (*b*) of an endothelial-lined sac-like structure and radially-disposed capillary-like tubes. Bar = 150 μm in (*a*) and 70 μm in (*b*). (From Pepper et al., J Cell Physiol 146: 170–179 (1991), with copyright permission from Wiley-Liss Inc.)

medium induces the production of a plasminogen activator inhibitor. We found that when added to microvascular endothelial cell monolayers, chondrocyte-conditioned medium increased the levels of PAI-1 and PAI-1 mRNA in these cells. To determine whether TGF-β might be the PAI-1 inducing factor, chondrocyte-conditioned medium was preincubated with anti-TGF-β antibodies before being added to endothelial monolayers. This significantly reduced the PAI-1 inductive properties of chondrocyte-conditioned medium [5].

To further assess whether an interrelationship might exist between protease activity and capillary sprout formation, we have taken advantage of a second in vitro model of angiogenesis. Endothelial cells cultivated on the surface of a three-dimensional collagen or fibrin gel, form a monolayer on the surface of the gel, and do not invade the underlying matrix. However, when they are treated with a physiological angiogenesis factor such as basic fibroblast growth factor (bFGF), the cells are induced to invade the underlying gel within which they form tube-like structures beneath the surface monolayer [6, 7]. This phenomenon is also induced by the pharmacological agents phorbol myristate acetate (PMA) [8] and sodium orthovanadate [9].

Knowing that these three agents induce invasion and the formation of capillary-like tubes in vitro, we next wished to determine their effect on the proteolytic properties of microvascular endothelial cells. Zymographic and Northern analyses revealed an increase in urokinase and urokinase mRNA in response to all three agents [6, 7, 9]. Using the technique of reverse zymography we also found greatly increased levels of PAI-1 in endothelial cell cultures treated with PMA, bFGF or vanadate. These results were confirmed by Northern analysis [7].

Thus, both during wound-induced migration and in response to angiogenic agents, capillary endothelial cells concomitantly produce urokinase and its physiological inhibitor, PAI-1. If extracellular proteolysis is necessary for angiogenesis, why are both protease and inhibitor produced together? To address this apparent paradox, we decided to assess in greater detail the effects of the two physiological factors bFGF and TGF-β1 on extracellular proteolysis and angiogenesis in our in vitro invasion system.

By Northern analysis we observed that although both PAI-1 and urokinase mRNA were induced by bFGF or TGF-β1, the magnitude and kinetics of these inductions were strikingly different. We have chosen to use the ratio of urokinase to PAI-1 mRNAs as a reflection of the potential proteolytic activity of the cells. In response to bFGF, there is a large increase in potential proteolysis. In contrast, in response to TGF-β1, there is a net anti-proteolytic effect. When both agents are added together, the potential proteolytic balance mimicks the levels seen in controls (Fig. 3) [7].

With these results in mind, we next determined how the changes in potential proteolysis might relate to angiogenesis in our in vitro invasion model. Whereas in controls the cells form a monolayer on the surface of the gel, in response to bFGF, they invade the underlying gel, forming tube-like structures with widely patent lumina. When TGF-β1 is added alone, the cells remain on the surface of the gel. When both bFGF and TGF-β1 are added together, the cells

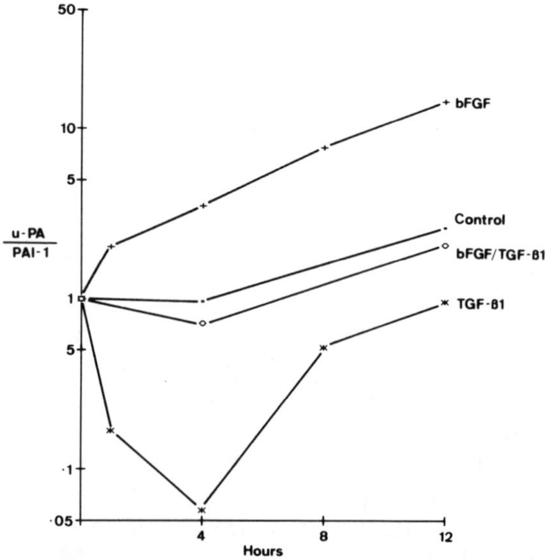

Figure 3. Potential proteolytic balance as reflected by the ratio of u-PA:PAI-1 mRNAs. The values for each mRNA, and hence the ratio, were arbitrarily taken to be 1 at time = 0 h. Following exposure to bFGF, an increase in potential proteolytic activity is observed over 12 h. In contrast, TGF-β1 decreases the u-PA:PAI-1 mRNA ratio, which is lowest after 4 h. In control cultures and in cultures exposed to both agents simultaneously, a small increase in u-PA:PAI-1 mRNA ratio is observed after 12 h. (From Pepper et al., J Cell Biol 111: 743–755 (1990), with copyright permission from Rockefeller University Press.)

invade the underlying gel as solid cell cords apparently devoid of lumina (Fig. 4) [7].

While recognising that the anti-proteolytic effect is only one of the many effects of TGF-β1 on endothelial cells, taken together, these results nonetheless allow us to construct a model of lumen formation during angiogenesis. In response to bFGF, proteolytic activity is increased, and the invading cells form tubes with widely patent lumina. When TGF-β1 is added at the same time, the increase in proteolytic activity is prevented, and the cells invade the underlying fibrin gel as solid cords. Although correlative, these results suggest that a net positive proteolytic balance is necessary for lumen formation.

What happens when there is excessive and uncontrolled proteolysis? Is this compatible with normal capillary morphogenesis? As described in the chapter by Montesano et al. in this book, we have found that when endothelial cells expressing the polyoma virus middle T oncogene are embedded in three-dimensional fibrin gels, they give rise to large cystic structures lined by a monolayer of endothelial cells [10]. Using a casein overlay technique which reveals the cell-associated PA activity of

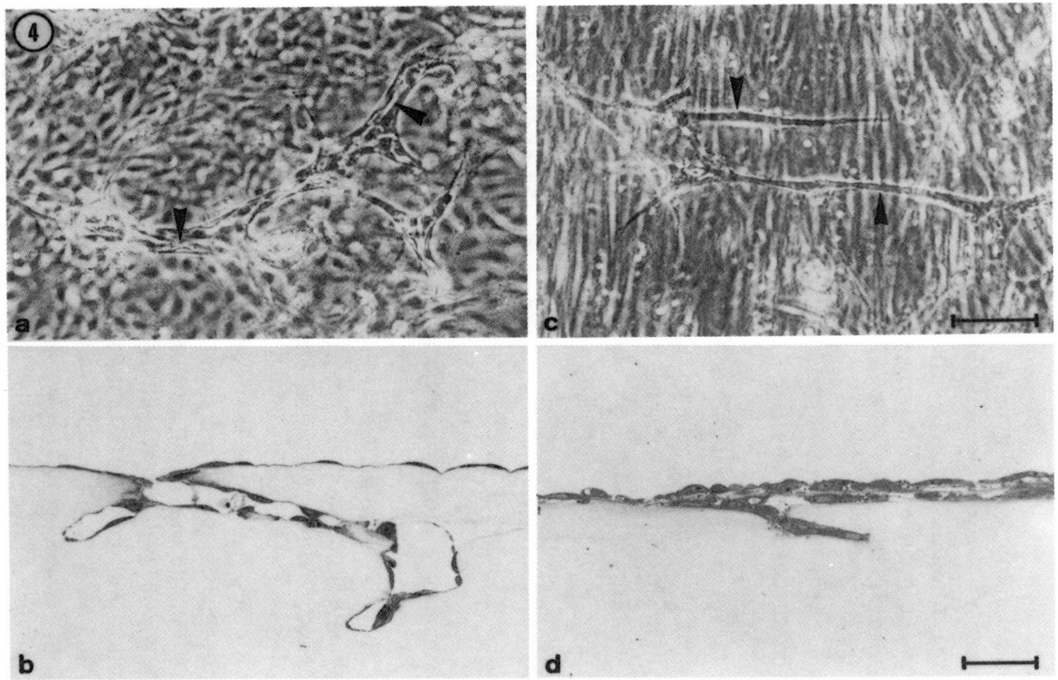

Figure 4. Bovine microvascular endothelial cell invasion of fibrin gels. Cells grown to confluence on the surface of fibrin gels in the presence of Trasylol were treated with bFGF or bFGF and TGF-β1 together for 7 days. (*a*,*c*, phase-contrast microscopy; *b*,*d*, semi-thin sections.) (*a* and *b*) Cultures treated with bFGF alone show widely patent lumina (arrowheads) which remain connected with the surface monolayer. (*c* and *d*) In marked contrast, in cultures treated with bFGF and TGF-β1 together, the cells invade the superficial parts of the fibrin gel as solid endothelial cell cords (arrowheads); the apparent multilayering is the result of superficial invasion of the fibrin gel – thin layers of gel can be seen between the cell layers. Bar = 10 μm in *a*, *c* and 50 μm in *b*, *d*. (From Pepper et al., J Cell Biol 111: 743–755 (1990), with copyright permission from Rockefeller University Press.)

individual cells, we also found that PA activity was greater in middle T-expressing endothelial cells when compared to their normal counterparts. Zymographic and Northern analyses revealed that middle T-expressing endothelial cell lines contained more urokinase and urokinase mRNA than normal endothelial cells. Furthermore, this effect appeared to be specific for endothelial cells, since the increase in urokinase mRNA was not seen in three non-endothelial cell lines expressing middle T. In addition, by reverse zymography we were unable to detect PAI-1 in the middle T-expressing endothelial cells, which was in contrast to the large amounts of PAI-1 present in three endothelial cell types not expressing middle T. Similarly, Northern analysis revealed a marked decrease in the levels of PAI-1 mRNA in the middle T-expressing cells [10].

Knowing then that the balance of extracellular proteolysis in middle-T-expressing endothelial cells was tilted to excessive proteolysis as a result of the increase in urokinase and the virtual absence of PAI-1, we asked whether the aberrant behaviour of these cells could be corrected by restoring their proteolytic balance to normal. As described by Montesano et al. in this book, we found that by adding serine protease inhibitors to the cultures, middle T-expressing endothelial cells instead of forming cysts, now formed branching capillary-like tubes within the fibrin gels [10].

These results demonstrate that excessive proteolytic activity is not compatible with normal capillary morphogenesis. Furthermore, by neutralizing this excessive activity by the addition of protease inhibitors, one can restore morphogenetic properties to the endothelial cells.

In summary, the studies which we have described demonstrate firstly that both urokinase and its physiological inhibitor PAI-1 are produced during endothelial cell migration and also in response to physiological and pharmacological agents which induce angiogenesis in vitro. Secondly, our studies have revealed a correlation between aberrant capillary morphogenesis and a spectrum of proteolytic properties of endothelial cells. On the one hand, reduced extracellular proteolysis is correlated with the formation of solid cell cords, or in other words, lumen formation is inhibited. On the other hand, too much proteolysis results in the formation of abnormal vascular structures.

These results therefore clearly suggest that a precisely regulated level of extracellular proteolysis resulting from the balance between proteases and their inhibitors is an essential requirement for normal capillary morphogenesis.

Acknowledgements. We would like to thank Drs. D. Belin, W. Risau and E. Wagner for their important contributions to our work. This work was supported by grants from the Swiss National Science Foundation, and by grants in aid from the Juvenile Diabetes Foundation International and the Sir Jules Thorn Charitable Overseas Trust.

References

1 Folkman J, Klagsbrun M: Angiogenesis factors. Science 1987; 235: 442–447.
2 Moscatelli D, Rifkin DB: Membrane and matrix localization of proteases: a common theme in tumour invasion and angiogenesis. Biochim Biophys Acta 1988; 948: 67–85.
3 Pepper MS, Vassalli J-D, Montesano R, Orci L: Urokinase-type plasminogen activator is induced in migrating capillary endothelial cells. J Cell Biol 1987; 105: 2535–2541.
4 Pepper MS, Montesano R, Vasalli J-D, Orci L: Chondrocytes inhibit endothelial sprout formation in vitro: evidence for involvement of a transforming growth factor-beta. J Cell Physiol 1991; 146: 170–179.
5 Pepper MS, Montesano R, Orci L, Vasalli J-D: Plasminogen activator inhibitor-1 is induced in microvascular endothelial cells by a chondrocyte-derived transforming growth factor-beta. Biochem Biophys Res Commun 1991; 176: 633–638.
6 Montesano R, Vassalli J-D, Baird A, Guillemin R, Orci L: Basic fibroblast growth factor induces angiogenesis in vitro. Proc Natl Acad Sci USA 1986; 83: 7297–7301.
7 Pepper MS, Belin D, Montesano R, Orci L, Vassalli J-D: Transforming growth factor beta 1 modulates basic fibroblast growth factor-induced proteolytic and angiogenic properties of endothelial cells in vitro. J Cell Biol 1990; 111: 743–755.
8 Montesano R, Orci L: Tumor-promoting phorbol esters induce angiogenesis in vitro. Cell 1985; 42: 469–477.
9 Montesano R, Pepper MS, Belin D, Vassalli J-D, Orci L: Induction of angiogenesis in vitro by vanadate, an inhibitor of phosphotyrosine phosphatases. J Cell Physiol 1988; 134: 460–466.
10 Montesano R, Pepper MS, Möhle-Steinlen U, Risau W, Wagner EF, Orci L: Increased proteolytic activity is responsible for the aberrant morphogenetic behaviour of endothelial cells expressing middle T oncogene. Cell 1990; 62: 435–445.

Angiogenesis: Key Principles - Science - Technology - Medicine
ed. by R. Steiner, P. B. Weisz & R. Langer

Metalloproteinase inhibition as a mechanism for the inhibition of angiogenesis

Marsha A. Moses and Robert Langer

Department of Surgery, The Children's Hospital Medical Center, Boston, MA 02115, and the Department of Chemical Engineering and Harvard-MIT Division of Health Science and Technology, Massachusetts Institute of Technology, Cambridge, MA 02139, USA

Metalloproteinases are a multigene family of metal-dependent enzymes (usually CA^{+2} or Zn^{+2}) whose activity is considered to be the key rate-limiting step in extracellular matrix degradation. Of the enzyme systems whose activity and control could have significant physiological ramifications, the metalloproteinase family has, in recent years, been shown to play a prominent role in the regulation of such phenomena as neovascularization, tumor growth and metastasis. An accumulating body of evidence has suggested that the control of metalloproteinases, specifically by their endogenous inhibitors, might provide an important strategy for the control of such phenomenon.

The metalloproteinase family of enzymes has been broadly subdivided into three categories on the basis of substrate specificity: the interstitial collagenases, the type IV collagenases (gelatinases), and the stromelysins [for further review see 1]. The degradation of extracellular matrix barriers, which is a prerequisite for those physiological events mentioned above, involves the proteolytic activity of a number of enzymes, the most important of which is the collagenase family [2].

Our laboratory has been particularly interested in the role that a collagenase inhibitor might play in the regulation of neovascularization. This objective has evolved from our laboratory's interest in the purification and characterization of an angiogenesis inhibitor from cartilage. The choice of cartilage as a potential source of angiogenesis inhibitors is a direct function of its avascularity. In addition to being avascular, normal cartilage is relatively tumor-resistant and chondrosarcoma, a tumor of the cartilage, is the least vascularized of all solid cancers. Since the pioneering work of Brem and Folkman in 1975 when they demonstrated that pieces of native cartilage could inhibit tumor-induced neovascularization [3] a number of different laboratories have attempted to purify and characterize angiogenesis inhibitors from cartilage [4–6]. Langer and coworkers showed that cartilage extracts containing a collagenase inhibitor could inhibit tumor-induced neovascularization when delivered both locally and systemically using two different animal models [7, 8]. In light of this observation, our laboratory chose to purify and characterize this collagenase inhibitor from cartilage in order to determine whether or not it was the active anti-angiogenic agent.

The cartilage-derived collagenase inhibitor (CDI) was purified by a combination of extraction in 2 M NaCl and precipitation with hydrochloric acid and ammonium sulfate followed by a series of conventional chromatography steps. The first of these chromatography steps was gel filtration on A-1.5 m Sepharose in the presence of 4 M guanidine-HCl. The inhibitor eluted at an apparent Mr of 28,000. Following gel filtration, the inhibitor was applied to a cation exchange column. The inhibitor bound to the resin low salt (0.05 M) but could be eluted by increasing the salt concentration to 0.2 M. Following ion exchange chromatography, a second gel filtration step was performed. The inhibitor was applied to a Sephadex G-75 (superfine) column where it eluted after the predominant peak of A_{280} nm absorbing material. Finally, the collagenase inhibitor was purified to a homogeneity using reverse phase high performance liquid chromatography. Using this purification protocol we were able to purify CDI approximately 1000-fold as well as to obtain amino terminal sequence data for this protein (Tab. 1).

Biochemical characterization of CDI reveals that it is a very stable molecule, withstanding exposure to pH 2, temperatures as high as 60 °C for up to 1 h and 4 M guanidine-HCl. Complete reduction and carboxymethylation do however result in the total loss of biological activity. CDI appears to be a specific inhibitor of mammalian collagenase. It inhibits both rabbit synovial cell collagenase and bovine corneal

Table 1. Samples were not reduced and alkylated, therefore, dashes (– – –) represent bona fide blanks that align by similarity with expected cysteine residues of previously reported collagenase inhibitors. Reprinted with permission [9]. Copyright 1990 by the AAAS.

NH_2-TERMINAL PROTEIN SEQUENCE OF CARTILAGE-DERIVED INHIBITOR

1	2	3	4	5	6	7	8	9	10
---	Thr	---	Val	Pro	Pro	His	Pro	Gln	Thr
11	12	13	14	15	16	17	18	19	20
Ala	Phe	---	Asn	Ser	Asp	Val	Val	Ile	Arg
21	22	23	24	25	26	27	28		
Ala	Lys	Phe	[Val][1]	Gly	Thr	Ala	Glu		

[1]Brackets indicate determination with less than full confidence

collagenase. It does not, however, inhibit trypsin or bacterial collagenase [9].

With the availability of purified CDI came the opportunity to determine how its activity might effect the process of neovascularization. However, in vivo assays for angiogenesis inhibition require large amounts of test material, are extremely expensive, and take long periods of time to complete. In light of these limitations, we chose to develop in vitro bioassays which could be utilized to test substances for their antiangiogenic potential. The foundation of the development of these assays lies in the knowledge of the sequence of events required for neovascularization. It is commonly accepted that virtually all adult angiogenesis begins as a sprouting process in which new capillaries bud from preexisting small venules. This first event requires the disruption of basement membrane surrounding the preexisting vessels and is correlated with the production of proteolytic enzymes such as collagenase and plasminogen activator by the endothelial cells stimulated by an angiogenic stimulus. Next the endothelial cells migrate out from the preexisting perivenule towards the angiogenic stimulus. The endothelial cells then proliferate in response to stimulator of angiogenesis. The final stage in the formation of new vessels is the process of lumen formation. The mature capillary bed is developed in which the capillaries are surrounded by basement membrane components and in some cases by a layer of pericytes or perivascular cells [10]. In summary, the steps in the formation of new capillary formation include basement membrane degradation, endothelial cell migration and proliferation, capillary tube formation, and differentiation into a mature blood vessel.

We began by studying the effect of CDI on growth factor-stimulated capillary endothelial cell (EC) proliferation. It occurred to us that it would be important to screen fractions from each of the column steps for an antiproliferative factor in the event that cartilage contained an inhibitor of EC proliferation that was not the collagenase inhibitor. In order to do this, we utilized a modification of an assay developed by Connolly and coworkers in 1985 [11]. They took advantage of the fact that capillary EC contain acid phosphatase activity and developed an in vitro assay which determines the number of EC cells in culture on the basis of the colorimetric measurement of acid phosphatase. This assay provided us with the opportunity to screen large numbers of fractions from each of our purification steps in a highly sensitive and reproducible fashion. By using the specific cells involved in angiogenesis, the capillary endothelial cells, and stimulating them with a known angiogenesis factor, in this case aFGF (acidic fibroblast growth factor), we could best mimic the angiogenic process in vitro. Using this assay, we found that an inhibitor of endothelial cell proliferation copurified with the collagenase inhibitor throughout the purification of CDI. Purified CDI was a potent inhibitor of aFGF-stimulated capillary EC proliferation with a concentration of 96 nM CDI causing 72% inhibition. Electronic cell counting assays supported these results. The antiproliferative effect of CDI appears to be an endothelial cell specific one. CDI does not inhibit smooth muscle cell proliferation nor Balb/c 3T3 cell proliferation. We also found in preliminary experiments that CDI could suppress the incorporation of tritiated thymidine by endothelial cells in response to FGF in a concentration-dependent manner with an IC_{50} of approximately 300 nM [9].

We next began to study the effect of CDI on another key component of the angiogenic process, the migration of capillary endothelial cells in response to an angiogenic stimulus. In order to do this we utilized a modification of the classic Boyden chamber technique. Into the lower wells of the Boyden chamber apparatus, we placed a known concentration of growth factor; the upper wells received a predetermined number of capillary endothelial cells and CDI. Over time, cells will attach to the upper surface of the coated polycarbonate membrane, migrate through, and attach to the lower membrane surface. The membrane can then be fixed and stained for counting. We found that an inhibitor of growth factor-stimulated capillary EC migration copurified with CDI throughout its purification, such that purified CDI inhibited capillary EC migration with an IC_{50} of 16 nM.

Given these results in our in vitro assays, were were very interested to learn whether or not CDI could inhibit neovascularization in vivo. It was first tested using the chick chorioallantoic membrane (CAM) as a third bioassay. We found that picomolar amounts of purified CDI implanted in methocellulose discs on the CAM surface caused significant inhibition of embryonic neovascularization, such that large avascular zones appeared on the surface of the CAM (Fig. 1a). In contrast, control CAMs implanted with empty methocellulose discs never developed avascular zones (Fig. 1b). Histological studies of these CDI-treated CAMs demonstrated a mesoderm that was thinner than normal and avascular relative to controls. Using this assay to monitor the antiangiogenic activity of CDI throughout its purification and developing dose dilution curves at each column step, we found that the two activities, the antiangiogenic activity and the anticollagenase activity did, in fact, copurify throughout the purification of CDI. CDI is one of the most potent inhibitors of angiogenesis in vivo when compared with the lowest reported doses of previously described angiogenesis inhibitors both tested alone in the CAM assay as well as those tested in conjunction with other factors [9].

Most recently, we have tested the effect of CDI on tumor-induced neovascularization in vivo. The rabbit corneal pocket model was utilized as our assay of choice, using V2 carcinoma as the angiogenic stimulus. In this in vivo model, we found that capillary growth towards the tumor was significantly inhibited by approximately 40 μg of highly purified CDI [12].

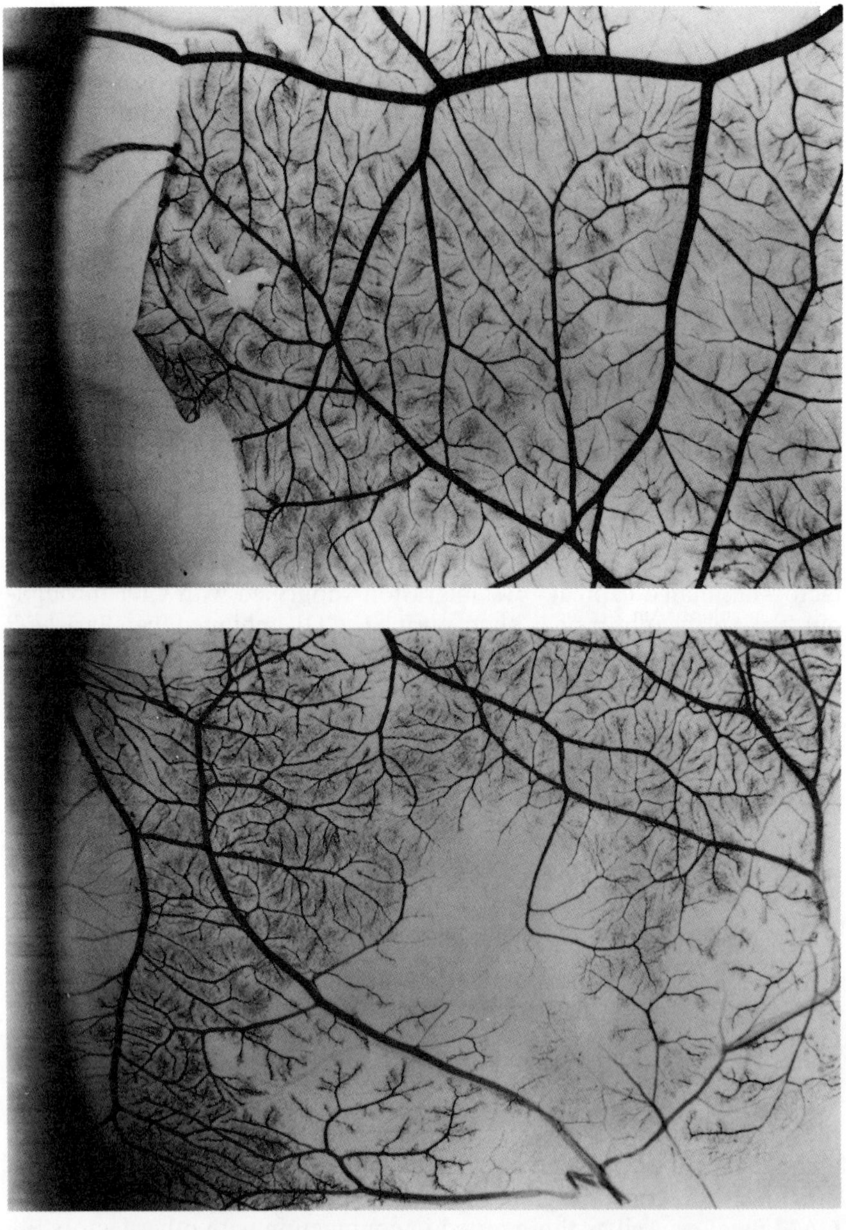

Figure 1. Inhibition of angiogenesis by CDI. *(a)* Control CAM *(b)* CDI-treated CAM. Reprinted with permission [9]. Copyright 1990 by the AAAS.

These results represent, to our knowledge, the first demonstration that a collagenase inhibitor, with no other known biological activity, can inhibit angiogenesis in vivo and in vitro. Taken together, these in vitro and in vivo studies demonstrate a key role for metalloproteinases in the process of neovascularization and suggest that one means of controlling the deregulated vascular growth characteristic of a number of 'angiogenic diseases' may be at the level of control of metalloproteinase activity.

References

1 Matrisian L: Metalloproteinases and their inhibitors in matrix remodelling. Trends Genet 1990; 6: 121–125.
2 Librach CL, Werb Z, Fitzerald ML, Chiu K, Corwin NM, Esteves RA, Grobelny D, Galardy R, Damsky C, Fisher SJ: 92-kD Type IV collagenase mediates invasion of human cytotrophoblasts. J Cell Biol 1991; 113: 437–449.
3 Brem H, Folkman J: Inhibition of tumor angiogenesis mediated by cartilage. J Expt Med 1975; 141: 427–439.
4 Eisenstein R, Kuettner KE, Neopolitan C, Soble L, Sorgente N: The resistance of certain tissues to invasion. III. Cartilage extracts inhibit the growth of fibroblasts and endothelial cells in culture. Am J Pathol 1975; 81: 337–348.
5 Sorgente N, Kuettner KE, Soble L, Eisenstein R: The resistance of certain tissues to invasion. II. Evidence for extractable factors in cartilage which inhibit invasion by vascularized mesenchyme. Am J Pathol 1975; 32: 217–222.
6 Shirai E: Production of an anti-angiogenesis factor by cultured chondrocytes and establishment of a cell line which produces the factor. J Osaka Univ Dent Soc 1987; 32: 163–180.
7 Langer R, Brem H, Falterman K, Klein M, Folkman J: Isolation of a cartilage factor that inhibits tumor neovascularization. Science 1976; 193: 70–72.
8 Langer R, Conn H, Vacanti J, Haudenschild C, Folkman J: Control of tumor growth in animals by infusion of an angiogenesis inhibitor. Proc Natl Acad Sci USA 1980; 7: 4331–4335.
9 Moses MA, Sudhalter J, Langer R: Identification of an inhibitor of neovascularization from cartilage. Science 1990; 248: 1408–1410.
10 Ausprunk D, Folkman J: Migration and proliferation of endothelial cells in preformed and newly formed blood vessels during tumor angiogenesis. Microvasc Res 1977; 14: 53–65.
11 Connolly DJ, Knight MB, Harakas NK, Wittwer AJ, Feder J: Determination of the number of endothelial cells in culture using an acid phosphatase assay. Anal Biochem 1986; 152: 136–140.
12 Langer R, Moses MA: Biocompatible controlled release polymers for delivery of polypeptides and growth factors. J Cell Biochem 1991; 45: 340–345.

Angiogenesis: Key Principles - Science - Technology - Medicine
ed. by R. Steiner, P. B. Weisz & R. Langer

TPA-induced differentiation of rat aortic endothelial cells is substrate-specific and receptor mediated

Marie P. Piechocki, Seema Kantak and James M. Onoda

Department of Radiation Oncology, 431 Chemistry, Wayne State University, Detroit, Michigan 48202, USA

Abstract. Under normal conditions, the morphology of cultured endothelial cells (EC) is characterized by a contact-inhibited monolayer with a distinct cobblestone appearance. However, when treated with phorbol esters, EC acquire fibroblast-like growth characteristics, are no longer contact-inhibited in growth, become invasive and form pre-capillary, tubular structures within collagen matrices. These events describe the basic processes of angiogenesis. We characterized the early effects of phorbol-12-myristate-13-acetate (TPA) on the attachment and spreading of rat aortic endothelial cells (RAEC) to purified extracellular matrix proteins by analyzing the distribution of fibronectin integrin receptors and the organization of cytoskeleton microfilaments during RAEC spreading on four different extracellular matrix peptides (i.e., Collagen Types I and IV, fibronectin and laminin). Type IV collagen appeared to be the best substrate for RAEC adhesion and spreading while laminin proved to be the poorest. TPA (0.1 μm) decreased the rate of RAEC spreading on Type IV collagen with a subsequent delay in cell-cell contact formation. In contrast, TPA increases the rate of spreading and cell-cell contact formation of RAEC plated on laminin.

Introduction

Recently, the extracellular matrix (ECM) has been implicated in the modulation of endothelial cell phenotype, growth and morphology during homeostasis and angiogenesis. Studies have examined the regulatory role(s) of the ECM from one of two perspectives: 1) that matrix bound resident growth factors (e.g., βFGF, heparan sulfate and plasminogen activators) serve as pluripotent angiogenic factors which modulate EC phenotype [1, 2] or 2) that collagenous and non-collagenous peptides of the matrix modulate endothelial cell growth, morphology, migration and angiogenesis [3]. The role of specific extracellular matrix proteins, integrin receptors and cytoskeleton in modulating endothelial cell attachment and spreading during angiogenesis have not been clearly elucidated. We have characterized the adhesion and spreading of RAEC on ECM peptides (e.g., Type IV collagen, laminin) in the presence or absence of TPA, a known mitogen and angiogenic agent [4]. In an attempt to elucidate the molecular modulation of these events, we examined the interactions among matrix peptides, adhesion integrin receptors and cytoskeleton in response to TPA during focal adhesion plaque formation and spreading.

Materials and methods

A) Isolation and culture of endothelial cells

Rat aortic endothelial cells (RAEC) were isolated from Sprague-Dawley Rats (Harlan Sprague, Indianapolis, IN) as previously described [5]. RAEC were cultured in sterile Dulbecco's Minimal Essential Media + 10% Fetal Bovine Serum (Gibco; Grand Island, NY) and maintained at 37°C, 5% CO_2. Media was changed every three days and cultures were passaged (4.5 mM EDTA and 0.25% trypsin) when a contact-inhibited monolayer was achieved (approximately once a week). RAEC between passage 10–19 were grown in T-75 culture flasks and used when confluent for experimental purposes.

B) Immunofluorescent staining of integrin receptors and cytoskeleton

1. Materials. Rhodamine-labeled phalloidin (Molecular Probes Inc., Eugene, Oregon) was used for the staining of F-actin filaments. Polyclonal Antisera against the fibronectin receptor (VLA-5/$\alpha_5\beta_1$) (Chemicon International, Inc. (Temecula, CA). Non-immune goat IgG, whole goat serum and fluorescein conjugated goat-anti-rabbit IgG $F_{(ab')}2$ fragment specific (Cappel Organon Teknika Corp.; West Chester, PA). Mouse Type IV Collagen and murine laminin (Collaborative Research; Bedford, MA).

2. Methods: Analysis of adhesion plaques (integrin receptors on the basal surface) and cytoskeleton of RAEC using fluorescent microscopy. RAEC (160,000 cells/chamber) in DMEM + 10% FBS were added to 35 mm Cell Wells™ (6-Well, Flat Bottom plates, Corning, NY) containing substrate-coated 18 mm² coverslips (10 μg per coverslip of Type IV collagen or laminin). After a 30-min incubation (37°C), 0.1 μM TPA or solvent (0.1% ethanol) was added. Samples were terminated 4 h after initial seeding. Lysis squirting was used to access the adhesion plaques formed at the abluminal surface of RAEC as previously described [6]. Samples were stained with primary (polyclonal antisera to the fibronectin receptor) and secondary (fluorescein conjugated $F_{(ab)}2'$ fragment) using conventional methods [6]. Rhodamine-labeled phalloidin was co-incubated with the FITC secondary for the staining of F-actin filaments. Samples mounted and analyzed at 600 × magnification under oil using a Nikon Optiphot-2 Microscope. Micrographs were recorded on Kodak T-MAX 400 Panchromatic film.

Results

RAEC attach and spread more rapidly on Type IV Collagen as compared to other extracellular matrix peptides (i.e., Type I collagen, fibronectin and laminin). Following initial adhesion (0.5 h) to Type IV collagen, RAEC undergo a rapid transition to begin spreading and appear flattened and circular within 1 hour. Once fully spread, RAEC undergo a rapid change in cell shape and adopt an elongated cell morphology with large, distinct, ruffled pseudopodia. Cell-cell contacts are also formed between adjacent polygonal shaped cells at this time (2 h). By 4 hours, RAEC juxtapose to organize as a monolayer. Treatment with 0.1 μM TPA at 0.5 h following RAEC addition to collagen-coated wells significantly inhibited their rate of spreading. RAEC still appear flattened and circular at 2 h, and do not initiate a shape change until 4 h. Cell-cell contacts are first observed at 8 h and there is a marked inhibition of monolayer formation.

RAEC spreading on laminin is significantly slower relative to RAEC plated on Type I or IV collagen or fibronectin. Moreover, the initial rate of adhesion to laminin is less than that observed for cells to collagen. No significant spreading is observed until 4 h and cell-cell contacts are not made until 8 h post-plating. After treatment with TPA, the rate of spreading of RAEC on laminin is significantly enhanced. A majority of the cells are flattened and fully spread by 2 h and cell-cell contacts are observed 4 h post-plating. By 8 h, cells are actively migrating as indicated by the presence of numerous vesicles in the perinuclear region.

Fig. 1 (*a–d*) demonstrates the effects of substrate on organization of F-actin microfilaments of the cytoskeleton (*a* and *c*) and distribution of fibronectin integrin receptors ($\alpha_5\beta_1$) (*b* and *d*). RAEC plated on Type IV collagen demonstrate maximum spreading at 4 h. The cytoskeleton is characterized by distinct stress fibers extending along the long axis of the cell. Microfilament bundles spanning these large areas are orientated in parallel as thick distinct stress fibers between which large areas of the ventral plasma membrane are visible (1*a*). Fibronectin receptors (FNR) are diffusely distributed without enrichment at stress fiber endings (1*b*). In contrast, RAEC plated on laminin are not fully spread (indicated by the relative small basal surface that is attached to the substrate) but a dense array of thin actin microfilaments are distributed in a circular pattern defining the cell periphery (1*c*). Short microfilament bundles, perpendicular to the circumference of the cell give rise to distinct stress fibers terminating at cell borders, and few, diffuse filaments are observed as stress fibers spanning the ventral plasma membrane (1*c*). Fibronectin receptors are distributed as discontinuous fibrillar streaks which follow the distribution of these stress fibers (1*d*).

Figure 1

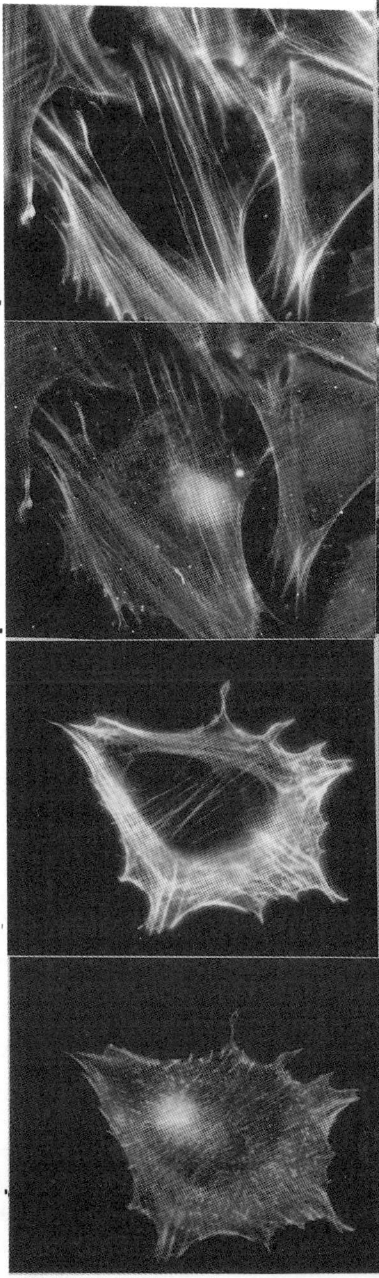

a. b. c d.

Figure 2

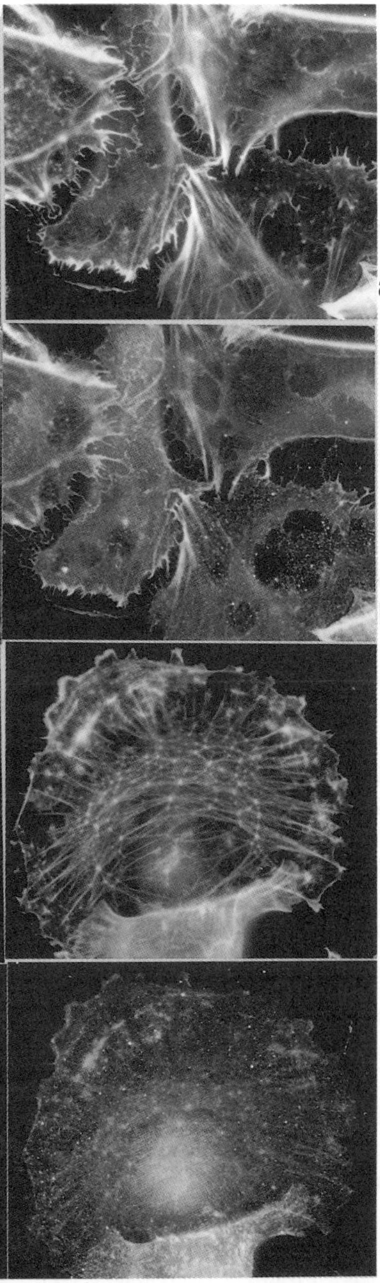

a. b. c. d.

Fig. 2 (*a–d*) demonstrates the effects of 0.1 μM TPA on microfilament organization (*a* and *c*) and FNR distribution (*b* and *d*) during RAEC spreading on collagen (*a,b*) or laminin (*c,d*). TPA causes a partial depolymerization of actin filaments in RAEC plated on Type IV collagen, giving rise to short disorganized filamentous actin indicated by a snowflake staining pattern (2*a*). A single, thick cortical filament extends along the long axis of the cell to define the cell boundary (2*a*). FNR are distributed in a diffuse pattern in the perinuclear region and a punctate distribution is observed in some instances (2*b*). FNR are also detected at the edges of plasmalemma coincident with isolated bundles of actin (2*b*). TPA causes partial depolymerization of actin in RAEC plated on laminin giving rise to an extensive rearrangement of the cytoskeleton. The cytoskeleton is characterized by an extensive network of thin microfilaments which extend along the axis of the cell with radical projections (2*c*). FNR follow the distribution of stress fibers in a uniformly diffuse pattern in the perinuclear region of the cell and are absent from the cell edge (2*d*).

Discussion

Aortic endothelial cells grown on collagen matrices (gels) exhibit angiogenic responses to TPA by forming tubular, capillary-like branching networks corresponding to enhanced collagenase activity and migration [4]. We hypothesized that exposure of EC to an agent with angiogenic activity (i.e., TPA) would allow us to define a role for receptors/cytoskeleton in EC interactions with discrete matrix peptides during the early angiogenic events of EC attachment, spreading and migration.

TPA treatment inhibited RAEC spreading on Type IV collagen but enhanced the rate of spreading on laminin. This appeared to be a function of adhesion plaque formation, which was present in RAEC on Type IV collagen but absent in RAEC on laminin. This lack of adhesion plaque formation following TPA treatment is facilitated by cytoskeletal reorganization and enhanced cell motility, spreading and cell-cell contact formation.

Figure 1 (*a–d*). Effects of substrate, Type IV Collagen (*a,b*) or Laminin (*c,d*) on organization of F-actin microfilaments of the cytoskeleton (*a* and *c*) and the distribution of fibronectin integrin receptors ($\alpha_5\beta_1$) (*b* and *d*) on the abluminal surface of RAEC.

Figure 2 (*a–d*). Effects of 0.1 μM TPA on microfilament organization (*a* and *c*) and FNR distribution (*b* and *d*) during RAEC spreading on collagen (*a,b*) or laminin (*c,d*). RAEC were treated with 0.1 μM TPA or solvent (0.1% ethanol, control) 0.5 h after seeding on substrate-coated coverslips.

The ability of TPA to alter RAEC spreading and cell-cell contact formation can be attributed to its direct effects on cytoskeletal arrangement and integrin receptor expression. However, the responses of RAEC to TPA are mediated by specific substrate proteins. Our studies suggest that during dynamic interactions (i.e., spreading) the cytoskeleton is more sensitive (i.e., prone to depolymerization) to a mitogenic agent such as TPA. Inasmuch as the spreading rates of RAEC on Type IV collagen and laminin differ significantly, the overall stability/organizational status of the cytoskeleton at the time of TPA treatment (0.5 h following exposure to substrate) may facilitate these differential responses. It appears that following exposure to discrete immobilized substrates (i.e., collagen Types I and IV, fibronectin and laminin) RAEC are predisposed to respond to TPA in a specific manner due to initial interactions between RAEC and substrate, and endogenous signals that mediate these activities.

References

1 Furcht LT: Critical factors controlling angiogenesis: Cell products, cell matrix, and growth factors. (Editorial) Lab Invest 1986; 55(5): 505–509.
2 Vlodavsky I, Korner G, Ishai-Michaeli R, Bashkin P, Bar-Shavit R, Fuks Z: Extracellular matrix-resident growth factors and enzymes: possible involvement in tumor metastasis and angiogenesis. Cancer Metastasis Rev 1990; 9: 203–226.
3 Madri JA, Williams SK: Capillary endothelial cell cultures: Phenotypic modulation by matrix components. J Cell Biol 1983; 97: 153–165.
4 Montesano R, Orci L: Tumor-promoting phorbol esters induce angiogenesis in vitro. Cell 1985; 42: 469–477.
5 Diglio CA, Grammas P, Giacomelli F, Weiner J: Rat-heart derived endothelial and smooth muscle cultures: isolation, cloning and characterization. Tissue Cell 1988; 20: 477–492.
6 Semich R, Robenek H: Organization of the cytoskeleton and the focal contacts of bovine aortic endothelial cells cultured on type I and III collagen. J Histochem Cytochem 1990; 38(1): 59–67.

Angiogenesis: Key Principles - Science - Technology - Medicine
ed. by R. Steiner, P. B. Weisz & R. Langer

Aortic endothelial cells in culture: factors which modulate their morphological and biosynthetic phenotypes

A. E. Canfield[a], A. B. Sutton[a], S. L. Schor[b], M. E. Grant[c] and A. M. Schor[a]

[a]*CRC Department of Medical Oncology, Christie Hosptial NHS Trust, Wilmslow Road, Manchester M20 9BX,* [b]*Department of Cell and Structural Biology, Coupland III Building, Coupland Street, University of Manchester, Manchester M13 9PL,* [c]*Department of Biochemistry and Molecular Biology, School of Biological Sciences, University of Manchester, Manchester M13 9PT, England*

Introduction

New capillaries arise from the pre-existing vasculature (arterioles, venules and capillaries) through a series of steps that involve the endothelium and the surrounding extracellular matrix (ECM). The principal morphological events involved in angiogenesis were described in the 1940s. Electron microscopic studies performed over 20 years later have confirmed these initial observations and added further detailed information [1–3].

Endothelial cells lining the blood vessels rest on a basement membrane. During the course of angiogenesis, endothelial cells penetrate this basement membrane and continue to migrate and proliferate into the surrounding connective tissue where they adopt an elongated, fibroblastoid morphology and self-associate in a head-to-tail fashion to form loops, sprouts and patent tubules. Endothelial cells in vitro retain the ability to display two distinct phenotypes (cobblestone and sprouting) and to undergo morphogenetic interactions leading to the formation of histiotypic structure. The nature of the surrounding ECM plays an important role in determining the endothelial cell phenotype [4, 5]. In this communication we examine the expression of cobblestone and sprouting phenotypes by cultured endothelial cells and the correlation between the morphological phenotype and the synthesis of ECM molecules by these cells.

Materials and methods

Isolation and culture of aortic endothelial cells

Bovine aortic endothelial cells were isolated from adult animals and characterised as previously described [4]. Both cloned and uncloned cell lines were examined. They were routinely cultured in Eagle's minimal essential medium (MEM) supplemented with 20% (v/v) donor calf serum, ascorbic acid (50 μg/ml), 2 mM-glutamine, 1 mM sodium pyruvate, non-essential amino acids, penicillin (100 U/ml) and streptomycin (0.1 mg/ml) (Gibco-Biocult Ltd, Uxbridge, England). This complete growth medium is referred to as 20% DCS-MEM. Stock cultures were maintained on gelatin-coated dishes [4] and were incubated at 37°C in a moist atmosphere of 5% CO_2 and 95% air.

For the various experimental investigations, cells were plated either onto a) gelatin-coated dishes, b) on the surface of 3-D gels of type I collagen, or c) homogeneously dispersed within the 3-D collagen gel matrix by mixing the cells with the collagen gelling solution before casting. These substrata were prepared as described previously [4, 6, 7].

Labelling of cell cultures, isolation and characterization of newly-synthesised proteins

Endothelial cells were incubated with [^{3}H]proline (Amersham Interantional, Amersham, England) for 20–24 h at 37°C, and the newly-synthesised proteins secreted into the medium and into the cell layer/matrix were collected and characterised as described in detail elsewhere [8–11]. Total RNA was prepared for endothelial cell cultures as previously described [10].

Results and Discussion

Expression of cobblestone and sprouting morphologies by aortic endothelial cells

We have studied the morphology displayed by 38 different lines of bovine aortic endothelial cells (18 cloned and 20 uncloned). All the cell lines examined (a) expressed factor VIII-related antigen, (b) were capable of displaying two distinct phenotypes (cobblestone and sprouting) and (c) could associate into tubule-like structures. On 2-D substrata (gelatin-coated dishes or the surface of collagen gels), aortic endothelial cells formed a compact, contact-inhibited cobblestone monolayer at confluence (Fig. 1A). Cells displaying the sprouting phenotype appeared

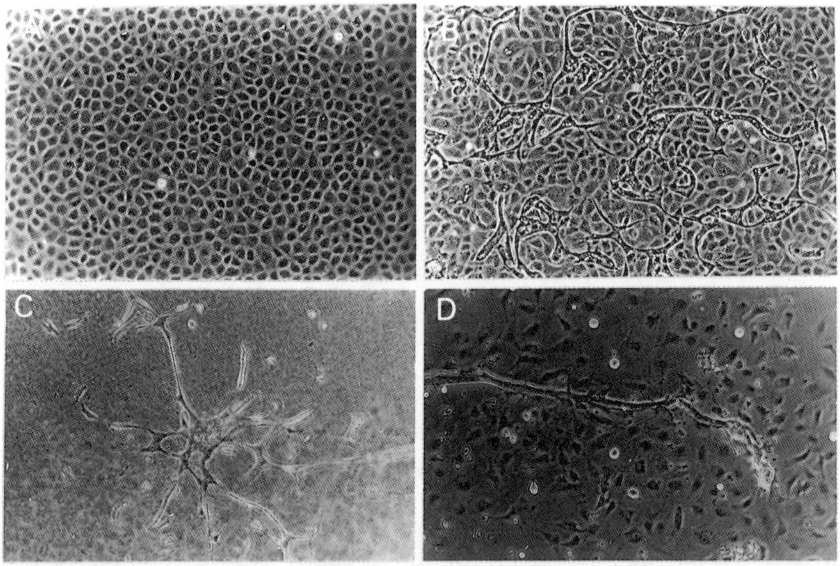

Figure 1. Morphology of aortic endothelial cells. *A* Confluent culture on gelatin-coated dish showing cobblestone morphology. *B* Post-confluent culture on gelatin-coated dish showing a network of sprouting cells under the cobblestone monolayer. *C* Sprouting endothelial cells within a 3-D collagen gel, 7days after plating. The sprouting cells have self-associated to form 3-D meshworks. *D* 3-month post-confluent culture on a gelatin-coated dish, showing tubules of endothelial cells surrounded by proliferating cells, Bar, 150 μm.

spontaneously under the cobblestone monolayer in post-confluent cultures on 2-D substrata (Fig. 1B). These cells formed in the presence of serum or platelet-poor plasma, without the addition of exogenous factors to the culture medium. In post-confluent cultures grown on the surface of 3-D collagen gels, the sprouting cells underneath the cobblestone monolayer did not remain at the surface but migrated down into the gel matrix forming dense cellular meshworks (similar to those shown in Fig. 1C).

Homogeneous populations of sprouting cells (without cobblestone cells) were obtained by plating endothelial cells directly within a 3-D collagen gel. Under these conditions, the cells elongate, migrate and self-associate to form complex networks (Fig. 1C) and tubule-like structures [4, 12]. The sprouting and cobblestone morphologies adopted by endothelial cells are reversible [4, 13].

Tubule-like structures can also be formed by tensile forces in post-confluent cultures of aortic endothelial cells which have been kept on plastic or gelatin-coated dishes for long periods (Fig. 1D). Under these conditions, the cell layer tends to roll up, occasionally leaving long cellular "bridges" or "tubules" that connect the rolled up cell layer to the tissue culture dish. These tubules remain attached to the culture dish

Table 1. Influence of cell density on sprouting cell formation by aortic endothelial cells

Serum concentration	Cell number at confluence (10^4 cells/cm^2)	Sprouting cells
20%	8–10	Yes
5%	3–7	No

Endothelial cells were plated on gelatin-coated dishes and grown in the presence of 20% or 5% DCS-MEM. At confluence, cell numbers were determined and the ability to form sprouting cells was noted. The results shown are from 5 separate experiments using 2 different cell lines.

for long periods, apparently increasing in length (Fig. 1D). They appeared to be formed by tensile forces and the tendency of the endothelial cells to self-associate. Histological and EM sections revealed that the majority of the tubules were solid cords of cells; lumeni were occasionally observed in some sections (not shown).

The appearance of sprouting cells in endothelial cell cultures may be modulated by culture conditions, such as cell density and the nature of the extracellular matrix (ECM). For example, under normal culture conditions, cells which appear confluent under the light microscope (betweeen 4 and 6×10^4 cells/cm^2) are still able to divide and form a more compact, cobblestone monolayer (between 8 and 20×10^4 cells/cm^2). If the cells are not allowed to reach this density (for example, by growing them in the presence of 5% serum – see Table 1), sprouting cells will not appear in the cultures. Similarly, if a post-confluent culture containing cobblestone and sprouting cells is incubated with either bacterial collagenase (1–2 mg/ml for 1 min) or EGTA (2–3 mM for 1 h), a small number of cells (between $2–3 \times 10^3$ cells /8 cm^2 dish) are removed from the cobblestone monolayer. After this initial incubation, sprouting cells are still present in the dish, but disappear within 3 h, indicating that they have migrated up into the monolayer and reverted back to the cobblestone morphology. The fact that identical results were obtained using collagenase and EGTA indicates that this phenomenon was due to removal of cells from the monolayer rather than hydrolysis of the ECM. Sprouting cells reapppear if collagenase- or EGTA-treated cultures are maintained to post-confluence.

The expression of the sprouting phenotype by endothelial cells can also be inhibited by incubation with alpha- and beta-xylosides [7]. This effect seems to be related to an inhibition of matrix deposition by the cells since the appearance of the sprouting phenotype was inhibited at xyloside concentrations that did not affect cell numbers. Table 2 shows the results of a typical experiment in which endothelial cells were cultured in the presence of alpha- and beta-xylosides (1.0 and 2.0 mM). These compounds inhibited both sprouting and incorporation of

Table 2. Effects of xylosides on sprouting cell formation and protein biosynthesis by endothelial cells

Sample	Presence of sprouting cells	Cts. min^{-1} incorporated/10^5 cells	% of control
Control	+ + +	135,800	100
1.0 mM β-xyl	+/–	62,000	46
2.0 mM β-xyl	–	41,500	31
1.0 mM α-xyl	+/–	79,300	58
2.0 mM α-xyl	–	73,200	54

Endothelial cells (9×10^4) were plated on gelatinised dishes (8 cm^2) in 20% DCS-MEM. Two days after plating, the medium was replaced by the control and test media and the cultures were incubated for a further 20-day period, with the medium being changed every second day. Duplicate cultures were incubated with [^{3}H]proline (10 μCi/ml) during the last 24 h of this period and the amount of radioactivity incorporated into newly-synthesised proteins was determined as described [14]. The morphology of the cells and the number of cells present were also determined. The results shown are from a typical experiment and are expressed as total cts min^{-1} incorporated/10^5 cells, and as a percentage of the control cultures.

Control cultures contained a network of sprouting cells (+ + +), whereas cultures grown in the presence of xylosides contained only a few (+/–) or no sprouting cells (–).

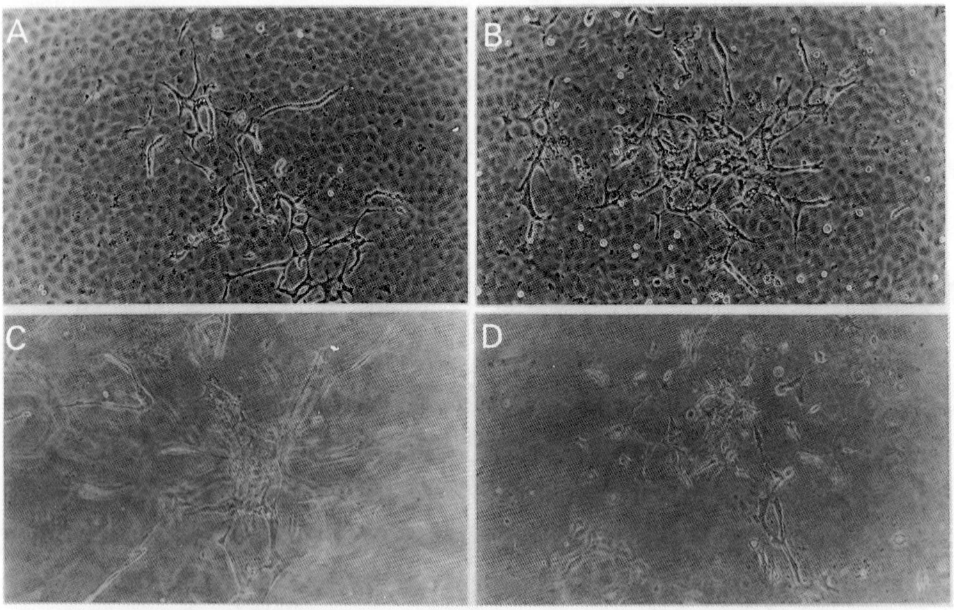

Figure 2. Sprouting cells embedded in an endothelial-produced ECM (*A*, *B*) and within a 3-D collagen gel (*C*, *D*) respond differently to TGFβ-1. *A* and *B*, post-confluent cultures of endothelial cells on gelatinised dishes were incubated with (*B*) or without (*A*) TGFβ-1 (1 ng ml^{-1}) in 20% DCS-MEM for 5 days. Neither the cobblestone nor the sprouting cells were affected by TGFβ-1 in terms of morphology or cell numbers. *C* and *D*, Endothelial cells were homogeneously dispersed within a 3-D collagen gel and allowed to adopt a sprouting morphology in 20% DCS-MEM over a 3 day period. At this time, TGFβ-1 (1 ng ml^{-1}) was added and the cultures incubated for a further 5 days. Sprouting cells in control medium (*C*) remained elongated and formed 3-D networks; the majority of cells maintained in the presence of TGFβ-1 (*D*) rounded up and failed to self-associate. Bar, 150 μm. Figures reproduced from [14].

[^{3}H]proline into newly-synthesised proteins in these cultures in a dose-dependent manner. A detailed analysis of the ECM deposited by cells cultured under these conditions will be presented elsewhere (Canfield A. E. and Schor A. M., manuscript in preparation).

Our recent studies suggest that cobblestone cells, sprouting cells underneath a cobblestone monolayer and sprouting cells within a 3-D collagen gel respond differently to TGFβ-1 [14]. When added to proliferating cultures on 2-D substrata (cobblestone), TGFβ-1 was both cytotoxic and cytostatic, caused marked changes in cell morphology and increased total protein synthesis. However, TGFβ-1 added to post-confluent cultures, had no effect on the cell numbers or on the morphology of the sprouting cells underneath the cobblestone monolayer (which are embedded within a complex endothelial-produced ECM) (Fig. 2A and B). On the other hand, TGFβ-1 was cytotoxic when added to sprouting endothelial cells embedded within a 3-D collagen gel (Fig. 2C and D). The observed lack of responsiveness to TGFβ-1 by sprouting cells embedded within a cell-produced ECM did not appear to be a consequence of growth factor inaccessibility since (a) the same cultures showed an increased level of protein synthesis in response to TGFβ-1 and (b) TGFβ-1 pre-incubated on these cultures was still active when added back to proliferating cultures [14].

Protein synthesis by aortic endothelial cells

Contradictory reports have been published regarding the types of collagen synthesised by vascular endothelial cells in vitro [8, 15–19]. We have identified the major species of matrix proteins (collagenous and non-collagneous) synthesised by six different lines of aortic endothelial cells and have found that fibronectin, thrombospondin and plasminogen activator inhibitor type I were consistently secreted by all six cell lines in vitro [9, 10 and unpublished data]. By contrast, the types of collagenous proteins synthesised by cobblestone endothelial cells were found to vary from line to line, but remained constant for each particular cell line. Table 3 shows the types of collagens synthesized by these six cell lines; three lines secreted different proportions of collagen types I, III and IV, one line secreted only collagen types I and III, another secreted only type IV collagen and another line secreted only type I collagen. A more detailed analysis of these results will be presented elsewhere (Canfield et al, manuscript submitted). It has been shown that aortic endothelial cells in monolayer cultures secrete type III collagen, and that type I collagen was produced when sprouting cells were present in the cultures [20, 21]. However, we show that *not all* endothelial cells synthesise type III collagen and, moreover, cells forming a cobblestone monolayer in

Table 3. Collagen synthesis by aortic endothelial cells

Endothelial cell line	Collagens synthesised
A, B, C,	I, III, IV
D	I, III
U10	IV
A1	I

Confluent cultures containing cobblestone endothelial cells (no sprouting cels) were incubated with [^{3}H]proline for 24 h at 37°C. Newly-synthesised proteins secreted into the medium were digested with pepsin (100 μg/ml, 4 h, 4°C) and analysed by SDS-PAGE as previously described [11]. Collagen types were identified (a) by their characteristic electrophoretic mobilities under non-reducing conditions and after interrupted reduction, (b) by comparison with collagen standards and (c) by immunoprecipitation with specific antibodies.

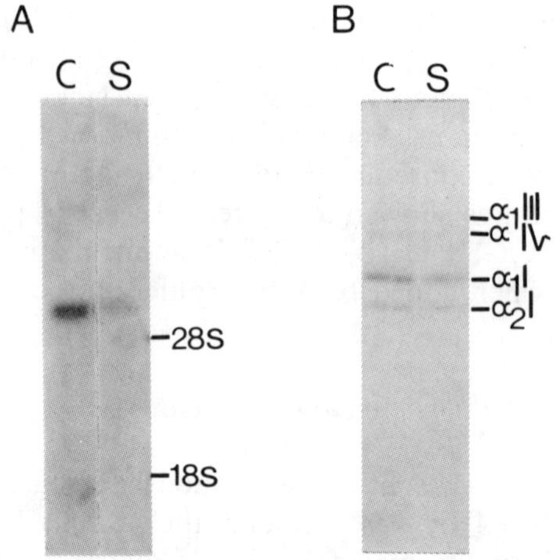

Figure 3. The effects of cell shape on matrix protein synthesis by aortic endothelial cells. *A* Total RNA (10 μg) from cobblestone (C) and sprouting (S) cells plated on or within 3-D collagen gels (respectively) was electrophoresed in a formaldehyde-containing agarose gel, blotted, probed with ^{32}P-labelled cDNA coding for thrombospondin, washed and exposed to X-ray film as described previously [10]. Sprouting cells contained 50% less mRNA for thrombospondin than did cobblestone cells (reproduced from [10]). *B* Endothelial cells plated on the surface of 3-D collagen gels (cobblestone, C) or within 3-D collagen gels (sprouting, S) were incubated with [^{3}H]proline (10 μCi/ml) for 24 h and the newly-synthesised collagenous proteins secreted into the medium were isolated by pepsin digestion as previously described [11]. Samples were electrophoresed after interrupted reduction on SDS-polyacrylamide gels and detected by fluorography. Both cobblestone and sprouting cells secreted a similar mixture of collagen types I, III and IV.

the absence of sprouting cells can synthesise type I collagen. It is clear that endothelial heterogeneity accounts for some of the discrepancies reported by different laboratories.

The influence of cell shape on matrix biosynthesis by aortic endothelial cells has been investigated previously by comparing confluent cultures of cobblestone cells with post-confluent cultures which contain cells displaying *both* the cobblestone and sprouting phenotypes [19–21]. Any changes in protein synthesis have generally been attributed to the presence of sprouting cells in these cultures [19–21]: however, they may also be due to the co-culture of sprouting and cobblestone cells or to the increased time in culture. We have overcome this potential problem by using the same substratum (type I collagen) and plating cells either on the gel surface (to obtain a homogeneous culture of cobblestone cells: as in Fig. 1A) or within the 3-D collagen matrix (to obtain a homogeneous culture of sprouting cells; Fig. 1C). These cells have been maintained in culture for the same length of time. Under these conditions, we have found that mRNA (Fig. 3A) and protein levels of thrombospondin were consistently down-regulated in sprouting compared with cobblestone cells in two endothelial cell lines examined [10]. However, the types of collagens synthesised by 5 different cell lines examined was the same whether they were cobblestone or sprouting. Thus, Fig. 3B shows that one particular cell line secreted collagen types I, III and IV whether they displayed the cobblestone or the sprouting phenotype. These results therefore demonstrate that the adoption of a sprouting phenotype by aortic endothelial cells is not necessarily associated with any change in collagen biosynthesis.

Conclusions

The results presented in this communication demonstrate that aortic endothelial cells in vitro can adopt two distinct, reversible phenotypes (cobblestone and sprouting) which are believed to represent the resting and angiogenic phenotypes (respectively) displayed by these cells in vivo. We show that six lines of cobblestone aortic cells secreted the same non-collagenous matrix proteins, fibronectin and thrombospondin; however, the types of collagen secreted by these cells varied from line to line but remained constant for each particular line of cells. Five lines of endothelial cells examined secreted the same types of collagen whether they displayed the cobblestone or sprouting phenotype. By contrast, thrombospondin secretion in two cell lines examined was down-regulated when the cells were sprouting. Whether similar changes in matrix protein biosynthesis occur when endothelial cells undergo angiogenesis in vivo remains to be elucidated. Discrepancies in findings from different laboratories may be due to endothelial cell heterogeneity; the function of this heterogeneity is not known.

Acknowledgements. The financial support of the Cancer Research Campaign and the Medical Research Council are gratefully acknowledged. We also thank Marjorie Evans for typing this manuscript.

References

1 Schor AM, Schor SL: Tumour angiogenesis. J Exp Path 1983; 141: 385–413.
2 Folkman J: Tumour angiogenesis. Adv Cancer Res 1985; 43: 175–203.
3 Blood CH, Zetter BR: Tumour interactions with the vasculature: angiogenesis and tumour metastasis. Biochim Biophys Acta 1990; 1032: 89–113.
4 Schor AM, Schor SL, Allen TD: Effects of culture conditions on the proliferation, morphology and migration of bovine endothelial cells. J Cell Sci 1983; 62: 267–285.
5 Madri JA, Pratt BM: Endothelial cell-matrix interactions: in vitro models of angiogenesis. J Histochem Cytochem 1986; 34: 85–91.
6 Schor AM, Schor SL: The isolation and culture of endothelial cells and pericytes from bovine retinal microvasculature: A comparative study with large vessel vascular cells. Microvas Res 1986; 32: 21–38.
7 Schor AM, Schor SL: Inhibition of endothelial cell morphogenetic interactions in vitro by alpha- and beta-xylosides. In Vitro 1988; 24: 659–668.
8 Canfield AE, Schor AM, Schor SL, Grant ME: The biosynthesis of extracellular matrix components by bovine retinal endothelial cells displaying distinctive morphological phenotypes. Biochem J 1986; 245: 375–383.
9 Canfield AE, Schor AM, Loskutoff DJ, Schor SL, Grant ME: Plasminogen activator inhibitor-type I is a major biosynthetic product of retinal microvascular endothelial cells and pericytes in culture. Biochem J 1989; 259: 529–535.
10 Canfield AE, Boot-Handford RP, Schor AM: Thrombospondin gene expression by endothelial cells in culture is modulated by cell proliferation, cell shape and the substratum. Biochem J 1990; 268: 225–230.
11 Canfield AE, Allen TD, Grant ME, Schor SL, Schor AM: Modulation of extracellular matrix biosynthesis by bovine retinal pericytes in vitro: effects of the substratum and cell density. J Cell Sci 1990; 96: 159–169.
12 Montesano R, Orci L, Vassali P: In vitro rapid organisation of endothelial cells into capillary-like networks is promoted by collagen matrices. J Cell Biol 1983; 97: 1648–1652.
13 Makarski JS: Evidence for reversible morphological phenotypes (sprouts) of cultured endothelial cells. Cell Biol Int Rep 1982; 6: 225–233.
14 Sutton AB, Canfield AE, Schor SL, Grant ME, Schor AM: The response of endothelial cells to TGFβ-1 is dependent upon cell shape, proliferative state and the nature of the substratum. J Cell Sci 1991 (in press).
15 Sage H, Crouch E, Bornstein P: Collagen synthesis by bovine aortic endothelial cells in culture. Biochemistry 1979; 18: 5433–5442.
16 Sage H, Pritzl P, Bornstein P: Secretory phenotypes of endothelial cells in culture: comparison of aortic, venous, capillary and corneal endothelium. Arteriosclerosis. 1981; 1: 427–442.
17 Sankey EA, Barnes MJ: Comparison of the collagenous products synthesised in culture by pig aortic endothelial and smooth muscle cells. Biochem J 1984; 218: 11–18.
18 Kramer RH, Fuh G-M, Karasek MA: Type IV collagen synthesis by cultured human microvascular endothelial cells and its deposition into the subendothelial basement membrane. Biochemistry 1985; 24: 7423–7430.
19 Levene CI, Bartlett C, Heale G; Phenotypic changes in morphology and collagen polymorphism of cultured bovine and porcine aortic endothelium. Atherosclerosis 1984; 52: 59–71.
20 Cott-Pereira G, Sage H, Bornstein P, Ross R, Schwartz S: Studies of morphologically atypical ("sprouting") cultures of bovine aortic endothelial cells. Growth characteristics and connective tissue protein systhesis. J Cell Physiol 1980; 102: 183–191.
21 Iruela-Arispe ML, Hasselaar P, Sage H: Differential expression of extracellular proteins is correlated with angiogenesis in vitro. Lab Invest 1991; 64: 174–186.

Angiogenesis: Key Principles - Science - Technology - Medicine
ed. by R. Steiner, P. B. Weisz & R. Langer

The behaviour of pericytes in vitro: relevance to angiogenesis and differentiation

A. M. Schor[a], A. E. Canfield[a], A. B. Sutton[a], T. D. Allen[b], P. Sloan[c] and S. L. Schor[d]

[a]*CRC Department of Medical Oncology, Christie Hospital NHS Trust, Wilmslow Road, Manchester M20 9BX*, [b]*Department of Ultrastructure, Paterson Institute for Cancer Research, Wilmslow Road, Manchester M20 9BX*, [c]*Department of Oral Medicine, Turner Dental School, Manchester University, Manchester M16 6FH, and* [d]*Department of Cell and Structural Biology, Coupland III Building, Coupland Street, Manchester University, Manchester M13 9PL, UK*

Introduction

Pericytes are defined by their characteristic morphology and location in vivo, i.e. stellate cells, embedded within the basement membrane of arterioles, venules and capillaries. Depending on the vessel and the age of the animal, pericytes have been reported to cover from 26 to 100% of the surface of the microvessels [1–5].

As an integral part of the microvascular wall, pericytes are believed to participate in the synthesis of the vascular basement membrane and in the formation of new blood vessels, although their specific role is not clear. During angiogenesis, the newly formed sprouts and cords are usually referred to as "endothelial"; pericytes only become apparent during the final stages of maturation. Because of this, it has been suggested that pericytes originate from the perivascular connective tissue as fibroblastic cells that migrate towards the new vessel and encircle it; therefore, pericytes have been suggested to be responsible for the arrest of vessel formation and the induction of vessel maturation [5]. Recent in vitro data lend support to this theory since it was found that active TGFβ (which inhibits endothelial cell proliferation and migration) is produced when adrenal cortex endothelial cells are co-cultured with retinal pericytes, whereas only latent TGFβ is produced by the same cells when cultured separately [6, 7]. However, it was recently demonstrated that pericytes are more numerous in vessels undergoing angiogenesis than in resting vessels [8]. Moreover, if pericytes were to signal the end of angiogenesis, it is difficult to explain why angiogenesis begins most commonly from capillaries rich in pericytes and why pericytes are more numerous during embryonic development (when neovascularisation is rampant) than in the adult animal [1]. In this communication we present data suggesting that during angiogenesis the newly formed sprouts are formed by both endothelial cells and pericytes. Furthermore, our data suggest that pericytes may represent primitive mesenchymal cells, able to differentiate along the osteogenic pathway.

Materials and methods

Cells and culture conditions

Pericytes were isolated from the retinal microvasculature of adult cows and cultured on plastic dishes in Eagle's Minimal Essential Medium (MEM) supplemented with 20% donor calf serum, 50 μg/ml ascorbic acid, 2 mM glutamine, 1 mM sodium pyruvate and non-essential amino acids (Gibco BRL, Paisley, Scotland) as described [4, 9]. This growth medium will be referred to as 20% DCS-MEM.

Bovine retinal endothelial cells were isolated as described [9] and maintained on gelatin-coated dishes in 20% DCS-MEM. Other cells used for comparison, such as aortic smooth muscle cells and bovine nuchal or human skin fibroblasts were isolated according to standard procedures [9, 10] and maintained on plastic dishes under the same conditions as the pericytes. The cells were cultured either on plastic culture dishes or on type I collagen gels. When using collagen gels as substratum the cells were plated either on the surface of the gels or within the 3-D gel matrices [9, 10].

Characterisation of newly synthesised proteins

Proteins synthesised by pericytes in culture were examined by discontinuous SDS/polyacrylamide gel electrophoresis, following labelling of the cultures with ^{3}H-proline. Specific proteins were identified by immunoprecipitation and by other biochemical properties as described [11–14]. Tenascin was identified by immunoblotting and laminin mRNA by Northern blot hybridisation [15].

Electron microscopy

Transmission electron microscopy (TEM) was performed as previously reported [4, 13]. Cells on collagen gels were infiltrated and embedded in Epon/Araldite via propylene oxide, and cells on plastic dishes were embedded via hydroxypropylmethacrylate and Lufts Epon.

Thin sections were stained with urnayl acetate and lead citrate and viewed at 80 kV in an AE801A TEM.

Localisation of matrix macromolecules by immunofluorescence

The localisation of matrix macromolecules in pericyte cultures was examined by indirect immunofluorescence using specific antibodies. Photographs were taken with an automatic camera; the exposure time was generally between 30 and 90 seconds; negative controls were stopped manually after 5 min exposure. The following antisera were used. Rabbit anti-mouse laminin from Gibco BRL (Paisley, Scotland), rabbit anti-human placental type IV collagen (kindly donated by Drs G. J. Harrap and W. Gibson, Unilever Research, Bedford, England), rabbit anti-chick tenascin (kindly donated by Dr R. Chiquet-Ehrismann, Friedrich Miescher Institute, Basle, Switzerland), and rabbit anti-chick type X collagen (a kind gift from Dr A. Kwan, Dept of Biochemistry and Molecular Biology, Manchester, England). As part of the characterisation of the pericyte and endothelial cultures we used rabbit anti-human von Willebrand Factor from Dako Ltd (Bucks, England), rabbit anti-cow glial fibrillary acidic protein (Dako Ltd), mouse monoclonal anti α-smooth muscle actin from Sigma (Dorset, England) and mouse monoclonal anti-desmin from Dako (clone D33). The second antisera, FITC conjugated, were rabbit anti-mouse and swine anti-rabbit (Dako Ltd). Controls included the substitution of the first antiserum by non-immune serum and the absorption of the specific antisera with the appropriate matrix molecule.

Results and discussion

Identification of retinal pericytes in vitro

Pericytes are identified by their tissue of origin (retinal microvessels) and by a series of characteristics that clearly distinguish them from other stromal cells such as fibroblasts, smooth muscle, endothelial and glial cells (summarised in Table 1). The lack of Factor-VIII-related antigen and glial fibrillary acidic protein excludes the presence of endothelial and glial cells in the pericyte cultures. The presence of α-smooth muscle actin and desmin excludes the presence of fibroblasts. Pericytes show various smooth muscle cell-like characteristics (Table 1) such as the presence of α-smooth muscle actin, the ability to proliferate within collagen matrices and the ability to contract such matrices. Both smooth muscle cells and pericytes form multilayers that retract into nodules when cultured on plastic dishes, but these cells differ in

Table 1. Characteristics of retinal pericytes and other stromal cells in culture

	Pericytes	Endothelial cells	Smooth muscle cells	Fibroblasts
Factor VIII	−	+	−	−
GFA protein	−	−	−	−
α-sm actin	+	−	+	−
Desmin	+	−	−	−
Cell retraction	++	−	+	−
Gell contraction	+	−	+	+
Gel invasion	−	+	+	+

behaviour when plated on the surface of 3-D collagen gels: under these culture conditions the smooth muscle cells form multilayers on the cell surface and migrate down into the gel matrix whereas pericytes form nodules and remain on the surface of the gel [4, 9].

Another feature which distinguishes aortic smooth muscle cells from retinal pericytes is the presence of desmin in the latter but not the former (Fig. 1). It is interesting that α-sm actin antibodies stain all the cells in both pericyte and smooth muscle cell cultures, whereas desmin antibodies stain approximately 30% of the pericytes and none of the smooth muscle cells. If the desmin-negative cells in the pericyte cultures were contaminating smooth muscle cells or fibroblasts, we would expect these cells to show different behaviour (for example to migrate into collagen gels). The fact that they all behave in an identical manner and

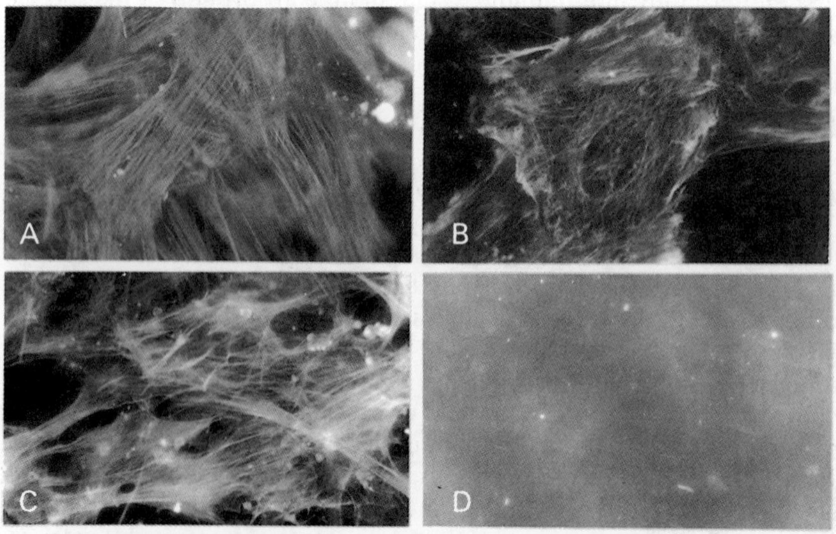

Figure 1. Immunofluorescence staining of bovine retinal pericytes (*A*, *B*) and aortic smooth muscle cells (*C*, *D*) with antibodies to α-smooth muscle actin (*A*, *C*) and desmin (B, D). Bar = 150 μm.

originate from retinal microvessels suggest that pericytes are heterogeneous regarding the presence of desmin. Both α-smooth muscle actin and desmin have been observed in pericytes in vivo [16, 17].

Morphology and protein biosynthesis by pericytes. Comparison with retinal endothelial cells

Pericytes may display two reversible phenotypes depending on culture conditions. When cultured on a two-dimensional substratum they display a characteristic stellate morphology (Fig. 2) with long fillamentous

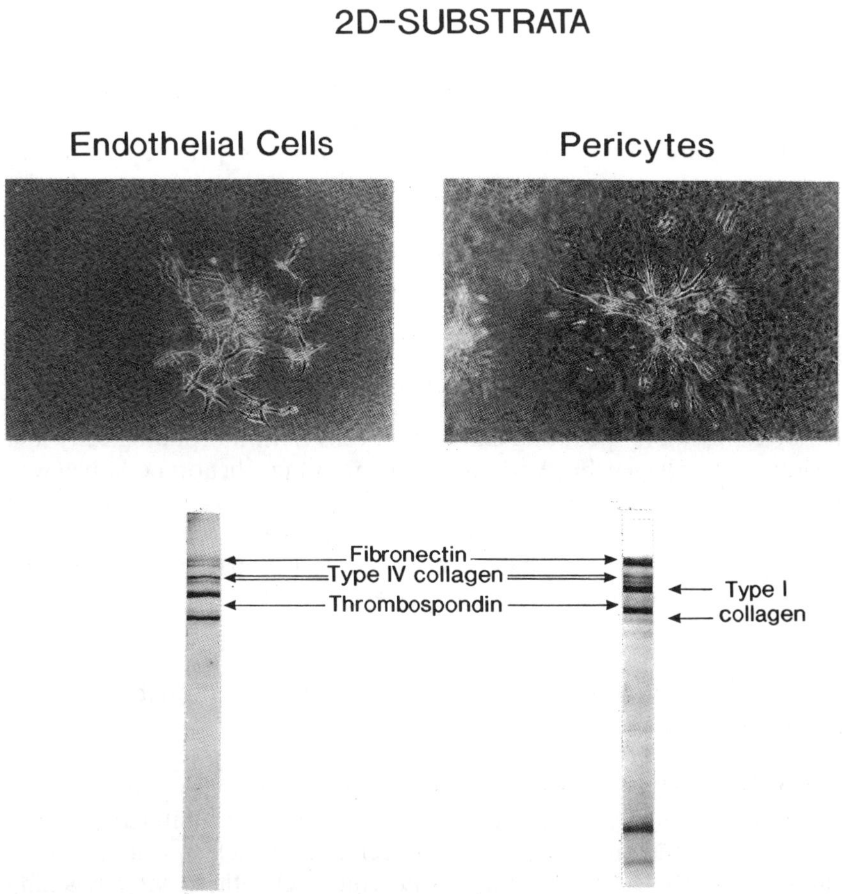

Figure 2. Retinal endothelial cells and pericytes plated on 2-D substrata (gelatin-coated or uncoated plastic dishes) display different morphologies and synthesise different collagen types. Bar = 150 μm. Cell cultures were incubated with ^{3}H-proline for 24 h; the proteins secreted in the medium were analysed by electrophoresis on 6.5% polyacrylamide gels under reducing conditions and detected by fluorography. The proteins present were identified as described [11–14].

processes [4]. This morphology is clearly different from the cobblestone morphology displayed by retinal endothelial cells on 2-D substrata (Fig. 2). Pericytes in vivo also appear as stellate cells with long processes [18, 19]. When plated within a 3-D collagen matrix, pericytes display an elongated, sprouting morphology, identical to that of endothelial cells plated under the same conditions (Fig. 3) [4, 9, 13]. Endothelial cells and pericytes plated within 3-D collagen gels as single cells show the same sprouting morphology but behave differently: namely, endothelial cells do not proliferate but they migrate and form aggregates and tubule-like structures. By contrast, pericytes proliferate within the gel and do not aggregate but remain mainly as single cells [9]. Pericytes tend to aggregate when cultured on 2-D substrata (see later). Such aggregates, recovered from 2-D cultures and plated within 3-D gels, look identical to the endothelial aggregates that form spontaneously within the gels (Fig. 3) [13]. The main proteins synthesised by retinal endothelial cells and pericytes are fibronectin, thrombospondin, plasminogen activator inhibitor (PAI-1) and collagens [11–13]. Different collagen types are synthesised by endothelial cells and pericytes. Under standard tissue culture conditions on 2-D substrata, "cobblestone" endothelial cells synthesise basement membrane-specific collagen (type IV) and "stellate" pericytes synthesise interstitial collagens (types I and III) (Fig. 2). When these cells are plated within a 3-D gel (and adopt a "sprouting" phenotype), both cell types synthesise significant amounts of type IV collagen in addition to type I collagen and minor amounts of type III collagen (Fig. 3). These results demonstrate that sprouting pericytes resemble sprouting endothelial cells both morphologically and in terms of collagen biosynthesis. The sprouting phenotype is believed to represent the angiogenic phenotype found in vivo. Accordingly, we have suggested that during angiogenesis, endothelial cells and pericytes may also be indistinguishable, both cell types participating in the formation of sprouts and new blood vessels [9, 11, 13].

Differentiation of pericytes in culture. Association with changes in matrix biosynthesis

Time-lapse video microscopy has shown that pericytes plated on 2-D substrata are highly migratory and are not contact-inhibited [4]. The cells overlap and form temporary associations which appear to break suddenly by the cells migrating away from each other. After reaching confluence, pericytes form multilayers and then nodules by a process involving (a) sudden retraction of a multilayer area into a small nodule or cell aggregate leaving empty patches in the substratum, (b) re-attachment of the nodule on the substratum followed by cell migration out of the nodule and cell proliferation. This process takes 4 to 5 h and results

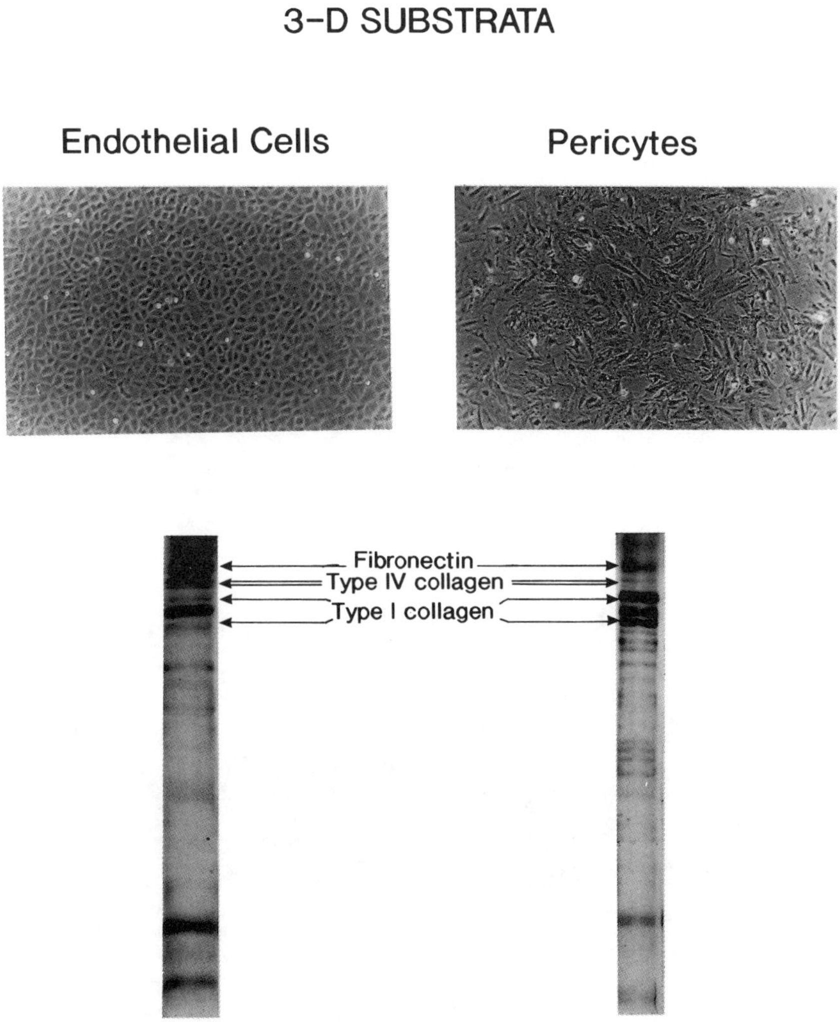

Figure 3. Retinal endothelial cells and pericytes plated within 3-D collagen gels show similar sprouting morphologies and synthesise similar relative amounts of collagen types I and IV. Bar = 150 μm. The specific proteins secreted into the medium were identified as for Fig. 2.

in the re-population of the cell-free areas of substratum, (c) new sudden retraction of a small nodule and surrounding multilayer leads to the formation of larger and denser nodules [4]. Nodules often appear connected by multicellular bridges or sprouts (Fig. 4); these apparently formed by the tensile forces involved in nodule formation.

Nodules are also formed by other cell types such as osteoblasts and smooth muscle cells. It is not known if the nodules in these cultures are formed by the same mechanisms as the pericyte nodules. Nodule formation by pericytes appears to be related to the differentiation of these

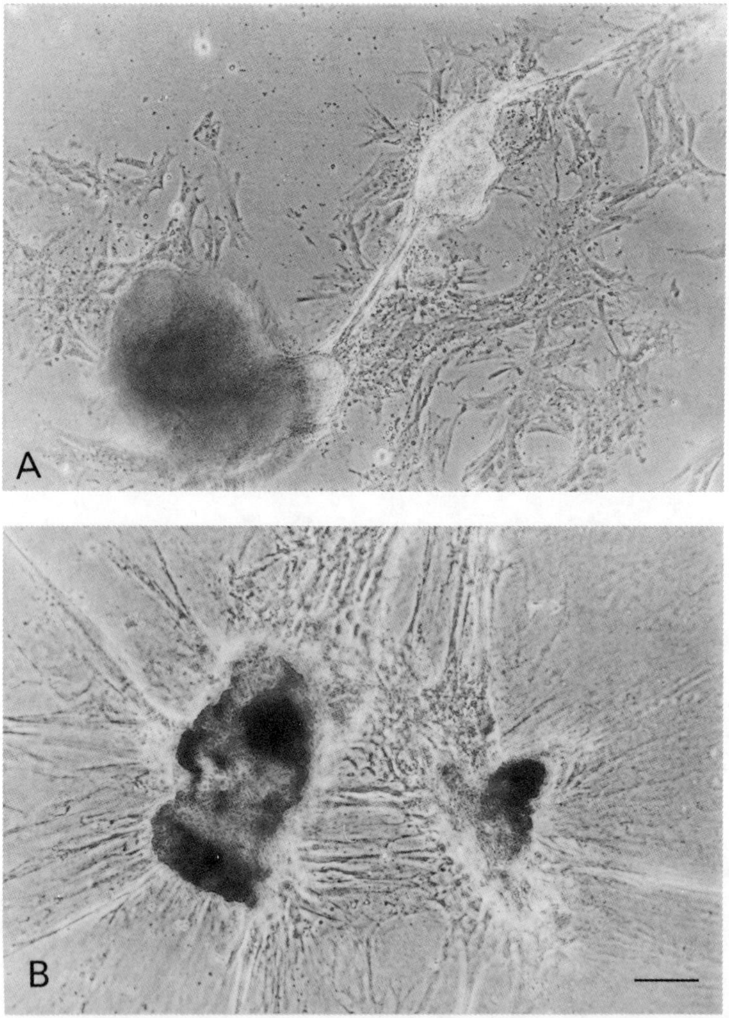

Figure 4. Post-confluent pericyte cultures on a plastic dish (*A*) and on the surface of a 3-D collagen gel (*B*) showing nodules interconnected by cellular bridges or sprouts. Isolated cells, confluent areas and cell-free substratum can also be observed. Calcification of the pericyte nodules occurs more rapidly on a collagen gel substratum than on plastic. Bar = 150 μm.

cells since (a) nodule formation is associated with changes in the synthesis and distribution of matrix macromolecules (see later) and (b) extracellular calcification occurs in the nodules (Fig. 4B; Fig. 5). The nodules contain cells and extracellular matrix. The cells are metabolically active and no signs of cell death have been observed within the nodules. The extracellular matrix contains striated fibrillar collagen and matrix-like vesicles; crystals of hydroxyapatite are first deposited within these structures (Fig. 5A), eventually the whole extracellular matrix

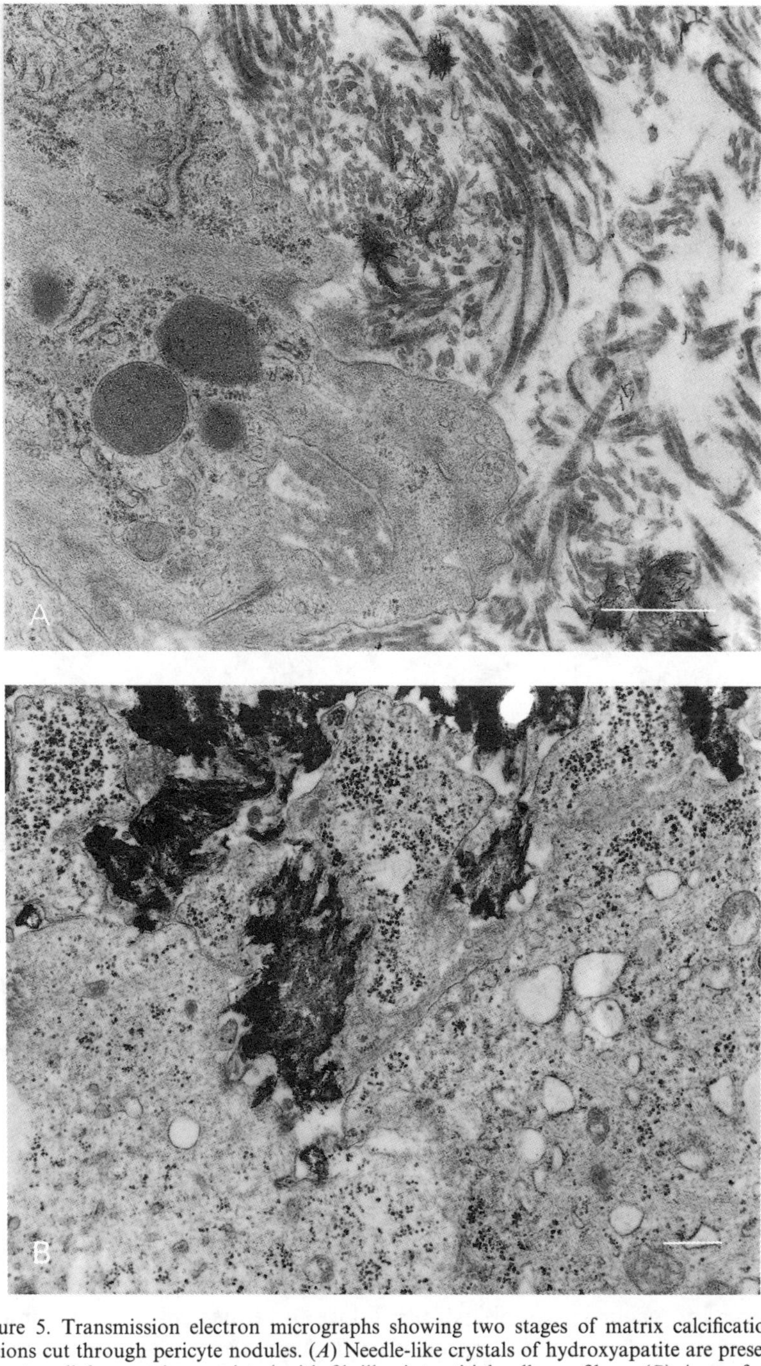

Figure 5. Transmission electron micrographs showing two stages of matrix calcification in sections cut through pericyte nodules. (*A*) Needle-like crystals of hydroxyapatite are present in the extracellular matrix associated with fibrillar interstitial collagen fibres. (*B*) Area of dense extracellular calcification. Bars = 1 μm.

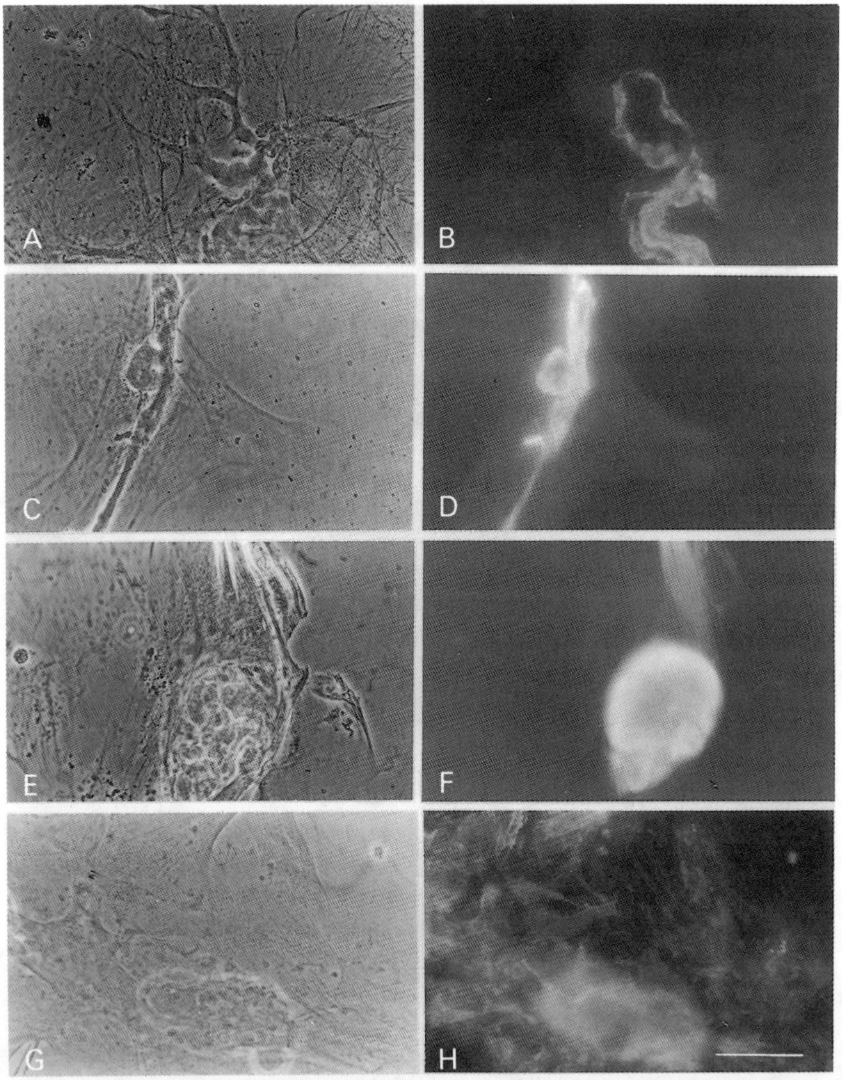

Figure 6. Phase (*A*, *C*, *E*, *G*) and immunofluorescence micrographs (*B*, *D*, *F*, *H*) of post-confluent pericyte cultures stained with specific antibodies for type IV collagen (*A*, *B*), laminin (*C*, *D*), type X collagen (*E*, *F*) and tenascin (*G*, *H*). Laminin, type IV collagen and type X collagen stain preferentially the sprouts and nodules. Tenascin is homogeneously distributed but becomes excluded from calcified nodules [15]. Figs. 6A and 6B are reproduced from [15]. Bar = 150 μm.

within the nodule becomes calcified (Fig. 5B). Calcification has been observed to occur in osteoblast, fibroblast and other cultures. Normally, the presence of serum and exogenous β-glycerophosphate are required [4, 20]. By contrast, pericyte nodules calcify in the presence of 5% or 20% serum and do not require the addition of β-glycerophosphate.

Levamisole, an inhibitor of alkaline phosphatase, inhibits calcification in the pericyte cultures [4].

Post-confluent pericyte cultures comprise cells at various stages of confluence, namely sparse areas, confluent (overlapping) monolayer, multilayer areas, sprouts, nodules and calcified nodules. The distribution and/or synthesis of certain matrix macromolecules varied on these different areas [15]. We have previously mentioned that, by comparison to stellate pericytes, an increase in type IV collagen synthesis was observed in pericyte aggregates plated within a 3-D collagen gel [13]. In aggreement with these results, staining with type IV collagen antibody (Fig. 6A, B) revealed the presence of this molecule on multilayer areas, nodules and sprouts but not in sparse or monolayer areas. Laminin (Fig. 6C, D) and a collagen similar to type X (BRP collagen, [14]) (Fig. 6E, F) were also found to be associated with nodules, sprouts and multilayer areas. Tenascin was homogeneously distributed throughout the cultures (Fig. 6G, H), but was excluded from the calcified nodules [15]. The synthesis of type IV collagen, laminin, BRP collagen and tenascin by pericytes has also been confirmed by biochemical analyses [13–15]. Tenascin and type X collagen have been previously associated with calcification, although their role is not known [21, 22]. An association between laminin, type IV collagen and calcification has not been reported previously.

Conclusions

We have demonstrated that pericytes can synthesise many components of the extracellular matrix including thrombospondin, fibronectin, laminin, tenascin, PAI-1, collagen types I, III and IV and a low Mr collagen which appears to be similar, but not identical, to type X collagen and which we have called BRP collagen [12–15]. We have shown that (a) under certain culture conditions pericytes can resemble endothelial cells both morphologically and in terms of protein biosynthesis, (b) pericytes can differentiate along the osteogenic pathway in vitro and (c) pericyte differentiation in vitro is associated with changes in the synthesis and/or distribution of laminin, type IV collagen, BRP collagen and tenascin. Our results suggest that pericytes in the adult represent pluripotential mesenchymal cells, as previously suggested [3, 23]. The similarities between sprouting endothelial cells and sprouting pericytes indicate that both cell types may participate in the formation of sprouts and new vessels during angiogenesis.

Acknowledgements. We thank Dr A Kwan, Dr W Gibson, Dr G Harrap and Dr R Chiquet-Ehrismann for the gifts of the various antisera used in our study. The financial support of the Cancer Research Campaign, the Medical Research Council and the Arthritis and Rheumatism Council is gratefully acknowledged. We also thank Marjorie Evans for typing the manuscript.

References

1 Bar TH, Wolff JR: The formation of capillary basement membranes during internal vascularisation of the rat's cerebral cortex. Z Zellforsch 1972; 133: 231–248.
2 Schor AM, Schor SL: Tumour angiogenesis. J Pathol 1983; 141: 385–413.
3 Sims DE: The pericyte. A review. Tissue and Cell 1986; 18: 153–174.
4 Schor AM, Allen TD, Canfield AE, Sloan P, Schor SL: Pericytes derived from the retinal microvasculature undergo calcification in vitro. J Cell Sci 1990; 97: 449–461.
5 Kuwabara T, Cogan D: Retinal vascular patterns. VI. Mural cells of the retinal capillaries. Archs Ophthalmol NY 1963; 69: 492–502.
6 Antonelli-Orlidge A, Saunders KB, Smith SR, D'Amore PA: An activated form of transforming growth factor β is produced by co-cultures of endothelial cells and pericytes. Proc Natl Acad Sci 1989; 86: 4544–4548.
7 Sato Y, Rifkin DB: Inhibition of endothelial cell movement by pericytes and smooth muscle cells: activation of a latent transforming growth factor-β1-like molecule by plasmin during co-culture. J Cell Biol 1989; 109: 309–315.
8 Schlingemann RO, Rietveld FJR, de Waal RMW, Ferrone S, Ruiter DJ: Expression of the high molecular weight melanoma-associated antigen by pericytes during angiogenesis in tumours and in healing wounds. Am J Pathol 1990; 136: 1393–1405.
9 Schor AM, Schor SL: The isolation and culture of endothelial cells and pericytes from the bovine retinal microvasculature. A comparative study with large vessel vascular cells. Microvas Res 1986; 32: 21–38.
10 Schor SL: Cell proliferation and migration on collagen substrata in vitro. J Cell Sci 1980; 41: 159–175.
11 Canfield AE, Schor AM, Schor SL, Grant ME: The biosynthesis of extracellular matrix components by bovine retinal endothelial cells displaying distinctive morphological phenotypes. Biochem J 1986; 235: 375–383.
12 Canfield AE, Schor AM, Loskutoff DJ, Schor SL, Grant ME: Plasminogen activator inhibitor type I is a major biosynthetic product of bovine retinal microvascular endothelial cells and pericytes in culture. Biochem J 1989; 259: 529–535.
13 Canfield AE, Allen TD, Grant ME, Schor SL, Schor AM: Modulation of extracellular matrix biosynthesis by bovine retinal pericytes in vitro: Effects of the substratum and cell density. J Cell Sci 1990; 96: 159–169.
14 Canfield AE, Schor AM: Identification and partial characterisation of a low Mr collagen synthesised by bovine retinal pericytes: Apparent relationship to type X collagen. FEBS Lett 1991 (in press).
15 Schor AM, Canfield AE, Sloan P, Schor SL: Differentiation of pericytes in culture is accompanied by changes in the extracellular matrix. In Vitro Cell Develop Biol 1991 (in press).
16 Skalli O, Pelte M-F, Peclet M-C, Gabbiani G, Gugliota P, Bussolati G, Ravazzola M, Orci L: α-smooth muscle actin, a differentiation marker of smooth muscle cells is present in microfilamentous bundles of pericytes. J Histochem Cytochem 1989; 37: 315–321.
17 Fujimoto T, Singer SJ: Immunocytochemical studies of desmin and vimentin in pericapillary cells of chicken. J Histochem Cytochem 1987; 35: 1105–1115.
18 Meyrick B, Reid L: Ultrastructural features of the distended pulmonary arteries of the normal rat. Anat Rec 1979; 193: 71–98.
19 Mazanet R, Franzini-Armstrong C: Scanning electron microscopy of pericytes in rat red muscle. Microvas Res 1982; 23: 361–369.
20 Kellerman O, Buc-Caron M-H, Marie PJ, Lamblin D, Jacob F: An immortalised osteogenic cell line derived from mouse teratocarcinoma is able to mineralise in vivo and in vitro. J Cell Biol 1990; 110: 123–132.
21 Mackie EJ, Thesleff I, Chiquet-Ehrismann R: Tenascin is associated with chondrogenic and osteogenic differentiation in vivo and promotes chondrogenesis in vitro. J Cell Biol 1987; 105: 2569–2579.
22 Poole AR, Pidoux I: Immunoelectron microscopic studies of type X collagen in endochondral ossification. J Cell Biol 1989; 109: 2547–2554.
23 Walter JB, Israel MS: General Pathology. 4th Edition. Churchill-Livingstone, Edinburgh and London 1974; 58.

Angiogenesis: Key Principles - Science - Technology - Medicine
ed. by R. Steiner, P. B. Weisz & R. Langer

Identification of non heparin-binding endothelial cell growth factor from rat myofibroblasts

Noboru Sato[a], Nobuo Tsuruoka[b], Mikio Yamamoto[a], Tatsuro Nishihara[b] and Tamotsu Goto[a]

[a]*Department of Biochemistry, National Defense Medical College, 3-2 Namiki, Tokorozawa 359, and* [b]*Suntory Institute for Biomedical Research, Mishima, Osaka 618, Japan*

Abstract. Myofibroblasts (Mfs) from rat fat tissues produced a potent endothelial cell growth factor (Mf-ECGF). The growth factor activity found in the conditioned media from primary cultures of Mfs, was labile to heat (80°C for 10 min) and proteinase (trypsin), and did not bind to heparin in the presence of 0.2 M NaCl. Mf-ECGF was partially purified 4760-fold with a recovery of 25% from serum-free conditioned media by sequential carboxymethyl (CM) ion-exchange column chromatography and gel filtration. This Mf-ECGF activity was recovered from the 40 kD region of a non-reducing SDS-PAGE, and from the pH region between 6.5 and 7 of isoelectric focusing, with recoveries of 20% and 65%, respectively. These results indicated that a major portion of ECGF activity in the conditioned media was clearly distinct from other well-known endothelial cell growth factors including fibroblast growth factors (FGFs).

Angiogenesis is involved in a wide range of biological and pathological processes including embryogenesis, luteinization, osteogenesis, wound healing, chronic inflammation, certain immune reactions, diabetic retinopathy, arthritis and solid tumor growth [1–5]. However, the mechanisms and factors influencing angiogenesis are not completely understood. We have recently developed a new in vitro angiogenesis model system useful for the study of capillary growth [6]. In this system, myofibroblasts (Mfs) from rat epididymal adipose tissues induce the formation of a capillary-like network from co-cultured microvascular fragments by producing an endothelial cell growth factor and by secreting an extracellular matrix which causes endothelial cells to form a capillary-like structure [6].

In the present study, we partially purified and characterized an endothelial cell growth factor from conditioned media of Mfs, and found that this factor is a 40 kD protein with non-heparin binding property which is distinct from many other endothelial cell mitogens including FGFs.

Materials and methods

Cells

Bovine capillary endothelial cells were isolated from adrenal cortex by the method of Folkman [7], and cultured in DMEN supplemented with 10% FBS. Myofibroblasts (Mfs) and microvascular fragments were obtained from rat epididymal fat pads by collagenase digestion and subsequent sequential fractionations using filter meshes and Percoll as described previously [6].

Preparation of conditioned media

For the preparation of conditioned media, freshly prepared rat Mfs were seeded in 25 roller bottles (Falcon 3035, 1750 cm^2) at a density of 7×10^6 cells per bottle. The cells were cultured until confluency in 500 ml of medium supplemented with 10% FBS for the first two weeks. Then the cells were maintained in 500 ml of serum-free GIT medium (Wako Pure Chemicals, Tokyo, Japan). The conditioned media were collected every 10 days, and twelve such changes were stored at -80°C until use.

Quantitative assay for Mf-ECGF

Mf-ECGF activity was assayed quantitatively using bovine capillary endothelial cells (see Fig. 1). Samples (0.1 ml) were serially diluted in 96-well plates (n = 3) with the GIT medium supplemented with 10 U/ml of heparin and 1 μg/ml of polymyxin B. Then capillary endothelial cells were seeded into each well at 1×10^4 cells per well in 0.1 ml of GIT medium supplemented with 10 U/ml of heparin and 1 μg/ml of polymyxin B. After 3 days' culture, cells were fixed with 0.1% glutaraldehyde and stained with 0.05% methylene blue for 10 min. After washing cells with water, the dye was extracted from the cells with 0.1 ml of 3% HCl, and the absorbance at 620 nm was measured. Activity was expressed in terms of the dilution to obtain 150% stimulation over control cultures. Control cells were cultured in a similar manner, but without addition of the sample. Standard deviations (n = 3) in this assay were <10%.

Characterization of Mf-ECGF

Mf-derived conditioned media were used for the following experiments: (i) Heat stability was tested by treating conditioned medium at

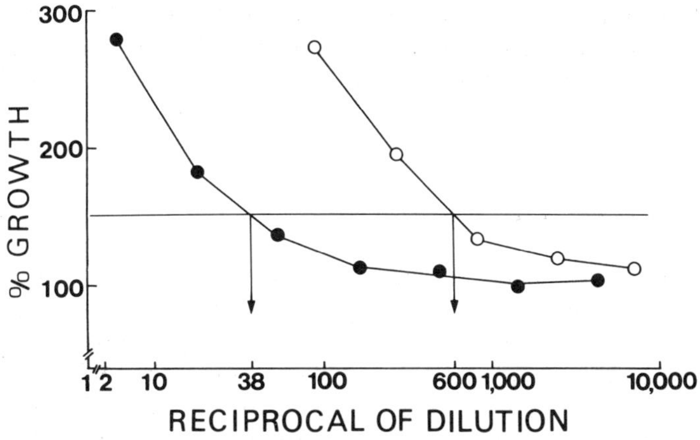

Figure 1. Assay of Mf-ECGF. The percent increase in endothelial cell growth compared to control cultures is shown. (●) Mf-conditioned medium prepared with serum free-GIT medium, (○) The same Mf-conditioned medium after twenty-fold concentration by an ultrafiltration and dialysis against 15 mM phosphate buffer (pH 6.5) overnight. Each point represents the mean of triplicate determinations. S.D. < 10%.

56°C for 30 min or at 80°C for 10 min. GIT medium heated in a similar manner was used as a control. (ii) Trypsin sensitivity was tested by treating medium with 400 μg/ml of the enzyme at 37°C for 2 h, terminating the reaction by the addition of 400 μg/ml of soy bean trypsin inhibitor. GIT medium containing only trypsin inhibitor served as a control. The effects of the variously treated samples on endothelial cell growth were quantitated using capillary endothelial cells as described above (n = 3).

Column chromatographies

Fifteen liters of the Mf-conditioned medium were concentrated with an ultrafiltration apparatus using a YM-10 membrane (Amicon, HP 10-20) at 4°C. The 20-fold concentrated medium (750 ml) was applied to a CM-column (CM-Toyopearl 650 M, Toyosoda Co., Tokyo, Japan), 4.4 × 20 cm, equilibrated with 15 mM phosphate buffer, pH 6.5. The bound materials were eluted with a linear gradient of NaCl (0–0.3 M) at a flow rate of 3 ml/min, and fractions (30 ml/fraction) were collected. Fractions having growth factor activity for capillary endothelial cells were pooled and concentrated 20-fold by ultrafiltration (Centricon 10, Amicon). The active sample (500 μl) was further resolved by gel filtration (Shim-Pack Diol-300, Shimazu Co., Tokyo, Japan), 7.9 × 500 mm, equilibrated with 20 mM Tris-HCl buffer (pH 8.3) containing 0.15 M NaCl and 0.1 M $MgCl_2$, eluting with the buffer at a flow rate of 0.5 ml/min, and collecting 1 ml fractions. The major active fractions obtained in the void-volume were collected and analyzed by SDS-PAGE and isoelectric focusing, and used for the heparin-binding test.

SDS-polyacrylamide gel electrophoresis

SDS-PAGE (10% polyacrylamide gel) was performed according to the method of Weber and Osborn [8]. The gel filtration-purified sample (50 U/10 μl) was allowed to stand at room temperature for 60 min in 0.01 M sodium phosphate buffer (pH 7) containing SDS without a reducing reagent, and then were applied to the gel after being mixed with glycerol and a tracking dye. Electrophoresis was performed at room temperature in a slab gel apparatus, then the gel was cut into 1-mm slices, and the proteins from each slice were eluted with 0.25 ml GIT medium overnight. The extracts were examined for ECGF activity.

Isoelectric focusing

The isoelectric point (pI) of Mf-ECGF was assessed by isoelectric focusing on 4% polyacrylamide gels (PAG plate, LKB) containing carrier ampholytes (pH 5.5–8.5) according to the method of LKB instruction manual. Namely, after applying the gel filtration-purified sample (40 U/8 μl), focusing was performed at a constant current (10 mA) for 4 h on an LKB multiphor electrofocusing apparatus cooled to 4°C. Cathode and anode buffers were 0.1 N NaOH and 0.4 M HEPES, respectively. After electrophoresis, the gel was cut into 5-mm slices, and the proteins were extracted from each slice by incubation in 0.25 ml GIT medium at room temperature for 3 h. Endothelial growth promoting activity extracted from the gel slices was quantitatively determined. Extracts from gel slices without protein served as controls.

Heparin binding

The gel filtration-purified preparation was used for a heparin-binding test after dialysis against the column equilibration buffer. The sample (1620 U) was applied to a heparin-5PW (7.5 × 7.5 mm, Toyosoda Co., Tokyo, Japan) column equilibrated with 15 mM phosphate buffer (pH 6.5) containing 0.2 M NaCl. After flow-through fractions were collected, the column was washed extensively with this buffer, and the bound materials were eluted with 2.0 M NaCl. Flow through, wash, and eluate fractions were quantitatively determined for ECGF activity.

Table 1. Effects of various treatments on Mf-ECGF activity

Treatments		% of residual activity
Control		100
Heat	(56°C, 30 min)	64
	(80°C, 10 min)	−9
Trypsin	(400 μg/ml, 37°C, 2 h)	6

Serum-free conditioned media were variously treated as described in Methods. Mf-ECGF activity was quantitated as described in Fig. 1 (n = 3). S.D. < 10%.

Results

Characterization and purification of Mf-ECGF

Exposure of Mf-conditioned media to heat (80°C for 10 min) or trypsin (2 h) completely destroyed its activity (Table 1), suggesting that the growth factor is a protein. To purify the growth factor, 15 l of the conditioned medium were concentrated 20-fold by ultrafiltration. The concentrate was dialyzed against 15 mM phosphate buffer (85% recovery of activity, see Fig. 1), then applied to a CM-column and eluted with a NaCl concentration gradient. The growth factor eluted at 0.2 M salt as a single peak containing 40% of the initial growth-promoting activity (Fig. 2 and Table 2). Fractions having growth factor activity were pooled and again concentrated by ultrafiltration, recovering 32% of the initial activity of conditioned media. In these procedures, Mf-ECGF

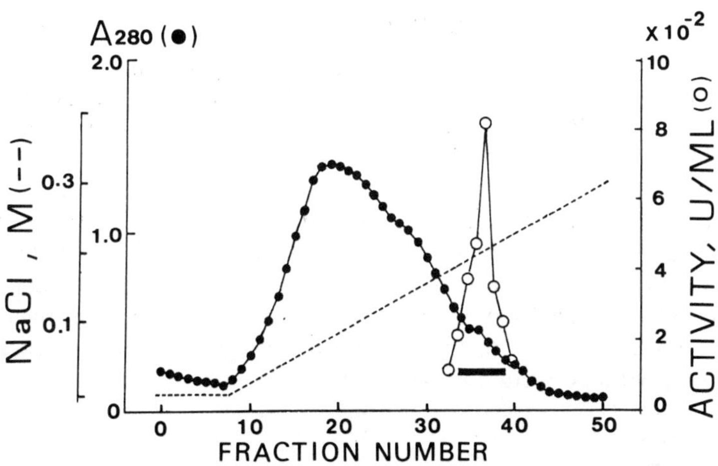

Figure 2. CM-column chromatography of Mf-ECGF. The twenty-fold concentrated Mf-conditioned medium (750 ml) was applied to a CM-column (4.4 × 20 cm) equilibrated with 15 mM phosphate buffer, pH 6.5. The bound materials were eluted with a linear gradient of NaCl (0–0.3 M). (—) indicates active fractions pooled for further studies.

Table 2. Summary of purification of Mf-ECGF

Step	Total activity (units)	Specific activity (units/mg)	Purification (fold)	Yield (%)
Conditioned medium	450,000	10	1	100
CM-column chromatography	383,000	3030	303	40
Gel filtration	117,000	47,600	4760	25

Mf-ECGF activity was determined by the quantitative method as described in Fig. 1. Protein concentration was estimated from the A280 nm (0.1%) = 1.

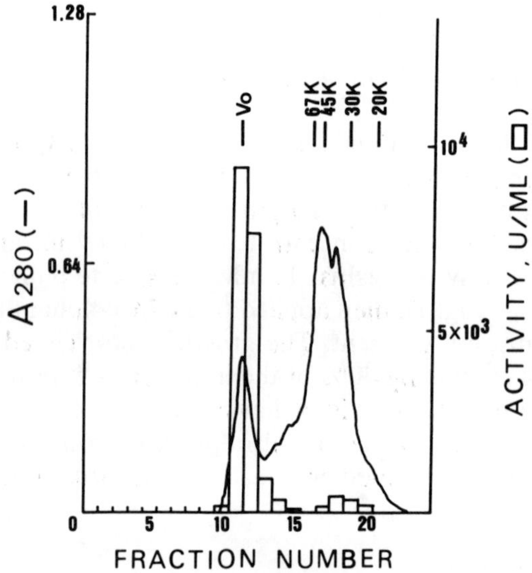

Figure 3. Gel filtration of Mf-ECGF. Active fractions (180 ml) from the CM-column were concentrated to 6 ml and dialyzed against 20 mM Tris-HCl buffer (pH 8.3) containing 0.15 M NaCl and 0.1 M $MgCl_2$. Aliquots (0.5 ml) were applied to Shim-Pack Diol-300 as described in Methods.

was purified about 280-fold. The concentrate was then fractionated by gel filtration. As shown in Fig. 3, a major peak of activity in the void volume and a minor one at 40 kD were recovered. In this step, the specific activity increased to 4.76×10^4 U/mg (4760-fold purification) with a recovery of 25% of the starting materials, when active fractions in the void volume were combined (Table 2).

SDS-PAGE and isoelectric focusing

The partially pure Mf-ECGF in the void-volume fractions of gel filtration was further analyzed by SDS-PAGE and isoelectric focusing.

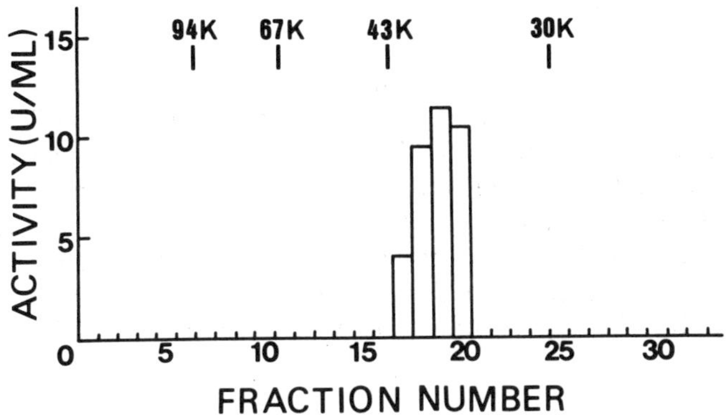

Figure 4. SDS-PAGE of Mf-ECGF. The sample (50 U/10 μl) of the active fraction from the gel filtration shown in Fig. 3 was subjected to SDS-PAGE. After electrophoresis, the gel was cut into 1-mm slices, and the proteins from each slice were eluted with 0.25 ml GIT medium overnight. The extracts were examined for ECGF activity.

Surprisingly, 20% of the applied Mf-ECGF activity survived after SDS-PAGE. And this activity migrated as a 40 kD protein on non-reducing SDS-PAGE (Fig. 4). After isoelectric focusing, proteins were extracted from gel slices and examined for growth factor activity. Mf-ECGF activity was detected only in the fractions with pHs ranging from 6.5–7 (65% recovery of activity, Fig. 5).

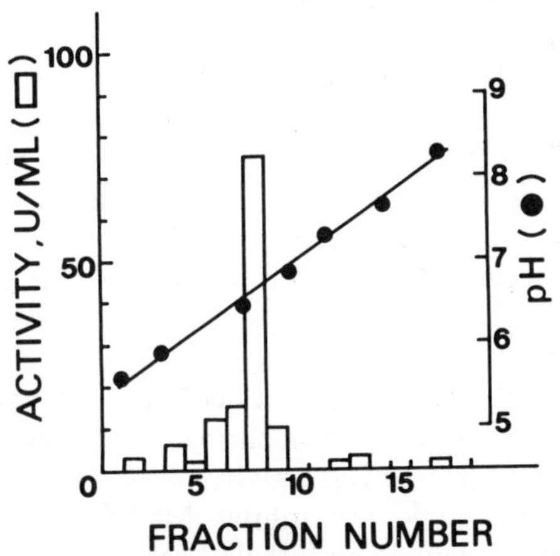

Figure 5. Isoelectric focusing of Mf-ECGF. The sample (40 U/8 μl) was applied to the PAG plate (LKB), and focused for 4 h at 10 mA with cooling. Then the gel was cut into 5-mm slices, and the proteins from each slice were extracted with GIT medium. The extracts were examined for ECGF activity. The isoelectric point was determined by use of pI calibration kits (LKB).

Heparin-binding test

To test if any of this activity might be due to an FGF-like protein released by Mfs, we tested its ability to bind to heparin-Sepharose. Using low ionic conditions (0.2 M NaCl in 0.01 M phosphate buffer) under which FGFs quantitatively bind to the affinity matrix, almost all (98%) Mf-ECGF activity (1580 units) was recovered in the flow-through fractions. On the other hand, no activity was detected in the wash, and eluate fractions with 2.0 M NaCl, effectively ruling out the possibility that Mf-ECGF might be a member of the FGF family of ECGFs.

Discussion

In the present study, we tried to characterize the endothelial cell growth factor from conditioned media of primary Mf cultures. By CM ion-exchange column chromatography and gel filtration, Mf-ECGF was purified 4760-fold with a yield of 25%. In this gel filtration, Mf-ECGF eluted in the void volume, suggesting a MW of >100 kD. However, on SDS-PAGE (non-reducing conditions), this activity migrated to the 40 kD region, indicating that Mf-ECGF is a 40 kD protein and aggregated in the gel filtration. Furthermore, this Mf-ECGF activity could be recovered in the pH region of 6.5–7 after isoelectric focusing. From these observations, we conclude that this factor is a 40 kD, non heparin-binding protein with a pI 6.5–7.

In the past few years, several peptide endothelial cell growth factors have been purified, their primary structures determined, and their genes cloned. These include acidic and basic FGFs [9–11], and endothelial cell growth factors from platelets [12, 13], pituitary folliculostellate cells [14, 15], and U 937 cells [16]. Other endothelial cell growth factors have been purified from fibroblasts [17], and chondrosarcoma [18]. Compared to these ECGFs, this factor has some unique characteristics. Namely, (i) acidic/basic FGFs [9–11] and ECGFs from fibroblasts [17], chondrosarcoma [18], pituitary folliculostellate cells [14, 15], and U 937 cells [16] are known to bind tightly to heparin and are eluted by 0.9-2 M NaCl from the heparin affinity column, whereas our Mf-ECGF did not display any binding to heparin. Moreover, heparin did not potentiate cell growth promoting activity of Mf-ECGF, and anti-bFGF neutralizing antibody could not inhibit Mf-ECGF activity (data not shown). (ii) A platelet-derived endothelial cell growth factor with a MW of 45 kD has been reported recently [13]. Although its MW appears close to ours, it is very acidic (pI 4.6), in contrast to our Mf-ECGF (pI 6.5–7). Moreover, the platelet-derived factor lacks a signal peptide sequence and therefore might not be secreted from cells [12]. (iii) The

MW of transforming growth factor-α, a non heparin-binding angiogenic factor and endothelial cell mitogen [19, 20], is also distinct from that of our factor. Thus Mf-ECGF has some different characteristics from those of other known endothelial mitogens.

It is not yet known whether Mf-ECGF serves as an angiogenic factor in vivo. However, it should be noted that our Mfs were derived from adipose tissues with well-developed microvascular networks. Furthermore, Mfs are generally observed in inflammatory, wounds and solid tumor tissues where capillary blood vessels are actively being formed [21, 22]. The development of antibodies to Mf-ECGF will enable us to understand its role in vivo.

Acknowledgements. The authors would like to thank Dr. A. R. Frackelton, Jr. at Brown University and Roger Williams Cancer Center for critical review of this manuscript, and Mrs. Nakami Sagara for technical assistance.

Abbreviations: CM, carboxymethyl; ECGF, endothelial cell growth factor; FBS, fetal bovine serum; FGF, fibroblast growth factor; kD, kilodalton; Mf, myofibroblast; SDS-PAGE, sodium dodecyl sulfate polyacrylamide gel electrophoresis.

References

1 Auerbach R: In: Lymphokines. E Pick (ed.), Academic Press Inc., New York, 4, 1981, pp 69–88.
2 Folkman J, Cotran R: Int Rev Exp Pathol 1976; 16: 207–248.
3 Folkman J, Klagsbrun M: Science 1987; 235: 442–447.
4 Gimbrone MA Jr, Martin BM, Baldwin WM, Unanue ER, Cotran RS: In: Pathology of the Endothelial cell. Nossal HL and Vogel HJ (eds.), Academic Press, Inc., New York, 1982, pp 3–17.
5 Gullino PM: In: Tissue Growth Factors. Baserga R (ed.), Springer-Verlag, Berlin, 1981, pp 427–447.
6 Sato N, Sawasaki Y, Senoo A, Fuse Y, Hirano Y, Goto T: Microvasc Res 1987; 33: 194–210.
7 Folkman J, Haudenschild C: Nature 1980; 288: 551–556.
8 Weber K, Osborn M: J Biol Chem 1969; 244: 4406–4412.
9 Baird A, Esch F, Mormede P, Ueno N, Ling N, Bohlen P, Ying S-Y, Wehrenberg WB, Guillemin R: Recent Prog Horm Res 1986; 42: 143–205.
10 Lobb RR, Harper JW, Fett JW: Anal Biochem 1986; 154: 1–14.
11 Thomas KA, Gimenez-Gallepo G: Trends Biochem Sci 1986; 11: 81–84.
12 Ishikawa F, Miyazono K, Helman U, Drexler H, Wernstedt C, Hasegawa K, Usuki K, Takaku F, Risau W, Heldin C-H: Nature 1989; 338: 557–562.
13 Miyazono K, Okabe T, Urabe A, Takaku F, Heldin C-H: J Biol Chem 1987; 262: 4098–4103.
14 Gospodarowicz D, Abraham JA, Schilling J: Proc Natl Acad Sci (USA) 1989; 86: 7311–7315.
15 Leung DW, Cachianes G, Kuang W-J, Goeddel DV, Ferrara N: Science 1989; 246: 1306–1309.
16 Keck PJ, Hauser SD, Krivi G, Sanzo K, Warren T, Feder J, Connolly DT: Science 1989; 246: 1309–1312.
17 Satoh T, Kan M, Kato M, Yamane I: Biochim Biophys Acta 1986; 887: 86–93.
18 Shing Y, Folkman J, Sullivan R, Butterfield C, Murray J, Klagsbrun M: Science 1984; 223: 1296–1298.
19 Marquarant H, Hunkapiller MW, Hood LE, Todaro GJ: Science 1984; 223: 1079–1082.
20 Schreiber AB, Winkler ME, Derynck R: Science 1986; 232: 1250–1252.
21 Lipper S, Kahn LB, Reddick RL: Pathol Annu 1980; 15: 409–441.
22 Seemayer TA, Lagace R, Schurch W, Thelmo WL: Pathol Ann 1980; 15: 443–470.

Angiogenesis: Key Principles - Science - Technology - Medicine
ed. by R. Steiner, P. B. Weisz & R. Langer

Differential expression of integrin cell-substratum adhesion receptors on endothelium

Steven M. Albelda

Pulmonary and Critical Care Section, University of Pennsylvania School of Medicine and The Wistar Institute, Philadelphia, Pennsylvania, USA

Abstract. The interactions of the endothelial cells with their underlying extracellular matrix is of fundamental importance to angiogenesis. Therefore, the distribution of integrin cell-substratum adhesion receptors on cultured and in situ endothelial cells was determined. Cultured endothelium from large and small vessels and large vessel endothelial cells in situ expressed integrins that bound to both basement membrane and inflammatory matrix components. In contrast, in situ microvessels primarily expressed basement membrane binding integrins. Alterations in the expression of the integrins on mircovessel endothelial cells may be important for microvascular remodeling during wound healing and/or angiogenesis.

Introduction

The interactions of endothelial cells with their underlying substratum is of fundamental importance to angiogenesis. Inber and Folkman have elegantly argued that the extracellular matrix provides capillary cells with "crucial information that directs the generation of three dimensional form" [1, 2]. Proliferation of microvessel endothelial cells [3] and the process of in vitro angiogenesis (i.e. the formation of capillary-like tubes by endothelial cells) can be strongly influenced by the matrix material on which endothelial cells are cultured [4, 5].

Among the most important molecular mediators of endothelial cell-matrix interactions are the integrins. These molecules represent a family of transmembrane cell adhesion receptors that exist as heterodimers consisting of noncovalently associated α and β subunits [6, 7]. The integrins have been divided into subfamilies on the basis of common β subunits [8]. Although the β subunit defines each subfamily, it is the combination of a particular α and β subunit that determines receptor specificity. There are at least seven receptors of the β_1 subfamily that bind primarily to extracellular matrix molecules such as laminin, collagen and fibronectin. Receptors of the β_2 subfamily are largely restricted to white blood cells and function in cell-cell adhesion and phagocytosis. The β_3 subfamily members interact with matrix proteins such as fibrinogen, fibronectin, vitronectin, and thrombospondin and include the platelet glycoprotein IIb/IIIa complex and the $\alpha_v\beta_3$ "vitronectin" receptor. The purpose of this study was to characterize the repertoire of integrin receptors on both cultured and in situ endothelial cells.

Methods

In order to study the distribution of integrins on cultured and in situ endothelium, a panel of monoclonal antibodies against specific integrin subunits was used. For immunoprecipitation, cultured endothelial cells were metabolically labeled with ^{35}S-methionine or surface labeled with ^{125}I followed by extraction in non-ionic detergent [9]. For immunohistochemistry, human tissue specimens were snap-frozen in liquid nitrogen at the time of resection, subsequently embedded in OCT Compound, and stored at -70°C. Serial frozen sections, 5–7 microns thick, were placed on poly-L-lysine (Sigma Chemical Company, St. Louis, Mo.) coated slides and fixed in cold (-20°C) acetone for 5 min. Following rehydration and blocking of non-specific binding sites with 1% normal serum, the sections were incubated in the first antibody for 60 min. Staining was performed using a three-step avidin biotin complex peroxidase technique (Vectastain Elite, Vector Laboratories, Burlingame, CA). The enzyme was developed using 3,3′-diaminobenzidine (DAB) as the chromogenic substrate.

Results

Endothelial cells derived from human umbilical veins, adult large vessels and fat capillaries expressed a very similar complement of adhesion receptors (Tab. 1) that included subunits from collagen/laminin binding integrins (α_2, α_3, and α_6) and subunits from fibronectin/fibrinogen binding integrins (α_5, α_v, and β_3). Although Kramer et al.

Table 1. Integrin expression

	Integrin subunit							
	Collagen-laminin receptors				Fibronectin-fibrinogen receptors			
Tissue	α_1	α_2	α_3	α_6	α_4	α_5	α_v	β_3
Cultured cells								
Large vessel endothelium	○	●	●	●	○	●	●	●
Capillary endothelium	○	●	●	●	○	●	●	●
In situ cells								
Large vessel endothelium	●	●	●	●	○	●	●	●
Microvessel endothelium	●	*	●	●	○	○	○	○

● = Readily detectable staining; * = Trace staining; ○ = No detectable staining.

[11] have reported that capillary endothelial cells in culture also express the $\alpha_1\beta_1$ laminin/collagen receptor, we have been unable to detect this integrin on cultured endothelium from large or small vessels [9].

Because we have previously observed differences in the types of integrins expressed on cultured cells compared with the same cell types observed in situ [10], the tissue distribution of integrin adhesion receptors on endothelial cells was analyzed with the same panel of monoclonal antibodies. A comparison of integrin expression by endothelial cells in culture and in various tissues (Tab. 1) illustrates a number of points:

1. There are differences in the pattern of integrins expressed by cultured cells when compared to those expressed by the same cells in tissues. For example, endothelial cells in situ express the α_1 laminin-binding subunit while cultured endothelial cells do not.

2. Endothelial cells in situ express multiple integrin subunits required for adhesion to collagen and laminin (α_1, α_2, α_3, and α_6) (Figs. 1 and 2).

3. The expression of the subunits involved in fibronectin/fibrinogen adhesion is quite different on the endothelium of large vessels compared with microvessels (Fig. 1). Large vessel endothelial cells have easily

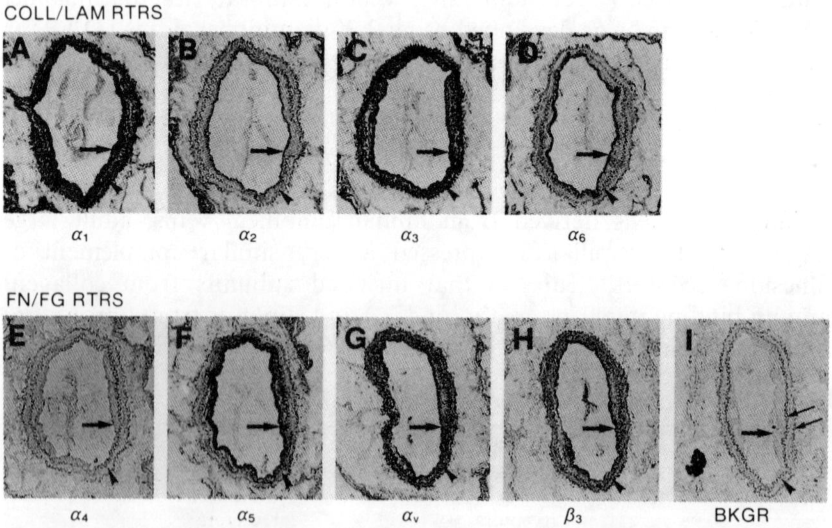

Figure 1. Distribution of integrins on the endothelium from a large blood vessel. Serial frozen sections of a medium sized pulmonary artery were immunostained with monoclonal antibodies directed against various integrin subunits. Top Panel: Distribution of integrin subunits α_1 (A), α_2 (B), α_3 (C) and α_6 (D). Endothelium (large arrows) expressed all these collagen/laminin binding receptors. Bottom Panel: Distribution of integrin subunits α_4 (E), α_5 (F), α_v (G) and β_3 (H). Panel I illustrates background staining of the vessel with a non-reacting monoclonal antibody – the small arrows mark the internal and external elastic lamina of the vessel. The smooth muscle layer is marked with arrowheads. Endothelium (arrows) expressed three out of four integrin subunits that bind primarily to fibronectin (FN) and fibrinogen (FG) – α_5 (F), α_v (G) and β_3 (H). Magnification: ×125.

Lymph Node

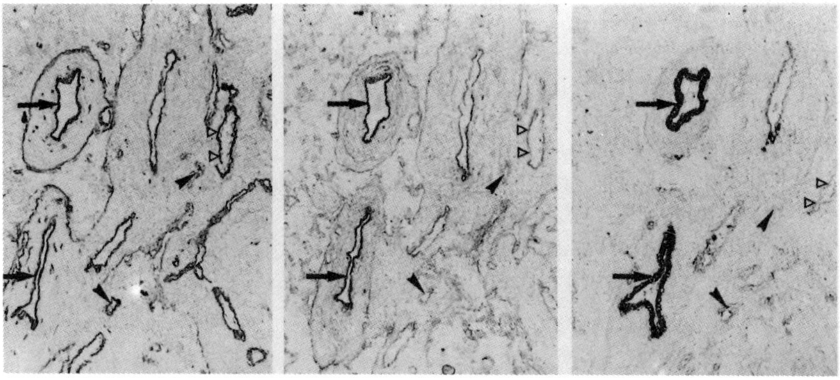

Figure 2. Distribution of integrins on microvessel endothelium. Serial frozen sections of a lymph node were immunostained with monoclonal antibodies against the α_6 (left panel), α_5 (middle panel), and β_3 (right panel) subunits of integrins. The endothelium of the microvessels (solid arrowheads), the larger arterioles (arrows), and high endothelial venules (open arrowheads) all stained strongly with the anti-α_6 laminin receptor antibody. In contrast, there were marked differences in staining intensity with the anti-α_5 and anti-β_3 antibodies. Only the endothelium of the larger vessels (arrows) strongly expressed α_5 and β_3, the subunits of integrins that bind primarily to fibronectin and fibrinogen. Magnification: $\times 125$.

detectable amounts of α_5, α_v, and β_3. In contrast, the expression of these receptors on the microvascular endothelial cells is much lower (Fig. 2).

Discussion

This study has shown that cultured endothelial cells derived from both large vessels and microvessels and the endothelium of large vessels in tissue contain a wide variety of integrin receptors that are able to bind both to the normal constituents of the basement membrane (laminin and Type IV collagen) and to matrix proteins found during inflammation and wound healing (fibronectin, fibrinogen, fibrin, thrombospondin). In contrast, normal microvessels, in situ, express much lower levels of fibronectin/fibrinogen receptors.

The consistently reduced expression of the fibronection/fibrinogen receptors in the in situ microvascular endothelium when compared with either the large vessel endothelium in tissues or the cultured endothelial cells is an especially interesting finding. The fact that capillary endothelial cells in culture produce large amounts of the $\alpha_5\beta_1$ and $\alpha_v\beta_3$ fibronectin/fibrinogen receptors suggests that microvascular cells have the potential to express these receptors under certain conditions. Angiogenesis may be one of these circumstances. The recent demonstration that peptides containing the RGD sequence (a sequence of amino acids known to block integrin mediated adhesion to fibronectin and vitronectin) can inhibit in vitro angiogenesis [12], supports the hypothesis that alterations in microvessel integrin expression may be an important step in new blood vessel formation. In preliminary experiments, we have seen an upregulation of the β_3 subunit on dermal microvessels in chronic inflammatory responses. The upregulation of at least one integrin subunit capable of binding to matrix proteins such as fibrinogen, fibrin, or fibronectin may enable capillary endothelial cells, which normally express these receptors at low levels, to migrate into areas of inflammation or tumour formation to participate in wound healing and angiogenesis.

References

1 Albelda SM, Daise M, Levine EN, Buck CA: Identification and characterization of cell-substratum adhesion receptors on cultured human adult large vessel endothelial cells. J Clin Invest 1989; 83: 1992–2002.
2 Albelda SM, Buck CA: Integrins and other cell adhesion molecules. FASEB J 1990; 4: 2868–2880.
3 Albelda SM: Endothelial and Epithelial Cell Adhesion Molecules. Am J Respir Cell Mol Biol 1991; 4: 195–203.
4 Hynes RO: Integrins: a family of cell surface receptors. Cell 1987; 48: 549–554.
5 Kramer RH, Cheng YF, Clyman R: Human microvascular cells use β_1 and β_3 integrin receptor complexes to attach to laminin. J Cell Biol 1990; 111: 1233–1243.
6 Albelda SM, Mette SA, Elder DA, Stewart R, Damjanovich L, Herlyn M, Buck CA: Integrin cell-substratum adhesion receptor distribution in malignant melanoma: Association of β_3 integrin subunit expression with tumor progression. Cancer Res 1990; 50: 6757–6744.
7 Form DM, Pratt BM, Madri JA: Endothelial cell proliferation during angiogenesis. Lab Invest 1986; 55: 521–530.
8 Ingber DE, Folkman J: How does extracellular matrix control capillary morphogenesis? Cell 1989; 58: 803–805.
9 Ingber DE, Folkman J: Mechanochemical switching between growth and differentiation during fibroblast growth factor-stimulated angiogenesis in vitro: Role of extracellular matrix. J Cell Biol 1989; 109: 317–330.
10 Kubota Y, Kleinman HK, Martin GR, Lawley TJ: Role of laminin and basement membrane in the morphological differentation of human endothelial cells into capillary-like structures. J Cell Biol 1988; 107: 1589–1598.
11 Montesano R, Orci L, Vassalli P: In vitro rapid organization of endothelial cells into capillary-like networks is promoted by collagen matrices. J Cell Biol 1983; 97: 1648–1652.
12 Nicosia RF, Bonanno E: Inhibiton of angiogenesis in vitro by arg-gly-asp-containing synthetic peptide. Amer J Pathol 1991; 138: 829–833.

Angiogenesis: Key Principles - Science - Technology - Medicine
ed. by R. Steiner, P. B. Weisz & R. Langer

Modulation of extracellular matrix receptors (integrins) on human endothelial cells by cytokines

Paola Defilippi, Chiarella Bozzo, Massimo Geuna[a], Paola Rossino, Lorenzo Silengo and Guido Tarone

Department of Genetics, Biology and Medical Chemistry, [a]Department of Biomedical Sciences and Human Oncology, Section of Pathology, University of Torino, Torino, Italy

Integrins are a family of receptor proteins that mediate adhesive interactions [1–3]. Human umbilical vein endothelial cells (HEC) express five distinct integrin complexes that mediate their interaction with basal membrane components [4–7]. We found that tumor necrosis factor alpha (TNFalpha) and interferon gamma (IFNgamma), two cytokines known to trigger the inflammatory response and to modify the HEC properties [8], induce selective alterations in the integrin expression. While TNFalpha induces the expression of a laminin-collagen receptor (integrin alpha-1/beta-1), not expressed in unstimulated HEC, a combination of TNFalpha and IFNgamma downregulates the vitronectin receptor (integrin alpha-v/beta-3). These modifications require a minimum of 24 h of exposure to cytokines and reach maximal levels between 48 and 72 h. Such alterations change the adhesive properties of the cells in a selective manner since the expression of other integrins (alpha-3/beta-1 and alpha-5/beta-1) are not altered, and are likely to play a role in the late phase of the inflammatory response.

Materials and methods

Cell culture and treatment

HEC were isolated from umbilical vein as described [9]. Confluent cultures were treated between passage 2 and 6 with 5 ng/ml TNFalpha, kind gift of Dr. M. Brockhaus (Hoffman La Roche, Basel), and/or 100 IU/ml IFNgamma, kind gift of Dr. G. Garotta (Hoffman La Roche, Basel). The treatments were performed in medium 199 supplemented with 20% FCS, in the presence of 5 ng/ml of recombinant basic fibroblast growth factor as trophic factor. For cytofluorimetric analysis, treated and untreated cells were detached with incubation in 5 mM ethylene diamine tetraacetate (EDTA) in PBS, and washed twice in culture medium. The cells were then incubated for 1 h at 4°C in PBS, 0.1 mM EDTA, 1 mg/ml bovine serum albumin, with the appropriate dilution of the specific antibodies. After washing, the cells were incubated 45 min with fluorescein-labeled affinity purified secondary antibodies, and analyzed on the flow cytometer Facs-Star Becton Dickinson, equipped with 5 watt argon laser at 488 nm.

Antibodies to integrins

The polyclonal antibodies to the fibronectin receptor, which recognizes beta-1 integrin, and to the beta-3 integrin were prepared in our laboratory as previously described in [10]. Mouse monoclonal antibodies to human integrin alpha-1 (TS2/7), alpha-5 (Mab B1E5), and alpha-3 (J143) were respectively kind gifts of Dr. M. Hemmler, Dr. C. Damsky and Dr. L. Old. Labeling of cells and immunoprecipitation were performed as described [10].

Results and discussion

Flow cytometry analysis indicated that HEC express high levels of alpha-3/beta-1, and alpha-5/beta-1 integrins, while alpha-1/beta-1 complex is absent. Inflammatory cytokines are known to activate HEC and to modify their properties [8]. By analyzing HEC cultures exposed to TNFalpha and IFNgamma, we found that treatment with these cytokines alters the expression of integrins on the cell surface. As shown in Fig. 1, exposure of HEC to 5 ng/ml of TNFalpha induces the expression of the integrin alpha-1/beta-1, a laminin and collagen receptor, which is not present in untreated HEC in vitro, and is also absent on umbilical vein endothelium in vivo [11]. In contrast, the treatment with TNFalpha does not modify the profile of alpha-3 and alpha-5 subunit on the cell surface (Fig. 1). The results obtained by flow cytometry were confirmed by immunoprecipitation of extracts of ^{35}S methionine-labeled HEC. As shown in Fig. 2A, HEC treated with TNFalpha express the alpha-1 subunit, which is absent on untreated HEC. The induction of alpha-1 subunit by TNFalpha expression was rather specific, since neither IFNgamma (Fig. 2A) nor IL-1beta (not shown) were active. These two latter cytokines have effects on the activation of HEC similar to those observed with TNFalpha [8]. The induction of the alpha-1 subunit was appreciable after 24 h of treatment and reached maximal levels at 48 h. To analyze the mechanisms by which TNFalpha leads to the induction of alpha-1, we tested substances known to trigger intracellular signalling

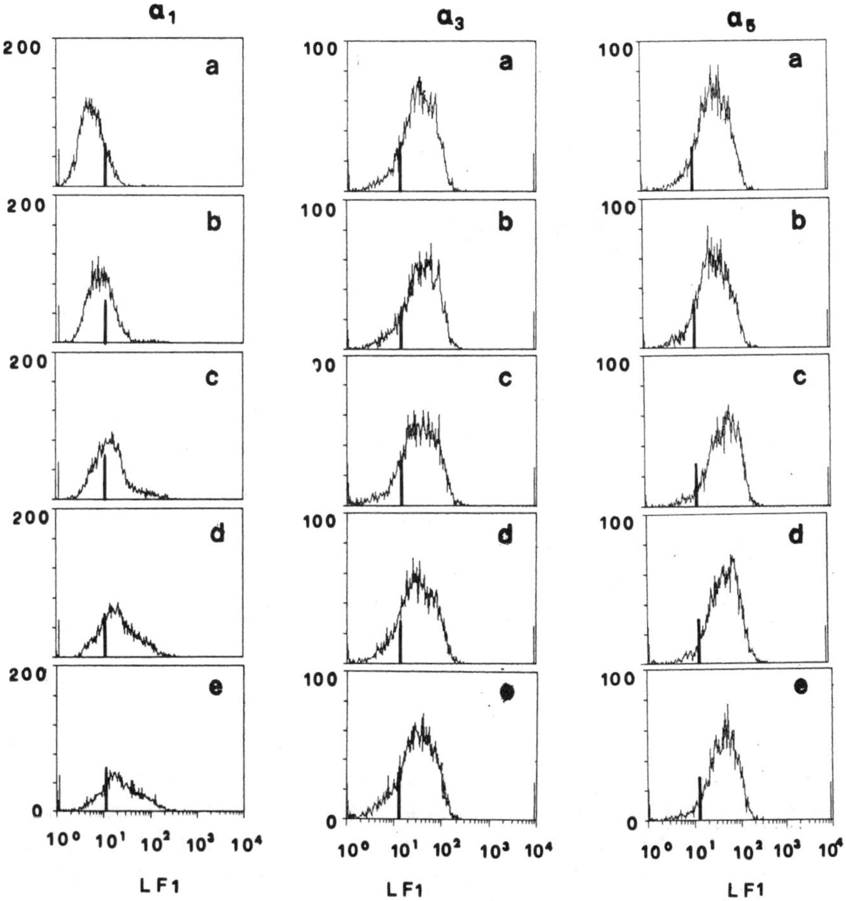

Figure 1. Flow cytometry analysis of integrin expression on HEC. Untreated HEC (*a*) and HEC treated with TNFalpha (5 ng/ml) for 24 h (*b*), 48 h (*c*), 72 h (*d*), 96 h (*e*), were detached, split in three samples and stained in suspension with the monoclonal antibodies TS2/7 (alpha-1), J143 (alpha-3), B1E5 (alpha-5), and analyzed as described in Material and Methods. The data are expressed as cell number (y axis) plotted as a function of fluorescene intensity (x axis) and are representative of three separate experiments. The vertical bar represents the boundary between the negative values (on the left), and the positive values (on the right).

pathways. We found that the tumor promoter PMA was active, suggesting the involvement of signalling mechanism through the protein kinase C [11]. In agreement with this hypothesis, the combination of TNF-alpha and PMA have a synergistic effect on alpha-1 induction [11]. Integrin alpha-1/beta-1 binds both collagen and laminin [12]; an interesting functional property of the alpha-1/beta-1 integrin is the ability to bind the P1 fragment of laminin, a region distinct from that recognized by other laminin receptors [13]. We have found that the induction of

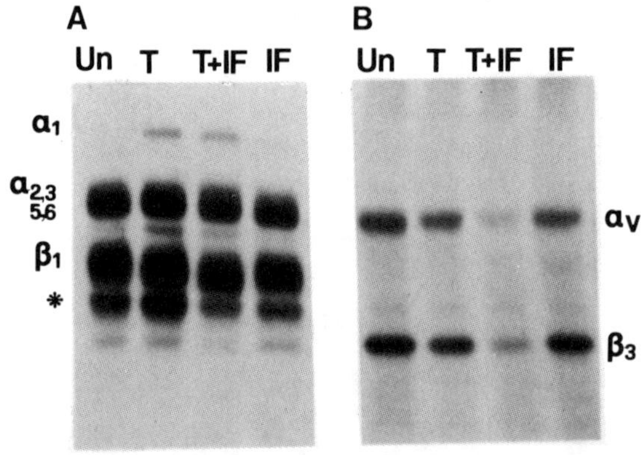

Figure 2. Immunoprecipitation of beta-1 and beta-3 integrins from HEC. Confluent HEC, untreated (UN), or treated 72 h with TNFalpha (T), IFNgamma (IF), and the combination of TNFalpha and IFNgamma (T + IF), were labeled for the last 15 h of culture with 40 μC/ml ^{35}S methionine. Cells were harvested, detergent extracted and each culture was split in two samples and subjected to immunoprecipitation with anti-fibronectin receptor (*A*), and beta-3 (*B*) antibodies. The anti-fibronectin receptor recognizes beta-1 integrin and coprecipitates all the alphas complexed to beta-1, which are indicated in the figure. The radioactive antigens were separated by 6% SDS-PAGE in non-reducing conditions. The asterisk indicates the immature form of the beta-1 subunit.

alpha-1 integrin changes the adhesive properties of HEC, by increasing the adhesiveness to collagen, laminin and laminin fragment P1, while adhesion to fibronectin and laminin fragment E8 is unchanged [11]. Furthermore, we have found a second specific effect of cytokines on integrin expressed on HEC. Treatment of HEC with a combination of TNFalpha and IFNgamma downregulates the expression of the integrin complex alpha-v/beta-3, also known as vitronectin receptor [10]. As shown in Fig. 2B, the vitronectin receptor is downregulated only by the combined treatment with TNFalpha and IFNgamma, while the two cytokines alone are inactive. In this case the two cytokines caused a downregulation of this receptor by specifically affecting the synthesis of the beta-3 subunit [10]. As a consequence of the decreased expression of the vitronectin receptor, cytokine-treated HEC show decreased adhesion to vitronectin, but adhere normally to fibronectin [10]. These results show that cytokines affect the expression of integrin on endothelial cells in a selective manner, indicating that the alteration of the endothelium-matrix interactions is likely to play an important role in the vascular response during inflammation.

Acknowledgements. We would like to thank Dr. M. Hemler, Dr. C. Damsky and Dr. L. Old for the gift of the monoclonal antibodies. We also thank G. Truffa and A. Navini for help in the experiments. This work was supported by the National Research Council. P.D. is a recipient of a fellowship from Fidia Pharmaceuticals Co.

References

1 Ruoslahti E, Pierschbacher MD: New perspectives in cell adhesion: RGD and integrins. Science 1987; 238: 491–497.
2 Hynes RO: Integrins: a family of cell surface receptors. Cell 1987; 48: 549–555.
3 Hemler ME: VLA proteins in the integrin family: structures, functions and their role on leukocytes. Ann Rev Immunol 1990; 8: 365–400.
4 Albelda SM, Daise M, Levine EM, Buck CA: Identification and characterization of cell-substratum adhesion receptors on cultured human endothelial cells. J Clin Invest 1989; 83: 1992–2002.
5 Cheng Y-F, Kramer RH: Human microvascular endothelial cells express integrin-related complexes that mediate adhesion to the extracellular matrix. J Cell Physiol 1989; 139: 275–286.
6 Tarone G, Stefanuto G, Mascarello P, Defilippi P, Altruda F, Silengo L: Expression of receptors for extracellular matrix proteins in human endothelial cells. J Lipids Mediators 1990; 2: S45–S53.
7 Kramer RH, Cheng YF, Clyman R: Human microvascular endothelial cells use beta-1 and beta-3 integrin receptor complexes to attach to laminin. J Cell Biol 1990; 111: 1233–1243.
8 Pober JS: Cytokine-mediated activation of vascular endothelium. Am J Pathol 1988; 133: 426–433.
9 Gimbrone MA: Culture of vascular endothelium. In T. H. Spaet (ed), Progress in Hemostasis and Thrombosis. Grune and Stratton, N.Y. 1976, Vol. 3, pp. 1–28.
10 Defilippi P, Truffa G, Stefanuto G, Altruda F, Silengo L, Tarone G: Tumor necrosis factor alpha and interferon gamma modulate the expression of the vitronectin receptor (integrin beta-3) in human endothelial cells. J Biol Chem 1991; 266, 12: 7638–7645.
11 Defilippi P, van Hinsbergh V, Bertolotto A, Rossino P, Silengo L, Tarone G: Differential distribution and modulation of expression of alpha-1/beta-1 integrin in human endothelial cells. J Cell Biol 1991, in press.
12 Clyman RI, Turner DC, Kramer RH: An alpha-1/beta-1 like integrin receptor on rat aortic muscle cells mediate adhesion to laminin and collagen types I and IV. Arterosclerosis 1990; 10: 402–410.
13 Hall D, Reichardt LF, Crowley E, Holley B, Moezzi H, Sonnenberg A, Damsky CH: The alpha-1/beta-1 and alpha-6/beta-1 integrin heterodimers mediate cell attachment to distinct sites on laminin. J Cell Biol 1990; 110: 2175–2184.

Angiogenesis: Key Principles - Science - Technology - Medicine
ed. by R. Steiner, P. B. Weisz & R. Langer

Possible mechanisms of type I collagen-induced vascular tube formation

Christopher J. Jackson, Kathryn Jenkins and Les Schrieber

Professorial Department of Rheumatology, Sutton Rheumatism Research Laboratory, Royal North Shore Hospital (RNSH), St Leonards, Sydney, Australia

Abstract. We investigated the effect of type I collagen on endothelial behaviour following its contact with the apical versus basal surface of cultured human endothelial cells. When endothelial cells were plated onto type I collagen they attached via their basal surface and formed a confluent monolayer. However, when type I collagen (100μg/ml) was added directly to the growth medium, so that it made contact with the apical surface of endothelial cells, it induced rapid capillary-like tube formation. Possible mechanisms were assessed using a) polyclonal (anti-VLA-2) and monoclonal (AK7) antibodies to different epitopes on the $\alpha_2\beta_1$ integrin receptor for collagen and b) drugs (chlorpromazine and trifluoperazine) that inhibit protein kinase C activity. Both anti-VLA-2 and AK7 (1–50 μg/ml) showed a dose-dependent inhibition of tube formation and cell attachement. At 50 μg/ml, anti-VLA-2 completely inhibited tube formation whereas AK7 caused only partial inhibition (< 50%). By contrast, AK7 was a more potent inhibitor of cell attachement than anti-VLA-2. Both chlorpromazine and trifluoperazine prevented tube formation. Conclusions: 1) The $\alpha_2\beta_1$ integrin receptor plays a role in both endothelial cell attachment and the induction of tube formation by type I collagen. 2) Protein kinase C may be involved in collagen-induced tube formation.

Introduction

The formation of vascular tubes in vitro from endothelia has been observed by many workers [1–3]. Ingber and Folkman [4] have suggested that at least two extracellular cues, one soluble and the other insoluble, are required for purified endothelial cells to form vascular tubes. The insoluble extracellular matrix component, collagen, can promote tube formation [5, 6], however, the mechanisms which control this process are poorly understood. We have demonstrated that endothelial cell polarity plays a prominent role in type I collagen-induced tube formation [7]. In this study, we investigated possible mechanisms of collagen-induced tube formation, involving the role of the $\alpha_2\beta_1$ integrin receptor on the endothelial cell surface and protein kinase C mediated signal transduction.

Methods

Cell isolation

Microvascular endothelial cells were obtained from human neonatal foreskins (FS) [8] and grown in BM-86 Wissler medium (Boehringer Mannhein, GmbH, Germany) supplemented with 40% bovine serum, 100 μg/ml endothelial cell growth supplement (ECGS) and 50 μg/ml heparin. ECGS was a kind gift of Mr. T. To (Sutton Laboratory, RNSH, Sydney, Australia). Large-vessel endothelium was derived from human umbilical vein (HUVE) as described by Jaffe [9] and grown in medium 199 containing 20% bovine serum plus 50 μg/ml ECGS and 50 μg/ml heparin.

Antibodies and drugs

Two IgG class antibodies, directed against the $\alpha_2\beta_1$ integrin receptor, were used at concentrations of 1–50 μg/ml. AK7 is a monoclonal antibody directed against a site on the α_2 chain and anti-VLA-2 is a polyclonal antibody binding to both chains. An IgG antibody without specificity to the $\alpha_2\beta_1$ integrin was used as the control. All antibodies were kindly provided by Dr. M. Berndt, Westmead Hospital, Sydney, Australia. Two drugs, chlorpromazine (Sigma, St. Louis, USA) and trifluoperazine (Sigma), were used at concentrations of 1–100 μg/ml, to inhibit the action of protein kinase C activity.

Tube formation protocol

Tube formation experiments were performed as previously described [7]. Briefly, endothelial cells were plated onto collagen- or gelatin-coated 35-mm dishes ($3-4 \times 10^5$ cells/dish) and incubated overnight in growth medium to form a confluent monolayer. The next day, soluble type I collagen (100 μg/ml) was added. Endothelial cells were pre-incubated with antibodies, drugs and/or phorbol myristate acetate (PMA) for 1–2 h before the addition of type I collagen. Tube formation was assessed using phase contrast microscopy (Nikon Diaphot, Japan).

Cell attachment assay

Confluent endothelial cultures were trypsinised, washed in HBSS, pelleted by centrifugation and resuspended in serum-free medium, containing 1% BSA. The cell suspension was incubated with the antibodies or protein kinase C inhibitors for 30 min, at 4°C, with end-to-end rotation. The cells were then seeded, in triplicate, into the wells of a 96-well plate, to give a final density of 2×10^4 cells per well. The wells were previously coated with fibronectin (control) or adsorbed type I collagen [7]. The cells were incubated for 2 h, in a humidified incubator at 37°C with 95% air/5% CO_2. They were then washed three times with HBSS, using gentle agitation, to remove the non-adherent cells. Cell attachment was quantitated using a methylene blue uptake assay [10]. Briefly, the attached cells were fixed in 10% formol saline, stained with 1% methylene blue for 30 min and washed four times in 0.1 M borate buffer. The colour was released from the cells by adding 0.1 ml ethanol: 0.1 M HCl (1:1(v/v)). Absorbance was read at 620 nm, against a blank column, in a microplate reader.

Results

Tube formation

When endothelial cells were plated onto type I collagen or gelatin they attached via their basal surface and formed a confluent monolayer. However, when type I collagen was subsequently added directly to the growth medium, so that it made contact with the apical surface of endothelial cells it induced rapid capillary-like tube formation, within 6–8 h. The addition of 10 ng/ml PMA accelerated tube formation (2–4 h) (Fig. 1). Similar observations were made for both HUVE and

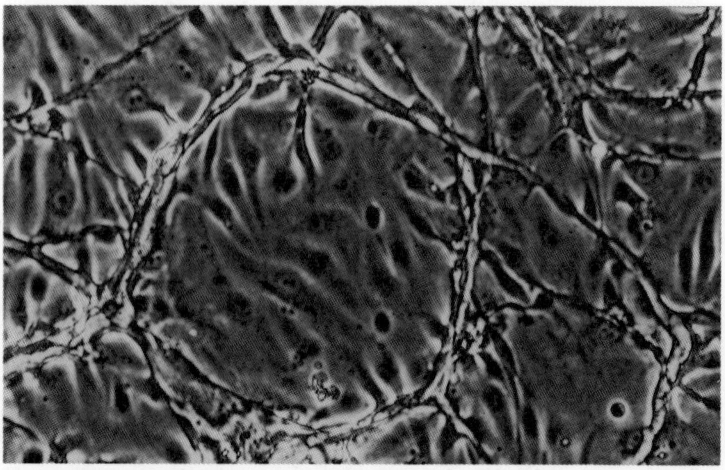

Figure 1. Foreskin endothelium cultured in the presence of 10 ng/ml PMA form tube-like structures 3 h after the addition of 100 μg/ml type I collagen. Note that the monolayer remains intact. (×140)

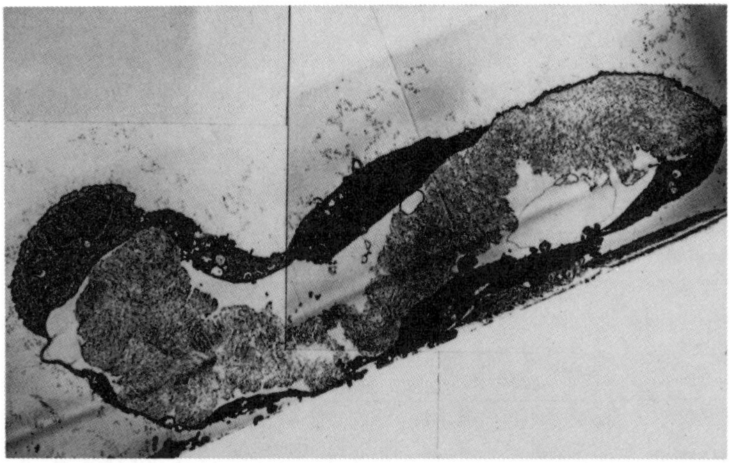

Figure 2. Transmission electron micrograph of an intact tube formed by human umbilical vein endothelium, 24 h after the addition of 100 μg/ml type I collagen to the growth medium. Note the presence of collagen fibrils within the lumen of the tube. (× 1400).

FS endothelium. Electron microscopy verified that the endothelial cells had formed intact tubes which enclosed a lumen and revealed that the lumina of these tubes contained collagen fibrils (Fig. 2.).

Effect of antibodies to the $\alpha_2\beta_1$ integrin

Tube formation. The polyclonal antibody, anti-VLA-2, demonstrated a dose-dependent inhibition of vascular tube formation between 1 and 50 μg/ml. At a concentration of 10 μg/ml, anti-VLA-2, was a potent inhibitor (>90%). The monoclonal antibody, AK7, demonstrated only weak inhibition of tube formation when used at concentrations of 5–50 μg/ml, with maximal inhibition of less than 20% at 50 μg/ml. These results are summarised in Tab. 1. The IgG control antibody did not affect tube formation when used at concentrations between 1 and 50 μg/ml.

Cell attachement

The monoclonal antibody, AK7, prevented endothelial cell attachment to type I collagen at low concentrations. At a concentration of 1 μg/ml, AK7 inhibited attachment of endothelium by 26.3 ± 7.1% (n = 3), compared to control antibody. No further inhibition was observed at higher concentrations. AK7 failed to prevent the attachment

Table 1. Inhibition of collagen-induced tube formation by the antibodies, AK7 and anti-VLA-2. FS endothelial monolayers were pre-incubated with antibody (1–50 μg/ml) for 1 h before the addition of type I collagen (100 μg/ml). Tube formation was assessed morphologically by phase microscopy and expressed semi-quantitatively as % tube formation compared to control (without antibody), represented as follows;
+ + + + >90% tube formation (<10% inhibition)
+ + + 50–90% tube formation
+ + 10–50% tube formation
+ <10% tube formation (>90% inhibition)

Conc. (μg/ml)	AK7	Anti-VLA-2
1	+ + + +	+ + + +
5	+ + +	+ +
10	+ + +	+
50	+ + +	+

of endothelium to fibronectin, which was used as a control. The polyclonal antibody, anti-VLA-2, did not prevent endothelial cell attachment to type I collagen at low concentrations (1–5 μg/ml). At a concentration of 50 μg/ml, anti-VLA-2 inhibited cell attachment by $27.3 \pm 7.8\%$ (n = 3). Inhibition did not occur on fibronectin coated plates.

Effect of protein kinase C inhibitors on tube formation

Both chlorpromazine and trifluoperazine inhibited tube formation by more than 50% of control when used at a concentration of 0.01 mM. At lower concentrations (0.001 mM), only very slight inhibition was seen. At higher concentrations (0.1 mM), no tube formation was observed. However, the endothelial monolayer was disrupted, suggesting that at high concentrations the drugs have a toxic effect on the cells.

Discussion

The ability of type I collagen to induce vascular tube formation is now established. Montesano et al. [6] have shown that capillary endothelial cells reorganised into a network of anastomosing capillary-like tubes when sandwiched between two layers of type I collagen. Cotta-Pereira et al. [11] have shown that "sprouting" cultures of bovine aortic endothelial cells produce a predominance of type I procollagen. Recently, Howard et al. [12] have reported that populations of capillary endothelial cells, which spontaneously form capillary-like tubes, exhibit type I collagen synthesis during the initial sprouting phase.

We have reported that type I collagen supports monolayer growth when endothelial cells attach via their basal surface, but induces tube formation when it contacts the apical side of the endothelial cell [7]. Here, we examine the role of the $\alpha_2\beta_1$ integrin receptor in these processes. $\alpha_2\beta_1$ is expressed on the cell surface of endothelial cells [13, 14] and binds to type I collagen.

Using two antibodies directed against different epitopes on the $\alpha_2\beta_1$ integrin receptor we have shown that this receptor participates in both endothelial cell attachment via the basal side and collagen-induced tube formation via the apical surface. Whereas the polyclonal antibody anti-VLA-2 was a potent inhibitor of collagen-induced tube formation, the monoclonal antibody AK7 was more effective in preventing endothelial cell attachment via the basal side of the cell. These paradoxical effects are difficult to explain. It is possible that different epitopes on the $\alpha_2\beta_1$ receptor are involved in the two processes. Further studies using monoclonal antibodies to different epitopes on the $\alpha_2\beta_1$ integrin should help resolve this question.

$\alpha_2\beta_1$ acts as a collagen receptor on some cells, e.g. fibroblasts and platelets, and a collagen/laminin receptor on other cells [15], including endothelial cells [16]. Because there is no obvious structural diversity in the $\alpha_2\beta_1$ molecule, Elices and Hemler [15] suggested that other factors such as cellular signalling or cell activation may contribute to the cell-type binding specificity. Whether such factors account for the differential effects of type I collagen on the apical versus basal side of endothelium, within the same cell, remains to be investigated.

Protein kinase C, an enzyme that is crucial for signal transduction, is thought to play a pivotal role in angiogenesis. Protein kinase C activation is involved in the migration and invasion of endothelial cells in vitro [17]. Montesano et al. [18] have suggested a role for protein kinase C in the formation of vascular tubes. Two lines of evidence from this study indicate that protein kinase C is involved in the signal transduction of collagen-induced tube formation. First, the addition of PMA accelerated tube formation. PMA is a potent stimulator of protein kinase C activity [19]. Second, tube formation was inhibited by the protein kinase inhibitors, chlorpromazine and trifluoperazine.

References

1 Ingber DE, Folkman J: Mechanochemical switching between growth and differentiation during fibroblast growth factor-stimulated angiogenesis in vitro: role of extracellular matrix. J Cell Biol 1989; 109: 317–330.

2 Maciag T, Kadish J, Wilkins L, Stemerman MB, Weinstein R: Organizational behavior of human umbilical vein endothelial cells. J Cell Biol 1982; 94: 511–520.

3 Folkman J, Haudenschild C: Angiogenesis in vitro. Nature 1980; 288: 551–556.

4 Ingber D, Folkman J: How does extracellular matrix control capillary morphogenesis?. Cell 1989; 58: 803–805.

5 Madri JA, Williams SK: Capillary endothelial cell cultures: phenotypic modulation by matrix components. J Cell Biol 1983; 97: 153–165.

6 Montesano R, Orci L, Vassalli P: In vitro rapid organization of endothelial cells into capillary-like networks is promoted by collagen matrices. J Cell Biol 1983; 97: 1648–1652.
7 Jackson CJ, Jenkins KL: Type 1 collagen fibrils promote rapid vascular tube formation upon contact with the apical side of cultured endothelium. Exp Cell Res 1991; 192: 319–323.
8 Jackson CJ, Garbett PK, Nissen B, Schrieber L: Binding of human endothelium to Ulex Europaeus I-coated dynabeads: application to the isolation of microvascular endothelium. J Cell Sci 1990; 96: 257–262.
9 Jaffe EA, Nachman RL, Becker CG, Minick CR: Culture of human endothelial cells derived from umbilical veins: identification by morphologic and immunologic criteria. J Clin Invest 1972; 52: 2745–2756.
10 Oliver MH, Harrison NK, Bishop JE, Cole PJ, Laurent GJ: A rapid and covenient assay for counting cells cultured in microwell plates: application for assessment of growth factors. J Cell Sci 1989; 92: 513–518.
11 Cotta-Pereira G, Sage H, Bornstein P, Ross R, Schwartz S: Studies of morphologically atypical ("sprouting") cultures of bovine aortic endothelial cells. Growth characteristics and connective tissue protein synthesis. J Cell Physiol 1980; 102: 183–191.
12 Howard PS, Myers JC, Gorfien SF, Macarak EJ: Progressive modulation of endothelial phenotype during in vitro angiogenesis. International Symposium on Angiogenesis 1991; 66(abstr).
13 Albelda SM, Daise M, Levine EM, Buck CA: Identification and characterization of cell-substratum adhesion receptors on cultured human endothelial cells. J Clin Invest 1989; 83: 1992–2002.
14 Giltay JC, Brinkman HJ, Modderman PW, von-dem-Borne AE, van-Mourik JA: Human vascular endothelial cells express a membrane protein complex immunochemically indistinguishable from the platelet VLA-2 (glycoprotein Ia-IIa) complex. Blood 1989; 73: 1235–1241.
15 Elices MJ, Hemler ME: The human integrin VLA-2 is a collagen receptor on some cells and a collagen/laminin receptor on others. Proc Natl Acad Sci USA 1989; 86: 9906–9910.
16 Languino LR, Gehlsen KR, Wayner E, Carter WG, Engvall E, Ruoslahti E: Endothelial cells use alpha 2 beta 1 integrin as a laminin receptor. J Cell Biol 1989; 109: 2455–2462.
17 Wolff S, Teuscher E: Studies on the influence of protein kinase-C modulators on the migration and protease expression of endothelial cells. International Symposium on Angiogenesis 1991; 113(abstr.).
18 Montesano R, Pepper MS, Orci L: Angiogenesis in vitro: Morphogenetic and invasive properties of endothelial cells. NIPS 1990; 5: 75–79.
19 Nishizuka Y: The family of protein kinase C for signal transduction. JAMA 1989; 262: 1826–1833.

Angiogenesis: Key Principles - Science - Technology - Medicine
ed. by R. Steiner, P. B. Weisz & R. Langer

Basic fibroblast growth factor bound to cell substrate promotes cell adhesion, proliferation, and protease production in cultured endothelial cells

Marco Presta, Marco Rusnati, Chiara Urbinati, Elena Tanghetti, Massimo Statuto, Ambra Pozzi, Anna Gualandris and Giovanni Ragnotti

Unit of General Pathology and Immunology, Department of Biomedical Sciences and Biotechnology, School of Medicine, University of Brescia, 25123 Brescia, Italy

Introduction

The extracellular matrix (ECM) plays a crucial role in the process of endothelial cell migration, proliferation, and differentiation during angiogenesis [1]. Several ECM proteins, like fibronectin (FN), affect cell behavior by binding to the various members of the integrin family of receptors, molecules that recognize the cell adhesion signal sequence Arg-Gly-Asp (RGD) [2].

Soluble growth factors have been purified which can induce neovascularization in vivo and stimulate endothelial cell proliferation in vitro [3]. Several data demonstrate that alteration of the adhesive capacity and of the integrity of ECM influences the response of endothelial cells to soluble growth factors [1]. Nevertheless, the mechanisms by which ECM proteins, adhesion receptors, and soluble growth factors interact to control angiogenesis remain poorly understood.

Examination of the primary sequence of the 155 amino acid form of the angiogenic molecule basic fibroblast growth factor (bFGF) shows the presence of the amino acid sequence Asp-Gly-Arg (DGR) repeated twice at positions bFGF(46–48) and bFGF(88–90) [4]. This sequence represents the inverse of the cell adhesion signal sequence RGD. Interestingly, both RGD- and DGR-containing peptides have been demonstrated to compete with adhesive proteins for integrin interaction [5].

We have observed that two synthetic peptides, corresponding to amino acid residues bFGF(38–61) and bFGF(82–101), exert a partial bFGF agonist and antagonist mitogenic activity in cultured endothelial cells without competing for the binding of bFGF to its cell membrane receptor [6]. Both peptides contain the DGR sequence. Moreover, synthetic tetra and heptapeptides containing the residues DGR or RGD specifically inhibit the mitogenic activity of bFGF [6]. These observations lead to the hypothesis that the two DGR sequences of the bFGF molecule may be involved in the modulation of the biological activity exerted by bFGF in cultured endothelial cells, possibly by interacting with adhesive RGD-dependent receptors.

In the present paper we summarize our findings on the properties of bFGF as a bioactive cell-adhesive molecule.

Materials and methods

Cells

Transformed fetal bovine aortic endothelial GM 7373 cells, corresponding to the described BFA-1c multilayered transformed clone [7], were obtained from the N.I.G.M.S. Human Genetic Mutant Cell Repository (Institute for Medical Research, Camden, NJ). Cells were grown in Eagle's minimal essential medium with 10% fetal calf serum (FCS), vitamins, essential and nonessential amino acids.

Adhesion assay

Confluent cultures of GM 7373 cells were trypsinized, washed, and resuspended in medium containing 1% FCS. Then, 50,000 cells were added to individual wells of polystyrene (non-tissue culture) microtiter plates that had been coated overnight at 4°C with bFGF (from 1 to 20 μg/ml in 100 mM carbonate buffer, pH 9.6). Bovine fibronectin (FN) at 20 μg/ml was used as a positive control for endothelial cell adhesion, while bovine serum albumin (BSA) was used as a negative control. Attachment of cells was allowed to occur for 2–4 h at 37°C. Microtiter wells were then washed once with 3 mM EDTA in PBS and once with medium containing 1% FCS. Washing procedure was repeated three times. The remaining attached cells were trypsinized and counted.

Results and discussion

bFGF has been reported to promote substrate adhesion of PC12 cells [8] and of capillary endothelial cells [9], but the mechanism(s) responsible for its adhesive capacity is not fully elucidated. Here, we have investigated the capacity of bFGF to promote the adhesion of endothelial GM 7373 cells to non-adhesive plastic.

As shown in Fig. 1A, coating of non-adhesive plastic with bFGF causes a dose-dependent increase in the number of attached cells and

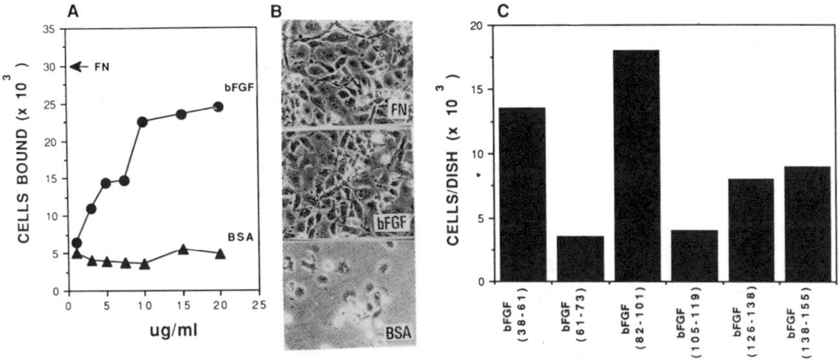

Figure 1. Effect of bFGF on adhesion and spreading of endothelial GM 7373 cells. *A* Dose-dependent cell-adhesion on bFGF-coated plastic. *B* Phase-contrast microscopy of the spreading of GM 7373 cells on plastic coated with 20 μg/ml bFGF, FN, or BSA. *C* Cell-adhesion on plastic coated with 300 μg/ml of different bFGF fragments.

promotes cell spreading. The cell-adhesive capacity of bFGF requires the presence of serum in the culture medium during the adhesion assay. Only a very limited cell-adhesion and spreading is in fact observed when the assay is performed in the absence of serum or in the presence of heat-inactivated serum (data not shown). Thus, bFGF requires a heat-labile serum component in order to promote cell adhesion and spreading. This component does not appear to be serum FN. Indeed, the cell-adhesive capacity of FN is resistant to heat inactivation and addition of 1 μg/ml of purified FN to serum-free medium during the adhesion assay does not restore the cell-adhesive capacity of bFGF. Interestingly, heat-inactivated bFGF retains, like FN, the capacity to mediate cell-adhesion and spreading. On the other hand, the basic heparin-binding protein histone III-S shows a cell-adhesive capacity much lower than bFGF (data not shown). Thus, the cell-adhesive capacity of bFGF does not seem to depend upon its tertiary structure nor its net basic charge, but rather upon some characteristics of its primary structure.

To assess this hypothesis, non-adhesive plastic was coated with synthetic peptides corresponding to different regions of the bFGF molecule. As shown in Fig. 1B, only peptides bFGF(38–61) and bFGF(82–101) significantly stimulate endothelial cell adhesion. As already mentioned, both peptides contain a core "DGR" sequence which appears to be involved in mediating the mitogenic activity of bFGF in endothelial cells [6]. These data suggest that DGR sequences may also mediate the cell-adhesive capacity of bFGF. Indeed, as observed for FN, bFGF-mediated cell-adhesion is inhibited by incubation of the cells with the synthetic heptapeptide G*RGD*SPK, but not by the negative control peptide G*RAD*SPK. Also, bFGF-mediated cell adhesion is calcium-dependent and it is inhibited by anti-$\alpha_5\beta_1$ integrin antibodies (data

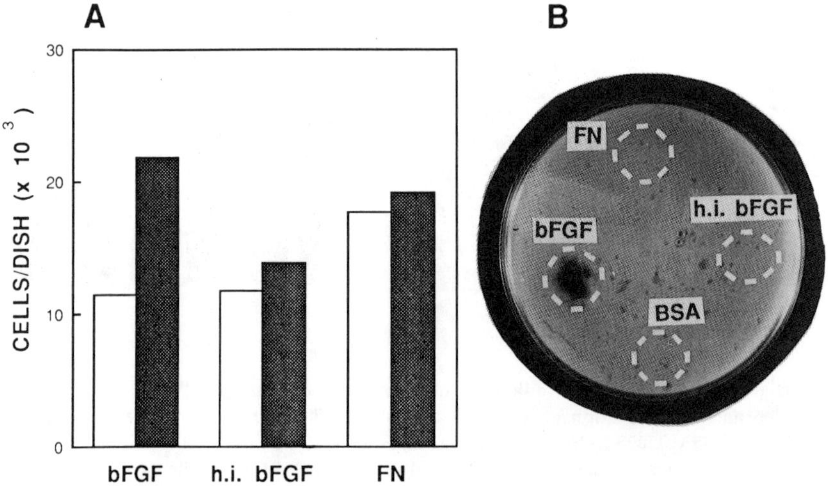

Figure 2. Biological activity of plastic-bound bFGF. *A* Mitogenic activity. Cells were seeded on plastic coated with 20 μg/ml bFGF, heat-inactivated bFGF (h.i. bFGF), or FN. After 2 h, floating cells were removed and adherent cells were counted immediately (open bar) or after 24 h of incubation in the presence of 0.4% FCS (black bar). *B* PA-inducing activity. Proteins to be tested were spotted onto a single plastic dish. Then, cells were seeded and allowed to adhere to coated plastic as in *A*. After removal of floating cells, adherent cells were overlayed with casein/plasminogen/agar. PA-dependent lysis was observed after 48 h of incubation at 37°C only for cells adherent to bFGF-coated plastic.

not shown). Thus, the capacity of bFGF to promote the adhesion of endothelial GM 7373 cells to non-tissue culture plastic requires a heat-labile serum component, depends upon specific sequences (possibly DGR sequences) in the bFGF molecule, and shows many of the characteristics of the integrin-mediated cell-protein interaction.

On this basis, we have addressed the question of whether surface-bound bFGF retains its biological activity. At low serum concentrations, GM 7373 cells proliferate when seeded on bFGF-coated plastic, but not when seeded on plastic coated with heat-inactivated bFGF or with FN (Fig. 2A). Also, surface-bound bFGF stimulates plasminogen activator (PA) production in GM 7373 cells, as shown by casein/agar overlay zymography (Fig. 2B). No PA production is instead observed in cells adherent to heat-inactivated bFGF or to FN.

In conclusion, surface-bound bFGF promotes endothelial cell adhesion and it is biologically active in situ. bFGF has been shown to be present in ECM both in vitro and in vivo [4]. Our findings shed a new light on the possible role of ECM-bound bFGF as a localized, persistent stimulus for endothelial cells during angiogenesis.

Acknowledgements. This work was supported by a grant from Associazione Italiana per la Ricerca sul Cancro and a grant from C.N.R. (Progetto Finalizzato Biotechnologie e Biostrumentazioni, Sottoprogetto Biofarmaci) to M.P.

References

1 Ingber DE, Flokman J: How does extracellular matrix control capillary morphogenesis? Cell 1989; 58: 803–805.
2 Ruoslahti E, Pierschbacher MD: New perspectives in cell adhesion: RGD and integrins. Science 1987; 238: 493–497.
3 Folkman J, Klagsbrun M: Angiogenic factors. Science 1987; 235: 442–447.
4 Burgess WH, Maciag T: The heparin-binding (fibroblast) growth factor family of proteins. Annu Rev Biochem 1989; 58: 575–606.
5 Akiyama SK, Yamada KM: Synthetic peptides competitively inhibit both direct binding to fibroblasts and functional biological assays for the purified cell-binding domain of fibronectin. J Biol Chem 1985; 260: 10,402–10,405.
6 Presta M, Rusnati M, Urbinati C, Statuto M, Ragnotti G: Functional domains of basic fibroblast growth factor (bFGF): possible role of Asp-Gly-Arg sequences in the mitogenic activity of bFGF. Ann NY Acad Sci 1991; in press.
7 Grinspan JB, Stephen NM, Levine EM: Bovine endothelial cells transformed in vitro with benzo(a)pyrene. J Cell Physiol 1983; 114: 328–338.
8 Schubert D, Ling N, Baird A: Multiple influences of a heparin-binding growth factor on neuronal development. J Cell Biol 1987; 104: 635–643.
9 Baird A, Schubert D, Ling N, Guillemin R: Receptor- and heparin-binding domains of basic fibroblast growth factor. Proc Natl Acad Sci USA 1988; 85: 2324–2328.

Angiogenesis: Key Principles - Science - Technology - Medicine
ed. by R. Steiner, P. B. Weisz & R. Langer

Thrombospondin modulates basic fibroblast growth factor activities on endothelial cells

Giulia Taraboletti, Dorina Belotti and Raffaella Giavazzi

Istituto di Ricerche Farmacologiche Mario Negri, Bergamo, Italy

Abstract. We previously reported that thrombospondin (TSP) induces endothelial cell (EC) adhesion, spreading and motility, suggesting that it can play a role in angiogenesis. We then studied whether TSP might modulate EC response to known angiogenic stimuli in vitro. Here we describe that TSP inhibits EC chemotactic response to basic fibroblast growth factor (bFGF). Furthermore, TSP and its 140 kD fragment reduce EC proliferative response to serum and bFGF. These data support the indicated role of TSP and its 140 kD fragment in angiogenesis and in related pathologies including tumor malignancy.

Introduction

Thrombospondin (TSP) is a multifunctional, multidomain extracellular matrix component, secreted by platelets and different cell types, including endothelial and tumor cells [1, 2].

Several in vivo and in vitro studies suggest a role for TSP in mediating processes related to tissue formation and remodeling, and particularly in angiogenesis. Histological studies have shown an increase in the amount of TSP at the site of vessel lesion [3] and in a late stage of vessel formation in the mouse embryo [4]. Direct evidence that TSP is involved in angiogenesis has been provided by Bouch and coworkers. They recently described a 140 kD protein, related to a cancer suppressor gene in hamster cells, that is immunologically and functionally identical to the carboxy-terminal region of TSP [5, 6]. This protein suppresses neovascularization in vivo in the rat cornea, and EC migration in vitro [5, 6]. The relationship of EC with TSP has been analyzed by several authors, who reported that EC produce TSP and interact with it through specific cell receptors [1, 2, 7].

We recently showed that TSP stimulates EC adhesion and spreading, and that it induces EC chemotactic and haptotactic migration [8]. We next investigated whether TSP could modulate EC response to angiogenic stimuli in vitro. Here we describe the effect of TSP on EC chemotactic and proliferative response to serum and bFGF.

Materials and methods

Purification of TSP from human platelets and production of its enzymatic fragments were performed as previously described [9]. Human recombinant basic fibroblast growth factor (bFGF) was kindly provided by Dr. F. Bertolero (Erbamont, Milano, Italy).

Cells

Bovine aortic EC (BAEC), kindly provided by Dr. E. Dejana (Istituto Mario Negri, Milano, Italy), and murine lung capillary cells (LE-II) were cultured in DMEM supplemented with glutamine and 10% FCS. Cells were tested and found free of mycoplasma infection.

Chemotaxis and proliferation

Chemotaxis assay was performed as described in detail [8, 9], using modified Boyden chambers, with 8 μm (LE-II) and 5 μm (BAEC) pore size gelatin-coated filters. Cell proliferation was measured as described [8].

Results

TSP modulation of bFGF-induced chemotaxis

Both bFGF and TSP are known inducers of chemotactic response in endothelial cells. When chemotactic concentrations of TSP (100 μg/ml) were added to bFGF (10 ng/ml) in the lower compartment of the chemotactic chamber, EC migration was higher compared to bFGF alone. However, the addition of non-chemotactic concentrations of TSP (0.1–1 μg/ml) to bFGF in the chamber resulted in a significant inhibition of endothelial cell motility (Tab. 1).

TSP modulation of endothelial cell proliferation

EC proliferation in vitro is stimulated by FCS and by bFGF. The addition of soluble TSP to EC together with either FCS or bFGF resulted in a dose-dependent reduction of endothelial cell proliferative response (Tab. 2). We also observed that the carboxy-terminal 140 kD fragment of TSP retained the ability to inhibit EC proliferation

Table 1. Effect of TSP on endothelial cell motility response to bFGF

Sample	Concentration (μg/ml)	Motility response to Control (no stimulus)	Motility response to bFGF (10 ng/ml)
Control	—	39.3 ± 2.0	84.7 ± 12.2
TSP	100	134.7 ± 21.7	260.3 ± 14.9
	10	102.0 ± 19.3	93.5 ± 5.5
	1	49.5 ± 3.5	75.0 ± 7.0
	0.1	49.5 ± 10.5	55.5 ± 2.5
	0.01	34.7 ± 13.2	50.0 ± 12.0

Data are expressed as migrated LE-II cells in 10 high power fields.

Table 2. Effect of TSP on endothelial cell proliferative response

Sample	Concentration (μg/ml)	Proliferative response to FCS (5%)	bFGF (10 ng/ml)
Control	—	100.0 ± 3.5	100.0 ± 3.7
TSP	100	43.4 ± 3.5***	30.8 ± 2.2***
	50	60.3 ± 3.6***	52.9 ± 4.9***
	10	97.0 ± 6.9	66.3 ± 0.3***
	1	103.2 ± 2.1	75.0 ± 7.7**
	0.1	102.1 ± 10.4	79.4 ± 3.4*
140 kD	5	45.0 ± 1.6***	nt
	2.5	62.7 ± 0.4***	36.1 ± 4.8***
	0.25	92.5 ± 4.3	nt
25 kD	5	70.0 ± 6.1***	nt
	2.5	81.0 ± 1.9***	71.8 ± 0.8**
	1	98.5 ± 4.0	nt

Data are percentage of BAEC proliferative response compared to control.
***$p \leq 0.001$; **$p \leq 0.01$; *$p \leq 0.05$.

(Tab. 2). A partial inhibitory activity of the 25 kD fragment was also observed, but this could be ascribed to the small amount of undigested TSP that co-purifies with this fragment.

TSP inhibition of endothelial cell motility and proliferative response was observed on both LE-II and BAEC cells.

Discussion

EC migration and proliferation are basic steps of the angiogenic process. Extracellular matrix components are known to play a role in angiogenesis, by functioning as a storage site for angiogenic stimuli, and by affecting cell structure and response to growth factors [10–12]. Our findings confirm that TSP is an angiogenesis regulatory factor, that acts by inducing EC adhesion and migration, and by inhibiting EC chemotactic and proliferative response to angiogenic stimuli. Further investigations are needed to understand the molecular mechanism involved.

The molecule of TSP contains different cell binding sites, which mediate several cellular functions. The 140 kD fragment has been reported to mediate tumor cell adhesion and haptotactic migration [9], and to inhibit angiogenesis in vivo and bFGF-induced EC migration in vitro [5, 6]. Our finding that this fragment also inhibits EC proliferation in response to serum and bFGF, confirms that this fragment contains the angiogenesis-suppressor site.

Acknowledgements. This work was supported by the CNR (National Research Council, Rome, Italy), contract No 90.00141.PF70, and by the Italian Association for Cancer Research, Milan, Italy. G.T. research is supported by the 'Fondazione per la Formazione Oncologica', and the European Institute of Oncology, Milano, Italy.

References

1 Frazier WA: Thrombospondin: a modular adhesive glycoprotein of platelets and nucleated cells. J Cell Biol 1987; 105: 625–632.
2 Mosher DF: Physiology of thrombospondin. Annu Rev Med 1990; 41: 85–97.
3 Wight TN, Raugi GJ, Mumby SM, Bornstein P: Light microscopic immunolocation of thrombospondin in human tissues. J Histochem Cytochem 1985; 33: 295–302.
4 O'Shea KS, Dixit VM: Unique distribution of the extracellular matrix component thrombospondin in the developing mouse embryo. J Cell Biol 1988; 107: 2737–2748.
5 Good DJ, Polverini PJ, Rastinejad F, Le Beau MM, Lemons RS, Frazier WA, Bouck NP: A tumor suppressor-dependent inhibitor of angiogenesis is immunologically and functionally indistinguishable from a fragment of thrombospondin. Proc Natl Acad Sci USA 1990; 87: 6624–6628.
6 Rastinejad F, Polverini PJ, Bouck NP: Regulation of the activity of a new inhibitor of angiogenesis by a cancer suppression gene. Cell 1989; 56: 345–355.
7 Murphy-Ullrich JE, Mosher DF: Interactions of thrombospondin with endothelial cells: receptor-mediated binding and degradation. J Cell Biol 1987; 105: 1603–1611.
8 Taraboletti G, Roberts DR, Liotta LA, Giavazzi R: Platelet thrombospondin modulates endothelial cell adhesion, motility, and growth: a potential angiogenesis regulatory factor. J Cell Biol 1990; 111: 765–772.
9 Taraboletti G, Roberts DR, Liotta LA: Thrombospondin-induced tumor cell migration: haptotaxis and chemotaxis are mediated by different molecular domains. J Cell Biol 1987; 105: 2409–2415.
10 Ingber DE, Folkman J: How does extracellular matrix control capillary morphogenesis? Cell 1989; 58: 803–805.
11 Rifkin DB, Moscatelli D: Recent developments in the cell biology of basic fibroblast growth factor. J Cell Biol 1989; 109: 1–6.
12 Burgess WH, Maciag T: The heparin binding (fibroblast) growth factor family of proteins. Ann Rev Biochem 1989; 58: 575–606.

Angiogenesis: Key Principles – Science – Technology – Medicine
ed. by R. Steiner, P. B. Weisz & R. Langer

Binding sites for basic fibroblast growth factor on solid tumors are associated with the vasculature

William F. Herblin and Janet L. Gross

The Dupont Merck Pharmaceutical Company, P. O. Box 80400, Wilmington, Delaware 19880-0400, USA

Introduction

Basic fibroblast growth factor (bFGF) is a member of a family of heparin binding growth factors that exhibit mitogenic and proliferative effects on many mesoderm- and ectoderm-derived cells [1, 2]. It has been shown to be a major angiogenic factor involved in the neovascularization of development and wound healing, as well as the abnormal proliferation of vasculature associated with tumor growth, metastases, diabetic retinopathy and atherosclerosis [3–5]. Multiple forms of the protein have been shown to exist and multiple classes of receptors have been identified [6].

bFGF has been localized by immunochemistry in many cell types and normal tissues [7], but not as much has been reported regarding the localization of high-affinity receptors. In this report we show that high affinity binding sites for bFGF are present in some normal tissues as well as in solid tumors. While the vasculature of the normal tissues are not densely labeled, the sites in the sections of solid tumors are associated with vascular-like structures that are also recognized by antibodies against Factor VIII and a lectin from *Dolicus biflorus*, two putative markers for endothelial cells [8].

Methods and materials

Recombinant basic fibroblast growth factor was provided by Synergen, Inc. and was iodinated with the lactoperoxidase method of Morrison et al. [9]. Antibody to Factor VIII was obtained in a kit from Accurate Chemical, Westbury, NY and used according to the manufacturer's directions. *Dolicus biflorus* agglutinin conjugated to horseradish peroxidase was from Sigma Chemcial, St. Louis, MO and was used according to the procedures given by Williams et al. [8].

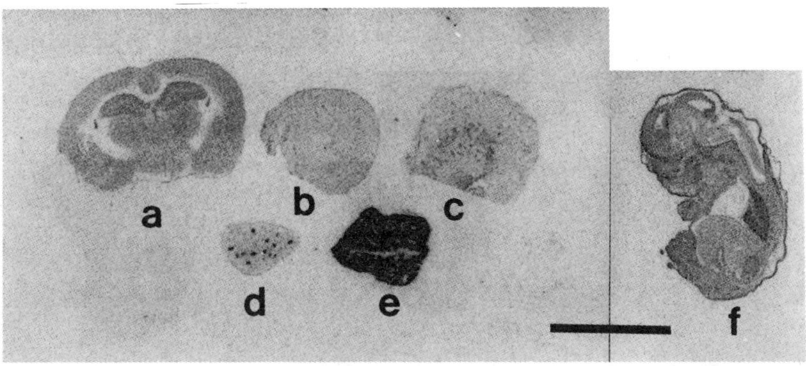

Figure 1. Labeling of normal adult and fetal tissues by ^{125}I-bFGF. Scale bar represents 2 cm. a) brain; b) heart; c) kidney; d) adrenal; e) spleen; f) fetus.

Normal tissues from rats or solid tumors from nude mice were rapidly removed, rinsed quickly in saline, frozen either on powdered dry ice or in hexane on dry ice and stored at −70°C until used. Autoradiography with ^{125}I-bFGF was performed using a modification of the binding assay reported by Moscatelli [10] as previously described [9].

Results

When normal tissues were examined by autoradiography with ^{125}I-bFGF, high-affinity binding sites were observed in several locations. The glomeruli of the renal cortex, the tubules of the renal medulla and the peri-arteriolar lymph nodes of the spleen were densely labeled (Fig. 1). The lung was the most heavily labeled tissue of all. In fetal tissue, labeling was associated with developing bone structures, consistent with the reported presence of bFGF receptors on chondrocytes [11], but the highest level of sites was in fetal lung tissue.

Although blood vessels were not densely labeled in normal tissues, there was concentrated ^{125}I-bFGF labeling to vascular-like structures in sections from various human xenografted solid tumors. The bright field veiw (Fig. 2A) of a section from a C6 rat glioma tumor propagated in nude mice illustrates the typical appearance of these structures. A dark field view in 2C demonstrates the concentration of grains over putative vessels. Figure 2B shows a similar section that was treated with *Dolicus biflorus* agglutinin conjugated to HRP. This lectin was reported to specifically recognize endothelial cells and labeled the same structures as ^{125}I-bFGF bound. Figure 2D shows another section from a C6 tumor that was immunostained with Factor VIII antibody.

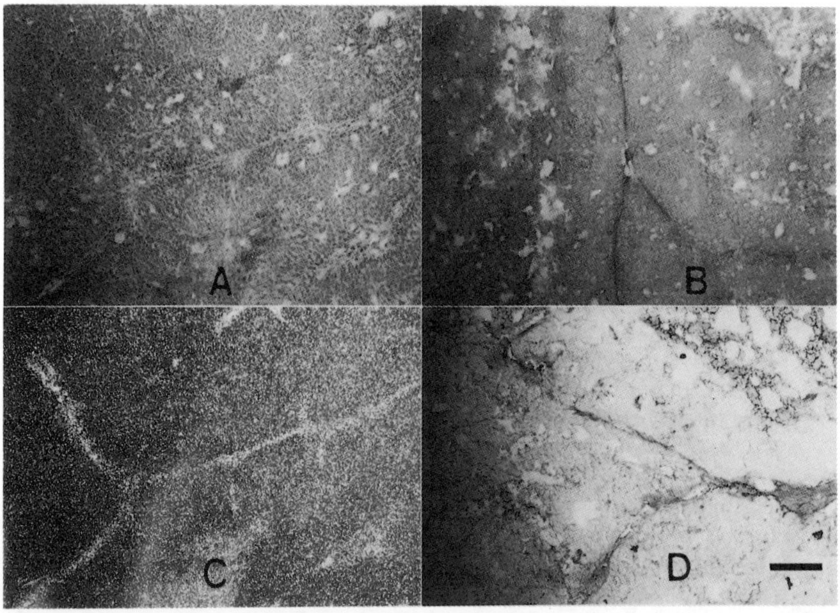

Figure 2. Labeling of secions from C6 tumor. *A*) Bright field view; *B*) Labeling by HRP conjugated *Dolicus biflorus* agglutinin; *C*) Dark field view of bFGF labeling; *D*) Labeling with antibody to Factor VIII. Scale bar represents 0.1 mm.

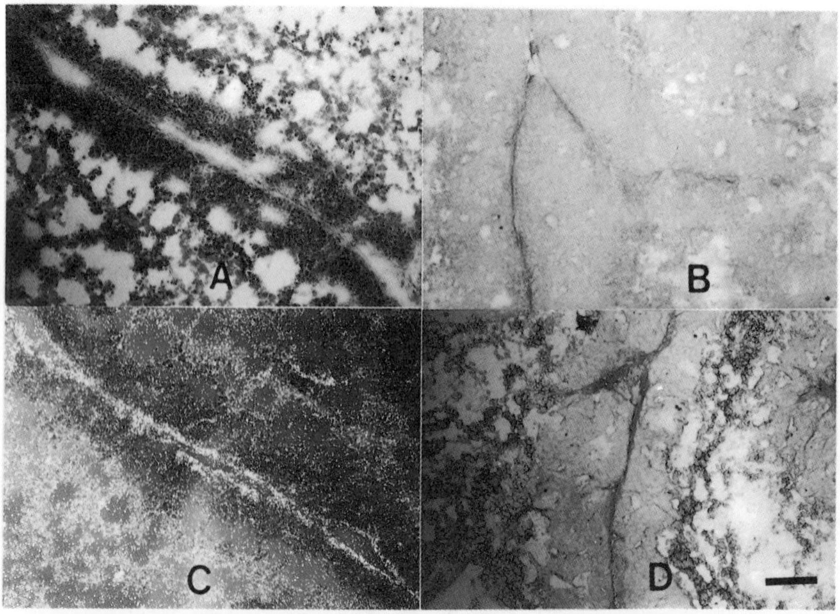

Figure 3. Labeling of sections from DLD-2 tumors. Panel identification as in Fig. 2. Scale bar represents 0.1 mm.

Although there was great variability within and between tumors, labeling by bFGF was consistently seen even in sections of tumors raised from cells which did not express high-affinity bFGF receptors in culture. Figure 3 shows sections from a different tumor, one raised from human colon carcinoma DLD-2 cells. DLD-2 cells did not express high affinity sites for bFGF in culture [12].

Discussion

A unique finding presented in this paper is that adult rat lung tissue possessed the highest level of high-affinity binding sites for bFGF; this has been a consistent finding in studies of fetal tissue and also in lung tissue from other species. In fetal tissue, the sites appeared to line the developing airways, while in adult tissue the level was so high that it was not possible to determine specifically labeled structures. Only structures that were not labeled, such as blood vessels, were clearly noticeable.

In the kidney, dense labeling was observed over the glomeruli and also over certain structures in the renal medulla. Should these medullary structures prove to be the vasa recta, then the labeling may correlate to capillaries in general. As in the lung, the mature vessels of the kidney were not labeled by bFGF. In studies of several tissues in various species, we have rarely observed any significant labeling of mature blood vessels by bFGF and have seen possible labeling of capillaries only in special cases in which the capillaries are 'massed', as in the renal glomeruli or vasa recta.

When sections from solid tumors were examined, there was clearly dense labeling associated with structures that have many of the characteristics of blood vessels. These structures traverse the tumor, branch into networks and are made up of cells with elongated nuclei that are clearly different from the cells. The structures stain positively for Factor VIII, although as Denekamp et al. observed, the staining was not as intense as would be expected form normal endothelial cells [13]. The same structures are also labeled by the lectin from *Dolicus biflorus*, which has been reported to recognize endothelial cells.

Several possibilities exist for the explanation of the differences between normal endothelial cells and those found in solid tumors: The absence in tumors of pericytes and smooth muscle cells, the rapid proliferation of the endothelial cells associated with tumor growth or the presence of the multiple growth factors and cytokines produced by the tumor cells themselves. Whatever the cause, the endothelial cells of the tumor vasculature exhibited a unique phenotype. This phenotype was associated with a more concentrated expression of bFGF sites than was found in other locations.

References

1 Burgess WH, Maciag T: The heparin-binding (Fibroblast) growth factor family of proteins. Ann Rev Biochem 1989; 58: 575–606.

2 Goldfarb M: The fibroblast growth factor family. Cell Growth Diff 1990; 1: 435–439.

3 Folkman J: The angiogenic activity of FGF and its possible clinical applications, Sara VR et al. (ed): Growth Factors: From Genes to Clinical Application. New York, Raven Press, 1990, pp 201–216.

4 Gospodarowicz D: Expression and control of vascular endothelial cells: Proliferation and differentiation by fibroblast growth factors. J Invest Dermatol 1989; 93: 39S–47S.

5 Schweigerer L: Fibroblast growth factor and angiogenesis. Z Kardiol 1989; 78 Supp 6: 12–15.

6 Robinson CJ: Multiple receptors found for the growing FGF family. TIPS 1991; 12: 123–124.

7 Rifkin DB, Moscatelli D: Recent developments in the cell biology of basic fibroblast growth factor. J Cell Biol 1989; 109: 1–6.

8 Williams RLL, Robertson D, Davies AJS: Identification of vascular endothelial cells in murine omentum using the lectin, *Dolicus biflorus* agglutinin: possible applications in the study of angiogenesis. Hist J 1989; 21: 271–278.

9 Morrison RS, Gross JL, Herblin WF, Reilly TM, LaSala PA, Alterman RL, Moskal JR, Kornblith PL, Dexter DL: Basic fibroblast growth factor-like activity and receptors are expressed in a human glioma line. Cancer Res 1990; 50: 2524–2529.

10 Moscatelli D: High and low affinity binding for fibroblast growth factors on cultured cells. J Cell Physiol 1987; 131: 123–130.

11 Chin JE, Hatfield CA, Krzesicki RF, Herblin WF: Interactions between interleukin-1 and basic fibroblast growth factor on articular chondrocytes. Arth Rheum 1991; 34: 314–324.

12 Gross JL, Morrison RS, Eidsvoog K, Herblin WF, Kornblith PL, Dexter DL: Basic fibroblast growth factor: A potential autocrine regulator of human glioma cell growth. J Neurosci Res 1990; 27: 689–696.

13 Denekamp J, Hill S: Angiogenic attack as a therapeutic strategy for cancer. Radiother Oncol 1991; Supp 20: 103–112.

Angiogenesis: Key Principles - Science - Technology - Medicine
ed. by R. Steiner, P. B. Weisz & R. Langer

Studies on basic fibroblast growth factor (FGF-β) gene expression in the rat and pig ovary using in situ hybridization and quantitative reverse transcriptase — Polymerase chain reaction techniques

Mark Guthridge, Jacqueline Schmitt, Joseph Bertolini, Jane Cowling, Andrew Runting, Sivi Katsahambas, Anne E. Drummond and Milton T. W. Hearn

Department of Biochemistry and Centre for Bioprocess Technology, Monash University, Wellington Road, Clayton, Victoria 3168 Australia

Abstract. In order to gain further understanding of the physiology of basic fibroblast growth factor (FGF-β) in the mammalian reproductive tract, the expression of FGF-β mRNA in the rat and porcine ovary has been examined by in situ hybridization and quantitative reverse transcriptase-polymerase chain reaction techniques during different stages of the estrus cycle. The results confirm that an increase in FGF-β mRNA levels occurs over the course of the estrus cycle. No FGF-β gene expression was detected during diestrus, the non-hormonal phase of the cycle, or at the early proestrus stage of the cycle. During late proestrus and estrus, FGF-β mRNA was predominantly localized to granulosa cells of the dominant follicles, and to a lesser extent, to secondary antral follicles not committed to ovulation. Theca cells also expressed FGF-β mRNA during this phase of follicular development, albeit in low abundance. During metestrus, after ovulation, in the newly formed corpora lutea FGF-β mRNA levels were maximal, however on entering the next cycle commencing at diestrus, no FGF-β mRNA was observed in the degenerating corpora lutea. These results indicate that expression of the FGF-β gene is differentially regulated during the estrus cycle. The biological significance of this expression and the potential role of FGF-β in local intra-ovarian regulation of the repetitive cycles of follicular differentiation, proliferation and maturation associated with ovarian revascularization are discussed.

Introduction

It is established that ovarian function is regulated by the secretion of gonadotrophins (lutropin, LH and follitropin, FSH) produced by the anterior pituitary in conjuction with locally produced steroid hormones (testosterone, T, estradiol, E_2 and progesterone, P). FSH promotes the growth and differentiation of follicles within the ovary and mediates the aromatization of androgens to estrogens by the granulosa cells. LH stimulates steroid production and induces ovulation in FSH-primed follicles. In addition, the development and maintenance of the corpus luteum requires LH stimulation.

Clearly LH and FSH play an essential role in ovarian regulation. However, it is also evident that other regulatory mechanisms must exist. The cycling ovary exhibits profound changes in tissue morphology during the estrus cycle which cannot be attributed solely to these hormonal influences. This point is illustrated by the vascularization of the theca interna during ovulation and corpus luteum development. Within the ovary, the selection and development of dominant follicles occurs adjacent to atretic and growing follicles, at various states of maturation. These discrete areas of growth and degeneration co-exist in an ovary exposed to circulating gonadotrophins. Thus, locally produced factors must play an active role in regulating ovarian function. This concept can be extended to regulatory mechanisms within individual follicles, whereby segregated cell populations influence each others function by the production of local factors. For example, the interaction of theca cells and granulosa cells in the production of estrogens, where androgens produced by the theca cells diffuse to the granulosa cells for aromatization to estrogens.

A number of growth regulating factors have been identified in the ovary or ovarian cell cultures. Epidermal growth factor (EGF) [1], insulin-like growth factors I and II (IGF-I, IGF-II) [2, 3], the transforming growth factors β and α (TGF-β, TGF-α) [2, 4, 5] the inhibins [6], and follistatins [7], are just a few of the factors which are postulated to regulate ovarian function in paracrine or autocrine interactions, mediating folliculogenesis, differentiation and steroidogensis.

Fibroblast growth factor-β (FGF-β), a potent angiogenic substance, has also been identified in the ovary and corpus luteum [8–10]. During ovulation, rapid and radical changes occur in the capillary network surrounding the follicle. Previously avascular granulosa cell layers are invaded by numerous proliferating capillaries. Similarly, the transformation of the collapsed follicle into a functional corpus luteum, following the expulsion of the ovum during ovulation, involves vascularization [11]. In the event that pregancy does not occur in a given cycle, the corpus luteum regresses and the estrus cycle recommences at diestrus. Putative in vivo roles for FGF-β in these angiogenic events can be envisaged.

In addition to promoting vascularization, FGF-β has the potential to regulate other aspects of ovarian function. Plasminogen activator (PA), a proteolytic enzyme involved in tissue restructuring, is produced by ovarian granulosa cells [12]. In the ovary, PA is postulated to be responsible for disruption of the follicle and thus the release of the ovum at ovulation. FGF-β has been shown to stimulate PA production by bovine capillary endothelial cells and to decrease the production of a PA inhibitor [13-15]. FGF-β may indirectly mediate ovulation and oocyte maturation via the ability of this growth factor to similarly stimulate PA production in the follicle. FGF-β has been found to exert inhibitory effects on the FSH-mediated processes of LH receptor induction (on granulosa cells) and aromatase activity [16, 17], thereby inhibiting E_2 production. Conflicting data has been reported regarding the action of FGF-β on granulosa cell PA production, with effects ranging from inhibition, to stimulation, to no effect, documented [18–20]. These discrepant data may reflect differences in animal species, culture conditions and cyclical variations depending on when the granulosa cells were harvested.

A proliferative capacity for FGF-β in the ovary is also likely since fibroblasts, mesenchymal cells and bovine aortic endothelial cells divide in response to FGF-β [21-24]. Granulosa cells are one of the most rapidly proliferating cell types in vivo and since these cells are mesoderm-derived they may respond to FGF-β stimulation. In vitro cultures of granulosa cells proliferate in response to FGF-like activity contained in granulosa cell extracts [21], demonstrating the potential for autocrine regulation by FGF-β. Other studies however indicate that FGFs are not mitogenic per se for rat granulosa cells, but nevertheless affect their function possibly via FGF receptors [17]. Furthermore, FGF-β delays the terminal differentiation of granulosa cells maintained in vitro, probably due to the ability of FGF-β to maintain granulosa cells in an active cycling phase, thereby preventing them from entering a quiescent state [25].

The potential thus exists for FGF-β to regulate ovarian function in vivo by exerting direct effects on vascularization, differentiation, steroidogenesis and gonadal cell maturation. Despite the documented in vitro actions of FGF-β, the in vivo mechanisms which regulate FGF-β synthesis and release are poorly understood. This is partly due to the fact that FGF-β, like other growth factors, is produced in small amounts by normal tissues, coupled with the short half-life of its mRNA transcript [26, 27]. It is also known that FGF-β is synthesized without a signal sequence indicating that FGF-β does not leave the cell by conventional pathways [28]. It has been postulated that FGF-β is co-secreted with extracellular matrix components providing a storage pool of immobilized growth factor. The affinity of FGF-like proteins for heparin strongly supports an association with heparan glycosaminoglycans of the extracellular matrix [29, 30].

In view of the current difficulties associated with monitoring FGF-β mRNA expression with cDNA probes [27], we recently developed sensitive in situ hybridization methods [32] with [α-^{35}S] dATP labelled oligonucleotides and a quantitative RT-PCR amplification assay [33] which allows low levels of mRNA from a small amount of tissue or a few cells to be measured. Significant modification of the in situ hybridization technique has been achieved [32] using oligonucleotide probes, specifically 3 labelled to high specific activity. These improvements have permitted highly stringent and sensitive methods to be employed for hybridization with FGF-β mRNA in tissue slices. Similarly, the reverse transcriptase-polymerase chain reaction procedure has been adapted to permit rapid quantitation. This latter procedure involves i) production of cDNA from mRNA; ii) amplification of the cDNA by PCR; iii) rapid and quantitative HPLC analysis of the PCR products with non porous microparticulate (<3 μm) ion exchangers; iv) restriction enzyme mapping of the cDNA product and v) dideoxy-nucleotide sequencing. The amount of FGF-β PCR product generated was shown [33] to be directly proportional to the initial amount of FGF-β mRNA in the purified sample. The simultaneous co-reverse transcription and co-amplification of a second, (endogenous or exogenous) and unrelated mRNA (distinct from the mRNA of interest), provided an internal standard which monitors assay efficiency and enabled quantification of the cDNA of interest.

These molecular biological techniques have permitted detection and localization of FGF-β gene transcripts in the cycling rat and pig ovary. The results confirm that the in vivo expression of FGF-β mRNA is differentially and repetitively regulated during the different stages of the estrus cycle.

Materials and methods

1. Tissue preparation

All studies were conducted in accordance with the Australian National Health and Medical Research Council guidelines for the care and use of laboratory animals. Wistar rats were maintained on a 12-h day/night cycle. Gilts (strain) were maintained at the Pig Research Centre, Werribee. Estrus cycles were determined by vaginal cytology. Only rats exhibiting 3 consecutive 4 day cycles were used in these investigations. The ovaries of rats ($n = 4$) at each stage of the estrus cycle were rapidly excised and snap frozen at −70°C in OCT compound. The reproductive tracts of cyclic or pregnant gilts were collected under aseptic conditions at local abattoirs and snap frozen at −70°C in OCT compound. Embedded tissues were cut into 8–10 μm sections at −20°C. Frozen tissue sections were thaw mounted onto acid washed, baked and poly-L-lysine coated slides and immediately placed on dry ice blocks for at least 30 minutes. Tissue sections were then subjected to the in situ hybridization procedure.

Tissue blocks (ca. 1 cm^3) were homogenized in 6 M guanidine thiocyanate −5 mM sodium citrate, pH 7.0, 0.1 M β-mercapto-ethanol using a PT10-35 Polytron. The homogenate was then centrifuged in a SS-34 rotor using a Sorval RC-5B refrigerated centrifuge. The guanidine thiocyanate supernatant was then recovered and the total mRNA extracted, fractionated on oligo-d-T cellulose and concentrated as described elsewhere [33].

2. Labelling of oligonucleotides

The design of the synthetic oligonucleotides complementary to FGF-β mRNA, the 18S rRNA primer, the cDNA primer B4, the PCR primers T1 and 2B, and the PCR detection oligonucleotide, Rex 1, have been described in our previous studies [32, 33]. Specific oligonucleotides, as appropriate, were 5′-[^{32}P]dATP labelled by T4 polynucleotide kinase or 3′-[α-^{35}S]dATP labelled with terminal deoxynucleotidyl-transferase as described previously [34]. The radio-labelled oligonucleotides were ethanol precipitated to ensure complete removal of the unincorporated [^{32}P] – or [α-^{35}S]dATP. Incorporation of the radiolabel into the oligonucleotides was monitored by trichloroacetic acid precipitation of aliquots onto glass fibre disks and scintillation counting.

3. In situ hybridization

All glassware was baked and solutions (except Tris-HCl) were DEPC treated and autoclaved. Cells and tissue sections were fixed for 10 min in 4% glutaraldeyde, 96 mM phosphate buffer pH 7.3 and 20% ethylene glycol, then washed 3 times over 15 min in 150 mM phosphate buffer, pH 7.3, followed by two washes in enzyme buffer, each for 10 min and twice for 10 min in enzyme buffer (15 mM Tris-HCl pH 7.5, 1 mM EDTA, 50 mM NaCl, 5% glycerol). Ribonuclease (RNase, 250 μg/ml in enzyme buffer) or enzyme buffer alone were applied alternately to tissue sections/cells, coverslipped and incubated at 37°C for 1 h in a humidified sealed container. Ribonuclease and ribonuclease-free slides were kept separate in all subsequent procedures. Coverslips were removed with 2 rinses in 150 mM phosphate buffer pH 7.3 and the slides were placed in hybridization buffer consisting of $5 \times$ SSC ($1 \times$ SSC = 0.15 M NaCl and 0.15 M trisodium citrate pH 7.0 and 10 mM β-mercaptoethanol), 50 mM phosphate buffer pH 8.0, 0.2% bovine serum albumin, 0.2% polyvinylpyrilodine, 0.2% ficol, 0.2% denatured and sheared Herring sperm DNA and 0.4% denatured *E-coli* tRNA for 1 h at room temperature and then fresh hybridization buffer at 37°C for 1 h. Slides not to be used immediately were washed twice for 5 min in 100% ethanol and stored at –70°C suspended over alcohol.

Stored slides were thawed and equilibrated for 10 min in hybridization buffer after which 30 μl of heat denatured [α-^{35}S] dATP tailed oligonucleotide (0.25×10^6 dpm in hybridization buffer/slide) was pipetted onto tissues or cells, coverslipped and placed in a sealed container humidified with hybridization buffer at 43°C for 24 h. Slides were then washed in decreasing salt concentrations at room temperature; 4 × SSC for 10 min to remove coverslips, then 3 × SSC for 20 min, 2 × SSC for 20 min and finally washed twice in 1 × SSC for 30 min each.

Ilford G5 nuclear track emulsion was melted at 42°C and diluted with an equivolume of distilled water. Slides were dipped in the emulsion, air dried and exposed at −70°C for up to 2 weeks. Slides were developed in Kodak D19 for 5 min, rinsed in 1% acetic acid for 1 min, fixed for 3 min in 24% thiosulphate, thoroughly rinsed in running tap water for 10 min and then air dried. Cells and tissues were stained in haematoxylin, fixed in Scotts Tap Water, counter stained in eosin, dehydrated by washing for 2 min each in 70% ethanol then 100% ethanol and finally cleared by a 2-min wash in Histoclear. The samples were mounted with Histomount onto polished glass coverslips.

4. Reverse transcriptase-polymerase chain reaction

FGF-βm RNA was first converted into complementary DNA (cDNA) using reverse transcriptase. Oligo-d-T-cellulose derived mRNA samples in distilled water were denatured by heating to 65°C for 10 min in a water bath. The 18S reverse transcription primer was denatured by boiling in a water bath for 10 min, followed by quenching on ice. AMV reverse transcriptase was diluted in enzyme dilution buffer and incubated at 4°C for 30 min prior to use. Reverse transcription was carried out at 42°C with 0.01 to 1 μg RNA, 0.3 μg primer, 6 mM $MgCl_2$ and PCR buffer (1 ×) or RT buffer (1 ×), 0.5 mM dNTPs (dATP, dTTP, dGRP, dCTP) and 20 units reverse transcriptase. In some cases, two cycles of reverse transcription were carried out by denaturing the cDNA/RNA mixture at 65°C for 10 min, quenching on ice, and the addition of further reverse transcriptase, followed by incubation at 42°C. Following production of cDNA, the FGF-β samples were either amplified immediately or stored at −20°C.

Prior to the addition of Taq polymerase, an aliquot of the FGF-β cDNA mixture was added to the transcription primer (B4 or T1) (1.0 nM), PCR buffer (1 ×), $MgCl_2$ (6 mM) and dNTPs (0.2 mM) and heat denatured at 96°C for 10 min then quenched on ice or a dry ice/ethanol bath. A cycle of PCR involved denaturation at 95–97°C for 1–1.5 min, primer annealing at 50–60°C for 1 min and primer extension at 70°C for 2.5 min. The extension time was increased by 1 s per cycle. Cycles were repeated using a Cetus thermal cycler up to 60 times in order to amplify the input cDNA material. Following PCR the Taq polymerase activity was inhibited by the addition of EDTA to 10 mM. The RT-generated cDNA was electrophoresed in 5% 18 × 20 cm polyacrylamide gels. 5% polyacrylamide, 1 × TBE (0.2 M Tris, 83 mM Boric acid, 0.02 M EDTA), was thoroughly degassed and 0.1% ammonium persulfate, 0.15% TEMED added. After pouring and allowing to set for 1 h, the DNA was applied to the gel in loading buffer (bromophenol blue and xylenecyanol in 0.02 M EDTA and glycerol) and electrophoresed at 100 mA until the dye front had progressed 3/4 through the gel. The gel was then dried on a gel dryer and autoradiographed between two high speed intensifying screens. Following electrophoretic separation, RT-PCR generated DNA fragments were capillary transferred to a Zeta Probe nylon membrane. To set up the transfer, orientation marks were made on the nylon membrane, which was then slowly wetted by dipping in H_2O followed by equilibration in 0.4 M NaOH for at least 15 min. Whatman 3 MM filter paper was also equilibrated for at least 15 min in the NaOH. The gel was then placed onto the saturated blotting paper in a tray containing 250 ml 0.4 M NaOH and overlaid with the membrane. The membrane was overlaid with two sheets of filter paper and a stack of paper towels was then placed on the top. The wet paper towels were removed 3–4 times over the 12–19-h period during which the transfer was carried out.

After transfer of RT-PCR generated DNA to the Zeta Probe, the membrane was dipped briefly in 5 × SSPE (0.74 M NaCl, 0.04 M $NaH_2PO_4 \cdot H_2O$, 0.005 M EDTA, pH 7.4), 0.1% sodium dodecyl sulphate (SDS) and then sealed in a heat sealable plastic bag with 5 ml of freshly prepared blocking solution, (5 × SSPE, 0.5% Blotto (Carnation Skim Milk Powder), 10% dextran sulphate, 7% SDS) per 5 cm^2 of membrane. Membranes were stored at −20% in blocking solution until used. Membranes were prehybridized for 6–12 h at 43, 55 or 65°C in a shaking water bath, and following the addition of ^{32}P-labelled oligonucleotide, probed for 19–24 h at the same temperature. The conditions for stringent hybridization were established using dot blot analysis. Following hybridization the membrane was removed from the bag and dipped briefly in 2 × SSC (0.3 M NaCl, 0.06 M Tri Sodium Citrate) then washed twice for 10-min periods with vigorous shaking in 2 × SSC, 0.1% SDS. This was followed by two 15-min washes in 1 × SSC, 1% SDS. The membranes were then washed for 5 min in 1 × SSC, 0.1% SDS and blotted dry using Whatman 3 MM filter paper. The membrane was then wrapped in Glad Wrap and autoradiographed between two high speed intensifying screens at −70°C until a signal was detected (12–72 h). Membranes could be probed with additional labelled oligonucleotides following removal of the hybridized probe immobilized on the membrane. This was achieved by boiling the membrane twice in a microwave oven in 0.1% SDS, 0.1 × SSC. Following washing in 5 × SSPE, the membrane was once again bagged and stored until rehybridization. Quantitation was achieved using a scanning densitometer or by high performance anion exchange chromatography (HPIEC) using 3 μm Tosoh TSK gel DEAE-NPR beads as described elsewhere [33].

Results and discussion

The ovary is one of the few tissues in mammals where repetitive cycles of angiogenesis occur during adult life. However, little is known about the physiological mechanism of expression and function of polypeptide growth factors implicated in ovarian vascularization. Most of the available knowledge arises from in vitro studies. In this context, the fibroblast (or heparin binding) family of growth factors have been shown to exhibit all the essential biological features of a profound physiological regulator, namely, chemotactic, angiogenic and proliferative properties which are highly regulated particularly at the post translational level. In our investigations the ability to monitor the expression of FGF-β mRNA in dynamic situations, namely in the cyclic rat and pig ovary, has been established using both in situ hybridization methods and quantitative reverse transcriptase-polymerase chain reaction procedures. Both techniques provided concordant results on the changes of FGF-β mRNA during the estrus cycle.

Figure 1 illustrates the in situ hybridization detection FGF-β mRNA of corpus luteum sections of a rat ovary (with and without RNase treatment) at the metestrus stage of the cycle. Under the in situ hybridization conditions employed, no specific localization of the FGF-β mRNA was observed when tissue was (1) incubated with RNase prior to hybridization; (2) incubated ^{35}S-labelled oligonucleotide in competition with an excess of unlabled antisense oligonucleotide and (3) incubated with a ^{35}S labelled sense oligonucleotide. The FGF-β mRNA expression in the rat ovary was observed to increase throughout the estrus cycle (diestrus, proestrus, estrus, metestrus). No detectable FGF-β mRNA levels were observed in follicles during diestrus. Following follicular recruitment during proestrus, FGF-β mRNA was observed to increase throughout the development course not only with the large antral follicles committed to ovulation at estrus, but also with the preantral follicles. The RT-PCR investigations indicated that there was a marked amplification (ca. 7-fold) of FGF-β mRNA expression in the ovary by metestrus. At this stage of the estrus cycle FGF-β mRNA expression was also very evident in specific populations of cells within the newly formed corpus luteum as determined in situ hybridization experiments. Consistently higher hybridization labelling of these cells in the corpus luteum was observed in both the rat and the pig ovary, compared to granulosa cells in developing follicles. With commencement of the next ovarian cycle (diestrus stage), very low levels of FGF-β mRNA persisted either initially in the atretic corpus luteum or

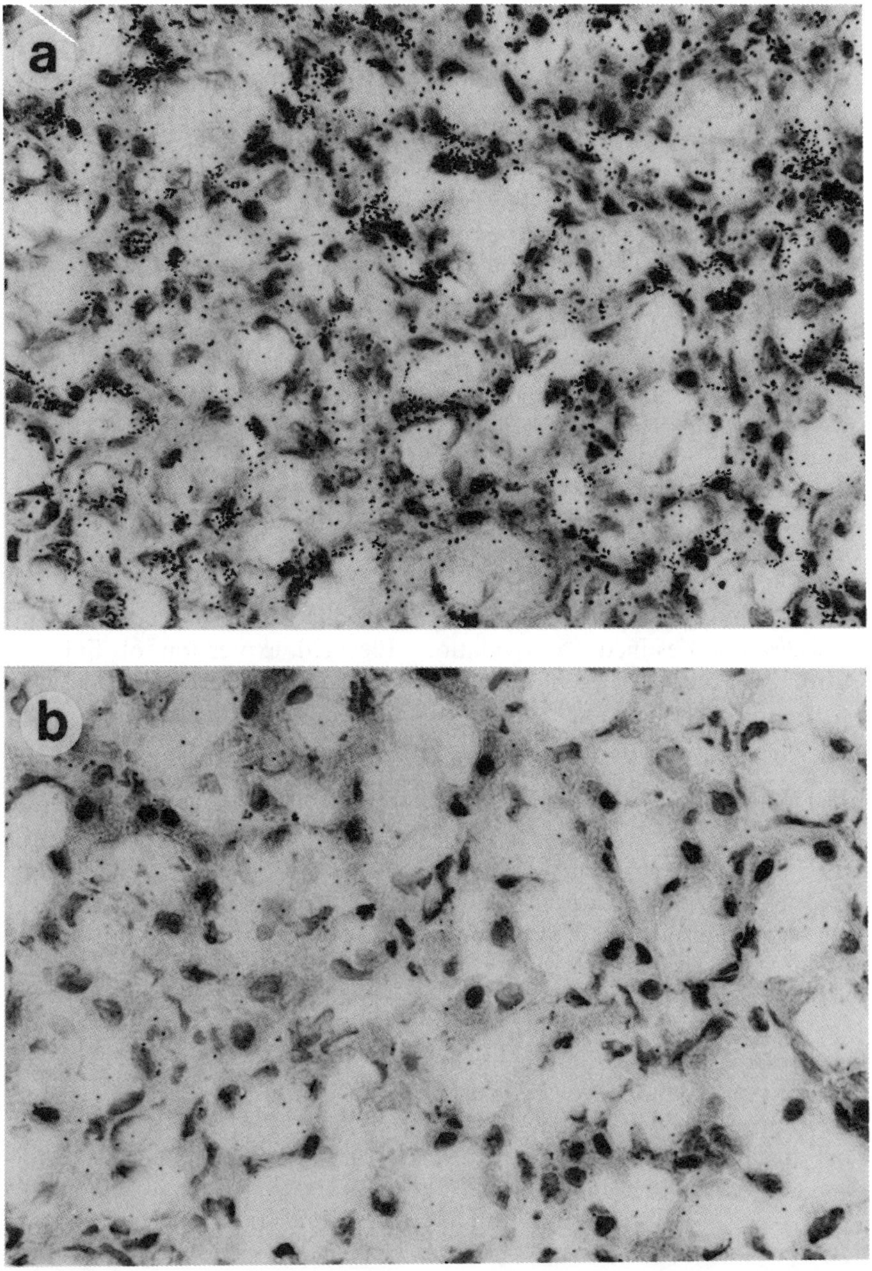

Figure 1. Photomicrograph of a corpus luteum at metestrus in the naturally cycling rat ovary. Serial sections were hybridized with [α-^{35}S]d ATP labelled FGF-β antisense oligonucleotide probes. The in situ hybridization procedures are given in the Materials and Methods section: *a*) corpus luteum; *b*) corpus luteum incubated with ribonuclease (250 μg/ml) prior to hybridization. Photographed at 400 ×.

in the senescent follicles, thereafter declining rapidly to the RNase control levels.

The combination of in situ hybridization techniques and quantitative reverse-transcriptase polymerase chain reaction procedures have permitted localization of FGF-β mRNA to the granulosa cells of developing follicles recruited during proestrus. Further, as documented in our associated studies (35), FGF-β mRNA expression can be modulated in vivo by the pleiotropic effects of the steroid hormone analogue, diethylstilbesterol. Estrogens are known to stimulate the synthesis of epidermal growth factor EGF in mouse uterine epithelial cells and IGF-1 in rat uterus in vivo [36, 37]. Previous studies have also shown [16, 17, 19] that pharmacological dosages of FGF-β will inhibit estradiol and progesterone production by granulosa cells in culture, suggesting the existence of a feed back loop. Thus, dual mediation of proliferation and differentiation by estradiol (and other steroids) and FGF-β can be proposed. Preliminary data obtained in this laboratory are consistent with this concept. Should further experimental results confirm this relationship, then the following scenario could well represent the estrogen–FGF-β axis.

Dominant follicles are known to secrete estrogen during their maturation, partly to promote their own growth but also to induce atresia in follicles not destined for ovulation. Elevated expression of FGF-β, through a negative feed back loop, could then influence estradiol biosynthesis, possible via (or as) a nuclear transcription regulator. The preferential advantage of steroid-mediation of FGF-β mRNA expression within the dominant follicles could further favour follicular selection and finally ovulation. Besides a mitogenic role on developing follicles, the expression of FGF-β by granulosa and theca cells is likely to selective promote vascularization of the theca interna surrounding dominant follicles prior to ovulation. Following estrus, neovascularization of the avascular granulosa cell layer, could then be accomplished by expansion of the capillary network of the theca interna, and the resultant newly formed corpus luteum. The marked decline in FGF-β mRNA levels at the onset of diestrus would then result in a diminished angiogenic signal and regression of the corpus luteum. Variations on this scenario could provide explanations for the regionalism of ovarian development under the influence of local regulators. As documented in this study, in situ hybridization and quantitative reverse transcriptase polymerase chain reaction procedures provided further evidence for the involvement of FGF-β in the in vivo modulation of ovarian function. The extraordinary changes in cell morphology, phenotype and dominance seen with the mammalian ovary clearly implicate the cooperative participation of local regulatory factors in addition to FGF-β. Our current studies are addressing the fascinating physiological and biochemical questions posed by the orchestration of this network of intra-ovarian regulators of ovarian function.

Acknowledgements. The support of the National Health and Medical Research Council of Australia, the Australian Pig Research Council and Monash University Research Fund is gratefully acknowledged.

References

1 Hsu, C-J, Holmes SD, Hammond JM: Biochem Biophys Res Commun 1987; 147: 242–247.
2 Hsu C-J, Hammond JM: Endocrinology 1987; 120: 198–206.
3 Ramasharma K, Li Ch: Proc Natl Acad Sci USA 1987; 84: 2643–2647.
4 Bendell JJ, Darrington J: Endocrinology 1988; 123: 941–948.
5 Skinner MK, Keski-Oja J, Osteen KG, Moses HL: Endocrinology 1987; 121: 786–792.
6 Robertson DM, de Vos FL, Foulds LM, McLachlan RI, Burger HG, Morgan FJ, Hearn MTW, de Kretser DM: J Mol Cell Endocr 1986; 44: 271–277.
7 Robertson DM, de Vos EL, McLachlan RI, Wettenhal REH, Hearn MTW, Burger HG, de Kretser DM: Biochem Biophys Res Commun 1988; 149: 744–751.
8 Bertolini J, Gutheridge M, Hearn MTW: J Chromatogr 1989; 491: 49–60.
9 Bertolini J, Hearn MWT: J Mol Cell Endocr 1987; 51: 187–193.
10 Gospodarowicz D, Cheng J, Lui GM, Barid A, Esch F, Bohlen P: Endocrinology 1985; 117: 2383–2387.
11 Bassett DL: Am J Anat 1943; 73: 251–262.
12 Beers WH, Strickland S, Reich E: Cell 1975; 6: 379–386.
13 Montesano R, Vassalli JD, Baird A, Guillemin R, Orci L: Proc Natl Acad Sci USA 1986; 83: 7297–7301.
14 Presta M, Moscatelli D, Silverstein JJ, Rifkin DB: Mol Cell Biol 1986; 6: 4060–4066.
15 Moscatelli D, Presta M, Rifkin DB: Proc Natl Acad Sci USA 1986; 83: 2091–2095.
16 Mondschein JS, Schomberg DW: Science 1987; 211: 1179–1180.
17 Adashi EY, Resnick CE, Croft CS, May JV, Gospodarowicz D: Mol Cell Endocr 1988; 55: 7–14.
18 Baird A, Hsueh AJW: Reg Peptides 1986; 16: 243–250.
19 Biswas SB, Hammond RW, Anderson LD: Endocrinology 1988; 123: 559–566.
20 Taponainen J, Leinonen PJ, Tapanainen P, Yamamoto M, Jaffe RB: Fert Ster 1987; 48: 576–580.
21 Neufield G, Ferrara N, Schweigerer L, Mitchell R, Gospodarowicz D: Endocrinology 1987; 121: 597–603.
22 Gospodarowicz D, Ill CR, Birdwell CR: Endocrinology 1977; 100: 1108–1120.
23 Gospodarowicz D, Neufield G, Schweigerer L: Mol Cell Endocr 1986; 46: 187–204.
24 Gospodarowicz D, Moran J: Proc Natl Acad Sci USA 1974; 71: 4648–4652.
25 Gospodarowicz D, Bialecki H: Endocrinology 1987; 103: 854–858.
26 Baird A, Esch F, Mormede P, Venu N, Ling N, Bohlen P, Wehrenberg WB, Guillemin R: Rec Prog Horm Res 1986; 42: 143–197.
27 Shimisaki S, Emoto N, Loba A, Mercado M, Shibata F, Cookesy K, Baird A, Ling N: Biochem Biophys Res Commun 1988; 157: 256–263.
28 Rogelj S, Weinberg RA, Fanning P, Klagsbrun M: J Cell Biochem 1989; 39: 13–23.
29 DiMario J, Buffinger N, Yamada S, Strohman RC: Science 1989; 244: 688–690.
30 Bashkin P, Dochrow S, Klagsbrun M, Svahn CM, Folkman J, Vlodavsky I: Biochemistry 1989; 28: 1737–1743.
31 Guthridge M, Bertolini J, Schmitt J, Hearn MTW: J Biochem Biophys Meth 1991; 22: 201–217.
32 Schmitt J, Guthridge M, Bertolini J, Hearn MTW: J Biochem Biophys Meth 1991; in press.
33 Cowling J, Guthridge M, Schmitt J, Bertolini J, Hearn MTW: Biochem Biophy Res Commun 1991, in press.
34 Guthridge M, Bertolini J, Cowling J, Hearn MTW: J Mol Endocr 1991, submitted.
35 Huet-Hudson YM, Chakraborty C, Suzuki Y, Andrews GK, Dey SK: Mol Endocr 1990; 4: 510–523.
36 Murphy LJ, Murphy LC, Friesen HG: Mol Endocr 1987; 1: 445–450.

Angiogenesis: Key Principles - Science - Technology - Medicine
ed. by R. Steiner, P. B. Weisz & R. Langer

Effects of platelet-activating factor on endothelial cells and fibroblasts in vitro

Rachel L. Smither and Tai-Ping D. Fan

Department of Pharmacology, University of Cambridge, Tennis Court Road, Cambridge CB2 1QJ, England

Introduction

Angiogenesis, the formation of new blood vessels, is normally under stringent control. Persistent angiogenesis can lead to pathological conditions such as atherosclerosis, diabetic retinopathy and rheumatoid arthritis. There are three major components of angiogenesis, namely protease production by endothelial cells (EC) and fibroblasts, migration and proliferation of EC, and EC tube formation. Recent studies, using both in vivo and in vitro models of angiogenesis, have established the role of several polypeptide growth factors in these processes [1]. Some, e.g. basic fibroblast growth factor, are involved in all three events, while others, e.g. transforming growth factor-β and tumour necrosis factor-α, are not. The role of small molecular weight inflammatory mediators, which are likely to be present at sites of tissue inflammation, is less well characterised.

One such mediator is platelet-activating factor (PAF). This phospholipid derivative has a unique repertoire of biological activities and target cells [2]. It is also known to be produced by EC [3] and to increase vascular permeability through a direct effect on vascular EC [4, 5]. We have recently shown that PAF enhances neovascularisation in an in vivo rat sponge model and that three structurally unrelated PAF antagonists inhibit neovascularisation in the same model [6]. Here we describe the effects of PAF on collagenase production by fibroblasts, and on EC migration and proliferation in vitro.

Methods

Rat aortic endothelial cells (RAEC) and bovine retinal endothelial cells (BREC) were obtained and cultured as described previously [7, 8]. Human gingival fibroblasts (HGF cells) obtained from human gingival explants were cultured as for RAEC. All cells were cultured in monolayers and maintained at 37°C, 5% CO_2.

For the collagenase assay, HGF cells in 24-well culture plates were incubated for 48 h with 10^{-14}–10^{-4} M PAF or 10^{-12}–10^{-6} M lyso-PAF in serum free medium. After incubation, collagenase activity in the culture medium was determined by the method of Cawston and Barret [9].

To study EC migration RAEC were grown to confluence on Thermanox tissue culture coverslips and then a single channel, 2 mm wide, wound was made in the monolayer using a perspex wounder previously developed in the laboratory. Wounded coverslips were incubated for 12 h with 10^{-11}–10^{-5} M PAF, or 10^{-10}–10^{-6} M lyso-PAF and then the width of the denuded area was measured using a computer assisted image analysis system (VIDS III, Analytical Measuring Systems).

Proliferation of RAEC and BREC grown in 96-well culture plates in the presence of 10^{-11}–10^{-5} M PAF or lyso-PAF was measured by a [^{3}H]-thymidine uptake assay.

Results

PAF (10^{-14}–10^{-4} M) caused a dose-dependent enhancement of collagenase production by HGF cells, as can be seen in Fig. 1. At 10^{-10}–10^{-6} M, PAF accelerated the closure of a 2-mm wound in RAEC monolayers resulting in smaller wound widths than controls 12 h after wounding (Tab. 1). PAF (10^{-11}–10^{-7} M) caused a statistically significant increase in [^{3}H]-thymidine uptake compared with controls in RAEC and BREC as can be seen in Fig. 2. Lyso-PAF, the biologically inactive metabolite of PAF, had no effect on fibroblast collagenase production, or on EC migration or proliferation (Figs. 1 and 2, Table 1).

Table 1. Effect of PAF and lyso-PAF on wound closure in RAEC[1]

Concentration (M)	PAF	Lyso-PAF
10^{-11}	95 ± 5	ND
10^{-10}	91 ± 4	98 ± 4
10^{-8}	78 ± 4*	106 ± 4
10^{-6}	76 ± 6*	96 ± 5
10^{-5}	84 ± 9	ND

[1]Cells were grown and wounded as described in the Methods section. Table shows wound width (% of control width) after 12 h incubation in the presence of 10^{-11}–10^{-5} M PAF or 10^{-10}–10^{-6} M lyso-PAF ± SEM. *$p < 0.005$, Student's t-test vs controls, n = 3 experiments. ND = not determined.

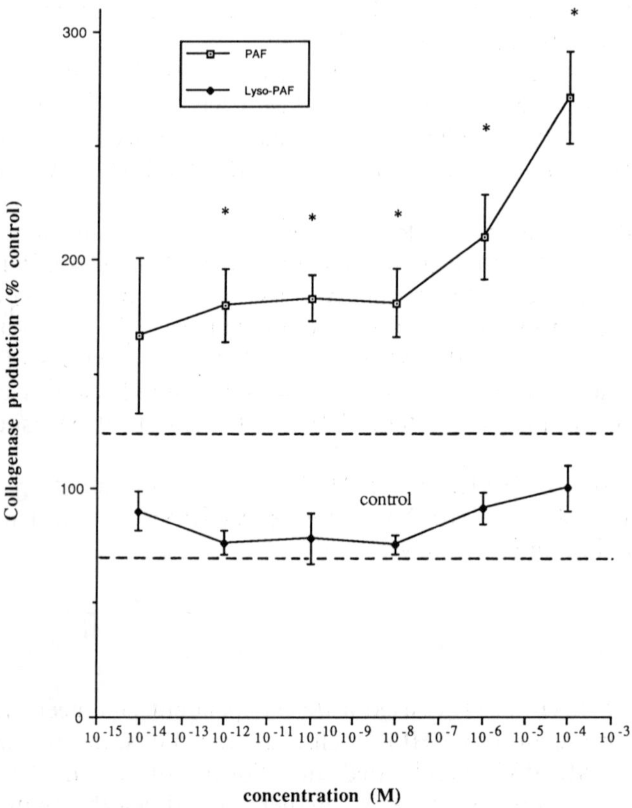

Figure 1. Effect of PAF and lyso-PAF on collagenase production by HGF. Graph shows concentration of PAF or lyso-PAF (log scale) against collagenase production expressed as a percentage of control collagenase activity in untreated cells (0.85 ± 0.05 U collagenase/ml). Bars represent s.e. mean. *$p < 0.005$. Student's t-test vs controls, $n = 3$.

Discussion

The observed effects of PAF on three of the sequential steps thought to occur during neovascularisation suggest it has a direct mechanism of action rather than one mediated through other cell types, enzymes or growth factors. The low concentrations at which PAF is effective, and the inability of lyso-PAF to have a similar effect over the same concentration range indicate the effects of PAF are the result of receptor-mediated processes and not due to non-specific membrane effects [10].

The magnitude of the collagenase production by HGF in response to exogenous PAF is comparable to that observed in EC stimulated by angiogenic preparations such as tumour cell lysates, bovine retinal extracts and adipocyte conditioned medium [11, 12]. Thus PAF-induced

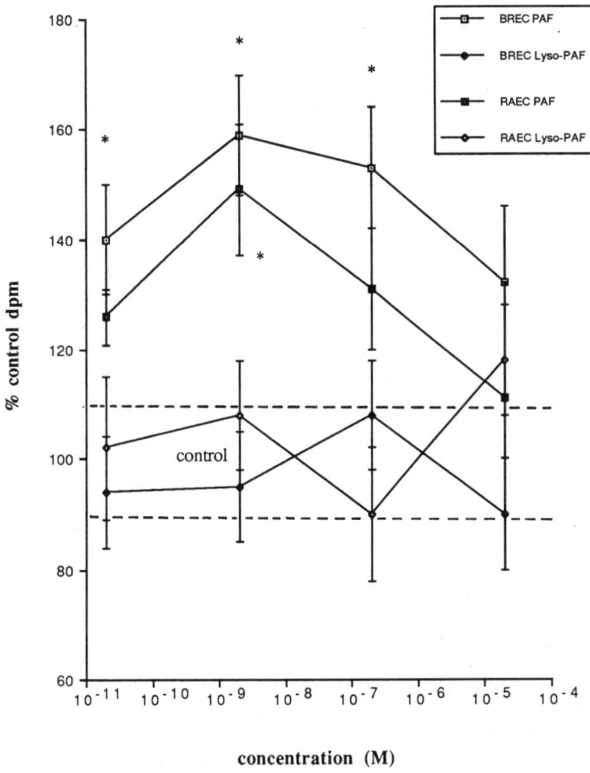

Figure 2. Effect of PAF and lyso-PAF on EC [^{3}H]-thymidine uptake. Graph shows [^{3}H]-thymidine uptake expressed as % control values against concentration PAF or lyso-PAF (M). bars represent s.e. mean. *$p < 0.005$, Student's t-test vs controls, $n = 3$ experiments for RAEC and $n = 12$ wells in representative experiments of 3 replicates for BREC.

fibroblast collagenase production may be of similar importance in angiogenesis.

Previous studies in cell wounding assays similar to the one employed here have shown that a significant increase in [^{3}H]-thymidine labelling does not occur until 36 h after wounding, suggesting that the response to denudation of the cell monolayer is predominantly migratory rather than proliferative in the first 12 h [13]. This was also the case in our model (data not shown). Thus we propose that PAF increases the migration of RAEC after wounding. This is consistent with the reported roles for PAF, at concentrations similar to those active in the present study, in changing EC morphology to a retracted, spindle-shaped form though to represent a migratory phenotype [10].

Preliminary studies in the laboratory have shown that [^{3}H]-thymidine uptake is a valid technique for assessing EC proliferation since the number of viable cells determined by the Trypan Blue exclusion method correlates well with [^{3}H]-thymidine uptake. Thus the [^{3}H]-thymidine studies described here reflect the ability of PAF to increase the proliferation of EC in vitro. The fact that RAEC proliferated in response to PAF is noteworthy since our in vivo observations with PAF and its antagonists were made in rats [6]. The enhancement of proliferation of BREC by PAF is also of note since the study of microvascular EC is pertinent to the study of angiogenesis [14].

In summary, PAF enhances three processes thought to be involved in neovascularisation, namely collagenase production by fibroblasts and the migration and proliferation of EC. The results of this study, together with previous observations of the effects of PAF in angiogenesis in vivo substantiate a role for this inflammatory mediator in the modulation of the angiogenic process.

References

1 Folkman J, Klagsbrun M: Angiogenic factors. Science 1987; 235: 442–447.
2 Braquet P, Touqui L, Shen TY, Vargaftig BB: Perspectives in Platelet-activating Factor Research. Pharmacol Rev 1987; 39(2): 97–1454.
3 Camussi G, Aglietta M, Malavast F, Tetta C, Piacibello W, Sanavio F, Bussolino F: The release of platelet-activating factor from human endothelial cells in culture. J Immunol 1983; 131(5): 2397–2403.
4 Wedmore CV, Williams TJ: Platelet-activating factor (PAF), a secretory product of polymorphonuclear leucocytes, increases vascular permeability in rabbit skin. Br J Pharmacol 1981; 74: 916–917P.
5 Humphrey DM, McMannus LM, Hanahan DJ, Pinckard RN: Morphological basis of increased vascular permeability induced by acetyl glyceryl ether phosphorylcholine. Lab Invest 1984; 50(1): 16–22.
6 Smither RL, Fan T-PD: PAF antagonists inhibit angiogenesis in a rat sponge model. Br J Pharmacol 1990; 99: 87P.
7 Cole OF, Fan T-PD, Lewis GP: Isolation, characterisation, growth and culture of endothelial cells from the rat aorta. Cell Biol Int Rep 1986; 10(6): 399–405.
8 McIntosh LC, Muckersie L, Forrester JV: Retinal capillary endothelial cells prefer different substrates for growth and migration. Tissue Cell 1988; 20(2): 193–209.
9 Cawston TE, Barret AJ: A rapid and reproducible assay for collagenase using [1-^{14}C]-acetylated collagen. Analyt Biochem 1979; 99: 340–345.
10 Bussolino F, Camussi G, Aglietta M, Braquet P, Bosia A, Pescarmona G, Sanavio F, D'Urso N, Marchisio PC: Human endothelial cells are target for platelet activating factor. I. Platelet activating factor induces changes in cytoskeleton structures. J Immunol 1987; 139: 2439–2446.
11 Gross JL, Moscatelli D, Rifkin DB: Increased capillary endothelial cell protease activity in response to angiogenic stimuli in vitro. Proc Natl Acad Sci USA 1983; 80: 2623–2627.
12 Moscatelli DA, Rifkin DB, Jaffe EA: Production of latent collagenase by human umbilical vein endothelial cells in response to angiogenic preparations. Exp Cell Res 1985; 156: 379–390.
13 Sholley MM, Gimbrone MA Jr, Cotran RS: Cellular migration and replication in endothelial regeneration. Lab Invest 1977; 36(1): 18–25.
14 Keegan A, Hill C, Kumar S, Phillips P, Schor A, Weiss J: Purified tumour angiogenesis factor enhances proliferation of capillary but not aortic endothelial cells in vitro. J Cell Sci 1982; 55: 261–276.

Angiogenesis: Key Principles - Science - Technology - Medicine
ed. by R. Steiner, P. B. Weisz & R. Langer

Angiogenin and endothelial cells

Josette Badet, Fabrice Soncin and Denis Barritault

Laboratoire de Biotechnologie des Cellules Eucaryotes, Université Paris, Val de Marne, Créteil, France

Antiogenin is a potent blood vessel-inducing polypeptide that has a unique ribonucleolytic activity. Initially identified from tumor cells conditioned medium through its ability to induce neovascularization in the chick embryo chorioallantoic membrane assay [1], angiogenin has also been purified from normal plasma [2] which suggested that it might be involved in endothelium homeostasis. Angiogenin is highly homologous to pancreatic ribonuclease A with 68% amino acid sequence homology and conserved essential active site residues [3]. However, angiogenin is inactive in standard ribonuclease assays but has a ribonucleolytic specificity toward ribosomal and transfer RNAs [4]. This enzymic activity might be relevant to the process of angiogenesis since a functional active site seems necessary for the expression of its angiogenic property [5]; however, it appears not to be sufficient [6].

Interactions of angiogenin with endothelial cells

Using recombinant human angiogenin produced by Rhône–Poulenc Santé and shown to be angiogenic, in vivo, both in the chick chorioallantoic membrane and the rabbit cornea [7, 8] we have demonstrated the presence of high affinity binding sites on endothelial cells from calf pulmonary artery (CPAE), bovine aorta, cornea and adrenal cortex capillary that could not be detected on Chinese hamster lung fibroblasts [9]. Receptors have also been reported on bovine brain capillary endothelial cells [10]. The interactions on CPAE cells displayed an apparent dissociation constant in the nanomolar range and involved about 200,000 angiogenin molecules per cell [9]. This number, that varies with the cell density, does not necessarily represent a number of receptors. However, it includes receptors since angiogenin, added to cultured cells, triggers off a series of intracellular events in endothelial cells such as a transient increase in 1,2-diacylglycerol [11] and activation of phospholipase A_2 [12], and in vascular smooth muscle cells, activation of phospholipase C and a rapid cholesterol esterification [13]. Besides, angiogenin might modulate mitogenic stimuli [10, 14].

Angiogenin has also been shown to interact with low-affinity/high capacity specific binding sites on CPAE cells as well as on extracellular matrix. This second family of interactions, with an apparent dissociation constant of 0.2 μM, involved millions of molecules [9].

An angiogenin-binding protein has been recently identified on bovine endothelial cells as a dissociable cell-surface component of 42 KD released by treating the cells with heparan sulfate [15].

Inhibition of angiogenin properties by placental ribonuclease inhibitor

Placental ribonuclease inhibitor (PRI) is a tight binding inhibitor of both the ribonucleolytic and angiogenic activities of angiogenin [16] with an inhibition constant value, Ki, in the femtomolar range [17]. This inhibitor also abolishes angiogenin binding to CPAE cells [9]. However, PRI is an intracellular regulatory protein and although immunoreactivity has been detected in normal human plasma [18], there is not yet any experimental evidence for angiogenin/inhibitor circulating complexes.

Effects of copper metal ions on angiogenin binding to endothelial cells and to ribonuclease inhibitor

Divalent copper, a modulator of angiogenesis [19], has been shown to induce a several-fold increase in specific cell-bound angiogenin and to reverse PRI-induced inhibition of angiogenin binding to CPAE cells [8, 9]. Furthermore, divalent copper metal ions inhibit angiogenin ribonucleolytic activity towards tRNA substrate [20].

Discussion

The variety of copper interactions with angiogenin properties is indeed puzzling. It might reflect a tight regulation of angiogenin functions necessary to balance its angiogenic potency with its high intravascular concentration. Copper also interacts with heparin and the angiogenic growth factor FGF [21], and appears, from other reported data, to modulate angiogenesis somehow [19]. At this point, it is worth noting that copper is required for other fundamental events such as central nervous system development, bone and connective tissue biosynthesis, electron transport and antioxidant protection where it acts as a prosthetic group [22]. Because of angiogenin peculiarities, studies on its mechanism of action should lead the way to a better knowledge of the biochemical events involved in angiogenesis.

In summary, modulation of angiogenin functions might involve interactions with cell-surface receptors, binding protein and extracellular matrix components, tight-binding inhibition of both its ribonucleolytic activity and cell binding by ribonuclease inhibitor under the overall influence of divalent copper, a modulator of angiogenesis.

Acknowledgements. We thank Dr. T. Cartwright and his collaborators for recombinant human angiogenin (Rhône-Poulenc Santé). This work was supported by funds from 'l'Association de la Recherche sur le Cancer' (grant no 6231) and 'le Ministère de l'Education Nationale'. J.B. was supported by 'l'institut National de la Santé et de la Recherche Médicale' and F.S. by 'le Ministère de la Recherche et de la Technologie'.

References

1 Fett JW, Strydom DJ, Lobb RR, Alderman EM, Bethune JL, Riordan JF, Vallee BL: Isolation and characterization of angiogenin, an angiogenic protein from human carcinoma cells. Biochemistry 1985; 24: 5480–5486.
2 Shapiro R, Strydom DJ, Olson KA, Vallee BL: Isolation of angiogenin from human plasma. Biochemistry 1987; 26: 5141–5146.
3 Vallee BL, Riordan JF: Chemical and biochemical properties of human angiogenin. Adv Exp Med Biol 1988; 234: 41–53.
4 Shapiro R, Riordan JF, Vallee BL: Characteristic ribonucleolytic activity of human angiogenin. Biochemistry 1986; 25: 3527–3532.
5 Riordan JF, Vallee BL: Human angiogenin, an organogenic protein. Br Cancer 1988; 57: 587–590.
6 Hallahan TW, Shapiro R, Vallee BL: Dual site model for the organogenic activity of angiogenin. Proc Natl Acad Sci USA 1991; 88: 2222–2226.
7 Denèfle P, Kovarik S, Guitton J-D, Cartwright T, Mayaux J-F: Chemical synthesis and expression of a gene coding for human angiogenin in *Escherichia coli* and conversion of the product into its active form. Gene 1987; 56: 61–70.
8 Badet J, Soncin F, N'Guyen T, Barritault D: In vivo and in vitro studies of angiogenin – a potent angiogenic factor. Blood Coag Fibrin 1990; 1: 721–724.
9 Badet J, Soncin F, Guitton J-D, Lamare O, Cartwright T, Barritault D: Specific binding of angiogenin to calf pulmonary artery endothelial cells. Proc Natl Acad Sci USA 1989; 86: 8427–8431.
10 Chamoux M, Dehouck MP, Fruchart JC, Spik G, Montreuil J, Cecchelli R: Characterization of angiogenin receptors on bovine brain capillary endothelial cells. Biochem Biophys Res Commun 1991; 176: 833–839.
11 Bicknell R, Vallee BL: Angiogenin activates endothelial cell phospholipase C. Proc Natl Acad Sci USA 1988; 85: 5961–5965.
12 Bicknell R, Vallee BL: Angiogenin stimulates endothelial cell prostacyclin secretion by activation of phospholipase A2. Proc Natl Acad Sci USA 1989; 86: 1573–1577.
13 Moore F, Riordan JF: Angiogenin activates phospholipase C and elicits a rapid incorporation of fatty acid into cholesterol esters in vascular smooth muscle cells. Biochemistry 1990; 29: 228–233.
14 Heath WF, Moore F, Bicknell R, Vallee BL: Modulation of mitogenic stimuli by angiogenin correlates with in vitro phosphatidylinositol biphosphate synthesis. Proc Natl Acad Sci USA 1989; 86: 2718–2722.
15 Hu G-F, Chang S-I, Riordan JF, Vallee BL: An angiogenin-binding protein from endothelial cells. Proc Natl Acad Sci USA 1991; 88: 2227–2231.
16 Shapiro R, Vallee BL: Human placental ribonuclease inhibitor abolishes both angiogenic and ribonucleolytic activities of angiogenin. Proc Natl Acad Sci USA 1987; 84: 2238–2241.

17 Lee FS, Shapiro R, Vallee BL: Tight-binding inhibition of angiogenin and ribonuclease A by placental ribonuclease inhibitor. Biochemistry 1989; 28: 225–230.
18 Feldman M, Kohtz DS, Kleinberg DL: Isolation and characterization of monoclonal antibodies against ribonuclease inhibitor. Biochem Biophys Res Commun 1988; 157: 286–294.
19 Folkman J, Klagsbrun M: Angiogenic factors. Science 1987; 235: 442–447.
20 Lee FS, Vallee BL: Characterization of ribonucleolytic activity of angiogenin towards tRNA. Biochem Biophys Res Commun 1989; 161: 121–126.
21 Shing Y: Heparin-cooper biaffinity chromatography of Fibroblast Growth Factors. J Biol Chem 1988; 263: 9059–9062.
22 Ettinger MJ: Cellular biochemistry of copper and inherited defects of copper metabolism. Life Chem Rep 1987; 5: 169–186.

Angiogenesis: Key Principles - Science - Technology - Medicine
ed. by R. Steiner, P. B. Weisz & R. Langer

Cultured endothelial cell mobilization induced by partially purified human uterine angiogenic factor

Nitza Mirsky[a] and Ella S. Lindenbaum[b]

[a]*Biology Department, Oranim, Haifa University and* [b]*Morphology Research Unit, Faculty of Medicine, Technion IIT, Haifa, Israel*

Introduction

Endothelial cell migration

The mobilization of endothelial cells may be induced by a number of external factors, and represents one step in a sequence of events which occurs during angiogenic stimulation. In vitro endothelial cells can organize and form a three-dimensional capillary network [1]. This process entails cellular motility.

In vivo studies have shown that some angiogenic factors induce directional migration but not necessarily cell proliferation [2]; therefore, endothelial cell migration during growth is independent of cell mitogenicity. In wound-healing experiments, the cell processes adhere to the fibrin matrix and anchor the cells, thus participating in the process of cell migration. At the capillary tip, the advancing edge grows by cell elongation, gliding over, and finally by reattachment. However, the mode of cell migration is not fully understood. Cell movement is an inherent characteristic of cells and the sessile state is caused by contact inhibition [3]. Released from inhibition, the cells seem to move while maintaining some contact with each other.

Human uterine angiogenic factor

The rapid vascular growth and proliferation associated with the normal menstrual cycle in general, and with pregnancy in particular, is unique among biological systems. The magnitude, density of vasculature and rate of growth point to the presence of specific factors controlling this process. The remodeling of the vascular bed in the endometrium of pregnancy – the decidua – commences prior to the establishment of the functional placenta, the extraembryonic membranes and the embryonic differentiation. Therefore, the decidua was selected as a likely source of angiogenic factor.

Materials and methods

Extraction procedure

Decidual tissue was obtained during legally induced first-trimester terminations of pregnancy at the Rambam Medical Center. The tissue was pooled, kept on ice and frozen in $-70°$ under sterile conditions. Batches of 100 g were extracted in our laboratory using ultrafiltration techniques. The frozen tissue was homogenized in cold, double-distilled water, centrifuged at 6000 rpm for 45 min, and the supernatant S1 retained. The sediment was resuspended and centrifuged as above. This supernatant, S2, was combined with S1 and acidified in 3% glacial acetic acid for 2 min, neutralized in NH_4OH, and recentrifuged. The supernatant, S3, was ultrafiltered through Millipore cartridge filters of 30,000 MW cutoff, yielding the F1 filtrate.

The F1 fraction was found to induce angiogenic activity on the chorioallantoic membrane (CAM) of the fertilized chick egg at 0.011 μg protein/disc [4]. In addition, F1 induced mitogenic activity at 10–100 ng protein/well in bovine brain capillary endothelial (BBCE) cell culture as measured by the thymidin incorporation method [5].

Chemotaxis of endothelial cells

Chemotactic activity was demonstrated under agarose using cultures of bovine brain capillary endothelial (BBCE) and bovine adrenal capillary endothelial (BACE) cells [6]. Each petri dish was coated with 7 ml agarose. Eight series of three wells, 2.4 mm in diameter and spaced 2.4 mm apart, were cut in each plate. The center well of each triplicate was seeded with 10 μl cell suspension (10^5 cells/well). The inner well contained 10 μl control medium (or water) and the outer well 10 μl of the tested material. The culture dishes were incubated at 37°C in 60% humidified atmoshpere containing 7.5% CO_2 [7]. Following overnight incubation, the plates were fixed in methanol for 10 min and in formalin for an additional 10 min. The gel was removed and the cells were stained in 10% Gimsa. Cell motility towards the tested reagents was measured and photographed using the Nikon SMZ 10 stereomicroscope, the Sony CCD camera and Sony Trinitron video monitor.

Cell migration was calculated as the differential between the linear distance which the cells migrated from the edge of the well towards the site of the chemotactic agent (A) and spontaneous migration (B).

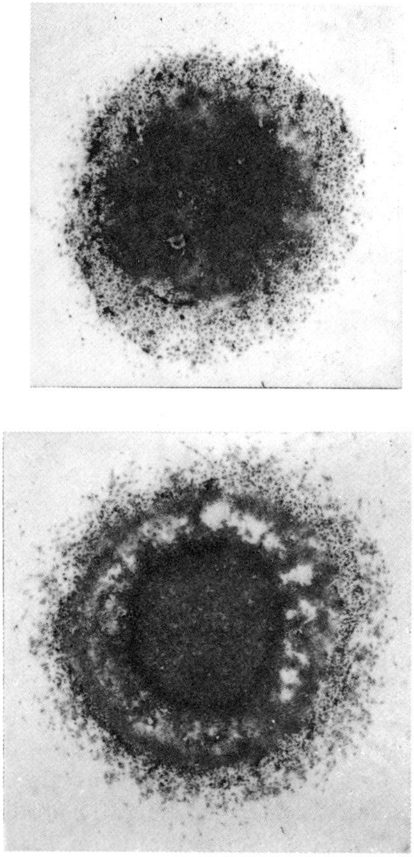

Figure 1. Photomicrograph of BACE cell culture treated with (*a*) 7.5 ng/well HUAF and (*b*) 10^{-7} M/well FMLP. Note the preferential migration pattern towards the site of the chemotactic agent. (× 20).

Results

Positive chemotactic response was demonstrated towards HUAF and FMLP using both BACE and BBCE cultures. Directional migration of the endothelial cells was expressed by cell elongation in the direction of movement and by endothelial cell mobilization (Fig. 1). Three concentrations of HUAF were tested – 15 ng/well, 7.5 ng/well and 3.75 ng/well. A dose response was observed; the optimal response occurred at 3.75 ng/well (Fig. 2a). Thus, the required dosage for inducing chemotactic activity is similar to that required for the mitogenic response which was active at 10–100 ng/ml [5]. BACE and BBCE migration was present also towards FMLP (Formal Methanyl Leucyl Phenyl Alanine) an analogue of a microbial peptide and a known chemotactic agent for

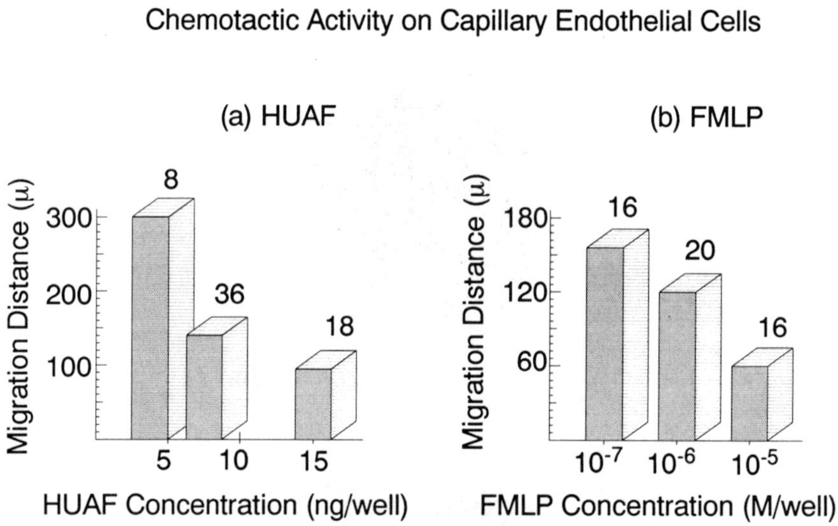

Figure 2. Chemotactic activity of HUAF (*a*) and FMLP (*b*) on capillary endothelial cells. Numbers above the bars denote the number of experiments.

neutrophils. This synthetic compound can bind to a special receptor a microbial peptide on the neturophil membrane. FMLP activity was measured using concentrations of 10^{-5} M, 10^{-6} M, and 10^{-7} M/well. A dose response was observed and the optimal chemotaxis was found at 10^{-7} M/well FMLP (Fig. 2b). The results show that the lower dosages were more effective than the higher dosages in producing cell motility. This fact might indicate that some inhibition to cell migration was present at the higher concentrations of both HUAF and FMLP.

Discussion

HUAF induced migration of 2 strains of endothelial cells after 12 h incubation. The response of the endothelial cells to both tested agents was dose dependent, i.e., cell migration distance increased with decreasing concentrations of tested agents.

For HUAF, the optimal concentration was noted at 3.75 ng/well; this compares favorably with a dose of 10–100 ng/well found active in inducing mitogenic response on BBCE culture [5]. Since the reaction of the endothelial cells to HUAF shows that with higher concentrations of protein, cell reactivity decreased, it may indicate that at higher concentrations, HUAF activity diminished cell migration. This effect is similar to the results obtained on cell mitogenicity which also decreased with increased dosages of HUAF [5]. In addition, since a similar dose response was present using increasing concentrations of FMLP, a gradual decrease in cell migration to FMLP was also demonstrated. The optimal migration was induced by FMLP at 10^{-7} M/well.

This report represents preliminary results. Further bioassays must be performed to confirm this data using additional agents of known chemotactic activity.

References

1 Folkman J, Haudenschild C: Angiogenesis in Vitro. Nature 1980; 288: 551–556.
2 Banda MJ, Knighton DR, Hunt TK, Werb Z: Isolation of a non-mitogenic angiogenesis factor from wound fluid. Proc Natl Acad Sci USA 1982; 79: 7773–7777.
3 Weiss P: Biological aspects of wound healing, in Patterson, Bradford W (eds): Wound Healing and Tissue Repair. University of Chicago Press, Chicago 1962.
4 Lindenbaum ES, Beach D: Angiogenesis induced on the Chorioallantoic Membrane by Partially Purified Human Uterine Factor. Proceedings of 5th World Congress for Microcirculation, Louisville, Kentucky, USA, August 1991.
5 Lindenbaum ES, Langer N, Beach D: Isolation and Culturing of Human Decidual Capillary Endothelial Cells in Serum-Free Medium Supplemented with Human Uterine Angiogenic Factor. Acta Anat 1991 (in press).
6 Nelson RD, Simmons LR, Quie PG: Chemotaxis under agarose: A method for measurement of spontaneous migration and chemotaxis of polymorphonuclear leukocytes and monocytes from human peripheral blood, in Bloom B, David JR (eds): Cell Mediated and Tumor Immunity. Academic Press, 1976, pp 663–675.
7 Haston WS, Shields JM: Neutrophil leukocyte chemotaxis: A simplified assay for measuring polarizing responses to chemotactic factors. J Immunol Meth 1986; 81: 229–237.

Angiogenesis: Key Principles – Science – Technology – Medicine
ed. by R. Steiner, P. B. Weisz & R. Langer

Two cloned cerebral endothelial cell phenotypes: an in vitro model for angiogenesis?

A. Amberger[a], U. Tontsch[a], P. Lemkin[b], G. Gabbiani[c] and H. C. Bauer[a]

[a]*Institut für Molekularbiologie, ÖAW, Billrothstr. 11, A-5020 Salzburg, Austria,*
[b]*Image Process. Section, Lab. of Math. Biol. NCI Frederick, Maryland, USA, and*
[c]*University of Geneva, 1 rue Michel-Servet, Geneva, Switzerland*

Abstract. Expression of smooth muscle α-actin and migratory behaviour of cloned cerebral endothelial cells (cEC) which exhibited two distinct phenotypes (type I, type II) were studied. Removal of mitogenic factors (αECGF, ECGS) and heparin from the culture medium resulted in a smooth muscle-like appearance (type II) of the cells, expression of smooth muscle α-actin protein and smooth muscle actin mRNA and in an increased migratory activity. In contrast, addition of growth factors and heparin led to a cobblestone-like phenotype (type I) which lacked the expression of smooth muscle α-actin but expressed other proteins as determined by 2-D-gel electrophoresis.

Introduction

We have previously shown that cloned cEC reversibly developed into two phenotypes depending on culture conditons: when endothelial cell growth supplement (ECGS) and heparin were added to the culture medium, cEC exhibited a cobblestone-like appearance (type I) whereas cells grown without these substances appeared elongated and spindle-shaped (type II) [1]. Apart from the phenomenological differences the two cell types differed also in their proliferation rate and in their ability to respond to glial and neuronal stimulation with an increase of blood-brain barrier (bbb) associated marker enzyme activities. Only type I cells displayed an increased activity of NaK-ATPase and gamma-glutamyl-transpeptidase (GGTP) after incubation with plasma membranes derived from glial cells or embryonic neurons [2, 3]. These findings led to the speculation that the two cloned cED phenotypes might be related to different endothelial functions in vivo, one being responsible for capillary formation and the other for bbb function.

In this study we have analysed cells of both phenotypes with 2-D-gel electrophoresis and the GELLAB II 2-D-gel analysis system. Furthermore, we studied the reversible expression of smooth muscle specific α-actin and the in vitro cell migration of type I and type II cEC.

Material and methods

cEC were prepared and established from capillaries derived from porcine brains as described by Tontsch and Bauer [1]. Two cEC phenotypes were obtained by either cultivation of the cells in medium 199 supplemented with 10% FCS, endothelial cell growth supplement (ECGS) or endothelial cell growth factor (ECGF-α) and heparin (cEC type I) or by cultivation in medium 199 without the mitogenic factors and heparin (cEC type II).

Immunofluorescence was carried out as described before [1]. Cells were labelled with ^{35}S-methionine overnight and the 2-D-gel electrophoresis was essentially carried out according to Bravo [4]. The fluorographs were then analysed using the GELLAB II system [5]. Immunoblots were performed as follows: extraction buffer (50 mM Tris HCl, pH 7.4; 0.154 M NaCl; 1 mM PMSF; 1 mM EDTA; 0.5% Triton-X-100 and 0.05% SDS) was added to cEC cultures and the cells were scraped off the petri dish. Immunoblots were done on PVDF membranes (Millipore) using a monoclonal anti smooth muscle α-actin antibody (Biomakor, Israel; 1:500) and alkaline phosphatase conjugated rabbit anti-mouse antibody from Dakopats (Denmark; 1:500). For Northern blot analysis total RNA was extracted as described elsewhere [6], denatured with formaldehyde, electrophoresed and transferred onto a nylon membrane as described by the manufacturer (Nytran, Schleicher and Schuell). The filters were hybridized with 5×10^6 cpm/filter of a ^{35}S-methionine labelled probe (pRAoαC) [7] for 17 h. The filters were then washed and exposed to Kodak XAO films at -80°C. Phagokinetic track assays were done according to Albrecht-Buehler [8].

Results

In Fig. 1, cEC are shown with (a) and without (b) ECGS or ECGF-α and heparin. Cells grown with these factors (type I) grew in monolayers and had a cobblestone-like appearance. Type II cEC cultured in medium only enriched with FCS were spindle-shaped and most of the cells stained positive for antibodies directed against smooth muscle α-actin (Fig. 1c,d). Both cEC types recognized *Bandeiraea simplicifolia lectin*, a specific marker for endothelial cells [9, 10]. These findings were confirmed by Immunoblot analysis as shown in Fig. 2. Only type II cells showed a band at 43 kD, the region of smooth muscle α-actin. Fig. 3 shows the time dependence of the reversible smooth muscle α-actin mRNA expression using Northern blot analysis. After 2 days a weak 1.7 kb signal was visible in cells deprived of mitogenic factors and heparin. One day later, a maximum of the 1.7 kb signal was reached. Addition of factors resulted in a disappearance of the signal after 12 h.

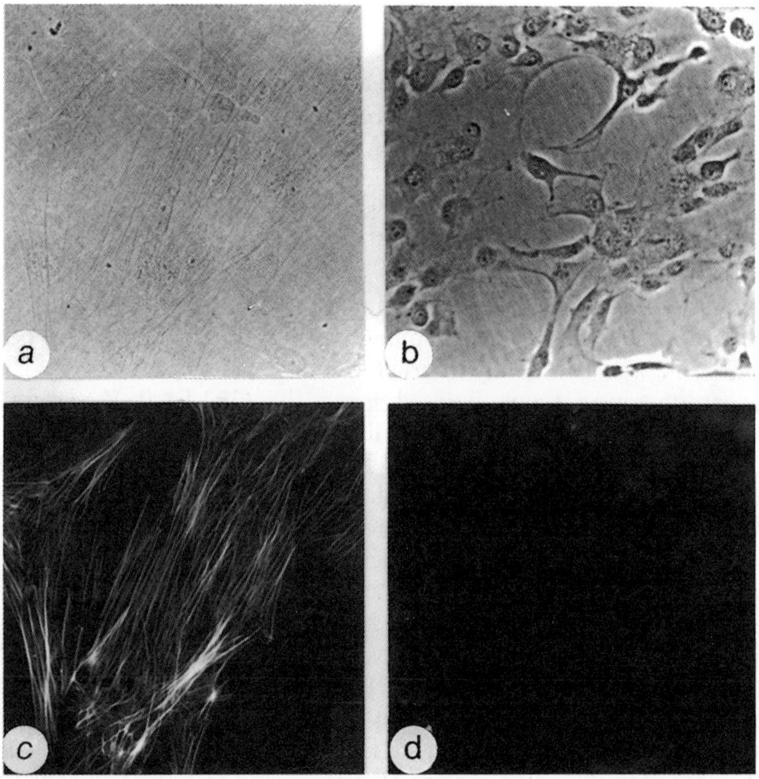

Figure 1. Double stain immunofluorescence of cEC grown on coverslips for 1 day, fixed with ethanol and glacial acetic acid. *a*) cEC type II phase photo; *b*) cEC type I phase photo; *c*) cEC type II labelled with smooth muscle-1 antibody (1:400 diluted), second antibody: FITC conjugated rabbit anti-mouse. *d*) cEC type I labelled with smooth muscle-1 antibody as described in *c*).

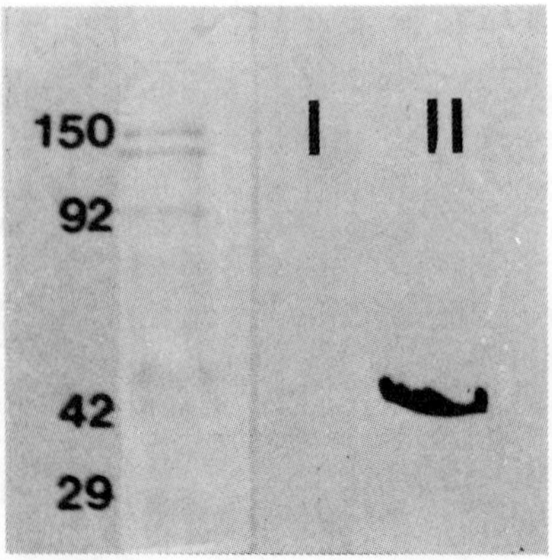

Figure 2. Smooth muscle α-actin expression in cEC detected by immunoblot technique. I) cEC type I; II) cEC type II.

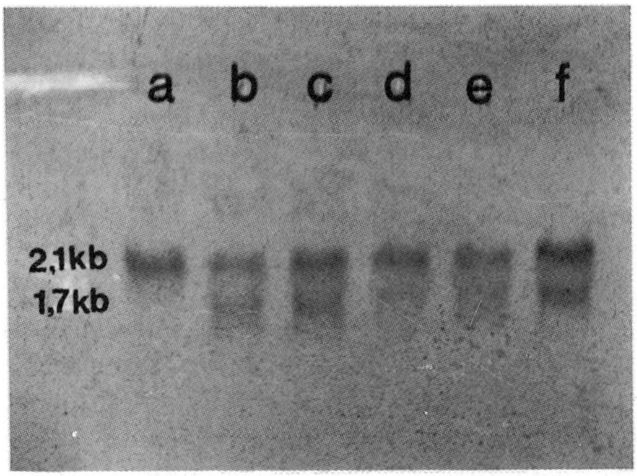

Figure 3. Northern blot of total RNA extracted from cEC; Probe: segment from the coding region of the smooth muscle α-actin gene from rat; ^{35}S-methionine labelled. 2.1 kb bands show the cytoplasmic actin mRNAs. (1.7 kb bands show the smooth muscle α-specific actin mRNAs. *a*) cEC type I, 3 days in culture; *b*)–*f*) cEC type II, 8, 6, 2, 10, 4 days in culture respectively.

The 2.1 kb signal, reflecting cytoplasmic actin isoform mRNA expression remained unchanged in type I and type II cells.

Both cell types were analysed with 2-D-gel electrophoresis using ^{35}S-methionine labelled samples and fluorography. The GELLAB II 2-D-gel analysis system was employed to determine unique proteins. Two proteins (MW 56 kD/pI 5.7 and 90 kD/pI 5.1) were only found in type I cEC (data not shown). The results of the phagokinetic track

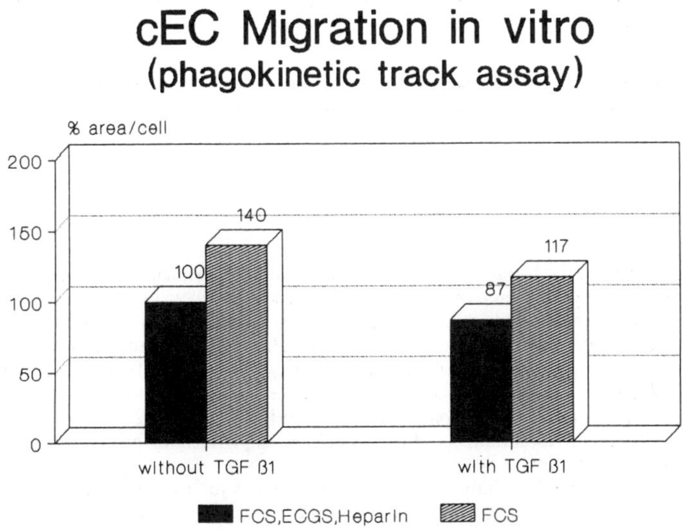

Figure 4. Phagokinetic track assay for assessment of cEC migration in vitro showing the migratory behaviour of type I cells (FCS, ECGS, Heparin) vs. type II cells (FCS)

experiments are summarized in Fig. 4 showing that type I and type II cells differed significantly in their migratory behaviour.

Discussion

In our study we have shown, that cloned cEC of different phenotypes are distinguishable by the epxression of at least three proteins. Two of them were identificed with 2-D-gel electrophoresis but could not be related so far to a distinct biological function. The expression of the third protein, smooth muscle α-actin, was studied with various methods revealing a time-dependent reversible mRNA expression. In a previous study we have demonstrated, that proliferation of type II cEC was significantly reduced compared to type I cells [1]. A similar phenomenon, i.e. the reciprocal relation of smooth muscle α-actin expression and cell replicative activity has only been reported for large vessel derived smooth muscle cells during normal and experimentally induced vascular intimal thickneing, obviously processes where enhanced migration of vascular tissue is involved [7]. Beside the smooth muscle-like appearance and the expression of smooth muscle α-actin, type II cEC also revealed increased migratory activity compared to type I cells. There is considerable evidence that the morphological and biochemical differences between cEC of the two phenotypes may be correlated to a functional diversity of the cells in vivo, such as bbb function (type I cEC) and capillary formation (type II cEC). Moreover, the question arises, whether there is a common origin of smooth muscle cells and endothelial cells.

Acknowledgement. This work was supported by the Austrian FWF, project number 7654, and the Swiss National Science Foundation, Grant Nr. 3.108–0.88.

References

1 Tontsch U, Bauer HC: Isolation, characterization and long-term cultivation of porcine and murine cerebral capillary endothelial cells. Microvasc Res 1989; 148–161.
2 Bauer HC, Tontsch U, Amberger A, Bauer H: Gamma-glutamyl-transpeptidase (GGTP) and $Na^{+}K^{+}$-ATPase activities in different subpopulations of cloned cerebral endothelial cells: responses to glial stimulation. Biochem Biophys Res Comm 1990; 168,1: 358–363.
3 Tontsch U, Bauer HC: Glial cells and neurons induce blood-brain barrier related enzymes in cultured cerebral endothelial cells. Brain Res 1991; 539: 247–252.
4 O'Farrell PH: High resolution two-dimensional electrophoresis of proteins. J Biol Chem 1975; 250: 4007–4021.
5 Lembkin PF, Lester EP: Database and search techniques for two-dimensional gel protein data: A comparison of paradigms for exploratory data analysis and prospects for biological modeling. Electrophoresis 1989; 10: 122–140.
6 Chomzcynski P, Sacchi N: Single step method of RNA isolation by acid guanidinium thiocyanate-phenol chloroform extraction. Analyt Biochem 1987; 162: 156–159.

7 Kocher O, Gabbiani G: Analysis of a-smooth-muscle actin mRNA expression in rat aortic smooth-muscle cells using a specific cDNA probe. Differentiation 1987; 34: 201–209.
8 Albrecht-Buehler G: The phagokinetic tracks of 3T3 cells. Cell 1977; 11: 395–404.
9 Laitinen L: Griffonia simplicifolia lectins bind specifically to endothelial cells in mouse tissues. Histochem J 1987; 19: 225–234.
10 Herman IM, D'Amore PA: Microvascular pericytes contain muscle and non-muscle actin. J Cell Biol 1985; 101: 43–52.

Angiogenesis: Key Principles - Science - Technology - Medicine
ed. by R. Steiner, P. B. Weisz & R. Langer

Bone endothelial cells: A tool for analyzing cell to cell interactions in the skeletal tissue

Maria Luisa Brandi

Department of Clinical Physiopathology, University of Florence Medical School, Viale Pieraccini 6, I-50139 Florence, Italy

Introduction

Bone is a complex tissue composed of a variety of cells of different nature, where osteoblasts, the bone-forming cells (of mesenchymal origin) and osteoclasts, the bone-resorptive cells (of hematopoietic origin) have been extensively characterized. However, the lack of in vitro models made it difficult to explore the role of bone endothelial cells in the process of bone formation and bone resorption. Indeed, the importance of bone vasculature in the fabrication of living bone has been proposed since 1763, when Albrecht von Haller stated that "the origin of bone is the artery carrying the blood and in it the mineral element". Vascularization is required before osteogenesis can occur, probably for a combination of factors including adequate oxygen tension, compression forces, nutrients and growth or differentiation agents. However, there is today little understanding of the role of bone endothelial cells in the bone remodeling process and of their interactions with stromal and hematopoietic systems of bone tissue. This general lack of interest may be explained by the extreme difficulty of investigating the vasculature of a tissue as hard as bone. The recent development of a clonal cell line of bone endothelial cells makes it possible today to analyze the possible interactions of bone endothelium with bone-forming and bone-resorbing cells [1].

Bone vasculature and bone formation

The possible relation between bone vasculature and bone formation was suggested in 1927 by Sir Arthur Keith, who expressed his suspicion that the bone-forming cells are derived from the endothelium of the capillary system [2]. After more than thirty years, Trueta, on the basis of the intimacy of endothelium-osteoblast contact, proposed that the endothelial cell itself is the osteoblast precursor [3]. Only recently these concepts have been developed by the use of in vivo as well as in vitro models. It is, in fact, well known that vascularized bone transfer represents a superior surgical methodology when compared with conventional bone grafts [4]. Moreover, a reduction in the sinusoidal and arterial capillary compartments is typical of osteoporosis [5] and electrical stimulation increases bone formation with a parallel increase in the number of capillaries [6]. The application of the bone chamber implant and horizontal intravital microscope supported the conclusion that neovascularization precedes neo-osteogenesis [7]. Furthermore, endothelial cells isolated from rat liver or bovine aorta can greatly enhance bone formation when implanted in diffusion chambers with fetal calvarial cells [8].

In vitro studies have led to the conclusion that bovine aortic endothelial cells in culture synthesize a potent bone active mitogen, establishing a potential link between initiation of bone formation and vascularization [9]. Moreover, transforming growth factor beta, a factor produced by osteoblastic cells and active on bone formation, is also angiogenic [10, 11], likewise basic and acidic fibroblast growth factors [12, 13].

Bone vasculature and bone resorption

Bone endothelial cells, despite providing the basic endothelial functions, have an essential role in the release of blood cells. Endothelial cells have also been shown to produce hemopoietic growth factors and other growth factors. The interactions of these humoral factors likely contribute to the regulation of hemopoiesis. Interestingly, the long-term marrow culture system has proved to be a useful model for the characterization of the hematopoietic environment in vitro [14]. Indeed, the maintenance of hematopoiesis is critically dependent upon a population of stromal cells and matrix. The precise lineage of the cells comprising the in vitro hematopoietic microenvironment is uncertain. Some workers have reported that the adherent cells in the bone marrow culture system are predominantly fibroblasts [15]. In contrast, others have found substantial numbers of endothelial-like cells [16].

Interestingly, during vascular invasion of the hypertrophied cartilage core, endothelial cells of the sprouting capillaries may have a direct or indirect role in resorption. In fact, osteoclasts are typically located on the tip of the sprouting capillary, while osteoblasts line the terminal wall of the capillary, like the rearguards ready to generate the organic matrix of bone (Fig. 1).

Indeed, clinical studies prove that vascular lesions of bone are associated with increased bone resorption. Diffuse cystic angiomatosis generally presents with multiple cystic areas of bony destruction [17].

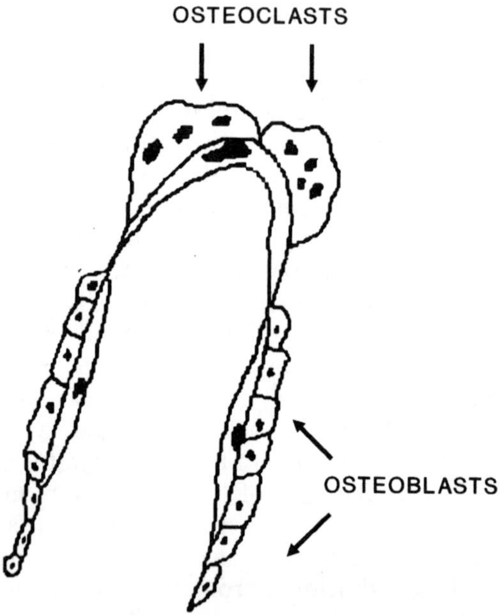

Figure 1. Schematic representation of the intimate relationship between a sprouting capillary invading the epiphyseal plate and osteoclastic and osteoblastic cells.

Radiographically and histopathologically, also lymphangioma of bone is lytic [18]. Vanishing bone disease, also described as disappearing bone disease, phantom bone, massive and idiopathic osteolysis, is also an osteolytic condition associated with a benign osseous angiomatosis and characterized by a progressive localized bone resorption [19–23]. The gross and microscopic features in this condition are not dissimilar from diffuse cystic angiomatosis. The demonstration of acid phosphatase activity in endothelial cells of the thin-walled vascular channels in this tissue suggests that these vessels might also be involved in the destructive process [24]. The presence of osteoclasts in vanishing bone disease is controversial. Rare multinucleated osteoclasts were a characteristic feature of the condition. However, numerous mononuclear perivascular cells rich in acid phosphatase activity were found at the erosion front of the bony matrix [24]. Interestingly, it is well known that mononuclear cells exhibit osteoclastic characteristics in vitro [25]. Therefore, it cannot be excluded that the abundant mononuclear cells at the resorption front in vanishing bone disease are in fact bone-resorbing elements of the heterogeneous osteoclast population. Perhaps, these cells are important in the resorptive process in this disorder.

Finally, Paget bone disease is characterized by important disturbances in bone vasculature with an increase in bone resorptive activity. The pathogenesis of this disorder and the possible correlations between the structural defects of the vasculature and the abnormalities of bone remodeling are still unclear.

Prospects

It is remarkable that the importance of bone endothelium in bone formation and resorption was postulated many years ago. The need for understanding the functional interactions between bone endothelium with the other stromal and hematopoietic systems of bone tissue is therefore urgently felt. The availability of bone endothelial cells as cultured clones [1], of preosteoclastic clonal cells [26] and of several osteoblastic cell lines offers a tool to address some of the basic questions using in vitro models. Is there a positive interaction between bone endothelium and osteoblasts? Can bone endothelium influence osteoclast recruitment and differentiation? Studies carried out in our laboratory are attempting to discover and characterize such interactions at a molecular level.

References

1 Streeten EA, Ornberg R, Curcio F, Sakaguchi K, Marx SJ, Aurbach GD, Brandi ML: Cloned endothelial cells from fetal bovine bone. Proc Natl Acad Sci USA 1989; 86: 916–920.
2 Keith, Sir A: Concerning the origin and nature of osteoblasts. Proc Royal Soc Med 1927; 21: 301–319.
3 Trueta J: The role of the vessels in osteogenesis. J Bone Joint Surg 1963; 45B: 402–418.
4 Cutting CB, McCarthy JG: Comparison of residual osseous mass between vascularized and nonvascularized onlay bone transfers. Plast Reconstr Surg 1983; 72: 672–675.
5 Burkhardt R, Kettner G, Bohm W, Schmidmeier M, Schlag R, Frisch B, Mallmann B, Eisenmenger W, Gilg TH: Changes in trabecular bone, hematopoiesis and bone marrow vessels in aplastic anemia, primary osteoporosis, and old age: a comparative histomorphometric study. Bone 1987; 8: 157–164.
6 Nanmark U, Buch F, Albrektsson T: Influence of direct currents of bone vascular supply. Scand J Plast Reconstr Surg 1988; 22: 113–115.
7 Winet H, Bao JY, Moffat R: A control model for tibial cortex neovascularization in the bone chamber. J Bone Miner Res 1990; 5: 19–30.
8 Villanueva JE, Nimni ME: Promotion of calvarial cell osteogenesis by endothelial cells. J Bone Miner Res 1990; 5: 722–739.
9 Guenther HL, Fleisch H, Sorgente N: Endothelial cells in culture synthesize a potent bone active mitogen. Endocrinology 1986; 119: 193–201.
10 Centrella M, Canalis E: Transforming and nontransforming growth factors are present in medium conditioned by fetal rat calvariae. Proc Natl Acad Sci USA 1985; 82: 7335–7339.
11 Merwin JR, Anderson J, Madri JA: Transforming growth factor-beta induces angiogenesis of microvascular endothelial cells in three-dimensional culture. J Cell Biol 1988; 107: 48A.
12 Klagsbrun M, Sullivan R, Smith S, Rybka R, Shing Y: Purification of endothelial cell growth factors by heparin affinity chromatography. Methods Enzymol 1987; 147: 95–105.
13 Canalis E, Centrella M, McCarthy T: Effects of basic fibroblast growth factor on bone formation in vitro. J Clin Invest 1988; 81: 1572–1577.
14 Dexter TM, Allen TD, Lajtha LG: Conditions controlling the proliferation of hemopoietic stem cells in vitro. J Cell Physiol 1977; 91: 335–344.

15 Bentley SA, Tralka TS: Characterization of marrow-derived adherent cells: evidence against an endothelial subpopulation. Scand J Haematol 1982; 28: 381–388.
16 Keating A, Singer JW, Killen PD, Striker GE, Salo AC, Sanders J, Thomas ED: Donor origin of the in vitro haematopoietic microenvironment after marrow transplantation in man. Nature 1982; 298: 280–283.
17 Brower AC, Culver JE, Keats TE: Diffuse cystic angiomatosis of bone: report of two cases. Am J Roentgen Radiother Nucl Med 1973; 118: 456–460.
18 Cohen J, Craig JM: Multiple lymphangiectasis of bone. J Bone Joint Surg 1955; 37: 585–590.
19 Johnson PM, McClure JG: Observations on massive osteolysis: a review of the literature and report of a case. Radiol 1958; 71: 28–42.
20 Gorham LW, Stout AP: Massive osteolysis (acute spontaneous absorption of bone, phantom bone, disappearing bone): its relation to haemoangiomatosis. J Bone Joint Surg 1955; 37A: 985–1004.
21 Heyden G, Kindblom LG, Nielsen JM: Disappearing bone disease: a clinical and histological study. J Bone Joint Surg 1977; 59A: 57–61.
22 Picault C, Comtet JJ, Imbert JC, Boyer JM: Surgical repair of extensive idiopathic osteolysis of the pelvic girdle (Jackson–Gorham disease). J Bone Joint Surg 1984; 66B: 148–149.
23 Cannon SR: Massive osteolysis. J Bone Joint Surg 1986; 68B: 24–28.
24 Dickson GR, Hamilton A, Hayes D, Carr KE, Davis R, Mollan RAB: An investigation of vanishing bone disease. Bone 1990; 11: 205–210.
25 Hattersley G, Chambers TJ: Generation of osteoclastic function in mouse bone marrow cultures: multinuclearity and tartrate-resistant acid phosphatase are unreliable markers for osteoclastic differentiation. Endocrinology 1989; 124: 1689–1696.
26 Brandi ML, Bezzini R, Gattei V, Pinto A, Tanini A, Bernabei PA: Development of an osteoclast phenotype in a human monoblastic cell line (FLG 29.1). J Bone Miner Res 1989; 4: S265.

Angiogenesis: Key Principles - Science - Technology - Medicine
ed. by R. Steiner, P. B. Weisz & R. Langer

Expression of angiogenic growth factors in the collateralized swine myocardium

Hari S. Sharma, M. Wünsch, M. Schmidt, R. J. Schott, R. Kandolf[a] and W. Schaper

Max-Planck Institute, Dept. of Experimental Cardiology, Bad Nauheim,
[a]*Max-Planck Institute for Biochemistry, Martinsried, Germany*

Introduction

Polypeptide growth factors contribute in the regulation of new blood vessel growth, a phenomenon called angiogenesis. In the adult, angiogenesis is an infrequent event. Exceptions are found during wound healing, fracture repair and in females, angiogenesis occurs during follicle development, ovulation and pregnancy [1]. Under normal physiological conditions, angiogenesis is tightly regulated, whereas, during pathological conditions such as tumour growth and retinopathies, it appears to be uncontrolled. In view of the physiologic and pathological importance of the angiogenesis, many efforts have been made to identify and characterize factors that can either stimulate or inhibit angiogenesis. In the last decade several angiogenic polypeptide growth factors have been purified, their amino acid sequences determined and their genes cloned [1, 3, 4].

During a slow coronary artery occlusion, small coronary arterioles and capillaries begin to grow in the ischemic region of the pig heart [5]. Polypeptide growth factors probably play an important role in this process of myocardial collateralization in response to ischemia. Potent angiogenic growth factors such as acidic and basic fibroblast growth factors (FGF) [6, 7] and transforming growth factor-β1 (TGF-β1) [8] have been demonstrated in the myocardium of various mammals. Endothelial cell growth factor-β (β-ECGF or aFGF) is an anionic polypeptide mitogen of 20 kD with nearly 55% amino acid identity to bFGF [1, 4]. TGF-β1 is a 25 kD homodimeric polypeptide differentiation factor with angiogenic properties [9]. Like β-ECGF, vascular endothelial growth factor (VEGF) is a heparin binding, dimeric peptide mitogen with target specificity to vascular endothelial cells [3]. The role of various growth factors has been understood with cell culture systems, relatively little is known about the physiological significance of these molecules in vivo. In a porcine model of progressive coronary artery occlusion, we examined the expression of β-ECGF, TGF-β1 and VEGF mRNAs using polymerase chain reaction (PCR), Northern and in situ hybridization techniques.

Materials and methods

Experimental model

Several domestic pigs were intubated endotracheally with room air enriched in oxygen under general anesthesia. After a lateral thoracotomy, a hygroscopic ameroid constrictor was implanted around the left circumflex coronary artery (LCx). These constricting rings slowly absorb body fluid and swell to compress the artery. Stenosis and occlusion of the coronary artery were determined by angiography after 2–3 weeks of ameroid implantation. At more than 90% of LCx stenosis and/or collateral growth from right coronary artery, pigs were sacrificed, heart was removed quickly and rinsed in ice-cold saline. Small tissue pieces from the LCx and normally perfused anterior wall were excised and snap frozen in liquid nitrogen and stored at −80°C until analyzed.

cDNA synthesis and PCR

Frozen tissue was pulverized in the presence of liquid nitrogen and homogenized in guanidinium thiocyanate buffer. Total cellular RNA was isolated according to the method of Chomczynski and Sacchi [10]. RNA concentration was measured by optical density measurement and usually the ratio OD_{260}/OD_{280} was >1.8. To perform PCR, single-stranded cDNA was synthesized from total RNA derived from normal and collateralized myocardia. An aliquot of above cDNA was used for PCR amplification. β-ECGF, TGF-β1 and VEGF-specific primers were designed from the published bovine, porcine and human cDNA sequences [3, 12–14]. cDNA products were amplified using each sense and antisense primers, in an automatic DNA thermal cycler [15]. PCR products were analyzed by poly-acrylamide or agarose gel electrophoresis, stained with ethidium bromide and visualized under UV light. For Southern hybridization, PCR products were electrophoresed on agarose gels, blotted on a nylon membrane and hybridized with radiolabeled 1.1 kb human cDNA insert encoding TGF-β1 (kindly provided by Dr. R. Derynck) or 2.1 kb human β-ECGF cDNA insert (kindly provided by Dr. T. Maciag) or a 930 bp cDNA insert encoding human VEGF (provided by Dr. Ferrara). Filters were washed under stringent conditions and auto-radiographed at −70°C for 1–2 h.

Northern and in situ hybridization

For Northern hybridization, total RNA from each tissue was size fractionated on a 1% agarose gel containing 2.2 M formaldehyde in the presence of ethidium bromide as described previously [11]. RNA was transferred to hybond-N membrane by vacuum blotting, UV cross-linked and ribosomal RNA bands were marked under UV light. Blots were hybridized with radiolabeled β-ECGF, TGF-β1 and VEGF cDNA probes at 42°C in a buffer containing 50% formamide. Filters were washed under stringent conditions and subsequently, exposed to Kodak XAR films at −70°C for 2–4 days. For in situ hybridization, porcine normal and collateralized tissue sections were hybridized with ^{35}S-labeled nick translated cDNA insert encoding β-ECGF and an anti-sense cRNA probe for TGF-β1.

Results

Angiogenesis leading to collateralization occurred in the ischemic region of the pig heart during left circumflex coronary artery occlusion by an ameroid constrictor. In a search for polypeptide growth factors which could be expressed during this myocardial angiogenesis, we

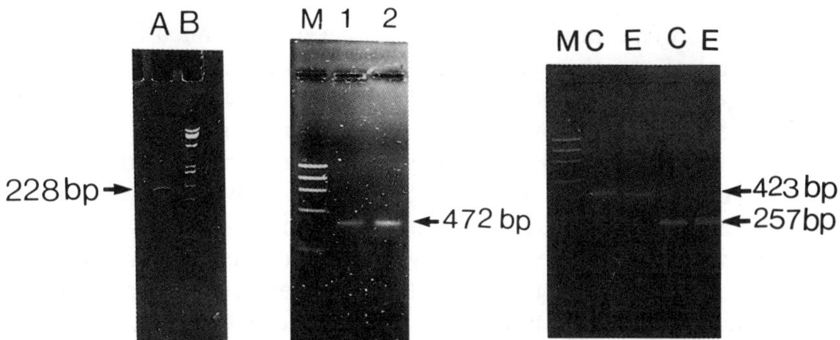

Figure 1. Detection of β-ECGF, TGF-β1 and VEGF transcripts in swine myocardium by PCR.
Panel A: VEGF PCR product was analyzed on a 8% polyacrylamide gel after ethidium bromide staining. A single band of 228 bp was amplified in the porcine myocardium. Size marker (B) is Hae III-digested PhiX DNA.
Panel B: PCR products amplified from total RNA derived from the collateralized (2) and normal myocardium (1) were analyzed on a 1.5% agarose gel by ethidium bromide staining. A major band of 472 bp (compared to the PhiX-Hae III marker) was amplified from both normal and collateralized myocardia.
Panel C: Agarose gel analysis of TGF-β1 PCR products by ethidium bromide staining revealed major bands of expected size of 423 and 257 bp in the collateralized (E) and normal (C) myocardium. M = size marker Hae III-cut PhiX DNA.

amplified by PCR DNA fragments encoding β-ECGF, TGF-β1 and VEGF from the normal and collateralized myocardia (Fig. 1). Furthermore, these PCR products were verified for specificity by Southern hybridization with their respective cDNA probe as well as by reamplification using internal primers. In Northern hybridization, we detected enhanced expression of two mRNA species of 2.5 and 1.9 kb hybridizing to a human TGF-β1 cDNA probe in collateralized tissue as compared to normal. In case of β-ECGF, we amplified a 472 bp DNA fragment and for VEGF, a fragment of 228 bp in the collateralized and normal myocardia (Fig. 1). However, by Northern hybridization, we observed higher levels of mRNAs encoding β-ECGF (4.6 kb) and VEGF (3.9 and 1.7 kb) in the collateralized myocardium as compared to the normal myocardium. We attempted to localize the β-ECGF and TGF-β1 mRNAs by in situ hybridization; surprisingly, β-ECGF transcripts were seen in the vascular walls, probably in smooth muscle cells bordering vascular endothelial cells in the collateralized myocardium. TGF-β1 mRNA could be localized predominantly in the cardiac myocytes. Similar to in situ localization of TGF-β1 mRNA, immunocytochemically, we observed the staining in the myocytes and a dense staining in the conducting Purkinje cells [8].

Discussion

It is now a well-established fact that pig heart is a much better model of the clinical situation of collateralization in response to chronic coronary occlusion in vivo [2, 5, 16]. This compensatory growth seems to be a very important event in the prevention of myocardial infarcation. At the cellular level, it is not yet clear how this adaptive mechanism of defense against ischemia is regulated. We believe that the myocardial response to a critical coronary stenosis in terms of adaptation and defense could be due to myocardial angiogenesis and development of collateral circulation. Recently, Görge et al. [5] reported that, in the pig heart, coronary blood flow was reduced to about 20%, and collateral flow rose to 60% of the normal 2–4 weeks after ameroid implantation. On the basis of coronary angiographic studies, we selected pigs with more than 90% stenosis and small vessels developed throughout the area at risk in a dense "blush"-like fashion without any macro-infarcation.

Employing PCR as a diagnostic tool, we searched for the mRNAs encoding β-ECGF, TGF-β1 and VEGF. PCR-amplified products were visualized on agarose or acrylamide gels by ethidium bromide staining and expected size of amplified DNA fragments were compared with Phi-X-HaeIII marker. Furthermore, Southern blot analysis of PCR products validated the specific amplification of β-ECGF, TGF-β1 and

VEGF. By PCR, we could not differentiate the expression pattern of these growth factors; however, Northern hybridization revealed enhanced expression of these molecules in the collateralized myocardium. This is the first time that, by Northern hybridization, β-ECGF mRNA was detected in the myocardial tissue. There are several reports showing the presence of β-ECGF protein in the normal myocardium [6, 7] and the mRNA expression in the primary cultures of isolated adult and neonatal rat myocytes [17]. It is clear that β-ECGF and TGF-β1 are present in the myocardium but the cell type(s) producing these factors under normal and experimental conditions such as collateralization are virtually unknown. By in situ hybridization, we observed β-ECGF transcripts in the vascular walls, and TGF-β1 in the cardiac myocytes in collateralized myocardium.

This study demonstrates the expression of β-ECGF, TGF-β1 and VEGF in the collateralized myocardium, suggesting their role in ischemia-induced angiogenesis. From these findings we postulate that TGF-β1 and/or a hitherto unidentified ischemia-related mitogen activator originate in the ischemic myocytes which occupy receptors on endothelial cells and switches on genes which are responsible for the mitosis of endothelial cells.

Acknowledgements. We thank Mrs. E. Neubauer for her excellent technical assistance.

References

1 Folkman J, Klagsburn M: Angiogenic factors. Science 1987; 235: 442–447.
2 Schaper W: The collateral circulation of the heart. Amsterdam, North Holland Publishing Co., 1971.
3 Leung DW, Cachianes G, Kuang WJ, Goeddel DV, Ferrara N: Vascular endothelial growth factor is a secreted angiogenic mitogen. Science 1989; 246: 1306–1309.
4 Klagsburn M, D'Amore PA: Regulators of angiogenesis. Ann Rev Physiol 1991; 53: 217–239.
5 Görge G, Schmidt T, Ito BR, Pantely GA, Schaper W: Microvascular and collateral adaptation in swine hearts following progressive coronary artery stenosis. Basic Res Cardiol 1989; 84: 524–535.
6 Quinckler W, Maasberg M, Bernotat-Danielowski S, Luthe N, Sharma HS, Schaper W: Isolation of heparin-binding growth factors from bovine, porcine and canine hearts. Eur J Biochem 1989; 181: 67–73.
7 Cassells W, Spier E, Sasse J, Klagsburn M, Allen P, Lee M, Calvo B, Chiba M, Haggroth L, Folkman J, Epstein S: Isolation, characterization, and localization of heparin-binding growth factors in the heart. J Clin Inv 1990; 85: 433–441.
8 Wünsch M, Sharma HS, Markert T, Bernotat-Danielowski S, Schott RJ, Kremer P, Bleese N, Shaper W: In situ localization of transforming growth factor-β1 in porcine heart: Enhanced expression after chronic coronary artery occlusion. J Mol Cell Cardiol, 1991. In press.
9 Roberts AB, Sporn MB, Assoian RK, Smith JM, Roche NS, Wakefield LM, Heine UI, Liotta LA, Falanga V, Kehrl JH, Fauci AS: TGF-β1: Rapid induction of fibrosis and angiogenesis in vivo and stimulation of collagen formation in vitro. Proc Natl Acad Sci (USA) 1986; 83: 4167–4171.
10 Chomczynski P, Sachhi N: Single-step method of RNA isolation by acid guanidinium thiocyanate-phenol-chloroform extraction. Anal Biochem 1987; 162: 156–159.

11 Rosen KM, Lamperti ED, Villa-Komaroff L: Optimizing the Northern blot procedure. Biotechniques 1990; 8: 398–403.
12 Jaye M, Howk R, Burgess W, Ricca GA, Chiu I-M, Ravera MW, O'Brien SJ, Modi WS, Maciag T, Drohan WN: Human endothelial cell growth factor: Cloning, nucleotide sequence and chromosome localization. Science 1986; 233: 541–545.
13 Esch F, Uenao N, Baired A, Hill F, Denoroy L, Ling N, Gospodarowicz D, Guillemin R: Primary structure of bovine brain acidic fibroblast growth factor (FGF). Biochem Biophys Res Comm 1985; 133: 554–562.
14 Derynck R, Rhee L: Sequence of the porcine TGF-β1 precursor. Nucleic Acid Res 1987; 15: 3187.
15 Saiki RK, Gelfand DH, Stoffel S, Scharf SJ, Higuchi RG, Horn GT, Mullis KB, Erlich HA: Primer-directed enzymatic amplification of DNA with a thermostable DNA polymerase. Science 1988; 239: 487–491.
16 Schaper W, Sharma HS, Quinkler W, Markert T, Wünsch M, Schaper J: Molecular biologic concepts of coronary anastomases. J Am Coll Cardiol 1990; 15: 513–518.
17 Weiner HL, Swain JL: Acidic fibroblast growth factor mRNA is expressed by cardiac myocytes in culture and the protein is localized to the extracellular matrix. Proc Natl Acad Sci (USA) 1989; 86: 2683–2687.

Angiogenesis: Key Principles - Science - Technology - Medicine
ed. by R. Steiner, P. B. Weisz & R. Langer

Tumor cell spheroids induce a mitogenic response in endothelial cells

Hera C. Lichtenbeld and Geert H. Blijham

University of Limburg, Department of Internal Medicine, P.O. Box 5800, NL-6202 AZ Maastricht, The Netherlands

Abstract. Conditioned media from spheroids of human colorectal tumor cell lines HT29, SW620, SW480 and CaCo-2 were tested for their mitogenic activity on human umbilical vein endothelial (HUVE) and bovine capillary endothelial (BCE) cells. All spheroid-conditioned media were capable of increasing the labeling indices in both endothelial cell populations, with the mitogenic activity decreasing in the following order: SW620, HT29/SW480 and CaCo-2. The endothelial cell mitogenic activity of the spheroids was not dependent on the presence of necrosis nor on their in vivo cytokinetic characteristics.

We conclude that tumor spheroids release mitogenic activity for endothelial cells, but this stimulation depends on the age of the tumor spheroid as well as on the type of tumor used.

Introduction

It has been suggested that the induction of angiogenesis is critical for tumors to grow successfully at the primary and metastatic sites [1]. However, very little is known about the nature of the tumor-endothelial cell interactions leading to migration, proliferation and differentiation of these latter cells. We chose spheroids of human colorectal cancer cell lines as a model for growing tumors in order to assess the kinetics and characteristics of the angiogenic response [2]. In this communication, we will focus on the proliferative response of endothelial cells to tumor spheroids as a function of (i) the type of cell line, (ii) the type of endothelial target and (iii) the size/age of the spheroids.

Materials and methods

Endothelial cells

Human venous endothelial cells were obtained from the umbilical cord (HUVE cells) according to a modified method of Jaffe [3]. Bovine Capillary Endothelial (BCE) cells were kindly provided by Dr. M. Furie (State University of New York, Stony Brook) through Dr. R. Montesano (University of Geneva, Department of Morphology, Geneva, Switzerland).

Table 1. Characteristics of human colorectal cancer cell lines

Cell line	Ploidy DNA index	Cytokinetics(*) L.I. (%)	TS (h)	Tpot (h)	Growth fraction	Differentiation
CaCo-2	1.90	16.9	6.5	40.2	0.50	moderately
SW480	1.20	23.6	8.2	34.7	0.68	poorly
HT29	1.53	16.4	7.2	46.3	0.58	poorly
SW620	1.27	17.7	6.3	35.9	0.32	poorly

(*if grown as xenografts in nude mice [6])

Primary human umbilical vein endothelial cells (HUVE) were routinely cultured in fibronectin-coated 25 cm^2 culturing flasks in standard culture medium: RPMI 1640, 10% human serum, 30 μg/ml Endothelial Cell Growth Supplement and heparin (100 IU/ml).

Bovine capillary endothelial cells (BCE) were cultured in α-MEM with 15% Donor Calf Serum.

Tumor cells

The tumor cell lines HT29, SW620, SW480 and CaCo-2 were used. The cells were cultured in Dulbecco's Modified Essential Medium (DMEM) with 10% Bovine Calf Serum with less than 10 pg endotoxins. Some characteristics of these tumor cell lines are shown in Tab. 1.

Tumor spheroids were grown in 5-cm culture dishes coated with 0.5% agarose in distilled water [4] to avoid adherence, with 5000 single cells on day 1.

Proliferation assay

Tumor-conditioned medium was collected 48 h after washing and endothelial cells were incubated with it for 24 h. Then the medium was refreshed and Bromodeoxyuridine (final concentration 10 μM) was added for 30 min, cells were harvested and cytospin preparations were made. They were stained according to the procedure of Schutte et al. [5] and the percentage of BrdU-positive cells was scored as the labeling index (LI). From each cytospin a total of at least 400 cells were counted. Statistical analyses was performed by using a Mann-Whitney test.

Results

Tumor-conditioned media of spheroids from the different colorectal cancer cell lines were tested for their mitogenic activity on HUVE and

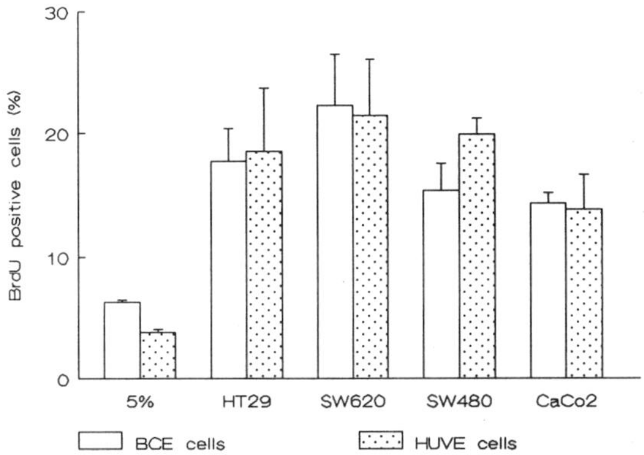

Figure 1. Labeling indices of endothelial cells when incubated 24 h with spheroid conditioned medium. All labeling indices are significantly different ($p \leq 0.05$) from control values.

BCE cells. For these experiments spheroids were at least 25 days of age (i.e. at maximal size). Results are expressed as endothelial cell LI with control versus conditioned media as shown in Fig. 1. Tumor spheroids obviously release mitogenic activity for both HUVE and BCE cells with results running in parallel. Of the four cell lines tested, SW620 yielded the highest and CaCo-2 the lowest LI, with the other two cell lines being of intermediate activity. These mitogenic differences were not correlated to ploidy levels or cytokinetics of these lines if grown as xenografts in nude mice, as can be seen by comparing these data with those of Tab. 1. It is of interest to note however, that of 13 cell lines tested the CaCo-2 xenograft was found to be unique in that it did not show in vivo growth despite a growth fraction of 50% and potential doubling time of 40 h. This points to the possibility that also in vivo this cell line is relatively deficient in angiogenic properties.

For the HT29 cell line the relationship between spheroid age and size and the release of mitogenic activity for endothelial cells (BCE) was investigated. The number of cells per spheroid as a function of age is given in Fig. 2, showing a doubling time of 1.5 days; a similar doubling time was observed for the growth in size (data not shown).

The mitogenic activity is depicted in Fig. 3, showing a declining activity in the first three weeks (the growth phase) and a plateau thereafter. Apparently with cell numbers increasing with a factor of at least 100 (from day 7 to day 21), the mitogenic activity decreases rather than increases.

Discussion

Human colorectal cancer cell lines differ in their capacity to release mitogenic activity for endothelial cells. One hypothesis to explain these

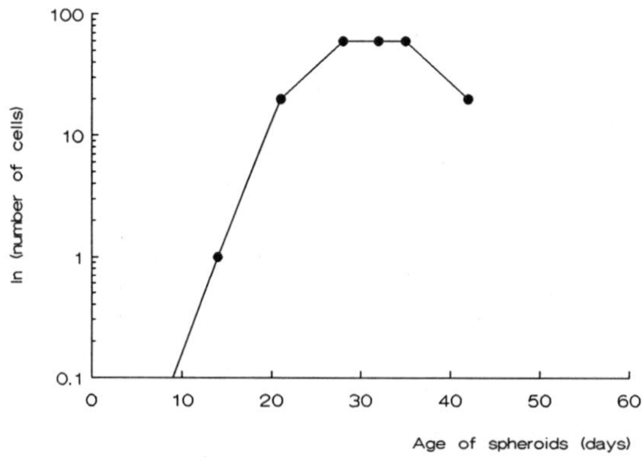

Figure 2. The number of cells per HT29 spheroid at different ages.

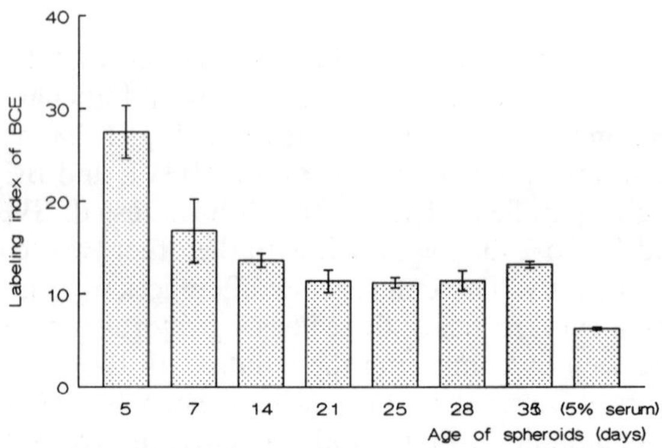

Figure 3. Labeling indices of capillary endothelial cells after incubation in conditioned medium of HT29 spheroids at different ages.

differences may be that this possibly antiogenic activity is related to the growth characteristics of the tumor, e.g. its proliferative potential or propensity to necrosis. We did not find a relation to the cytokinetics of the colorectal cancer cell lines if grown as xenografts in nude mice (Tab. 1) or in vitro (data not shown). A more direct comparison between the proliferation within the spheroids and their mitogenic potential for endothelial cells is currently being performed. If the mitogenic activity reflects the occurrence of oxygen limitation, one would expect this activity to be a function of the size of the spheroids with a particular critical timepoint at around day 21, when necrosis starts to occur. Our results do not favor this hypothesis as the mitogenic activity appears to decrease at higher cell numbers per spheroid and reaches its nadir at day 21.

In summary, the tumor spheroid model may be useful to determine tumor-endothelial cell interactions and their regulation in a direct way.

So far we have described a mitogenic effect of spheroids on endothelial cells and found this activity not to be dependent on the occurrence of necrosis or inherent proliferative activity of the tumor cell line. Experiments to characterize this mitogenic activity and to expand our observations of invasive endothelial cell properties are in progress.

References

1 Folkman J: What is the evidence that tumors are angiogenesis dependent. J Nat Cancer Inst 1990; 82(1): 4–6.
2 Sutherland RM: Cell and environment interactions in tumor microregions: the multicell spheroid model. Science 1988; 240: 177–184.
3 Jaffe EA, Nachman RL, Becker CG, Minick CR: Culture of human endothelial cells derived from unbilical veins. J Clin Invest 1973; 52: 2757–2764.
4 Carlsson J, Yuhas JM: Liquid-overlay culture of cellular spheroids. Cancer Res 1984; 95: 1–23.
5 Schutte B, Reynders MMJ, Bosman FT, Blijham GH: The effect of tissue fixation on anti-bromodeoxyuridine immunohistochemistry. J Histochem Cytochem 1987; 35: 1343–1345.
6 Blijham GH, Schutte B, Verstijnen C, Arends JW, Reynders M, Bosman F: In vivo and in vitro cytokinetic studies on 13 human colorectal (HCC) cell lines. Am Soc Clin Oncol 1987; 6: 18.

Angiogenesis: Key Principles - Science - Technology - Medicine
ed. by R. Steiner, P. B. Weisz & R. Langer

A monoclonal antibody (E-9) binds preferentially to the vasculatures of human tumors, embryonic and regenerating tissues

Ji Min Wang, Robin D. Hunter and Shant Kumar

Christie Hospital, Manchester, England

Angiogenesis plays an important role in a variety of physiological and pathological processes such as embryogenesis, wound healing and tumour growth. Although vascular smooth muscle cells and pericytes take part in angiogenesis, endothelial cells are considered to be the critical component in neovascularisation [1]. The migration and proliferation of endothelial cells are the two key events that occur in angiogenesis [2–4]. Furthermore, monoclonal antibodies against endothelial cells have been raised in an attempt to identify the process of angiogenesis [5–7]. A clearer understanding of the role of endothelial cells in angiogenesis will contribute to the control of angiogenic diseases [8, 9]. In the present report, we describe a monoclonal antibody E-9, which binds preferentially to the vasculature in human tumours, embryonic and regenerating tissues.

Materials and methods

Production of monoclonal antibodies

Human umbilical vein endothelial cells used for immunisation were cultured with medium 199 and MCDB 104 (1:1) plus 20% foetal calf serum and endothelial cell growth supplements (Sigma). The cells were detached from culture flasks with rubber policeman after 3 washes in PBS. BALB/C mice were immunised by 3 intraperitoneal injections of 2×10^7 HUVE cells. Fusion of the spleen cells and NS_1 myeloma cells was performed 4 days after the last booster injection.

The hybridoma supernatants were tested for specific activity with cultured HUVE cells by ELISA and with tumour and normal tissue sections using indirect immunohistochemistry. Hybridomas which showed reactivity with cultured HUVE cells or the vasculature of tumours were subsequently cloned by limiting dilution. One hybridoma

(E-9) was selected for further characterisation with a panel of tumours, embryonic, regenerating and normal adult tissues. Analysis of the antibody subclass was carried out using a mouse monoclonal antibody isotyping kit (Serotec).

Immunostaining

Indirect immunofluorescence was performed either on cell suspensions or on frozen tissue sections. Briefly, viable cells or tissue sections were incubated with the hybridoma supernatant for 30 min at room temperature and binding was revealed by fluorscein-labelled $F(ab)_2$ fragment of rabbit anti-mouse immunoglobulins (DAKO).

Immunohistochemistry was performed by an indirect peroxidase method. The frozen sections were incubated with the hybridoma supernatant for 1 h at room temperature, washed and reacted with peroxidase-labelled rabbit anti-mouse immunoglobulin at 1:100 dilution (DAKO). The slides were stained with DAB solution (Sigma) and counterstained briefly with Gill's haematoxylin. The staining of E-9 antibody was compared with that of a pan-endothelial cell monoclonal antibody 5.6E (developed by Drs P. Dial and S. Kumar – PECAM/CD31).

Results

Immunocytochemical reaction of E-9 monoclonal antibody with human and bovine cells

Analysis of E-9 antibody reactivity was carried out on human peripheral blood smears from seven persons using indirect immunoperoxidase staining. No reactivity was observed in any case.

E-9 antibody was positive to HUVE cells in suspension at different passages (from passage 2 to 17) and more than 90% of the cells were stained. The antibody did not stain human vascular smooth muscle cells or bovine aortic endothelial cells.

Specificity of E-9 antibody on human tumour, embryonic and regenerating tissues

E-9 antibody stained blood vessels in tumours (32/33), foetal tissues (7/7) and regenerating tissues (3/3) [Tabs. 1 and 2, Fig. 1]. The number of blood vessels staining in the tumours is related to the mass of tumour cells. The blood vessels in and around tumours show stronger staining than blood vessels further away from tumour cell masses. Blood vessels in foetal tissues always exhibit strong staining.

Table 1. Immunostaining of the vasculature of tumours with E-9 antibody

Sources of tumour	Number positive/total
Breast carcinoma	11/11
Colon carcinoma	4/4
Rhabdomyosarcoma	3/3
Neuroblastoma	2/2
Lymphoma	2/2
Wilms' tumour	2/2
Renal carcinoma	1/1
Stomach carcinoma	1/1
Cervix carcinoma	2/2
Ovary carcinoma	4/4

Table 2. Reactivity of E-9 antibody against foetal and regeneration tissues

Source of tissue	Number positive/total
Foetal kidney	2/2
brain	1/1
skin	1/1
intestine	1/1
striated muscle	2/2
Healing ulcer (legs)	3/3
Synovium (rheumatoid)	2/2

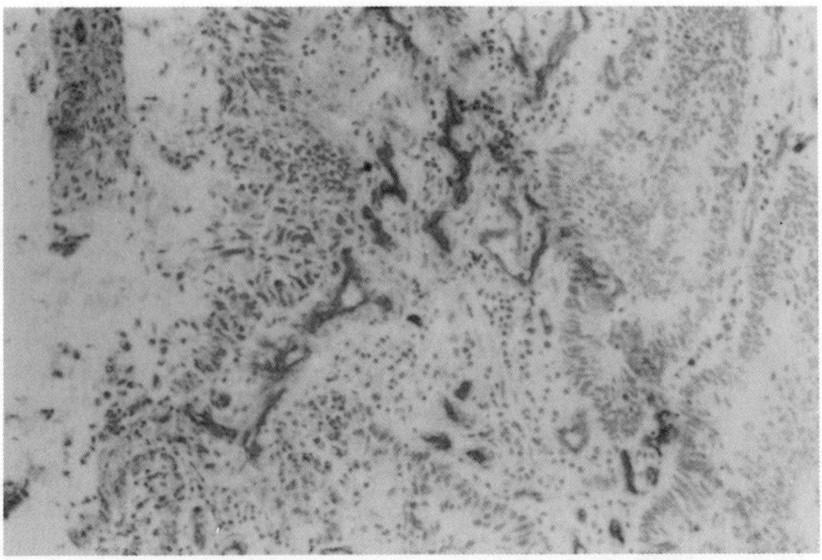

Figure 1. Cryostat section of a colon carcinoma stained with E–9 monoclonal antibody – only the tumour blood vessels are staining.

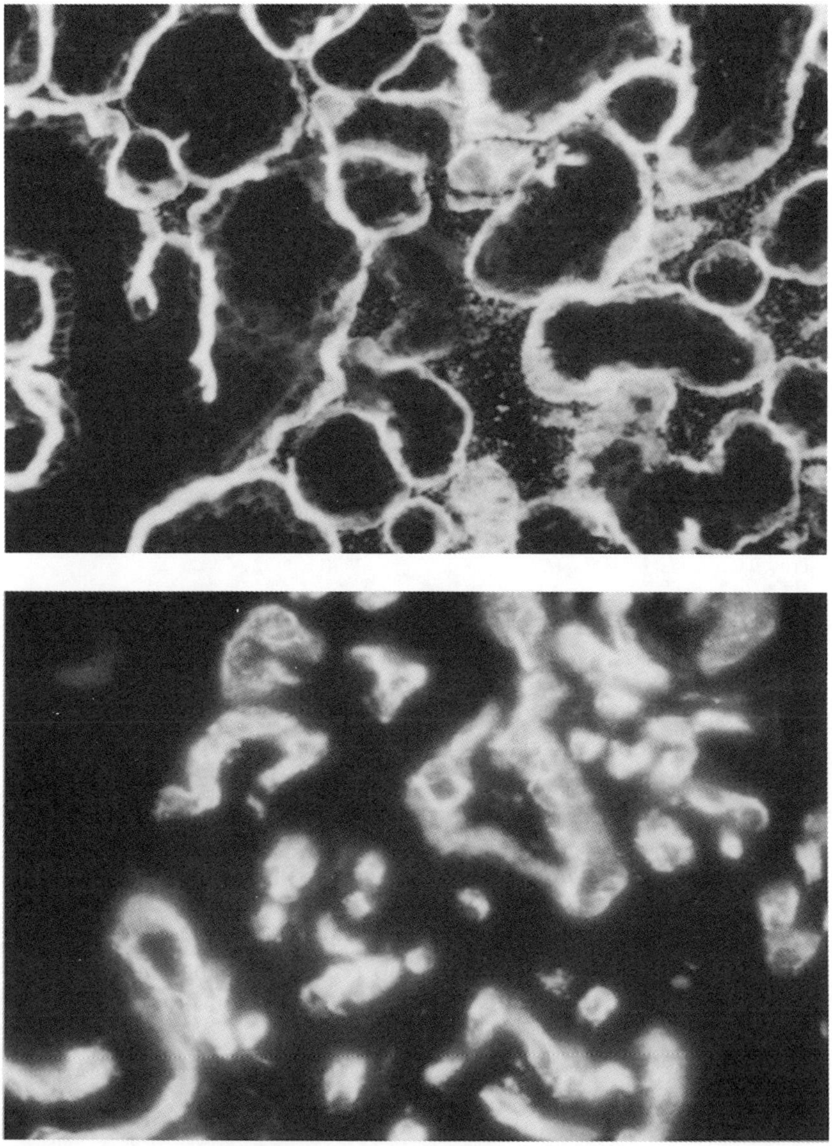

Figure 2. Two serial cryostat sections of human placenta stained with: *A* E-9 – only outer layer of trophoblast stains; *B* 5.6E – blood vessel within the placental villi show staining.

Reactivity of E-9 antibody to normal adult tissues

The blood vessels in 14 to 62 normal adult tissues tested showed positive staining with E-9 antibody. The positively staining tissues were mainly mucosa (2), skin (3) and tonsil (4) [Tab. 3]. Only a small

Table 3. Reactivity of E-9 antibody against normal adult tissues

Source of tissue	Number positive/total
Colon	0/8
Stomach	0/3
Striated muscle	0/6
Heart	0/3
Liver	1/5
Lung	1/5
Tonsil	4/4
Pancreas	0/3
Skin	3/6
Mucosa	2/3
Kidney	1/4
Brain	1/2
Bronchus	0/1
Tongue	0/1
Artery	0/1
Oesophagus	0/1
Lymph node	0/1
Nerve	0/2
Spleen	0/1
Uterus	1/2
Overall	14/62

percentage (from 5% to 30%) of blood vessels in skin and mucosa were stained. The blood vessels of 4 tonsils stained as strongly as blood vessels in tumours.

The staining of E-9 antibody with placenta tissues

Eleven placentas were stained. The E-9 antibody stained only the outer layer of trophoblast and had no reactivity with blood vessels in placental villi (Fig. 2). In contrast, 5.6E stained the blood vessels but not the outer layer of trophoblast.

Discussion

Using endothelial cell growth supplement "activated" human umbilical vein endothelial cells as immunogen, we raised a monoclonal antibody E-9 against endothelial cells. The results of immuno-staining with E-9 antibody show that it binds preferentially to the vasculature in human tumours, embryonic and regenerating tissues. There is a qualitative and quantitative difference in the staining of tumour and normal tissues. This preliminary report has raised some important questions regarding angiogenesis in tumours, embryonic and regenerating tissues.

Although E-9 antibody stained the vasculature of all tumours, embryonic and regenerating tissues tested, it also stained some normal tissues. Why some normal organs (skin, mucosa and tonsil) stained with E-9 antibody (9/13) whereas normal colon is consistently negative (0/8) is not clear. E-9 antibody has positive staining with foetal tissues and has no reactivity with blood vessels of placental villi. Why that is so is not known either. The possible function of the antigen recognised by E-9 cells will be elucidated by further characterisation of the antigen.

References

1 Folkman J: What is the role of endothelial cells in angiogensis? Lab Invest 1984; 51: 601–604.
2 Ausprunk DH, Folkman J: Migration and proliferation of endothelial cells in preformed and newly formed blood vessels during tumor angiogenesis. Microvasc Res 1977; 14: 53–65.
3 Denekamp J, Hobson B: Endothelial cell proliferation in experimental tumours. Br J Cancer 1982; 46: 711–720.
4 Hirst DG, Denekamp J, Hobson B: Proliferation kinetics of endothelial and tumour cells in three mouse mammary carcinomas. Cell Tissue Kinet 1982; 15: 251–261.
5 Hagemeier HH, Vollmer E, Goerdt S, Osthoff KS, Sorg C: A monoclonal antibody reacting with endothelial cells of budding vessels in tumours and inflammatory tissues, and non-reactive with normal adult tissues. Int J Cancer 1989; 38: 481–488.
6 Schlingemann RO, Dingjan GM, Emeis JJ, Blok J, Warnaar SO, Ruiter DJ: Monoclonal antibody PAL-E specific for endothelium. Lab Invest 1985; 52: 71–76.
7 Ramani P, Bradley NJ, Fletcher CDM: QBEB/10, a new monoclonal antibody to endothlium: assessment of its diagostic utility in paraffin sections. Histopathology 1990; 17: 237–242.
8 Folkman J: What is the evidence that tumors are angiogenesis dependent? JNCL 1990; 82: 4–6.
9 Denekamp J: Endothelial cell proliferation as a novel approach to targeting tumour therapy: Br J Cancer 1982; 45: 136–139.

Angiogenesis: Key Principles – Science – Technology – Medicine
ed. by R. Steiner, P. B. Weisz & R. Langer

Liver or lung colonization by F9 teratocarcinoma cells follows specific interactions with the target organ

Dario Rusciano, Patrizia Lorenzoni and Max M. Burger

Friedrich Miescher Institute, P.O. Box 2543, CH-4002 Basel, Switzerland

Introduction

It is widely accepted today that, in many instances, cancer metastases do not occur at random, and specific organs represent the preferential target of given tumors. According to the two main theories that have been proposed to explain the organ distribution pattern of metastases, either the vascular connections of the primary tumor [1] and/or specific interactions between the malignant cells and the target organ [2] will decide the localization of metastatic colonies.

F9 is a murine teratocarcinoma cell line that has been defined as quasi-nullipotent, since in vivo it only gives rise to undifferentiated tumors, while in vitro, under the induction of retinoic acid and dibutyrryl-cyclic-AMP (RA/DBA), it can differentiate to give rise to a limited type of progeny, namely endodermal cells [3].

In the following paper we show that the organ colonization pattern of tail vein injected F9 cells (differentiated or undifferentiated) may correlate to either the vascular distribution of the injected cells, and - most importantly - to the specific interactions of these cells with the organ microenvironment.

Materials and methods

Cell line

F9 cells were cultured and induced to differentiate as described [3, 4].

Adhesion assays

Extracellular matrix (ECM) was prepared from rat organ homogenates, and coated on 24-well plates (Falcon), as described [5]. Laminin and fibronectin were obtained pure from Sigma.

The adhesion assays were performed essentially as described [5]. Briefly, cells were trypsinized and resuspended in serum-free DMEM (SFM) at the desired cell concentration (when differentiated F9 cells were used, they were trypsinized on the third day of induction with RA/DBA). Then 1.0 ml of the cell suspension was plated in sextuplicates wells (24-well plates, Falcon) that have been precoated with either ECM, or the pure glycoproteins.

When assaying cell adhesion onto the ECM, F9 were radioactively labeled with ^{3}H-TdR, and the percentage of adhesion was evaluated estimating the radioactivity adherent to plastic 6 h after the additon of the cell, with respect to the total added radioactivity. When assaying cell adhesion to laminin or fibronectin, the amount of adherent cells was estimated by cell staining with crystal violet [6], and the percentage of adherent cells was then expressed with respect to the amount of cell that adhered after 22 h to uncoated plastic in the presence of 10% FCS.

Growth promotion

Organ-membrane extracts were obtained from mouse organs as described [5]. F9 cells, either differentiated or three days after induction with RA/DBA, were plated at 2×10^3/well in 24-well plates in DMEM containing 5% FCS (SCM). On day 1 and 2 after seeding, 300 mg of proteins/well were added, together with fresh SCM, and on day 3 cell growth was evaluated by crystal violet staining [6].

Results and discussion

It was previously shown that tail-vein inoculation of undifferentiated F9 cells resulted mainly in liver colonization [4]. The lungs, that represent the first capillary bed encountered by tail-vein injected cells, were only very rarely colonized [4].

The lodgement and subsequent growth of metastatic cells into an organ may be facilitated by specific interactions with the local extracellular and pericellular matrix [7, 8], and by the presence in the target organ of paracrine growth factors that may specifically stimulate the growth of the malignant cells [9, 10]. We found that the organ colonization pattern of undifferentiated F9 cells correlated with both their adhesive properties, and their organ specific growth response. In fact F9 cells adhered very poorly to laminin-coated surfaces, and to lung-derived ECM, while they adhered much better to fibronectin and to liver-derived ECM (Tab. 1). It is interesting to note that laminin is a prominent component of the lung endothelial basement membrane, and that fibronectin is present at the level of the Disse's spaces in the liver. We also found that, soon after tail-vein inoculation, radioactively labeled undifferentiated F9 cells do not accumulate in the lungs, and, according to their poor adhesiveness to laminin and lung-derived ECM, they recirculate very fast through the lungs, and arrive at the liver, where they can eventually find a better adhesive substrate [5]. Moreover, while organ-membrane extracts derived from mouse lungs do not have any effect on the growth of F9 cells, liver-derived extracts have a growth promoting effect on these cells (Fig. 1), thereby suggesting that a combination of adhesion mechanisms and specific growth stimulation may contribute to the specific organ colonization pattern of undifferentiated F9 cells [5].

Induction of differentiation of F9 cells by treatment with RA/DBA has been shown to change dramatically their organ colonization pattern after tail-vein injection. Differentiated F9 cells will colonize the lungs and not the liver [4, 11].

We also investigated how adhesiveness and organ-specific growth response might influence the in vivo behavior of RA/DBA treated F9 cells. Comparing the results with those obtained with undifferentiated F9, we found a decrease of the general adhesiveness to each substrate tested (laminin and fibronectin, lung- and liver-derived ECM), although the relative ratio of adhesion did not change significantly, so that differentiated F9 still adhered better to fibronectin over laminin, and to liver-derived ECM over lung-derived ECM (Tab. 1). In agreement with this behavior, tail-vein inoculation of radioactively labeled differentiated F9 showed that also in this case the cell are quickly released from the

Table 1. Specific cellular adhesion

Substrate	F9 #	F9 RA/DBA #
Laminin 1 $\mu g/cm^2$	17.2 ± 3.0	14.9 ± 0.7
Fibronectin 1 $\mu g/cm^2$	44.8 ± 1.4	19.9 ± 1.1
Lung ECM 3 mg/well	19.8 ± 3.6	9.3 ± 2.1
Liver ECM 3 mg/well	56.2 ± 15.0	24.7 ± 2.6

#: % of adherent cells. 2×10^5 F9 cells, either differentiated or not, were plated onto the indicated coating. The percentage of adherent cells was evaluated 6 h after plating.

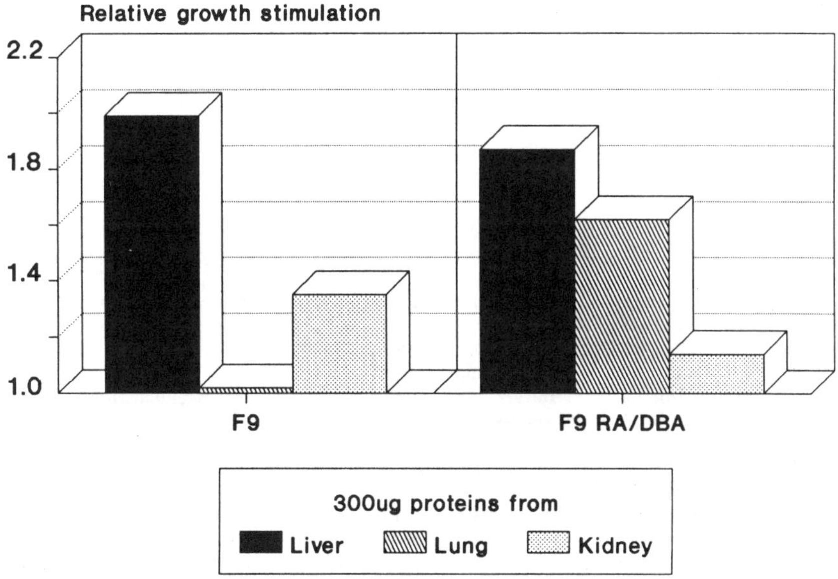

Figure 1: Organ-specific growth promotion F9 cells, either differentiated or not, were treated with the indicated amount of proteins extracted from mouse organ-membranes (materials and methods). Relative growth stimulation was evaluated on the third day after seeding, as the ratio between the growth obtained in the presence of added proteins, and the growth of control wells, that did not receive further additions.

lungs, with a kinetics very similar to that obtained with undifferentiated F9 (not shown). But, contrarily to what happens with untreated cells, RA/DBA treated F9 accumulate in the lungs to a higher extent during the first minutes after the injection (90 to 95% for treated F9 vs 50 to 60% for untreated F9). And, interestingly enough, only after induction of differentiation there is at least a subpopulation of cells that is able to respond to lung-mediated growth stimulation (Fig. 1).

From these data we may speculate that differentiated F9 cells are more likely to be mechanically trapped in the lungs (maybe because of their bigger size and/or a lesser deformability), although they are not efficiently retained at that site because of their low adhesiveness to the local ECM. But now those cells that could remain and survive in the lungs are also growth stimulated by some locally produced factor, so that they can develop into a visible colony.

In conclusion, we think that the specific liver colonization ability of undifferentiated F9 cells follows mainly specific interactions between the target organ and the circulating cells, either in terms of adhesion and growth promotion. On the contrary, the lung colonization ability of differentiated F9 cells might also depend on their circulation route, since they are more likely to be trapped in the first capillary bed encountered.

However, it may then be the presence of specific growth factor(s) in the lungs that gives a determinant contribution to the development of tumor colonies at that site.

References

1 Ewing J: A treatise on tumors. 3rd edition, 1928, W.B. Saunders, Philadelphia.
2 Paget S: Distribution of secondary growths in cancer of the breast. Lancet 1889; 1: 571–573.
3 Strickland S, Mahdavi V: The induction of differentiation in teratocarcinoma stem cells by retinoic acid. Cell 1978; 15; 393–403.
4 Terrana B, Rusciano D, Pacenti L: Organ colonization pattern of retinoic acid treated and untreated mouse embryonal carcinoma F9 cells. Cancer Res 1987; 47: 3791–3797.
5 Rusciano D, Lorenzoni P, Burger MM: The role of both specific cellular adhesion and growth promotion in liver colonization by F9 embryonal carcinoma cells. Int J Cancer 1991; 48, in press.
6 Gillies RJ, Didier N, Denton M: Determination of cell number in monolayer cultures. Analyt Biochem 1986; 159: 109–113.
7 Liotta LA, Nageswara Rao C, Wewer UM: Biochemical interactions of tumor cells with the basement membrane. Ann Rev Biochem 1986; 55: 1037–1057.
8 Doerr R, Zvibel I, Chiuten D, D'Olimpio J, Reid LM: Clonal growth of tumors on tissue-specific biomatrices and correlation with organ site specificity of metastasis. Cancer Res 1989; 49: 384–392.
9 Burger MM and Madnick HM: General and liver specific metastasis: Possible mechanisms, in Bianchi L, Gerok W, Landmann L, Sickinger K, Stalder GA (eds); Liver in metabolic diseases (Proceedings of the 35th Falk Symposium). MTP Press Ltd. 1983; pp 351–365.
10 Sargent NSE, Oestreicher M, Haidvogl H, Madnick HM, Burger MM: Growth regulation of cancer metastases by their host organ. Proc Natl Acad Sci (Wash) 1988; 85: 7251–7255.
11 Leoncini L, Pacenti L, Rusciano D, Burroni D, Garbisa S, Cintorino M, Terrana B: Correlation between differentiation and lung colonization by retinoic acid treated F9 cells as revealed by the expression pattern of extracellular matrix and cell surface antigens. Amer J Pathol 1988; 130: 505–514.

Angiogenesis: Key Principles - Science - Technology - Medicine
ed. by R. Steiner, P. B. Weisz & R. Langer

Induction and altered distribution of tenascin in the basal lamina of colorectal adenomas and carcinomas

St. Riedl[a], P. Möller[b], A. Faissner[c] and P. Schlag[a]

Department of Surgery[a], Department of Pathology[b], Department of Neurobiology[c], University of Heidelberg, Germany

Introduction

Tenascin is an extracellular matrix molecule with isoforms ranging from 190 kD to 260 kD. Its molecular structure is six-armed with a central core containing disulfide junctions. The six subunits of tenascin contain paracentral EGF-like repeats, a segment resembling the type III repeats of fibronectin with RGD cell attachment signals and a terminal nodular domain resembling fibrinogen β and γ chains including a calcium-binding segment [1]. Tenascin is synthesized by fibroblasts. In vitro TGF-β induces tenascin synthesis. During embryonic development tenascin is found at specific times and locations. In adult tissues it is present only in very restricted locations and in low tissue concentrations. Tenascin is prominently expressed in wound healing and growth of a variety of tumors [2]. We have studied tenascin expression in normal colon and colorectal neoplasias to investigate the change of tenascin pattern under pathological conditions.

Materials and methods

26 specimens of normal colon mucosa originating from all parts of the large intestine, 38 adenomas and 52 adenocarcinomas of the colorectal mucosa were investigated. Representative samples of fresh frozen tissue were stored at $-70°C$. Sections of 4 to 6 μm were dried, fixed in acetone and immunostained. A polyclonal rabbit IgG antibody to human tenascin(pJ1) was used in dilution of 1:250 with the same binding characteristics as a monoclonal mouse antibody (608). As a secondary antibody, affinity pured polyclonal goat antibody (Jackson Immuno Research, West Grove, PA, USA) in dilution of 1:20, and as

third antibody, a soluble complex of horseradish peroxidase in dilution of 1:100 (Dakopatts, Glostrup, Denmark), were used [3].

All dilution and washing steps were carried out in PBS pH 7.4. Using 3-amino-ethyl-carbazole as chromogen the peroxidase reaction resulted in an intense red precipitate. The sections were weakly counter-stained with Harris hematoxylin and mounted with glycerol gelatin.

Results

In normal colon mucosa a very regular tenascin pattern emerged: the basal lamina was tenascin negative at the base of the crypt, displayed a discontinuous pattern in the middle part and developed into a continuous tenascin positive band at the upper part, broadening steadily towards the mucosal surface. A very sparce interstitial meshwork of tenascin could be observed in the upper parts of the lamina propria (Fig. 1). In all 38 adenomas a loss of the above-described polarity in tenascin content of the basal lamina was obvious. Tenascin was present throughout the basal lamina in a mostly smooth and continuous staining. Tenascin induction was also detected in the whole lamina propria (Fig. 2). Loss of polarity with extensive positivity of the basal lamina was a constant feature in colon carcinomas (Fig. 3). Unlike adenomas, carcinomas more frequently exhibited a rough ($p = 0.012$) and discontinuous ($p = 0.006$) basal lamina (Fig. 4). As to the clinical stage at the time of resection, presence of lymph node metastases (Dukes C, N1-3, respectively) was correlated with roughness ($p = 0.04$) (Tab. 1). The tumor stroma of the carcinomas regularly showed a tenascin neoformation.

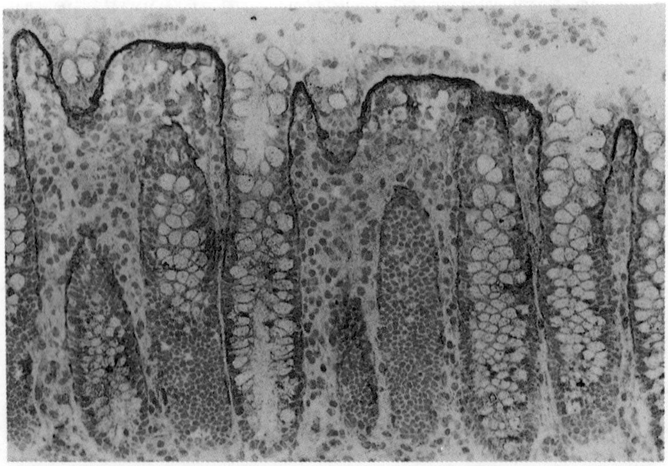

Figure 1. Tenascin staining in normal colon mucosa. Tenascin content of the basal lamina is high at the mucosal surface and steadily decreasing towards the bottom of the crypt. Tenascin positive fibrils are very scarce and barely viable within the apical part of the lamina propria.

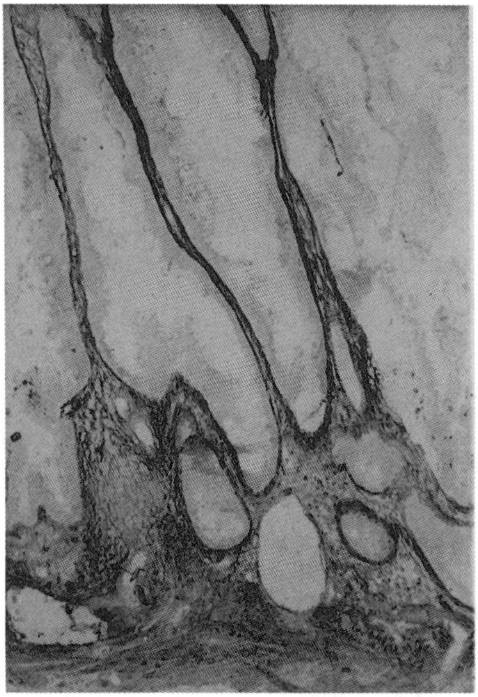

Figure 2. Tenascin induction in a colon adenoma. Loss of polarity and increase of tenascin content of the basal lamina and tenascin positivity in the lamina propria of a tubular adenoma.

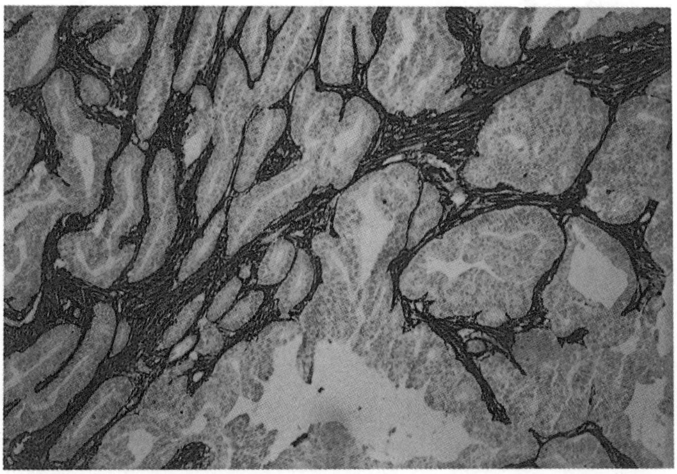

Figure 3. Tenascin induction in a colon carcinoma. Loss of polarity and tenascin positivity in the basal lamina and the stroma of an invasive adenocarcinoma of the colon.

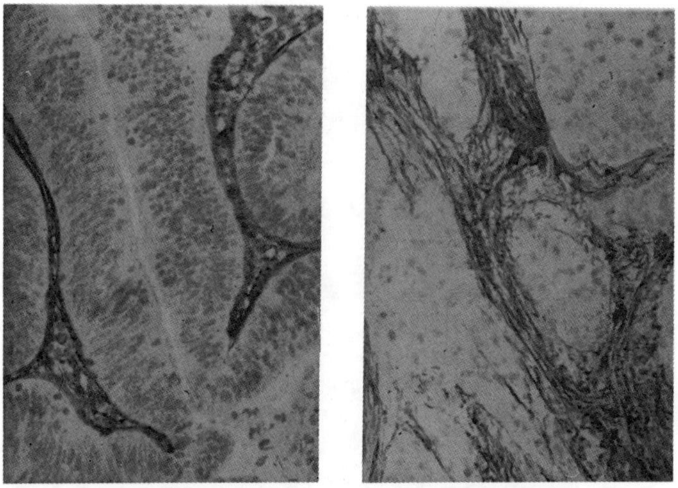

Figure 4. Structural alterations in the basal lamina of different carcinomas in tenascin staining. Even and continuous basal lamina of an adenocarcinoma and discontinuous basal lamina with curly alterations in another case.

Table 1. Structural alterations of the basal lamina as revealed by tenascin staining

		Surface qualities			Continuity	
Tissue	n	Smooth	Mixed	Rough	Continuous	Discontinuous
Adenomas	38	24	10	4	33	5
Carcinomas						
Total	52	17	20	15	27	25
Dukes B	22	12	6	4	12	10
Dukes C	15	2	4	9	5	10

Discussion

There is increasing evidence that tenascin plays an important role in embryogenesis but also in injured tissues and malignant tumors by modulating growth and differentation of different cell types. In physiological situations tenascin is frequently found in the interface of epithelial and mesenchymal tissues. Tenascin is associated with the basal lamina of several steadily regenerating normal epithelia, such as murine intestinal mucosa, rodent and human skin and human preovulatory glands. As in murine intestinal mucosa we found tenascin in the upper parts of the basal lamina of colon mucosa. Regarding its fibronectin antagonizing effect, tenascin is thought to promote the shedding of intestinal epithelia into the gut lumen [4]. The complete loss of polarity in tenascin content of the basal lamina in combination with tenascin precipitates in the adjacent stroma was observed only in adenomas and carcinomas. This fact may reflect the loss of epithelial organization depending on cell departments in regard to cell differentation. Under normal conditions, continuously dividing cells, so-called enteroblasts, are situated on tenascin negative basal lamina. In contrast, adenomatous and carcinomatous neoplastic cells in cell cycle dwell upon a tenascin positive basal lamina. Ultrastructurally the contact between intestinal epithelial cells is microvillous-like in tenascin positive areas and smooth and continuous in tenascin negative areas. The loss of the micro-heterogenity of the polarized basal lamina might cause an overall decreased adherence of neoplastic epithelia. It seems possible that tenascin promotes invasive growth of neoplastically transformed epithelial cells by disturbing the microarchitecture of the basal lamina.

References

1 Ericson HP, Bourdon MA: Tenascin: an extracellular matrix protein prominent in specialized embryonic tissues and tumors. Ann Rev Cell Biol 1989; 5: 71–92.

2 Natali PG, Zardi L: Tenascin: a hexameric adhesive glycoprotein. Int J Cancer 1989; 4: 66–68.

3 Faissner A, Kruse J: J1/Tenascin is a repulsive substrate for central nervous system neurons. Neuron 1990; 5: 627–637.

4 Probstmeier R, Martini R, Schachner M: Expression of J1/Tenascin in the crypt-villus unit of adult mouse small intestine: implications for its role in epithelial cell shedding. Development 1990; 109: 313–321.

Angiogenesis: Key Principles - Science - Technology - Medicine
ed. by R. Steiner, P. B. Weisz & R. Langer

On the quantitative rat mesenteric-window angiogenesis assay

Klas Norrby

Department of Pathology, University of Göteborg, Sahlgren Hospital, S-413 45 Göteborg, Sweden

Abstract. The rodent mesenteric-window angiogenesis assay permits the quantitative assessment of angiogenesis in adult, normally-vascularized mammalian tissue in the intact animal. Using appropriate optical magnifications, even very small, newly-formed vessels are recorded. The ultimate value of the model will, of course, depend on how pertinent it is to the features which govern angiogenesis in the tissues that are involved in clinically-relevant, angiogenesis-related diseases.

Introduction

A number of angiogenesis assays have been developed [1]. In order to put the rodent mesenteric-window assay in perspective, some features of the currently most widely-used assays – the rabbit cornea implantation technique [2, 3] and the chick chorioallantoic membrane (CAM) [4] assays – as well as the more recently introduced subcutaneous rodent sponge angiogenesis models [5, 6], will be discussed briefly. Any critical points raised in relation to these assays by no means imply that the results they produce are necessarily unreliable: each system provides reliable information of some kind.

The rabbit cornea and the chick embryo chorioallantoic membrane angiogenesis assays

These assays, which were introduced by Folkman and co-workers some 15–20 years ago, have been extremely useful and successful from many points of view. In fact, virtually all our present knowledge of experimental angiogenesis has been obtained using these techniques. Like most methods, they do, however, have limitations, some of which will be discussed.

In the rodent cornea assay, in which the test material is placed in a pocket created by microsurgery at a distance from the limbus, a positive test results in the directional growth of vessels from the limbus towards the pocket. Since the cornea is normally strictly avascular, the angiogenesis-regulating mechanisms of the tissue may be specific, however. One point in favour of this technique is that it permits continuous observations and the use of one eye as an internal control.

In the CAM assay, the acceleration or retardation of organogenic angiogenesis is examined in the vicinity of the test material which is applied to the membrane in the opened egg or in a Petri dish. The test substance is usually adsorbed to a polymer disc from which it is released slowly; similar discs are used in the cornea micropocket. The CAM technique does not lend itself to the study of true de novo angiogenesis since the vascular system is triggered proliferatively by a score of unidentified endogeneous factors. One distinct point in favour of this assay is its relative simplicity and usefulness for screening purposes, but the observation period is relatively short, usually ≤ 3 days.

Unfortunately, the quantification of angiogenesis in the CAM and the cornea assays has proved difficult [7, 8]. Consequently there has long been a call for other angiogenesis assays that could be used as a complement to or instead of the CAM and the cornea implantation methods. For instance, Vallee et al. [9] expressed the view that "the design and verification of specific, reliable, repeatable and precise methodology to determine angiogenesis remains an imperative of high priority". Since most human angiogenesis-related disorders occur in adult, normally-vascularized tissues which lack physiological angiogenesis, it is, furthermore, essential to develop angiogenesis assays which are applicable for the study of angiogenesis in adult, normally-vascularized mammalian tissue.

The subcutaneous rodent sponge models

A method for objective, continuous and reproducible studies of the blood flow in newly-formed vessels that invade subcutaneous implants of polyester discs in rats, using the ^{133}Xe clearance technique, was introduced by Andrade, Fan and Lewis in 1987 [5]. The surgical procedure creates a wound-healing-like lesion, as evidenced by the fact that the sponges are initially encapsulated by granulation tissue, before they are gradually infiltrated by host blood vessels. By implanting subcutaneous discs made of polyvinyl alcohol foam in mice, a somewhat similar technique was introduced by Fajardo and co-workers in 1988 [6]. Projected images of medial plane histological paraffin sections of the disc are analysed. The most reliable parameter to be measured is said to be the "centripetal vessel growth", i.e. the radial distance between the most central portion of the vessel which is situated closest to the centre of the disc and the edge of the disc.

Whether angiogenesis-regulating features which operate in normally-vascularized mammalian tissue actually operate in these sponges is still an open question, however. Most probably the functional and proliferative characteristics of the newly-formed microvasculature are significantly affected by the microenvironment and it is noteworthy that the central portions of the polyvinyl alcohol foam discs lack not only mature collagen and elastic fibres but certain cell types as well. Moreover, surgical procedures introduce uncontrollable effects in the tissue which is affected, be it primarily in the extracellular matrix or the cells, not to mention the involvement of additional angiogenic factors produced or released locally, or derived from the blood.

The rodent mesenteric-window angiogenesis assay

Although originally published in 1986 by Norrby, Jakobsson and Sörbo [10], the assay may not be familiar to many students of angiogenesis. Some features of the method will be described in brief.

The mesenteric window, which measures 5–10 μm in thickness, is perhaps the thinnest tissue in the mammalian body. The virtually two-dimensional vasculature of the tissue is quantifiable by two technically independent methods, i.e. in histological sections cut perpendicular to the surface [10, 11] and in spread preparations of intact, unsectioned windows where the entire vascular tree is visualized [12].

The *number of vessel profiles per unit length of tissue* is counted microscopically in stained sections of plastic-embedded mesenteric membrane. The smallest discernible vessels have a cross-sectional luminal area corresponding to a circular diameter of about 3.0 μm; the rat erythrocyte is ≥ 6.3 μm in diameter. The number of discernible vessels is, furthermore, related to the optical magnification used: only some 40% of all the vessels recorded at $\times 1000$ are discernible at $\times 100$ [11]. The reading at $\geq \times 400$, which is our standard, is straightforward; the inter- and intra-personal reproducibility for blind readings is very high, $r \geq 0.99$ using the linear regression analysis.

In spreads of intact windows the *vascularized area* (VA), as a percentage of the whole window area or in μm^2, and the *vascular density* (VD) within the vascularized area are measured morphometrically, whereas the *total vascular quantity* is computed from VA $\times$ VD [12]. Moreover, the number of vessels invading the test tissue from the surrounding fatty tissue, in terms of the *number of circumferential vessels per unit length of perimeter*, is computed from the number of vessels crossing the border of the window and its perimeter [13]. These parameters are assessed with a high degree of reproducibility, usually $r \geq 0.98$.

Clearly, both modes of the assay are quantitative in terms of the microvasculature present in the tissue at any given time point. The idea of continuously observing the mesenteric-window vessels using a transparent chamber has not been pursued because even an aseptic laparotomy deranges the homeostasis of the test tissue significantly, as evidenced by mast-cell secretion and increased cell proliferation [14].

One drawback of the mesenteric-membrane assay is that comparatively large quantities of test material are needed.

The model is well suited for testing the systemic effects of drugs on angiogenesis. The sc injection of testosterone in prepubescent male rats [11] and of protamine in adult rats [15] atoxically suppresses mast-cell-mediated angiogenesis elicited by the ip injection of a highly-selective mast-cell secretagogue. Moreover, the sc administration of a histamine H_2-receptor antagonist suppresses mast-cell-mediated angiogenesis whereas the sc administration of heparin, but not of an oligosaccharide fragment thereof, stimulates angiogenesis elicited by ip saline injection in rats. Since the mast cell is not only the sole source of heparin but also the repository of histamine in the body, these findings strongly suggest that the activated mast cell is capable of directly initiating angiogenesis, in agreement with the previous finding of mast-cell-mediated angiogenesis in the same test tissue [10, 16]. In passing, structure-function studies are needed to elucidate the critical molecular difference between the standard heparin and its oligosaccharide fragment with regard to the angiogenesis-stimulating effect, and the quantitative mesenteric-window assay appears to be uniquely suited to this type of study.

Acknowledgement. Supported by the Swedish Medical Research Council, project no. 5942.

References

1 Gullino PM: Angiogenesis factor(s), in Baserga R (ed): Tissue Growth Factors. New York, Springer-Verlag, 1981, pp 427–449.
2 Zauberman H, Michaelson IC, Bergmann F, Maurice DM: Stimulation of neovascularization of the cornea by biogenic amines. Exp Eye Res 1969; 8: 77–78.
3 Gimbrone MA, Jr, Cotran RS, Leapman SB, Folkman J: Tumor growth and neovascularization: An experimental model using the rabbit cornea. J Natl Cancer Inst 1974; 52: 413–427.
4 Auerbach R, Arensman R, Kubai L, Folkman J: Tumor induced angiogenesis: lack of inhibition by irradiation. Int J Cancer 1975; 15: 241–245.
5 Andrade SP, Fan T-PD, Lewis GP: Quantitative in vivo studies on angiogenesis in a rat sponge model. Br J Pathol 1987; 68: 755–766.
6 Fajardo LF, Kowalski J, Kwan HH, Prionas SD, Allison AC: The disc angiogenesis system. Lab Invest 1988; 58: 718–724.
7 Vu MT, Smith CF, Burger PC, Klintworth GK: An evaluation of methods to quantitate the chick chorioallantoic membrane assay in angiogenesis. Lab Invest 1985; 53: 499–508.
8 Blood CH, Zetter BR: Tumor interactions with the vasculature: angiogenesis and tumor metastasis. Biochem Biophys Acta 1990; 1032: 89–118.
9 Vallee BL, Riordan JF, Lobb RR, Higachi N, Fett JW, Crossley G, Buhler R, Budzik G, Breddam K, Bethune JL, Alderman EM: Tumor-derived angiogenesis factors from rat Walker 256 carcinoma: an experimental investigation and review. Experientia 1985; 41: 1–15.
10 Norrby K, Jakobsson A, Sörbo J: Mast-cell-mediated angiogenesis: a novel experimental model using the rat mesentery. Virchows Arch [Cell Pathol] 1986; 52: 195–206.

11 *Idem*: On mast-cell-mediated angiogenesis in the rat mesenteric-window assay. Agent Actions 1990; 30: 231–233.
12 *Idem*: Quantitative angiogenesis in spreads of intact rat mesenteric windows. Microvasc Res 1990; 39: 341–348.
13 *Idem*: Age-dependent mast-cell-mediated angiogenesis. APMIS 1988; Suppl 2: 251–261.
14 Franzén L, Norrby K: A tissue model for quantitative studies on time course of healing, rate of healing, and cell proliferation after wounding. APMIS (A) 1983; 91: 281–289.
15 Jakobsson A, Sörbo J, Norrby K: Protamine and mast-cell-mediated angiogenesis in the rat. Int J Exp Pathol 1990; 71: 209–217.
16 Norrby K, Jakobsson A, Sörbo J: Mast-cell secretion and angiogenesis, a quantitative study in rats and mice. Virchows Arch [Cell Pathol] 1989; 57: 251–256.

Angiogenesis: Key Principles - Science - Technology - Medicine
ed. by R. Steiner, P. B. Weisz & R. Langer

Wound healing in aged animals – Effects of locally applied Transforming Growth Factor beta 2 in different model systems

David A. Cox, Samuel Kunz*, Nico Cerletti, Gary K. McMaster and Robert R. Burk

CIBA-GEIGY Ltd., Pharmaceuticals Research Division, Molecular Biology Department and Experimental Infectious Models Group, CH-4002 Basel, Switzerland*

Abstract. Local application of a growth factor which could stimulate cell turnover, extracellular matrix synthesis and blood vessel formation in the skin should improve and accelerate wound healing processes which are often impaired in old age. We demonstrate the effects of TGF-β2 in promoting wound repair in old animals where normal healing responses are shown to be naturally slower. The potential use of TGF-βs for the treatment of wound injuries, including chronic non-healing ulcers in the elderly, is discussed.

Introduction

Wound healing is a complex interactive repair process characterised by sequential phases of inflammation, cell migration and proliferation, extracellular matrix (ECM) synthesis, angiogenesis and remodelling [1]. It is recognised that wound healing becomes slower and is often impaired with advancing age [2]; a phenomenon which can be directly related to a number of cutaneous changes which occur as ageing progresses. These alterations include a decrease in the rates of cell turnover, ECM synthesis and blood vessel formation [3], processes that are normally regulated by a range of soluble secreted growth factors, some of which are potential candidates for use in wound healing therapies. Growth factor-based therapeutics that stimulate and accelerate these processes may prove especially useful for the treatment of wounds that heal slowly due to natural reasons, or which fail to respond to existing therapies, as is often the case with severe burns, venous stasis-, diabetic-, or decubitus ulcers [4], injuries which collectively represent a major problem in geriatric medicine today.

Transforming Growth Factor beta (TGF-β) is one such candidate as a therapeutic agent for tissue repair and three highly homologous isoforms of this molecule are known to exist in man and other mammals. TGF-βs are stored in platelets and are released and activated from precursor forms at sites of injury. At physiological concentrations, these active peptides are not only potent chemoattractants and activators of macrophages and fibroblasts, but also regulators of ECM synthesis [reviewed in ref. 5]. TGF-β1, the original member of this family of molecules, has already been shown to have beneficial effects on the healing of primary and secondary intention wounds [6] and in stimulating neovascular formation [7]. Furthermore, its greater efficacy over other growth factors (e.g. EGF or PDGF) and its effects in reversing wound healing deficits in diabetic animals [8], or in animals treated with glucocorticoids [9] or doxorubicin [10], suggest a therapeutic use for this peptide not only in the treatment of chronic, non-healing wounds but also in surgical wounds where repair processes are potentially at risk due to steroid treatment, radiation therapy or chemotherapy. This present study investigates the in vivo effects of a second isoform, TGF-β2, following topical or local application in three age-impaired wound healing models and also in an established angiogenesis model.

Materials and models

Reagents

TGF-β2 was isolated from bovine milk as described previously [11]. For use in each of the wound healing models, TGF-β2 was dissolved at the concentrations indicated in a viscous vehicle buffer consisting of a 0.5% or 0.8% w/v solution of hydroxypropyl cellulose (Hercules, Wilmington) in 10 mM histidine, 140 mM NaCl, pH 7.4. In the angiogenesis assay, TGF-β2 was dissolved in filter-sterilised vehicle buffer (0.5%) at the appropriate concentration. 5 μl of this solution was applied directly onto 1% methylcellulose discs (10 × 10 mm) which were then air-dried in a sterile air stream. All solutions and materials used in the following models were sterile, endotoxin free and pyrogen free.

Wound healing and angiogenesis models

Partial-thickness wound model in old mice. Single 1 × 1 cm thermal injuries were performed by application of a brass block, equilibrated to 80°C, to the depilated dorsal thorax of fully anaesthetised 450–500-day-old male C57/BL6J mice. Following surgical removal of the blister, TGF-β2 was applied daily, in 0.8% vehicle buffer (25 μl) and at the indicated dosage, for 5 days onto the open wound. Animals were caged separately for the duration of the experiment and sacrificed on day 6, at which time the healing response was evaluated by planimetry and histological analyses. Control wounds in the old animals were treated with buffer alone or were left untreated for the duration of the experiment. Results were also compared with the epithelial regeneration process in young (56–84-day-old) C57/BL6 mice with identical middermal burns which were left untreated for the 5-day experimental period. Areas of burns that had regenerated epithelium were outlined onto uniform thickness transparent overhead projector film and the percentage of each original burn area that had healed was calculated by planimetry. Measurements were not made on wounds showing evidence of infection or excessive haemorrhaging (less than 3% of all wounds).

Full-thickness wound model in old rats. 5-cm linear incisions were performed with surgical scissors 1.5 cm on each side of the dorsal midline on the depilated thorax of fully anaesthetised 450-day-old Wistar rats. A single application of TGF-β2, in 0.8% vehicle buffer (100 μl) and at the indicated dosage, was applied directly to the edges of the left side incision and a placebo protein (BSA), at the corresponding protein concentration, was applied at the same time to the edges of the contralateral right side incision. Control animals received vehicle buffer only in left side incisions, whilst contralateral right side incisions received no treatment prior to suturing. Edges of each wound were then coated with 5 evenly placed horizontal mattress sutures of 5-0 Ethilon. Animals were caged separately and the wounds were left to heal for varying periods up to 21 days post treatment. After sacrifice the entire dorsal skin was removed and all subcutaneous fat was dissected from the underside using a surgical scalpel. A template consisting of two parallel surgical blades (8 mm apart) was then used to excise strips of skin between sutures on each incision for tensile strength measurements. Samples were taken from one end of each incision for histological analysis. The maximum load tolerated by each excised skin sample was measured with a Universal Tensile Strength Machine Model 144501 (Zwick, Ulm, FRG). Measurements were made on 30 × 8 mm strips which were secured between hydraulic clamps and then stretched to breaking point at a rate of 10 mm per minute, with the maximum load recorded on a chart recorder. Measurements were made on triplicate samples from each wound. Breaking strength was not measured on wounds showing evidence of infection or excessive haemorrhaging (less than 3% of all wounds).

Porous Chamber Implant model in old rats. Empty rigid polytetrafluoroethylene tubes (internal and external diameters, 10 and 12 mm respectively; length 32 mm), each perforated by approximately 250 regularly spaced holes (diameter 1 mm) and sealed at each end with a removable cap of identical material, were gas sterilized and surgically inserted sub-cutaneously, in symmetrical fashion, through small incisions into the dorsal flanks of fully anaesthetized 450-day-old Wistar rats. One gas-sterilized tissue cage was implanted into each flank and the incisions were closed with single surgical clips (Clay-Adams Auto-Clips, 9 mm) which were removed 5 days after surgery. Animals were used for experimentation 14 days after implantation of the chambers. At this time, TGF-β2 was injected daily, in 0.5% vehicle buffer (100 μl) and at the indicated dosage, for 5 days directly into the left side chamber through the body wall. Contralateral right side chambers received corresponding equal amounts of a placebo control (BSA). Control animals received vehicle buffer alone in the left side chambers whereas right side chambers remain untreated for the duration of the experiment. All animals were individually caged for the duration of the experiment and were sacrificed 24 h after the last series of injections. Chambers were then removed from each animal by aseptic technique, the fibrous tissue from inside each chamber was 'wet' weighted and the total amount of serous protein in the sterile chamber fluid was measured by using the Pierce BCA protein assay reagent according to manufacturers instructions. Samples of fibrous tissue removed from inside and outside each chamber were prepared for histological analysis. Measurements were not made on chambers showing evidence of infection or rejection (less than 3% of all chambers).

Chicken Chorio-allantoic Membrane (CAM) assay. Methycellulose discs containing 10–100 ng TGF-β2 were placed onto the exposed CAMs of fertilised chicken eggs on day 10 of incubation. After 72 h the vascular rearrangements in response to each disc were qualitatively scored according to the following scale:

+2 Clear, unequivoval vascular rearrangement
+1 Unclear, slight difference in vascularity
0 No effect visible
−1 Unclear, slight zone of inhibition in the capillary vascularisation
−2 Clear, inhibitory zone in the capillary vascularisation

Control discs contained vehicle buffer alone (0 negative control) or 10 pg oleic acid (+2 Positive Control). Care must be taken to avoid fragments of shell or membrane falling onto the CAM since these often become inflammatory foci which modify the vascularisation.

Results

Partial-thickness wound model in old mice

Planimetrical analyses of wounds demonstrate that topical application of TGF-β2 for 5 days, stimulated and accelerated epithelial regeneration in partial-thickness wounds on old mice in a dose-dependent fashion when compared with vehicle buffer only or untreated wounds (Fig. 1). The enhanced rate of healing resembled that observed in young animals with wounds which had remained untreated for the duration of the experiment. Comparsion of the healing rates of untreated wounds

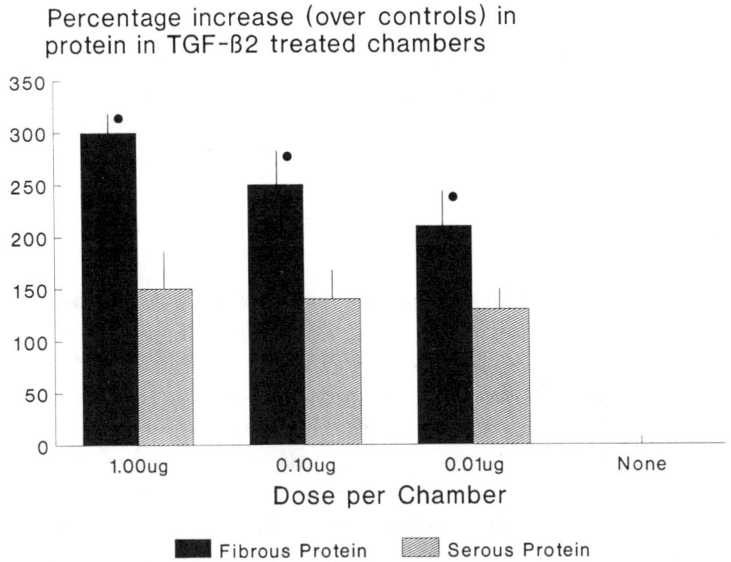

Figure 1. Planimetrical analyses of TGF-β2 treated partial-thickness (PT) wounds, vehicle treated PT wounds and untreated PT wounds in old mice (solid bars), and untreated PT wounds in young mice (hatched bars) on Day 6. Results are shown as mean ± SEM of 16 estimations. ● $p < 0.01$ compared with controls. ●● $p < 0.01$ comparing untreated wounds in old vs young animals.

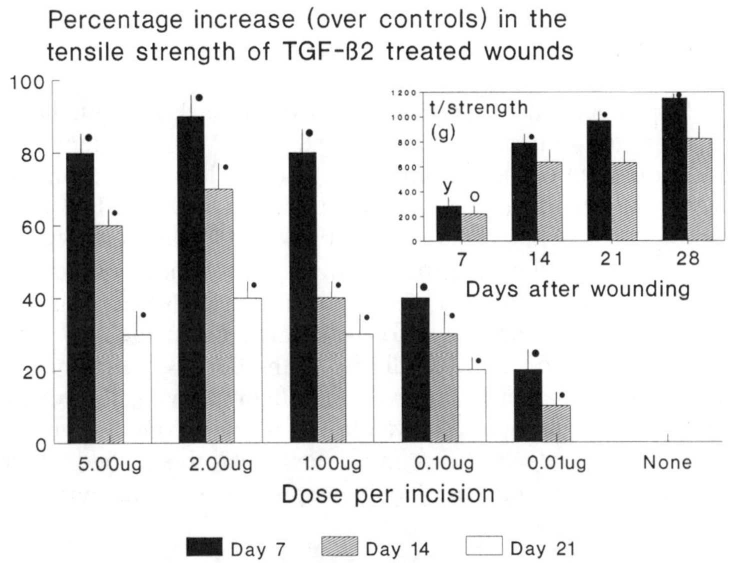

Figure 2. Tensile strength analyses of TGF-β2 treated incisional (I) wounds, vehicle treated I wounds and untreated I wounds in old rats over a 21-day period. Results are shown as mean ± SEM of 16 estimations. ● $p < 0.01$ compared with controls. Inset: Tensile strength analyses of untreated I wounds in old (O) and young (Y) rats over a 28-day period. Results are shown as mean ± SEM of 16 estimations. ●● $p < 0.01$ comparing untreated wounds in old vs young animals.

on old and young mice showed a markedly slower healing process in the old animals. Histological analyses of old mice wound biopsies revealed the extent of the regenerating epidermis and also a transient hyperkeratosis at the wound periphery on Day 6 following treatment with TGF-β2.

Full-thickness wound model in old rats

Tensile strength measurements of wound biopsies demonstrate that a single topical application of TGF-β2 markedly enhanced the breaking strength of full-thickness incisional wounds in old rats in a dose-dependent fashion over a 21-day time period when compared against control wounds (Fig. 2). Comparsion of the healing responses of untreated incisional wounds in old and young rats clearly shows a markedly slower healing rate in the old animals (Fig. 2; inset), histological analyses revealed the high degree of cellularity of TGF-β2-treated wounds; evidenced by a marked increase in the influx of mononuclear cells and fibroblasts and an increase in collagen production over the 21-day period as compared to control wounds. A transient hyperkeratosis was also evident in TGF-β2-treated wounds which persisted up to 14 days after the treatment.

Porous chamber implant model in old rats

Daily local injection of TGF-β2 for 5 days markedly enhanced the accumulation of total fibrous tissue and (to a lesser extent) serous chamber protein, in a dose-dependent fashion in left-sided chambers as compared to the right-sided contralateral chambers which had received equal amounts of a placebo protein (Fig. 3). No differences between left-sided and right-sided chambers in the control group were observed. Histological analyses revealed an influx of migrating fibroblasts and mononuclear cells within the TGF-β2-treated chambers and also an increase in the thickness and vascularity of the fibrous matrix surrounding the chambers. A sterile infiltrate of inflammatory cells, consisting predominantly of macrophages, was found in the serous fluid of TGF-β2-treated chambers whereas contralateral placebo-treated chamber fluid showed a predominance of polymorphonuclear leukocytes.

Chicken chorio-allantoic membrane (CAM) assay

Application of TGF-β2 (10–100 ng) to the chicken CAM resulted in a dose-dependent rearrangement of the membrane vasculature (score +2; Fig. 4). Positive effects were evidenced by a radial orientation of

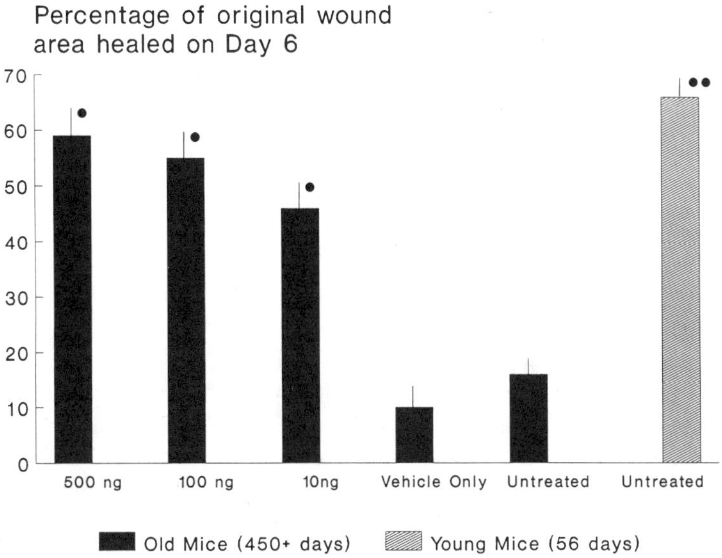

Figure 3. Analyses of fibrous protein and serous protein in TGF-$\beta 2$ treated-, placebo treated-, vehicle only- or untreated chambers on Day 6. Results are shown as mean $\pm$ SEM of 8 estimations. ● $p < 0.01$ compared with placebo treatments.

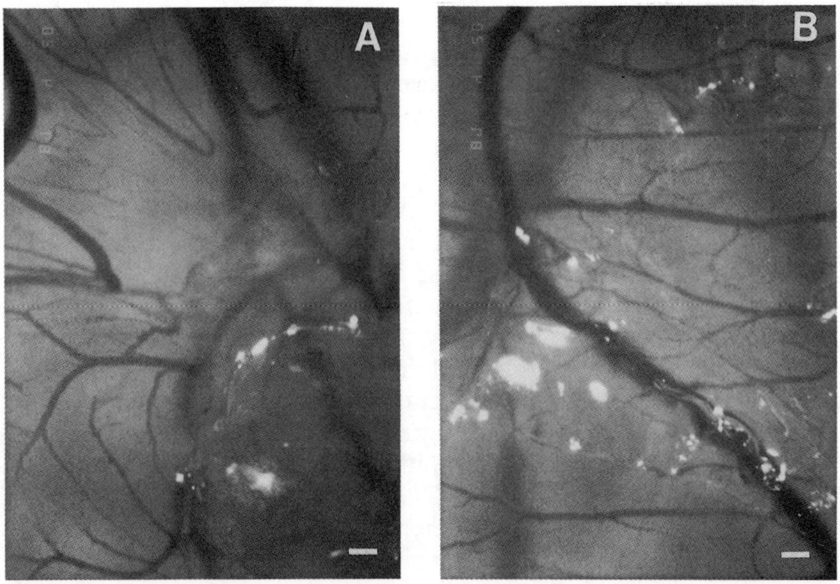

Figure 4. Vascular rearrangements in the chicken CAM assay after 72 h. TGF-$\beta 2$ on a methylcellulose disc induced a marked vascular rearrangement in the CAM (*A*) as compared to a control disc alone (*B*). Bars, 1 mm.

newly formed capillaries towards the disc and a formation of larger blood vessels which was clearly apparent after 72 h. Discs alone had no influence on the membrane vasculature whereas 10 pg oleic acid caused a clear positive effect (score +2) in this model.

Discussion

The structure of the skin enables wounds to be categorized according to the depth of the injury and the tissues affected [1]. This present study describes the animal models which have been used to study, in a defined way, the effects of a second TGF-β isoform on the healing of different types of wounds. Whereas the efficacy of TGF-β1 has been previously demonstrated in animals where defective wound healing has been artificially induced, our studies with TGF-β2 have focussed on aged animals where repair processes are naturally impaired.

The results obtained from these experiments demonstrate that TGF-β2 accelerates wound dealing processes, firstly in full-thickness wounds to the dermis (which heal by an initial granulation phase followed by phases of re-epithelialisation and contraction) and secondly, in partial-thickness wounds to the epidermis (which heal essentially by a process of re-epithelialisation). Our findings demonstrate the chemoattractive effects of TGF-β2 on the recruitment of macrophages and fibroblasts and the stimulatory effects on the subsequent production of fibrous ECM in incisional wounds and porous implants. The increase in matrix vascularity together with the angiogenic effects produced in the CAM assay suggests a further important role for TGF-β2 in increasing blood flow to a wound. TGF-β1 has already been shown to be a migration factor for human keratinocyte cells [12] and to initiate the outgrowth of epithelial cells from skin explant cultures [13]. We have previously reported the migration of normal human skin epithelial cells (NCTC 2544; ATCC CRL 19) in response to TGF-β2 [11] and this present study now demonstrates that this factor enhances the re-epithelialisation of partial-thickness wounds. It may therefore be that TGF-βs have two different effects on epidermal cells during wound healing; firstly, an early stimulatory effect on cell migration and secondly, as reported for different epithelial cell types [5], a growth inhibitory effect which would result in a termination of the re-epithelialisation process. Whether this first effect is directly mediated by TGF-βs or is produced as a result of interactions between cells and other cytokines present at the wound site is currently unknown. An impairment in the migration of epidermal cells from decubitus ulcer biopsy explant cultures has recently been suggested as a cause of the protracted epithelialisation in these chronic wounds which fail of respond to conventional treatments [14].

Careful clinical investigation of the efficacy of TGF-βs in impaired or non-healing wounds is the next important step in the development of these peptides as wound healing agents.

Acknowledgements. The authors would like to thank C. Ardines, J. Vaxelaire, S. Tobler, and P. Schad, for their excellent technical assistance, and Drs. G. Bilbe and W. Seiler for their critical reading of the manuscript.

References

1 Peacock EE: Wound Repair. Saunders, Philadelphia, 1984.
2 Grove GL: Age-related differences in healing of superficial skin wounds in humans. Arch Dermatol Res 1982; 272: 381–385.
3 Carter DM, Balin AK: Dermatological aspects of aging. Med Clin N Am 1983; 67: 531–543.
4 Seiler WO, Stähelin HB: Decubitus ulceration in the elderly. Principles and practice of nursing home care. Springer, New York, 1989: 328–349.
5 Massagué J: The Transforming Growth Factor β Family. A Rev Cell Biol 1990; 6: 597–641.
6 Amento EP, Beck LS: TGF-β and wound healing. Clinical applications of TGF-β. CIBA Foundation Symposium No 157. Wiley, London, 1991: 115–129.
7 Yang EY, Moses HL: Transforming growth factor β1-induced changes in cell migration, proliferation, and angiogenesis in the chicken chorioallantoic membrane. J Cell Biol 1990; 111: 731–741.
8 Broadley KN, Aquino AM, Hicks B, Ditesheim JA, McGee GS, Demetriou AA, Woodward SC, Davidson JM: Growth factors bFGF and TGF-β accelerate the rate of wound repair in normal and diabetic rats. Int J Tiss Reac 1988; X: 345–353.
9 Pierce GF, Mustoe TA, Lingelbach J, Masakowski VR, Gramates P, Deuel TF: Transforming growth factor beta reverses the glucocorticoid-induced wound-healing deficit in rats: Possible regulation in macrophages by platelet-derived growth factor. Proc Natl Acad Sci USA 1989; 86: 2229–2233.
10 Curtsinger LJ, Pietsch JD, Brown GL, von Frauenhofer A, Ackerman D, Polk HC, Schultz GS: Reversal of adriamycin-impaired wound healing by transforming growth factor beta. Surg Gynaecol Obstet 1989; 168: 517–522.
11 Cox DA, Burk RR: Isolation and characterisation of milk growth factor, a transforming growth factor β2-related polypeptide, from bovine milk. Eur J Biochem 1991; 197: 353–358.
12 Nickoloff BJ, Mitra RS, Riser BL, Dixit VM, Varani J: Modulation of keratinocyte motility. Correlation with production of extracellular matrix molecules in response to growth promoting and antiproliferative factors. Am J Pathol 1988; 132: 543–551.
13 Hebda PA: Stimulatory effects of transforming growth factor-beta and epidermal growth factor on epidermal cell outgrowth from porcine skin explant cultures. J Invest Dermatol 1988; 91: 440–445.
14 Seiler WO, Stähelin HB, Zolliker R, Kallenberger A, Lüscher NJ: Impaired migration of epidermal cells from decubitus ulcers in cell cultures. A cause of protracted wound healing? Am J Clin Pathol 1989; 92: 430–434.

Angiogenesis: Key Principles - Science - Technology - Medicine
ed. by R. Steiner, P. B. Weisz & R. Langer

Intradermal angiogenesis in nude mice induced by human tumor cells or b-FGF

Gisbert Weckbecker, Rong Liu and Levente Tolcsvai

Preclinical Research, Sandoz Pharma Ltd., CH-4002 Basle, Switzerland

Tumor-induced angiogenesis is a promising target for anti-cancer drugs. However, testing of angiostatic drugs in vivo requires adequate models. Thus, our study was aimed at establishing angiogenesis models in the skin of nude mice which allow use of both human cancer cells and growth factors for inducing blood vessel formation. The angiogenic activity of the two human tumor cell lines SiHa (cervix) and MDA-MB-231 (breast) was evaluated following intradermal injection. Angiogenesis was demonstrated to be a time- and cell concentration-dependent process. Alternatively, bFGF (1 μg) adsorbed to microspheres was effective in eliciting an angiogenic response under similar conditions. Both models were validated by use of the standard angiostatic combination cortisone plus heparin. Moreover, of a series of compounds tested, suramin, compound 62-759 (platelet-activating factor antagonist) and 89-443 (aromatase inhibitor) were found to exert significant angiostatic effects. In conclusion, tumor- and growth factor-induced angiogenesis in the skin of nude mice may be an interesting alternative for studying blood vessel formation in vivo.

Introduction

As demonstrated by Folkman's group [1–3] angiogenesis is an important permissive step in tumor progression. Models to study blood vessel formation include the CAM, the rabbit or rat cornea and, more recently, in vitro systems that allow to track migration and tube formation of endothelial cells (cf. this book). However, there is still a demand for in vivo models that allow evaluation of potential angiogenesis modulators. Based on angiogenesis studies in normal mice [4, 5] we assessed the utility of intradermally induced angiogenesis in immune-deficient nude mice.

Materials and methods

Female Balb/A nu/nu mice (17 to 22 g, 6–8 weeks old) from IFFA Credo (Lyon) were used for the experiments. SiHa cells (human squamous carcinoma of the cervix) and MDA-MB-231 cells (human breast adenocarcinoma cells) came from ATCC. The cells were grown in RPMI 1640 medium supplemented with 5% FCS at 37°C and 5% CO_2. For inoculation, subconfluent monolayer cultures were trypsinized and washed 3 times in PBS using centrifugation at 1800 rpm for 5 min. Final cell suspensions were prepared in 0.5% of trypan blue/PBS (1:25, v/v) to help identify the injection sites. Mice were anesthetized by 400 mg/kg chloral hydrate (Sigma) i.p. Tumor cells (0.1 to 10×10^6 in 0.1 ml) were inoculated intracutaneously using a 26G needle. Usually two midventral sites/mouse were injected which were located distant from the main ventral skin vessels so that the background vessel count was low. Control groups received 0.1 ml 0.02% trypan blue in PBS. At the indicated times following injection, anesthetized mice were sacrificed by CO_2 inhalation. The skin was mounted onto a plastic ring (40 mm diameter) for evaluation by an inverted microscope (Zeiss IM) at 12.5- and 25-fold magnification. As a measure of angiogenesis, vessels were photographed and those were counted that were directly connected with the tumor. In control animals those vessels were counted that were connected to a defined area around the injection site. This area corresponded to the mean area of the dermal tumors. The latter was determined by use of a caliper according to the equation $3.14 \times r^2$. Microsphere-induced angiogenesis: Matrix Gel Blue A (Amicon) diluted with saline (1:1) was incubated with bFGF (bovine) for 30 min at 37°C. This suspension (0.05 ml corresponding to 1 μg bFGF) was injected i.d. Control animals received either the microspheres without FGF or Evans blue/0.9% NaCl. For statistical analysis Student's t-test was used; data are shown as mean $\pm$ SEM.

Results

Injection of human tumor cells such as SiHa (cervix) or MDA-MB-231 (breast) into the skin of athymic nude mice led to the formation of dermal blood vessels. New tumor-associated vessels were tortuous, exhibited various lengths and diameters and grew in radial patterns. The angiogenic response was dependent on the number of cells injected intradermally (Fig. 1). Moreover, a clear time-dependence was apparent in that, e.g., with 10^7 SiHa cells per injection the vessel counts were already significantly elevated at day 2 and reached a maximum at day 10 after injection. In a similar experiment, B16F10 murine melanoma

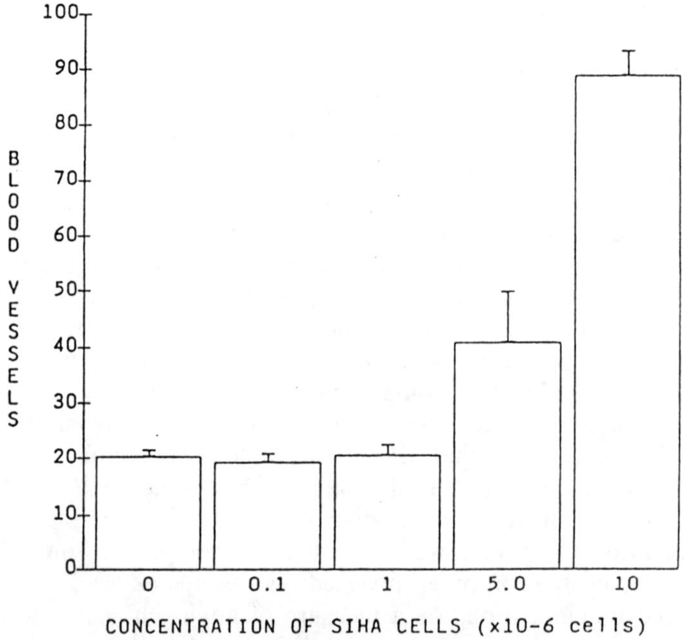

Figure 1. Angiogenesis induced in the skin of nude mice as a function of increasing amounts of SiHa tumor cells injected intradermally.

cells gave rise to an aggressively growing tumor that could be evaluated by this method only for a few days following inoculation.

The model of tumor cell-induced angiogenesis in nude mice was validated by use of the well-established angiostatic combination cortisone (75 mg/kg/day, s.c.) plus heparin (p.o. or s.c.). Oral (500 U/ml drinking water) and subcutaneous (630 U/kg, bid) administration of heparin gave similar results, moreover, no difference in the inhibitory potency was seen when treatment was started on the day of tumor inoculation or 3 days later. In animals treated with this combination, angiogenesis was significantly ($p < 0.001$) inhibited by up to 72% as compared to vehicle-treated controls. However, there were very serious side effects including extensive subcutaneous bleeding with s.c. heparin and massive weight loss averaging 3 g per animal. In an effort to identify compounds that are better tolerated, suramin was tested and found to exert highly significant, dose-dependent angiostatic effects without causing, under the conditions used, overt signs of toxicity (Fig. 2). Interestingly, compound 62-759, a receptor antagonist of PAF (platelet-activating factor) also suppressed dose-dependently blood vessel formation in this model.

In order to evaluate the contribution of growth factors to tumor-induced angiogenesis, preliminary experiments using microspheres coated

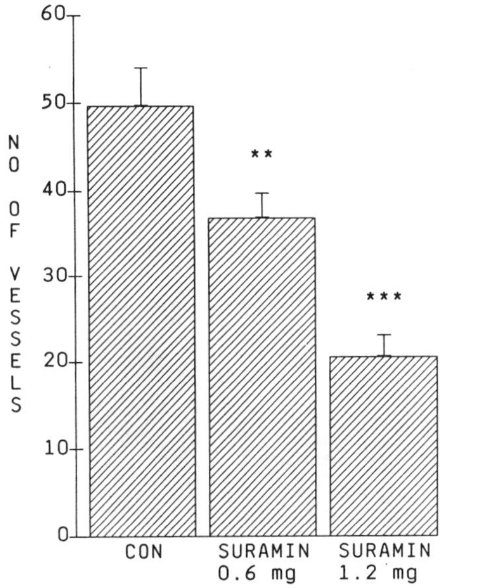

Figure 2. Inhibitory effect of intraperitoneally administered suramin on SiHa cells-induced angiogenesis.

with bFGF were carried out. Analogous to tumor cells, the intradermal injection of 1 μg bFGF adsorbed to the solid carriers elicited a marked increase in blood vessels around the injection site. It should be noted that the uncoated microspheres used were already nonspecifically angiogenic as compared to the injection of vehicle alone. However, the additional increase in vessel counts caused by bFGF was always significantly different from that induced by uncoated microspheres. Again, growth-factor-dependent angiogenesis was highly susceptible to treatment with heparin/cortisone (Fig. 3). Of the various compounds tested in this model, aromatase inhibitors appeared to be interesting: compound 89-443 (1.8 mg/kg, p.o., b.i.d.) totally suppressed bFGF-dependent angiogenesis while equimolar CGS 16-949A lacked any inhibitory activity (both compounds exerted similar inhibitory effects on aromatase).

Discussion

The in vivo angiogenesis models described here combine several advantages in that tumor cells of human origin or adequately administered bFGF can be used for the intradermal induction of blood vessels;

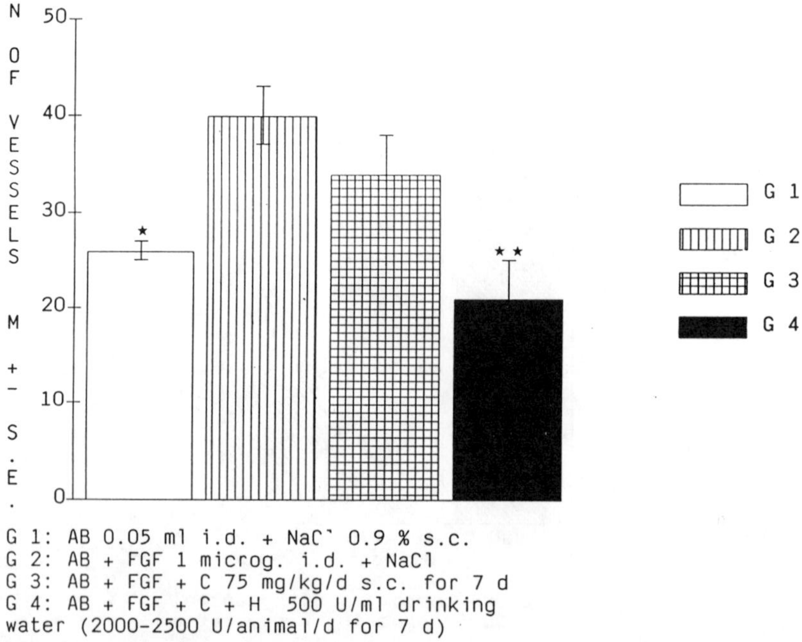

Figure 3. bFGF-induced angiogenesis in the skin of nude mice and inhibition by cortisone/heparin. (AB, microspheres; C, cortisone; H, heparin).

this setup may allow to investigate the role of growth factors in human tumor-induced angiogenesis and to address the question as to the mode of action of angiogenesis modulators. Moreover, the hairless transparent skin of nude mice allows ready evaluation by light microscopy. Since human tumors can be used, this model may be relevant for the clinical assessment of the angiogenic potential of a given tumor and to determine the feasibility of an angiostatic therapy. Our studies using various known and novel inhibitors of angiogenesis demonstrate the utility of the models in compound screening. However, we are currently looking into the possibility of developing a further refined model that allows continuous monitoring of vessel development without sacrificing the animal.

References

1 Folkman J: Tumor angiogenesis. Adv Cancer Res 1985; 43: 175–203.
2 Folkman J, Langer R, Linhardt RJ, Haudenschild C, Taylor S: Angiogenesis inhibition and tumor regression caused by heparin or a heparin fragment in the presence of cortisone. Science 1983; 221: 719–725.

3 Folkman J: Die Gefässversorgung von Tumoren, in Schirrmacher E (ed): Krebs – Tumoren, Zellen, Gene. Spektrum Wissenschaft, Verlagsgesellschaft Heidelberg, 1987, pp 134–146.
4 Sidky YA, Borden EC: Inhibition of angiogenesis by interferons: Effects on tumor- and lymphocyte-induced vascular responses. Cancer Res 1987; 47: 5155–5161.
5 Kreisle RA, Ershler WB: Investigation of tumor angiogenesis in an Id mouse model: Role of host-tumor interactions. J Nat Cancer Inst 1988; 80: 849–853.

Angiogenesis: Key Principles - Science - Technology - Medicine
ed. by R. Steiner, P. B. Weisz & R. Langer

Angiogenic activity of PAF and inhibition of blood flow by *Bothrops Jararaca* venom in a mouse sponge model

L. B. G. B. Vieira, J. A. A. Rodrigues, V. H. R. Leite and S. P. Andrade[a]

Department of Physiology and Biophysics, Federal University of Minas Gerais, Belo Horizonte, MG, Brazil and [a]Department of Pharmacology, Royal College of Surgeons of England, London WC2A 3PN, England

Introduction

Angiogenesis is a natural and necessary response to injury in wound healing and in endometrial regeneration. However, in tumor growth and a variety of vascular diseases, angiogenesis can be seen as contributing to the pathology of these conditions. A number of in vitro [1, 2] and in vivo [3–5] systems in which blood vessel formation can be analyzed have been described. Recently, we have developed a new, quantitative method of measuring angiogenesis in a sponge implant in rats [6], which detects neovascularization of an implanted sponge by measuring ^{133}Xe washout as an indication of blood flow. Parallel histological studies of the implants showed a good correlation between morphological signs of neovasculariation and Xenon washout. The purpose of the present study was to determine if the angiogenesis model in rat, modified for use in mice, could be used to detect potential angiogenic factors and inhibitors of blood flow. Platelet-activating factor (PAF), a low molecular weight inflammatory mediator, was investigated for its ability to promote neovascularization in the implants. *Bothrops Jararaca* venom, causes necrotic lesions in man and was used to demonstrate the ability of our model to detect changes in vascular tone.

Materials and methods

Mice

Male Balb/c mice (20–30 g) were used for all experiments.

Chemicals

^{133}Xenon injection (10 mCi in 3 ml) was obtained from Amersham International, Platelet-activating factor (PAF 1-0-alkyl-2-acetyl-sn-glycero-3-phosphorylcholine) from Sigma; *Bothrops Jararaca* venom was a gift from Department of Biophysics, Federal University of Minas Gerais, Brazil.

Sponge implants

Sterile polyester discs (8 mm × 4 mm) with a cannula attached were used as a matrix for vessel growth [6]. The implants were introduced through a 1 cm incision in the skin at the back of the anesthetized mouse. The cannula was exteriorized and plugged with a smaller sealed polythene tubing. Animals were housed individually and provided with a normal diet and water.

Blood flow measurement

For blood flow measurements the animals were anaesthetized, Xenon (^{133}Xe) solution (10 μl) injected into the sponge, the cannula plugged and the washout of radioactivity from the implant monitored [6]. Washout of ^{133}Xe is expressed in terms of its half-time ($t_{1/2}$; time taken for the radioactivity in the implant to fall to 50% of its original value).

Histology of the sponge implants

The implants were removed for formalin fixation and processed by standard histological methods. Indirect immunofluorescence for laminin was performed using anti-laminin antibodies as previously described [7]. Images were recorded on Tri X Pan film (Kodak).

Effects of PAF and Lyso-PAF

On day 4 postimplantation PAF or Lyso-PAF (its biologically inactive metabolite, 2 μg in 50 μl) was injected into the sponge and this treatment repeated daily for another 4 days (5 injections in total), in groups of 6 animals. A control group (n = 7) received 50 μl of PBS (phosphate buffered saline) in the same schedule. These injections were made immediately after the measurement of ^{133}Xe washout.

Fibrovascular infiltration

Estimates of fibrovascular infiltration of the implant were derived from the wet weight of the sponge disc and its content of protein [8] and (GAG) glycosaminoglycan [9]. For this the implants were removed on day 7 postimplantation in the control and PAF treated groups.

Effects of B. Jararaca venom

On day 14 after implantation 5, 10, 20 μg of venom in 50 μl of saline were given into implants immediately before ^{133}Xe.

Statistical analysis

Results are given as mean SEM from n animals. Comparison between groups were made with Student's t-test for unpaired data.

Results

Normal angiogenesis in the implant

At various times after implantation ^{133}Xe washout from untreated sponges was measured expressed as the $t_{1/2}$ and the mean values shown in Fig. 1. At 4 days after implantation, $t_{1/2}$ was high and fell progressively to within the range of $t_{1/2}$ for normal untreated skin 5 ± 1 min by 13 days and remained at this value during the subsequent days. Sequential histology of sponges excised from animals at different times showed a clear progression of events leading to capillary formation after early infiltration of neutrophils, macrophages, fibroblasts and endothelial cells laid down on an amorphous extracellular matrix. Blood vessels were detected histologically either using hematoxilin-eosin staining or by immunofluorescence for laminin, an excellent marker for developing capillaries within normal and neoplastic tissues [10]; neuronal structures have also been identified in sponges removed 14 days postimplantation.

PAF and angiogenesis

The effects of PAF and lyso-PAF on angiogenesis were measured following 5 daily doses of these agents (Fig. 2). On day 7 postimplantation, animals that had received PAF showed significantly lower $t_{1/2}$ values than those given lyso-PAF or PBS. The latter two groups were not different at any time.

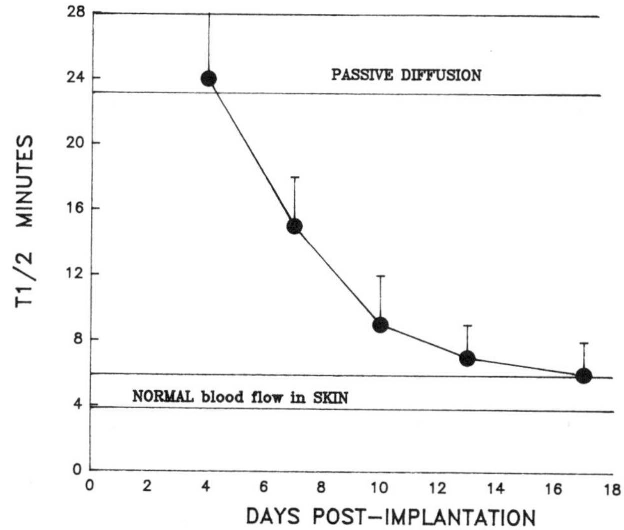

Figure 1. Establishment of blood flow in sponge implants expressed as $t_{1/2}$ plotted against number of days postimplantation. ^{133}Xe washout due to diffusion and from normal mouse skin are indicated. Data represent mean ± SEM from 20 mice.

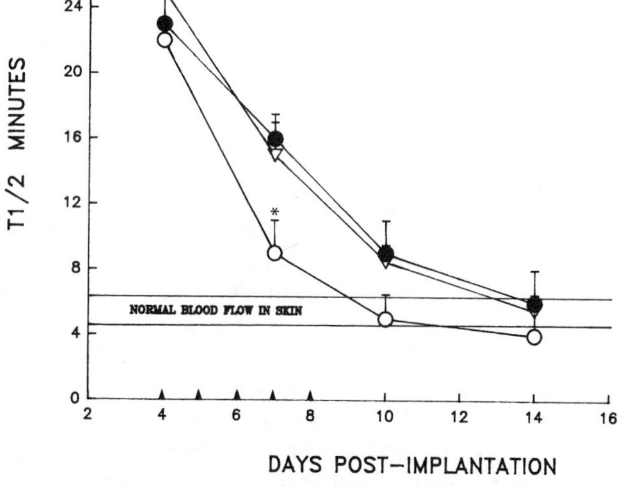

Figure 2. Enhanced angiogenesis elicited by PAF (○) indicated by lower $t_{1/2}$ values compared to lyso-PAF (▽) or PBS (●). Data represent mean ± SEM from 6–8 mice per group. $^{*}p < 0.05$, ▲ injection.

Evaluation of fibrovascular infiltration

Results of the variables measured as indicators of tissue infiltration in the group treated with PAF were significantly higher than in the control group; wet weight for PAF 254 ± 18 mg vs 224 ± 19 mg control; protein for PAF were 216 ± 10 μg/mg sponge vs 128 ± 11 and GAG for PAF 0.370 ± 0.05 μg/mg of sponge vs 0.244 ± 0.03 for the control (n = 7–10 in each group per parameter).

Effects of B. Jararaca venom on implant blood flow

At 14 days postimplantation, the sponge implants were fully vascularized; $t_{1/2}$ was at the normal skin value and was stable. At this time *B. Jararaca* venom was injected and its acute effect on ^{133}Xe washout measured. The results showed an increase in $t_{1/2}$ from a control value of 5 ± 2 min to 10 ± 3, 19 ± 3 and 26 ± 4 min respectively after injection of 5, 10, 20 μg of the venom.

Discussion

The technique used for determining angiogenesis in mice in the present study seems to offer some advantages over existing methods as it provides a non-destructive assay of blood flow which can be repeated in the same animal over several days during development of new vessels in the sponge. Some recently developed in vivo models of angiogenesis, although quantitative [4, 5], do not allow sequential observation of blood vessels as each animal provides information for only one time point. The development of blood flow demonstrated by progressive decrease in ^{133}Xe $t_{1/2}$ must represent neovascularization since the sponge originally contained no blood vessels and, furthermore, the vessels formed in such sponges have been shown to react to vasodilator PGE_2 and vasoconstrictor noradrenaline in rats [6], and to endothelin-1 and PAF in mice (Andrade, S. P., unpublished results). The presence of blood vessels in the implants was confirmed by histological results using antibodies to laminin. Identification of neuronal structures in the implant also suggest that blood vessel formation and nerve growth are intimately associated in this system. It may be possible to use this model to study specifically the innervation of newly formed vasculature.

PAF, but not lyso-PAF, caused maximum vascularization of the sponges on day 7 postimplantation as indicated by a reduction in $t_{1/2}$ and increases in wet weight, protein and GAG contents, suggesting a positive angiogenic effect of PAF in this system. This finding adds a lipid compound to the varied list of angiogenic agents. Many mediators of inflammation promoting venoconstriction with consequent capillary congestion may synergise to induce neovascularization [11].

The responsiveness of the newly formed blood vessels in the implants to vasoactive agents was shown by *B. Jararaca* venom. The local vasoconstriction following the venom was clearly expressed as a slower washout of ^{133}Xe.

In summary, the sponge implant model in mice provides a reproducible quantitative measure of angiogenesis capable of sequential sampling. It has been used to show the angiogenic effects of PAF and vasoconstrictor effects of *B. Jararaca* venom.

Acknowledgements. We thank Dr Y. S. Bakhle for helpful discussions and assistance.

References

1 Folkman J, Haudenschild C: Angiogenesis in vitro. Nature 1980; 288: 551–556.
2 Nicosia RF, Ottinetti A: Growth of microvessels in serum-free matrix culture of rat aorta: A quantitative assay for angiogenesis in vitro. Lab Invest 1990; 63: 115–122.
3 Folkman J: Angiogenesis; initiation and control, in Fishman AP (ed): Endothelium. Ann NY Acad Sci 1982; 401: 212–227.
4 Fajardo LF, Kowalski J, Kwan HH, Prionas SD, Allison AC: The disc angiogenesis system. Lab Invest 1988; 58: 718–724.
5 Plunkett ML, Halley JA: An in vivo quantitative angiogenesis model using tumor cells entrapped in alginate. Lab Invest 1990; 62: 510–517.
6 Andrade SP, Fan T-PD, Lewis GP: Quantitative in vivo studies on angiogenesis in a rat sponge model. Br J exp Path 1987; 68: 755–766.
7 Timpl R, Rohde H, Robey PG, Rennard SI, Foidart JM, Martin GR: Laminin – a glycoprotein from basement membrane. J Biol Chem 1979; 254: 9933–9937.
8 Lowry OH, Rosebrough JN, Farr AL, Randall RJ: Protein measurement with the folin phenol reagent. J Biol Chem 1951; 193: 265–275.
9 Farndale RN, Sayers CA, Barret AJ: A direct spectrophotometric microassay for sulfated glycosaminoglycans in cartilage cultures. Connec Tiss Res 1982; 9: 247–248.
10 Folkman J, Watson K, Ingber D, Hanahan D: Induction of angiogenesis during the transition from hyperplasia to neoplasia. Nature 1989; 339: 58–61.
11 Chiarugi V, Ruggiero M, Porciatti F, Vannuchi S, Ziche M: Cooperation of heparin with other angiogenic effectors. Int J Tiss Reac 1986; 8: 129–133.

Angiogenesis: Key Principles – Science – Technology – Medicine
ed. by R. Steiner, P. B. Weisz & R. Langer

Further studies on angiogenesis in a rat sponge model

Tai-Ping D. Fan[a], De-En Hu[a], Rachel L. Smither[a] and G. Austin Gresham[b]

[a]*Department of Pharmacology, University of Cambridge, Tennis Court Road, Cambridge CB2 1QJ, and* [b]*Department of Histopathology, Addenbrooke's Hospital, Cambridge CB2 2QQ, England*

Introduction

In recent years, techniques involving subcutaneous implanation of sponges have become a popular means for studying angiogenesis. For example, polyvinyl alcohol sponges have been used to quantitate the effects of cartilage derived growth factor [1] and epidermal growth factor [2] on the DNA, protein, collagen and hydroxyproline content of the implants. In another study, double velour Dacron sponges were used to induced neovascularisation which was then assessed by light microscopy, autoradiography, electron microscopy and angiography [3]. More recently, a disc angiogenesis system was devised, incorporating ethylene-vinyl acetate co-polymer to facilitate sustained release of angiogenic agonists and antagonists [4]. These various sponge models are therefore amenable to biochemical and morphological analyses to provide valuable information on the control mechanisms on angiogenesis. However, such analyses can only be carried out post mortem.

In 1987, we described a sponge model which not only allows histological and biochemical studies to be made but also has the major advantage of using a simple ^{133}Xe washout technique for repeated quantitation of blood flow changes through the sponges for periods of several weeks [5]. Here we discuss the rationale for the development of this method and its validation.

Blood flow changes in sponge implants measured by ^{133}Xe clearance

The ^{133}Xe clearance technique was originally introduced to provide optimum measurement of cutaneous blood flow in man [6], based on the suggestion that the clearance of a freely diffusible tracer from the

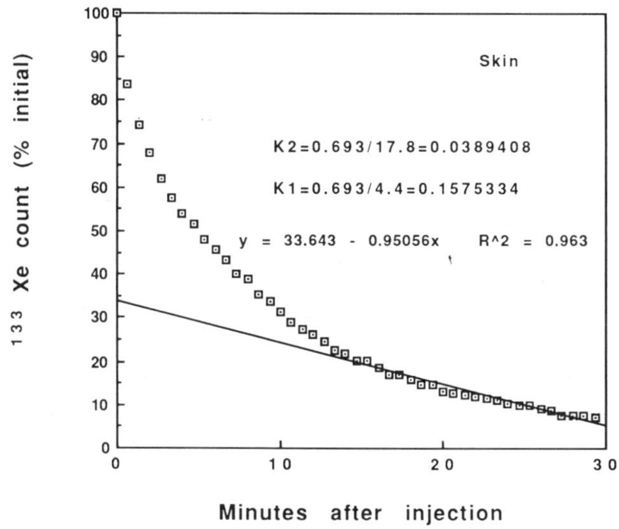

Figure 1. ^{133}Xe clearance curve in normal rat skin. Data are expressed as % of the initial radioactivity injected intradermally. n = 4.

site of injection would be a measure of the local tissue circulation. This technique has been adapted to measure blood flow in skin grafts [7, 8]. As shown in Fig. 1, when injected into rat skin, ^{133}Xe-saline produces biexponential washout curves over 30 min. By curve-peeling, the $t_{1/2}$ of the fast (cutaneous) and slow (subcutaneous) components and their respective clearance constants K_1 and K_2 can be calculated [7]. If the tissue to blood partition coefficient is known, *absolute blood flow* (ml/min/g) can be determined. However, this technique requires continous monitoring of the washout of radioactivity for at least 30 min, thus limiting the number of observations that can be made in each experiment. Furthermore, derivation of the clearance constants from the washout curves is laborious and time-consuming.

To overcome this problem, the washout of radioactivity from sponge implants was measured every 40 s for 30–45 min. The data were then treated in two ways. First, $t_{1/2}$ values and clearance constants of the fast and slow component were derived by curve-peeling. Second, the clearance at 6 min was calculated as a percentage of the initial count [8]. The two sets of results were compared so as to establish if measurements of 6 min ^{133}Xe clearance would provide a simple and fast alternative to detect *relative blood flow changes* in sponge implants during angiogenesis.

Fig. 2 shows ^{133}Xe clearance curves from sponge implants over 14 days. From Day 8 onwards, two components can be resolved reasonably well to obtain $t_{1/2}$ and clearance constants. Thus, $t_{1/2}$ values decrease from 66.5 min on Day 4 to 20.7 min, 10.2 min, 7.6 min and

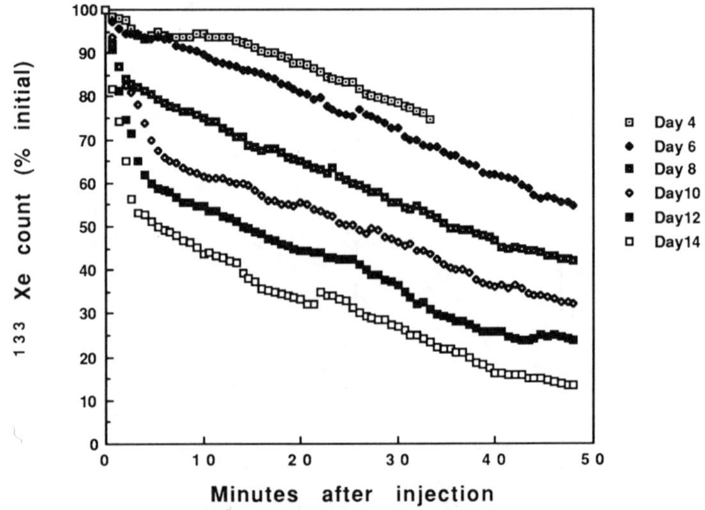

Figure 2. ^{133}Xe clearance curves in sponges 4, 6, 8, 10, 12 and 14 days after implantation. Data are expressed as % of initial radioactivity injected into the sponge implants. n = 8 for each time point.

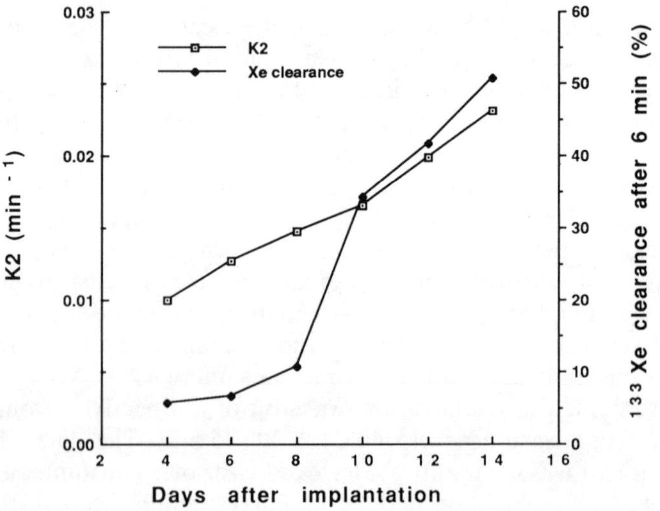

Figure 3. Correlation between K_2 values and 6 min ^{133}Xe clearance values. To calculate 6 min ^{133}Xe clearance, the formula [(initial count − count at 6 min)/initial count] × 100%, is used. These data are derived from Fig. 2.

6.1 min on Days 8, 10, 12 and 14, respectively. K_1 values before Day 6 are too small to be determined accurately. However, it reaches 0.231 on Day 6 and remains relatively stable until Day 14. Fig. 3 shows a time-dependent increase in K_2 values, indicating that it is probably a function of tissue infiltration and vascularity of sponge implants. Fig. 3

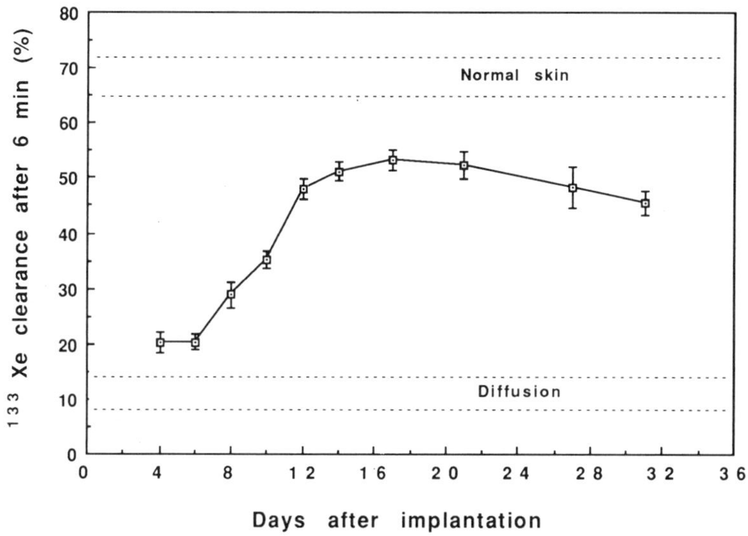

Figure 4. Time of course of sponge neovascularisation as measured by 6 min ^{133}Xe clearance technique. Data represent mean $\pm$ SEM. $n = 40$.

also shows K_2 values to be well correlated with 6 min ^{133}Xe clearance values (correlation coefficient = 0.902). Thus, 6 min ^{133}Xe clearance values represent a good estimation of the clearance capacity of the neovasculature induced by sponge implants. Fig. 4 shows that ^{133}Xe clearance from the sponges is initially due to passive diffusion but increases from Day 8 as they become vascularised until it approaches that of normal skin by Day 16. Since the sponges originally contained no blood vessels, the development of blood flow must represent a neovascularisation.

Preliminary morphometric analysis of sponge implants

A series of experiments was carried out to establish the relationship between ^{133}Xe clearance and tissue infiltration of sponge implants. Groups of animals were killed after ^{133}Xe measurements and their sponges processed for histology. Serial sections (10 μm) were prepared at eight levels and stained with H&E. Computer-assisted image analysis shows that there is a good correlation between the extent of tissue infiltration (as % of total sponge area) and the capacity of sponge neovasculature to clear ^{133}Xe (Fig. 5). Furthermore, five daily doses of 5 μg of HA3, an angiogenic fragment of hyaluronic acid [9], increased ^{133}Xe clearance relative to controls ($59.1 \pm 2.7\%$ ^{133}Xe cleared in 6 min vs $49.0 \pm 2.7\%$ for controls on Day 9 after implantation, $p < 0.05$,

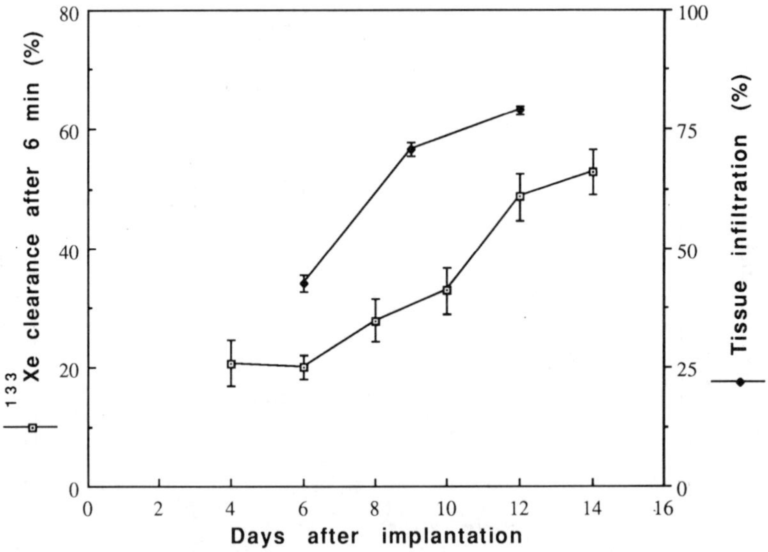

Figure 5. Correlaton between 6 min ^{133}Xe clearance and tissue infiltration of sponge implants. Data represent mean $\pm$ SEM. $n = 4$.

$n = 7-8$). In contrast, 10 μg/sponge/day of the platelet-activating factor (PAF) antagonist WEB 2086 [10] caused a decrease in ^{133}Xe clearance relative to controls on Day 8 after implantation ($22.3 \pm 6.0\%$ vs $40.6 \pm 5.3\%$, $p < 0.025$, $n = 7-9$). Tissue infiltration into the sponges was shown to be greater in HA3-treated sponges compared to controls ($75.8 \pm 4.5\%$ and $69.1 \pm 1.48\%$ of total section area respectively, $n = 3$) and less in those treated with WEB 2086 ($63.2 \pm 2.2\%$ vs $73.5 \pm 3.2\%$ for controls, $p < 0.05$, $n = 3-4$).

Recently, we used an endothelial cell marker, *Bandeiraea simplicifolia* lectin I isolectin B_4 [11], to visualise blood vessels in sponge sections and to quantitate their density by image analysis. Preliminary analysis of the infiltrated areas under higher magnification suggests that HA3 increased the vascular density of the sponges while WEB 2086 diminished it.

Assessment of angiogenic factors or inhibitors

Potential modulators of angiogenesis are administered daily through attached cannulae into the sponge implants. It might be argued that blood vessel reactivity to these agents might interfere with the assessment of angiogenesis by ^{133}Xe measurements. However, the new vessels will have to be developed before being influenced by vasoactive agents. Furthermore, the test substances are always injected via the cannulae into the sponges 24 h before ^{133}Xe measurements so as to exclude their possible acute effects on the microvasculature (dilation or constriction). Control studies using laser Doppler flowmetry and ^{133}Xe clearance technique established that, at the doses studied, drugs, fatty acids and peptides did not produce significant blood flow changes in normal rat skin at 1–60 min, 6 h, 12 h and 24 h; nor did they show any vasoactivity in 14-day-old sponges.

Experiments with cyclo-oxygenase and/or lipoxygenase inhibitors, and angiostatic steroids (hydrocortisone and tetrahydro-S) indicate that prostaglandins and leukotrienes do not play a significant role in this model [12]. However, three structurally distinct PAF antagonists did inhibit the angiogenic response, suggesting PAF may be involved in some aspects of angiogenesis [10]. Furthermore, daily administration of 3 pmol of human interleukin 1 alpha (IL-1α), or 1 nmol of the neuropeptides substance P (SP), calcitonin gene related peptide (CGRP), vasoactive intestinal peptide (VIP) or the inflammatory autacoid bradykinin (BK) caused an enhancement of angiogenesis. When combined, subthreshold doses of IL-1α (0.3 pmol) and the neuropeptides or BK (10 pmol) were sufficient to promote neovascularisation. Preliminary histological studies of the implants have shown increased vascularity in the test sponges as compared to controls. Futhermore, the effects of SP, VIP, and BK were abolished by selective receptor antagonist [D-Pro^2, D-Phe^7, D-Trp^9]-SP, VIP-(10–28) and des-Arg^9, [Leu^8]-BK respectively [13]. Thus, the interactions among fatty acid derivatives, neuropeptides and cytokines could lead to the excessive neovascularisation seen in rheumatoid arthritis.

Perspective of sponge models

Because of the low cost and technical simplicity, sponge implants have been used as a matrix to study would healing [14], inflammatory reactions [15–17], allograft rejection [18, 19] and angiogenesis [1–5, 20]. It is noteworthy that this model offers the possibility to measure the levels of eicosanoids [10], interleukins [19, Hori and Fan, unpublished data] and other cytokines/growth factors in the sponge exudate fluid. Such studies are important in the elucidation of the role of these factors in angiogenesis.

Unlike the rabbit ear chamber, hamster cheek pouch, corneal implant assay and chick choriollantoic membrane, external or gross inspection cannot be performed on the sponges. To circumvent this problem, we incorporated cannulae into sponge implants so that their functional vascularity could be assessed by ^{133}Xe clearance technique. Thus, our sponge model and its variants [21, 22] are suitable for routine examination of angiogenic factors or inhibitors. Subsequently, histochemical and morphometric techniques are applied to confirm and quantitate neovascular growth.

Acknowledgements. This work was supported by British Heart Foundation and the Wellcome Trust. We thank Drs J Saklatvala, D West and S Kumar for the gifts of IL-1α and hyaluronic acid fragments. We also thank Drs LH Smaje, I Gartside and PE Thorpe for discussions on the ^{133}Xe clearance technique.

References

1 Davidson JM, Klagsbrun M, Hill KE, Buckley A, Sullivan R, Brewer PA, Woodward SC: Accelerated wound repair, cell proliferation, and collagen accumulation are produced by a cartilage-derived growth factor. J Cell Biol 1985; 100: 1219–1227.
3 Buckley A, Davidson JM, Kamerath CD, Woodword SC: Epidermal growth factor increases granulation tissue formation dose dependently. J Surg Res 1987; 43: 322–328.
3 Richardson TC, Humphreys JAH, Townsend KMS: Subcutaneous implantation of double velour Dacron into the mouse: infiltration and angiogenesis. Br J exp Path 1987; 68: 359–366.
4 Fajardo LF, Kowalski J, Kwan HH, Prionas SD, Allison AC: The disc angiogenesis system. Lab Invest 1988; 58: 718–724.
5 Andrade SP, Fan T-PD, Lewis GP: Quantitative in-vivo studies on angiogenesis in a rat sponge model. Br J exp Path 1987; 68: 755–766.
6 Sejrsen P: Blood flow in cutaneous tissue in man studied by washout of radioactive xenon. Circ Res 1969; 25: 215–229.
7 Lewis GP, Peck MJ, Williams TJ, Young BA: Measurement of blood flow in rabbit skin homografts and autografts using a ^{133}Xe clearance technique. J Physiol 1976; 254: 32–33P.
8 Fan T-PD, Lewis GP: Blood flow, histamine content and histidine decarboxylase activity in rat skin grafts and their modification by cyclosporin-A. Br J Pharmacol 1982; 76: 491–497.
9 West DC, Hampson IN, Arnold F, Kumar S: Angiogenesis induced by degradation products of hyaluronic acid. Science 1985; 228: 1324–1326.
10 Smither RL, Fan T-PD: PAF antagonists inhibit angiogenesis in a rat sponge model. Br J Pharmacol 1990; 99: 87P.
11 Laitinen L: *Griffonia simplicifolia* lectins bind specifically to endothelial cells and some epithelial cells in mouse tissues. Histochem J 1987; 19: 225–234.
12 Smither RL, Fan T-PD: Role of arachidonic acid metabolites in a rat sponge model of angiogenesis. Int J Rad Biol 1991; (in press).
13 Fan T-PD, Hu D-E: Modulation of angiogenesis by inflammatory polypeptides. Int J Rad Biol 1991; (in press).
14 Edwards LC, Pernokas LN, Dunphy JE: The use of plastic sponge to sample regenerating tissue in healing wounds. Surgery, Gynaecology & Obstetrics 1957; 105: 303–309.
15 Higgs GA, Flower RJ, Vane JR: A new approach to anti-inflammatory drugs. Biochem Pharmacol 1979; 28: 1959–1961.
16 Parnham MJ, De Leve LD, Saxena PR: Development of enhanced blood flow responses to prostaglandin E_1 in carrageenan-induced granulation tissue. Agents Actions 1979; 9: 510–515.
17 Smaje LN, Noor NM, Clough GF: Changing sensitivity to H_1 and H_2 receptor agonists in the growing vasculature, in Shu Chien (ed): Vascular Endothelium in Health and Disease. Advances in Experimental Medicine and Biology. New York, Plenum, 1988, vol 242, pp 145–150.
18 Roberts PJ, Hayry P: Sponge matrix allografts: a model for analysis of killer cells infiltrating mouse allografts. Transplantation 1976; 21: 437–442.
19 Ford HR, Hoffman RA, Wing EJ, Magee DM, McIntyre LA, Simmons RL: Tumour necrosis factor, macrophage colony-stimulating factor, and interleukin 1 production within sponge matrix allografts. Transplantation 1990; 50: 460–466.
20 Mahadevan V, Hart IR, Lewis GP: Factors influencing blood supply in wound granuloma quantitated by a new in vivo technique. Cancer Res 1989; 49: 415–419.
21 Mahadevan V, Hart IR. Flavone acetic acid-induced tumour vasculature shutdown. Int J Rad Biol 1991.
22 Vieira LBGB, Andrade SP, Rodrigues JAA, Leite VHR: Angiogenic acitivity of PAF and inhibition of blood flow by *Bothrops Jararaca* venom in mice sponge model. This meeting.

Angiogenesis: Key Principles - Science - Technology - Medicine
ed. by R. Steiner, P. B. Weisz & R. Langer

A multichannel wounding device for the study of vascular repair in vitro

Tai-Ping D. Fan, Emma E. Frost and Amanda D. Wren

Department of Pharmacology, University of Cambridge, Tennis Court Road, Cambridge CB2 1QJ, England

Introduction

Vascular endothelial cells (EC) in vivo usually have a turnover measurable in years [1], however, they can be stimulated to proliferate much more rapidly, i.e., within 2–3 days, in response to injury or an angiogenic stimulus. The processes involved in capillary formation: (i) degradation of the existing basement membrane by metalloproteinases, (ii) migration of EC towards the angiogenic stimulus and (iii) proliferation of these EC to form the new capillary loops [2], are normally strictly regulated. However, a breakdown in the controlling mechanism is thought to be responsible for aberrant neovascularisation in solid tumor growth, rheumatoid arthritis, diabetic retinopathy and hypertension.

Several in vitro models have been developed to study angiogenesis. Migration assays, commonly based on the removal of a defined proportion of EC from a confluent monolayer, involve the application of a filter disk [3], or the use of a mechanical scraper [4–8]. Techniques for growing EC in 3-dimensional extracellular matrix gels [9–11] are used to monitor the organisation of EC into capillary-like networks.

We have modified the earlier wounding techniques [3–8] to produce an assay for the study of wound repair in vitro [12], which is rapid, sensitive and versatile. It can be used with several cell types such as rat aortic smooth muscle cells (SMC) and human fibroblasts from patients with gingivitis (HGF), as well as human umbilical vein endothelial cells (HUVEC). Most recovery measurement experiments are complete in 8–16 h. The multiple injury of a monolayer by this technique amplifies any biochemical signals released by the cells to a detectable level.

To test the sensitivity of the assay, we choose to examine the effects of three distinct growth factors/cytokines on repair of these cells after injury.

Materials and methods

Basic FGF was purchased from Penninsula Laboratories, TGFβ and IL-1α were gifts from Genentech Inc and Dr. J. Saklatvala of Strangeways Research Laboratory, Cambridge, respectively. All culture media were purchased from ICN-Flow and heat-inactivated foetal calf serum (FCS) was from Sera-Labs.

Cells were grown on 13-mm round Thermanox® coverslips (Miles Laboratories). At confluence the cells were incubated for 24 h in serum-free medium prior to wounding [11]. The coverslips were incubated in media containing 5% FCS and various concentrations of test substance, immediately post wounding. After the predesignated time, the cells were washed ×3 with Ca^{2+}/Mg^{2+}-free phosphate buffered saline and fixed for 3–5 min in 100% ethanol. The wounded area of each coverslip was measured using a computer-assisted image analysis system (Solitaire, Seescan; VIDS III, Analytical Measuring Systems). The results were expressed as % recovery using the formula:

$$\% \text{ recovery} = (1 - \text{wound area at } T_n/\text{wound area at } T_0) \times 100\%$$

where T_n is the number of hours post-injury and T_0 is immediately post-injury.

Prostacyclin production by the wounded monolayers was measured as its stable metabolite 6-oxo-$PGF_{1\alpha}$ using a specific radioimmunoassay with minor modifications [13].

Results

Image analysis showed that 37.1 ± 1.1% (n = 28) of the HUVEC monolayer was removed, giving a total wound area of 49 mm^2, 45.9 ± 3.7% (n = 24) HGF were removed, a wound area of 61 mm^2 and SMC by 54.5 ± 0.53% (n = 12), a wound area of 72 mm^2. Studies on the time course of migration established that 8 h post-injury was, with 50% recovery, the optimum time point for examining the effects of exogenous factors on HUVEC and HGF (Fig. 1). SMC, however, show a biphasic response and the optimum time point chosen for this cell type was 18 h, when 25% recovery had occurred.

The results obtained with bFGF (1–10 ng/ml) indicated enhanced wound repair of HGF and HUVEC in a dose-dependent manner but did not affect PGI_2 release (Fig. 2), while 24-h pre-treatment with the same concentrations of TGFβ caused a dose-related inhibition (Fig. 3), again, there was no effect on PGI_2 release. In contrast, IL-1α (0.1–10 ng/ml) appeared to exert no effect on HUVEC or HGF wound regeneration and no effect on PGI_2 release (Fig. 4).

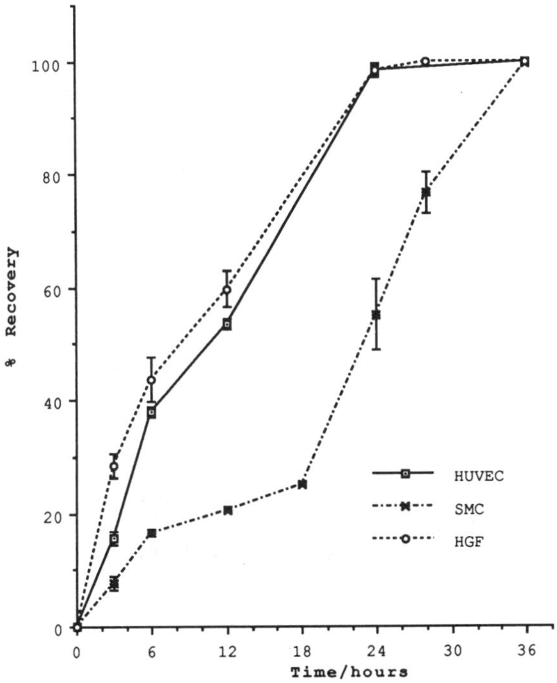

Figure 1. The % recovery of HUVEC (n = 28), HGF (n = 14) and SMC (n = 12), with time.

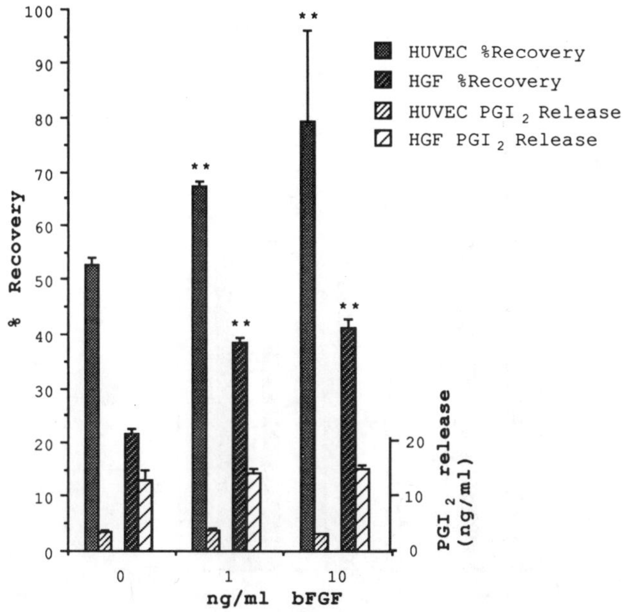

Figure 2. The effects of a post wounding, bolus dose of bFGF on HUVEC (n = 12) and HGF (n = 8) wound repair and 6-oxo-PGF1_{α} release, after 8 h incubation. ** = $p < 0.005$ vs control.

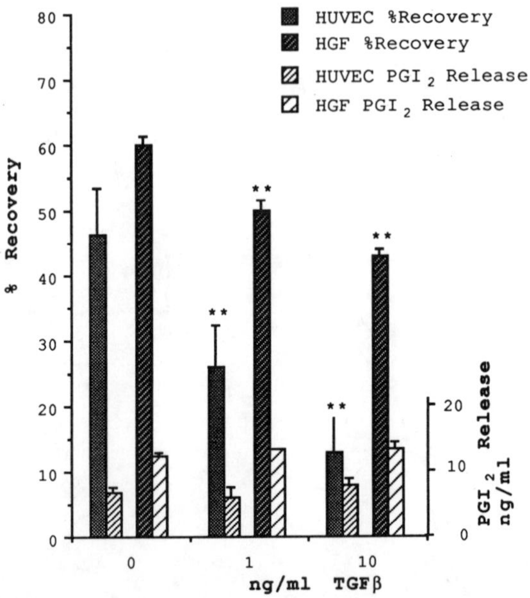

Figure 3. The effects of 24-h pretreatment followed by a post wounding, bolus dose of TGFβ on HUVEC ($n = 12$) and HGF ($n = 8$) wound repair and 6-oxo-PGF1$_{\alpha}$ release, after 8 h incubation. ** = $p < 0.005$ vs control.

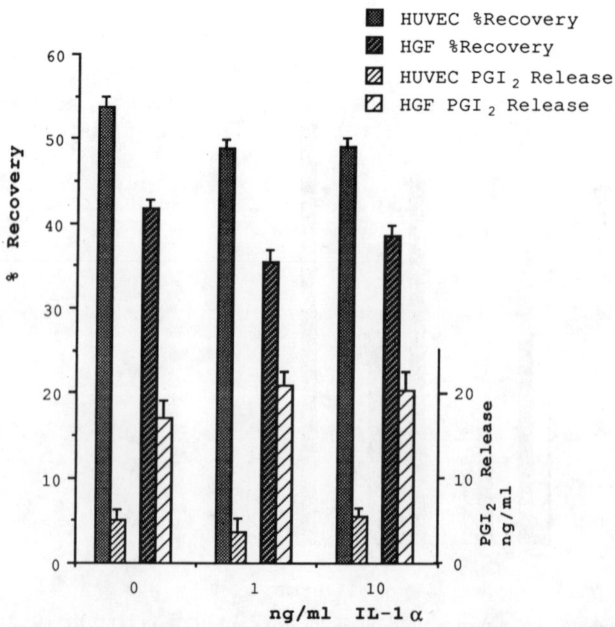

Figure 4. The effects of a post wounding, bolus dose of IL-1α on HUVEC ($n = 12$) and HGF ($n = 8$) wound repair and 6-oxo-PGF1$_{\alpha}$ release, after 8 h incubation.

Discussion

These results demonstrate that this wounding device is versatile and reproducible and can be used to study the effects of physical injury on several different cell types involved in vascular repair. The use of coverslips in a 24-well plate not only means that the amounts of material (cells, foetal calf serum, growth medium, and peptides/drugs under investigation) can be scaled down substantially, but also the number of measurements per test can be increased compared with other systems.

The results obtained with a known mitogen, bFGF, and a known growth inhibitor, TGFβ, are consistent with earlier observations [4, 14, 15]. IL-1α is known to stimulate collagenase release by fibroblasts and play an important role in mediating the activity of leucocytes in response to inflammation [16]. The apparent lack of its effects in this model does not, however, rule out its possible role in wound repair. It would appear that this method can be used to quantitatively study the effects of various factors on angiogenesis in vitro.

Further studies using agents which inhibit cell migration and/or division will allow us to assess the differential effects of growth factors and/or inhibitors on these two components of wound repair. A better understanding of the biochemical and physical mechanisms involved in the formation of capillary sprouts should help in the identification of potential targets for pharmacological intervention to combat angiogenic diseases.

Acknowledgements. This work was supported by the British Heart Foundation and the Human Research Trust. We wish to thank Jim Sharpe for making the wounders for us and Dr P Weissberg, Department of Biochemistry, University of Cambridge and the CMB Group at Strangeways Research Laboratory for supplying the SMC and HGF, respectively. ADW is an MRC Scholar.

References

1 Denekamp J: Vasculature as a target for tumour therapy. Prog Appl Micro Circ 1984; 4: 28–38.
2 Folkman J, Ausprunk DH, Langer R: Connective tissue: small blood vessels and capillaries, in Kelly WN (ed): Textbook of Rheumatology, 1st edition. Philadelphia, Saunders, 1981. vol. 1, pp 210–220.
3 Joyce NC, Matkin ED, Neufeld AH: Corneal endothelial wound closure in vitro. Effects of EGF and/or indomethacin. Invest Ophthalmol Vis Sci 1989; 30: 1548–1559.
4 Sato Y, Rifkin DB: Autocrine activities of basic fibroblast growth factor: Regulation of endothelial cell movement, plasminogen activator synthesis, and DNA synthesis. J Cell Biol 1988; 107: 1199–1204.
5 Gotlieb AI, Spector W: Migration into an in vitro experimental wound. A comparison of porcine aortic endothelial and smooth muscle cells and the effects of culture irradiation. Am J Path 1981; 103: 271–282.
6 Klein-Soyer C, Beretz A, Millon-Collard R, Abecassis J, Cazenave J-P: A simple in vitro model of mechanical injury of confluent cultured endothelial cells to study quantitatively the repair process. Thrombos Haemost 1986; 56: 232–235.

7 Madri JA, Stenn KS: Aortic endothelial cell migration. Am J Path 1982; 106: 180–186.
8 Kinsella MG, Wight TN: Modulation of sulfated proteoglycan synthesis by bovine aortic endothelial cells during migration. J Cell Biol 1986; 102: 679–687.
9 Folkman J, Haudenschild C: Angiogenesis in vitro. Nature 1980; 288: 551–556.
10 Madri JA, Pratt BM: Endothelial cell matrix interactions: In vitro model of angiogenesis. J Histochemistry and Cytochemistry 1986; 34: 85–91.
11 Montesano R, Orci L, Vassalli JD: In vitro rapid organisation of endothelial cells into capillary-like networks is promoted by collagen matrices. J Cell Biol 1983; 97: 1648–1652.
12 Fan T-PD, Frost EE: A new in vitro assay for the study of endothelial cell wound repair. Eur J Pharm 1990; 183: 1809.
13 Salmon JA: A radioimmunoassay for 6-keto-prostaglandin $F_{1\alpha}$. Prostaglandins 1978; 15: 383–395.
14 Montesano R, Vassalli J-D, Baird A, Guillemin R, Orci L: Basic fibroblast growth factor induced angiogenesis in vitro. Proc Natl Acad Sci USA 1986; 83: 7297–7301.
15 Heimark RL, Twardzik DT, Schwartz SM: Inhibition of endothelial regeneration by type-β transforming growth factor from platelets. Science 1986; 233: 1078–1080.
16 Martin M, Resch K: Interleukin-1: more than a mediator between leukocytes. TiPS 1988; 9: 171–177.

Angiogenesis: Key Principles - Science - Technology - Medicine
ed. by R. Steiner, P. B. Weisz & R. Langer

Assessment of angiogenic potential—the use of AIDS-KS cell supernatants as an in vitro model

R. Adatia[a], L. Poggi[a], E. W. Thompson[b], R. C. Gallo[c], G. F. Fassina[d] and A. Albini[a]

[a]*Istituto Nazionale per la Ricerca sul Cancro Genova, Italy,* [b]*Vince Lombardi Cancer Center, Georgetown, Washington D.C., USA, the* [c]*National Cancer Institute, Bethesda, Maryland, USA, and* [d]*Centro de Studio per la Neurofisiologia Cerebrale C.N.R., Genova, Italy*

Introduction

Metastasis, the passage of primary tumour cells throughout the body via the vascular system and their subsequent proliferation into secondary lesions in distant organs, represents a poor prognosis and therefore an understandably feared event for cancer patients. Despite considerable advances in cancer diagnosis and treatment, most deaths are the result of metastases resistant to conventional treatment [1]. Rather than being a random process, metastasis involves a series of organised steps leading to the growth of a secondary tumour. Malignant tumours stimulate the production of new vessels by the host, and this process is a prerequisite for the increase in size of a new tumour [2]. Angiogenesis, not only permits tumour expansion but also allows the entry of tumour cells into the circulation and is probably the most vital event for the metastatic process [3]. Metastasis and angiogenesis [4] have received much attention in recent years. A biological understanding of both phenomena seems to be an urgent priority towards the search for an effective prevention and treatment of tumour progression. Studies in vitro and in vivo have shown that one of the most important barriers to the passage of malignant cells is the basement membrane. The crossing of such barriers is a vital step in the formation of a metastasis [5].

Basement membranes consist mainly of laminin, collagen IV, entactin and heparan sulphate proteoglycan [6]. A study of tumour cell invasiveness through such membranes is considered to be an effective way of assessing malignant potential. In the past, animal models have been used to study metastasis, however one could not account for the influence of the organism as a whole on the metastatic process in situ. It seemed more expedient to use isolated membranes such as from lens, bladder wall, or amnion and chorioallantoic membranes [5]. Fragments of heart and muscle or cultures of epithelial cells and hepatocytes were also found effective. However, the difficulty in obtaining biological membranes made the development of an artificial membrane device an invaluable asset. Such a 'synthetic' membrane was prepared by the reconstitution of laminin, collagen IV, heparan sulphate proteoglycan and entactin. The final substance, Matrigel [7], is a liquid at 4°C which can easily be coated onto porous filters and forms a gel-like layer at a higher temperature. Such filters can be used to separate cells between the two compartments of a Boyden chamber [8]. Cells are seeded in the upper chamber, the lower chamber containing only chemoattractant or serum-free medium. Cells capable of migrating across the filter barrier are quantitated on the lower surface of the filter. Chemotaxis tests using gelatin-covered filters are employed as a measure of cell migratory ability under the influence of specific factors.

Metastasis involves both the degradation of basement membrane by proteolytic enzymes and the migration of cells in response to chemoattractants such as growth factors. Tumour neo-angiogenesis resembles metastasis in that it requires the migration of endothelial cells and invasion of their own basement membrane. Since neo-vascularisation resembles tumour cell invasion, it seems likely that similar mechanisms underlie both phenomena. Invasion of tumour cells as well as endothelial cells through basement membrane involves their attachment to the membrane surface, probably mediated by laminin or fibronectin [9]. Subsequently, proteolytic enzymes are produced usually in reponse to growth factors such as bFGF [10], ECGF, PDGF, interleukin-1 [11] and TNF. All these events may be studied using the Boyden chamber assay.

The current particular interest in Kaposi's sarcoma is due to its prevalence in patients with acquired immune deficiency syndrome (AIDS) [12]. AIDS-KS is an aggressive tumour lesion and has been shown to be highly angiogenic, producing vascularised KS-like lesions in mice injected with AIDS-KS cells from long-term cultures [13]. These lesions consist largely of cells recruited from the host. KS cells also induce angiogenesis in the chick chorioallantoic membrane assay [13]. KS lesions appear to be of vascular origin where paracrine growth stimulation is important for their maintenance and progression [14]. Evidence suggests that activated KS cells make several factors including interleukins-1 and 6, bFGF, chemoattractants and chemosuppressors and in addition novel factors. These factors may stimulate angiogenesis [15].

The correlation of malignant cell invasiveness and endothelial cell invasiveness in the initial phases of neo-angiogenesis led us to test our invasion assay as in vitro model of angiogenesis and angiogenic potential.

Results and discussion

HUVEC (human umbilical vein endothelial cells) and BAEC (bovine aorta endothelial cells) are not generally capable of invading reconstituted membranes. They have a low capacity for migrating through gelatin-coated filters without the presence of chemoattractants. However, both migration and invasion of these cells in the Boyden chamber is positively stimulated under the influence of AIDS-KS cell serum-free supernatants. On the contrary, fibroblast and endothelial cell conditioned media were unable to promote invasive behaviour. The production of an angiogenic factor by the KS cells is strongly suggested by these data [16]. Inhibitors of cell-laminin binding and proteolytic enzyme action prevented the significant invasion of AIDS-KS cells themselves. The degradation of basement membrane components, in particular type IV collagen, by proteolytic enzymes, has aroused much interest [17]. Two separate type IV collagenases with molecular weights of 72 kD and 92 kD are now known. Tumour cell invasion and type IV collagenase activity seem to be linked and recent work has shown evidence for this positive correlation. There is a notable correlation between the 72 kD type IV collagenase and metastatic potential. In vivo activation of these enzymes is poorly understood, however plasmin and plasminogen activator are thought to play a role in producing active enzyme species. Various highly aggressive human tumours such as carcinomas, melanomas, hepatomas, fibrosarcomas and reticulum cell sarcomas have shown raised levels of type IV collagenase activity as compared to benign controls. Malignant cell lines produce high levels of 92 and 72 kD type IV and collagenases/gelatinases. In certain systems it seems that the 72 kD collagenase form is particularly associated with the invasive phenotype. We have recently shown that TIMP-2, a tissue inhibitor of metallaproteinase which binds more to the 72 kD of collagenase IV, is a inhibitor of tumour cell invasion [18]. Here we show that endothelial cells are also able to produce a collagenolytic activity of the same molecular weight as in metastatic cells (Fig. 1). We hypothesize that the production of this enzyme might be permissive for invasion under the appropriate chemotactic stimulus. Expression of collagenases has been shown in both malignant cell lines (A205, HT1080) and in BAEC and HUVEC. It is of future interest to assess whether activation by AIDS-KS cell supernatants may produce enhanced expression and activity of these enzymes in endothelial cells. Inhibitors of degradative enzyme action prevented the significant invasion of AIDS-KS cells themselves [16], this may prove an important factor in anti-metastasis treatment. IL-1, a cytokine produced by AIDS-KS cells does not appear to induce endothelial cell (BAEC) invasiveness and antibodies against IL-1 reduced HUVEC invasion by approximately 30%. Basic FGF (and to a lesser extent acidic FGF) stimulates invasion of HUVEC (Fig. 2).

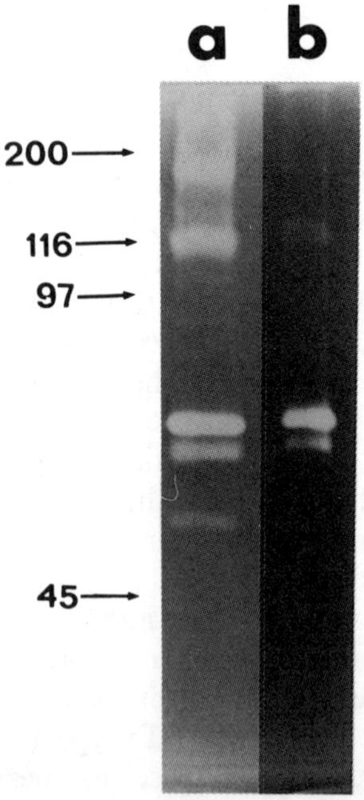

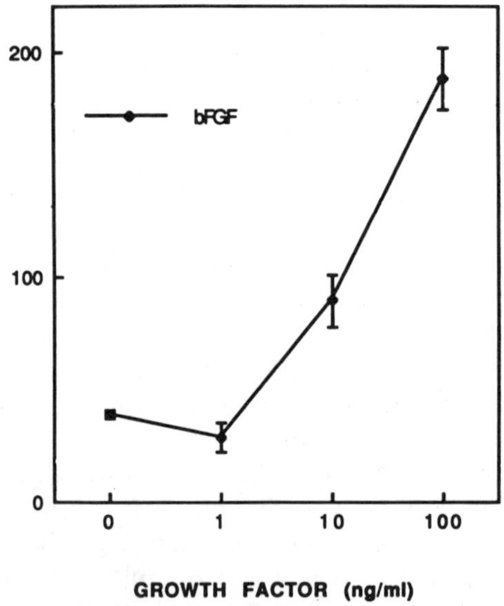

AIDS-KS cells were cultivated as monolayers in RPMI 1640 with 15% FCS and the addition of 20% medium conditioned with HTLV-II positively transformed T4 lymphocytes to each culture vessel. HUVEC and BAEC were maintained in culture using RPMI 1640 and DMEM specified media respectively. Cells were incubated with appropriate serum-free medium for 24 h, and the medium centrifuged to remove cell debris.

Zymograms: Ethanol-precipitated protein form 6-h serum- and cell-free supernatants was loaded onto 7% polyacrylamide gels containing 0.19 mg/ml gelatin. Developed gels were stained with Coomassie blue and enzyme-containing regions observed as contrasting white bands.

Invasion assay: Polycarbonate filters were coated with a fine layer of reconstituted basement membrane, Matrigel. The lower compartment of the Boyden chamber was filled (200 microlitres) with conditioned medium or medium containing growth factors as chemoattractant, the top of the chamber was filled with cell suspension (300000 cells/ml). The two compartments were separated by the filters (8 micron pore) and the chambers incubated at 37°C for up to 6 h. The filters were then removed, fixed and stained, cells that had migrated were observed microscopically on the filter underside. For chemotaxis assessment the procedure was similar except that the filters were coated with gelatine as a cell adhesion substance rather than using Matrigel.

Figure 1 shows the presence of collagenases in: a) A2058 melanoma cells and b) BAEC. Figure 2 shows the response of HUVEC to various concentrations of bFGF.

Noticeably though, antibodies against bFGF inhibited only 10% of KS-CM stimulated HUVEC and BAEC invasion [16]. Therefore, although AIDS-KS cells produce IL-1 and significant quantities of a bFGF-like molecule, they are probably not the major contributors to the angiogenic properties of this tumour.

To summarise, these data appear to indicate the production of an in vitro angiogenic factor by AIDS-KS cells which differs from IL-1 and bFGF, the other cytokines under investigation. Furthermore we have preliminary evidence that a type IV collagenase might be involved in endothelial cell invasiveness of basement membrane. Our experiments also denote the usefulness of AIDS-KS cell supernatants as a model for angiogenic effects. A recent study of the inhibition of invasion of endothelial cells by sulphated chitin derivatives was also demonstrated using the Boyden chamber chemoinvasion assay [19]. The Boyden chamber, chemoinvasion and chemotaxis assays and the use of various endothelial cell lines are invaluable systems to study in relative detail angiogenic potential and hence metastatic events.

Acknowledgements. This work was financed by a grant from the Minestero della Sanità, IV Progetto AIDS, no. 6206002.

References

1 Fidler IJ: Tumor heterogeneity and the biology of cancer invasion and metastasis. Cancer Res 1978; 38: 2651–2660.
2 Folkman J, Klagsbrun M: Angiogenic factors. Science 1987; 235: 442–447.
3 Liotta L, Kleinerman J, Saidel G: Quantitative relationships of tumor cells, tumor vessels and pulmonary metastases following tumor implantation. Cancer Res 1974; 38: 997–1004.
4 Folkman J: Tumor angiogenesis. Adv Cancer Res 1985; 43: 175–203.
5 Terranova VP, Hujanen ES, Martin GR: Basement membrane and the invasive activity of metastatic tumour cells. J Natl Cancer Inst 1986; 77: 311–316.
6 Timpl R, Rohde H, Gehron-Robey P, Rennard SI, Foidart JM, Martin GR: Laminin—a glycoprotein from basement membranes. J Biol Chem 1979; 254: 9933–9937.
7 Kleinman HK, McGarvey ML, Hassell JR, Star VL, Cannon FB, Laurie GW, Martin GR: Basement membrane complexes with biological activity. Biochemistry 1986; 25: 312–318.
8 Albini A, Aukerman SL, Melchiori A, Thompson EW, Reich R, Shima TB, Martin GR, Iwamoto Y: Basement membranes, reconstituted to assess the invasiveness of tumor cells. Tumor Prog Metastasis 1988; 12: 261–269.
9 Kleinman HK, Cannon FB, Laurie GW, Hassel JR, Aumailley M, Terranova VP, Martin GR, Dubois-Dalcq M: Biological activities of laminin. J Cell Biochem 1985; 27: 317–325.
10 Tsuboi R, Sata Y, Rifkin DB: Correlation of cell migration, cell invasion, receptor number, proteinase production and basic fibroblast growth factor levels in endothelial cells. J Cell Biol 1990; 110: 511–517.
11 Orr FW, Buchanan MR, Tron VA, Guy D, Lauri D, Sauder DN: Chemotactic activity of endothelial cell-derived interleukin-1 for human tumour cells. Cancer Res 1988; 48: 6758–6763.
12 Armes J: A review of Kaposi's sarcoma. Adv Cancer Res 1989; 53: 73–87.
13 Salahuddin SZ, Nakamura S, Biberfeld P, Kaplan MH, Markham PD, Larsson L, Gallo RC: Angiogenic properties of Kaposi's sarcoma-derived cells after long-term culture in-vitro. Science 1988; 242: 430–433.
14 Werner S, Hofschneider PH, Roth WK: Cells derived from sporadic and AIDS-related Kaposi's sarcoma reveal identical cytochemical and molecular properties in-vitro. Int J Cancer 1989; 43: 1137–1144.
15 Barnes DM: New clues about Kaposi's sarcoma. Science 1988; 242: 376–377.
16 Thompson EW, Nakamura S, Shima TB, Melchiori A, Martin GR, Salahuddin SZ, Gallo RC, Albini A: Supernatants of acquired immunodeficiency syndrome-related Kaposi's sarcoma cells induce endothelial cell chemotaxis and invasiveness. Cancer Res 1991; 51: 2670–2677.

17 Stetler-Stevenson WG: Type IV collagenases in tumor invasion and metastasis. Cancer Metast Rev 1990; 9: 289–303.
18 Albini A, Melchiori A, Santi L, Liotta L, Stetler Stevenson W: TIMP-2 inhibits tumor cell invasion. J Natl Cancer Inst 1991; 83: 775–779.
19 Murata J, Saiki I, Makabe T, Tsuta Y, Tokura S, Azuma I: Inhibition of tumor-induced angiogenesis by sulfated chitin derivatives. Cancer Res 1991; 51: 22–26.

Angiogenesis: Key Principles - Science - Technology - Medicine
ed. by R. Steiner, P. B. Weisz & R. Langer

Delivery systems for angiogenesis stimulators and inhibitors

Robert Langer

MIT E25–342, Department of Chemical Engineering, 77 Mass Ave, Cambridge, MA 02139, USA.

Abstract. The development of biocompatible, controlled release systems for macromolecules has provided the opportunity to deliver angiogenesis stimulators and inhibitors. These systems have also facilitated the purification and characterization of a number of these substances. In this paper, we discuss controlled release delivery systems which release angiogenesis factors through either porous polymer matrices, degradable polymeric delivery systems or modulated polymer release systems.

Introduction

Polymeric release systems release their contents continuously over long time periods – from days to years. Over the past fifteen years, controlled release systems have been developed for drugs, pesticides, nutrients, agricultural products and fragrances. However, nearly all of the systems that have been developed have not been capable of slowly releasing molecules of large molecular weight ($MW > 600$) such as proteins, e.g., angiogenesis stimulators and inhibitors. In fact, up until 1976 it was a fairly common conception in the field of controlled release that effective systems could not be developed for macromolecules [1]. However, after several years of effort we discovered an approach that permitted the continuous release of biologically active macromolecules as large as 2,000,000 daltons from normally impermeable, yet biocompatible, polymers for over 100 days [2]. In this paper we review three areas of our research, 1) systems that release large molecules through porous polymer matrices, 2) novel degradable polymeric delivery systems, 3) pulsatile controlled release polymer systems.

Controlled delivery systems that release proteins by diffusion through pores

Relatively rapid release of proteins can be achieved using gels such as polyacrylamide or highly porous membranes (e.g. millipore filters), however, in these cases diffusion is generally too fast to be of value for either biological testing or therapeutic applications. The first approach that permitted controlled release of large molecules from biocompatible polymers was based on the discovery that mixtures of solid proteins and hydrophobic polymers could release the proteins for hundreds of days. The earliest methods to prepare these controlled release systems involved dissolving the polymer in an appropriate solvent and adding the macromolecule (protein) in powder form [2]. The resulting mixture can be cast in a mold and dried. When the pellets are placed in water, they release the molecules trapped within the polymer matrix. Subsequently, methods of forming these polymer-macromolecule delivery systems which require no solvent at all, enable microspheres to be formed, or permit microgram quantities to be released, were developed [for review, see 3].

A number of polymer systems were examined both for tissue biocompatibility and release kinetics. Polymers that function best include non-degradable ethylene-vinyl acetate copolymers or degradable lactic/glycolic acid copolymers. Certain hydrogels like polyhydroxyethylmethacrylate (Hydron) or polyvinyachol also performed effectively, but released proteins for shorter time periods [2].

The release mechanism in these systems involves movement of the polypeptide through a complex porous path in the polymer matrix. If the polymer erodes, this will increase the porosity which will also effect the release rate. There are a number of design parameters that can be used to control release rates; these include protein particle size and loading, protein solubility and molecular weight, polymer composition and molecular weight, and the dimensions and shape of the matrix. Both angiogenesis stimulators [4–6] and inhibitors [7] have been slowly released using this approach.

Degradable polymers

Degradable controlled release systems have an advantage over other systems in obviating the need to surgically remove the drug-depleted device. In many cases, however, the release is augmented by diffusion through the matrix, rendering the process difficult to control – particularly if the matrix is hydrophilic and therby absorbs water, promoting degradation in the interior of the matrix. To maximize control over the release process, it is desirable to have a polymeric system which degrades only from the surface and deters the permeation of the drug molecules. Achieving such a heterogeneous degradation requires the rate of hydrolytic degradation on the polymer matrix surface to be much faster than the rate of water penetration into the bulk. With this in mind, we proposed that an ideal polymer would have a hydrophobic backbone, but with a water labile linkage. Many classes of polymers, including polyesters, polyamides, polyurethanes, polyorthoesters, and polyacrylonitriles have been studied for controlled delivery applications, but only polyorthoesters erode from surface and then only if additives were included in the matrix. In designing a biodegradable system that would erode in a controlled heterogeneous manner without requiring any additives, we have suggested that, due to the high lability of the anhydride linkage, polyanhydrides may be a promising candidate.

In a number of studies, a model polyanhydride – a copolymer of carbophenoxy propane (CPP) and sebacic acid (SA) – was used. We discovered that 1-mm thick discs of polyCPP will completely erode in over 3 years in aqueous media. The degradation rates can be enhanced by copolymerization with sebacic acid (SA). An increase of 800 times was observed when the sebacic acid concentration reached 80%. By altering the CPP/SA ratio nearly any degradation rate between 1 day and 3 years can be achieved [8].

The release behavior of drugs incorporated into these polymers depends on both the polymer and the formulation procedure (solvent casting, compression molding, injection molding). Release of proteins can be achieved using this approach, but it is important to use formulation methods that involve minimal heating [9]. These systems have also been used to deliver angiogenesis inhibitors [10].

Modulated release systems

We have also developed several polymeric systems capable of delivering drugs at increased rates on demand. The first system consists of drug powder dispersed within a polymeric matrix (generally ethylene-vinyl acetate copolymer, EVAc) together with magnetic beads. Release rates are controlled by an oscillating external magnetic field, which is generated by a device that rotates permanent magnets beneath the vials. By placing small plastic cages containing animals on the top of the device, it can also be used for in vivo studies. Polymer matrices containing drug and magnets can release up to 30 times more drug when exposed to the magnetic field compared to baseline release, and release rates return to normal when the magnetic field is discontinued. The response time to peak release rate is nearly immediate. The magnetically controlled implant does not cause inflammation in vivo. This was confirmed by the lack of edema, cellular infiltrate or neovascularization as judged by gross and histologic examination in animals. A variety of proteins including insulin and albumin have been released using this magnetic approach [11, 12]. We also discovered that ultrasound could affect the release of substances from polymers [13].

It is hoped that the delivery systems discussed in this paper will have an ever increasing impact – first on the testing, and second on the eventual use – of angiogenesis stimulators and inhibitors.

References

1 Stannett VT, Koros WJ, Paul DR, Lonsdale HK, Baker RW: Recent advances in membrane science and technology. Adv Polym Sci 1979; 32: 71.
2 Langer R, Folkman J: Polymers for the sustained release of proteins and other macromolecules. Nature 1976; 263: 797.
3 Langer R, Brown L, Edelman E: Controlled release and magnetically modulated release systems for macromolecules, in: Drug and Enzyme Targeting. Meth Enzym 1985; 112: 399.
4 Gospodarowicz P, Bialecki H, Thakral TK: The angiogenic activity of the fibroblast and epidermal growth factor. Exp Eye Res 1979; 28: 501.
5 Polverini P, Cotran R, Gimbrone M, Unanue E: Activated macrophages induce vascular proliferation. Nature 1977; 269: 904.
6 Silberstein GB, Daniel CW: Reversible inhibition of mammary gland growth by transforming growth factor beta. Science 1987; 237: 291.
7 Langer R, Brem H, Falterman K, Klein M, Folkman J: Isolation of a cartilage factor that inhibits tumor neovascularization. Science 1976; 193: 70–72.
8 Leong KW, Brott BC, Langer R: Bioerodible polyanhydrides as drug-carrier matrices: I. Characterization, degradation and release characteristics. J Biomed Mat Res 1985; 19: 941.
9 Chasin M, Domb A, Ron E, Mathiowitz E, Leong K, Laurencin C, Brem H, Grossman B, Langer R: Polyanhydrides as drug delivery systems, in: Biodegradable Polymers for Drug Delivery, Langer R, Chasin M (Eds), Marcel Dekker Inc, NY 1990; pp. 43–70.
10 Langer R, Lund D, Leong K, Folkman J: Controlled release of macromolecules: Biological studies. J Controlled Release 1985; 2: 331.
11 Hsieh DST, Langer R, Folkman J: Magnetic modulation of release of macromolecules from polymers. Proc Natl Acad Sci 1981; 78: 1863.
12 Edelman E, Brown L, Langer R: In vitro and in vivo kinetics of regulated drug release from polymer matrices by oscillating magnetic fields. J Biomed Mat Sci 1987; 21: 339.
13 Kost J, Leong K, Langer R, Effects of ultrasound on polymer degradation and release of incorporated substances. Proc Natl Acad Sci 1989; 86: 7663.

Angiogenesis: Key Principles - Science - Technology - Medicine
ed. by R. Steiner, P. B. Weisz & R. Langer

Angiopolarity of cell carriers: directional angiogenesis in resorbable liver cell transplantation devices

E. Wintermantel[a,b], L. Cima[a], B. Schloo[c] and R. Langer[a]

[a]*Massachusetts Institute of Technology, Cambridge, MA, USA,*
[b]*Swiss Federal Institute of Technology, Zürich, Switzerland,*
[c]*Deborah Heart and Lung Center, Brown Mills, NJ, USA*

Abstract. The purpose of this study was to obtain directional angiogenesis of small blood vessels and capillaries to an implant made from a resorbable polymer for hepatocyte transplantation. It was intended to mimic the native acinar structure of the liver in order to facilitate replication of the cells and organ growth. The implant device structure was designed for injection to minimize surgical trauma. Hollow microspheres with an open porous wall structure and one large central opening were made from poly(d,l-lactic-co-glycolic acid) (85:15 lactic:glycolic). This polymeric scaffold was seeded with hepatocytes and implanted into the abdominal wall muscle of syngeneic Fisher rats. Specimens explanted up to 56 days p.o. showed hepatocyte survival and the development of a directional blood supply. This phenomenon is coined "angiopolarity". The study should help in addressing the issue as to whether avascular cell implants with post-transplantation organ growth should be attempted. Processing options in applying heat to the polymer solution allow manufacturing of larger microspheres with different diameters of central openings. This would allow the use of the scaffold for other cell transplantations than hepatocytes.

Introduction

Transplantation using cells isolated from a living relative of the patient and transplanted through injection using a resorbable polymer carrier could be an attractive means of therapy for malfunctioning or damaged organs. Vacanti et al. [1] reported an implantable system requiring surgery enabling selective hepatocyte transplantation using bioabsorbable polymers processed into small branches. Demetriou et al. [2] describe a novel approach in which hepatocytes were attached to non-degradable collagen-coated microcarrier beads prior to injection into the peritoneal cavity. We now report a new polymer design using the potential to combine some of the desirable properties of these approaches

-degradibility,
-injectability, as well as potentially permitting
-angiogenesis to occur.

Angiogenesis could provide a high degree of structural integration as needed for a critical cell mass. Provoquing angiogenesis around the polymer implant should result in a blood vessel distribution similar to the vascular pattern of a liver acinus: the intended polymer cell carrier was therefore processed to approach the design of a functional unit of the target organ. Its blood supply is characterized by a hepatic arteriole feeding a bulk mass of hepatocytes by diffusion from the acinar axis to the acinar periphery [3]. A polymer replicating this navel like natural design would be named angiopolar as a vascular pole could be distinguished. These requirements led to the following polymer design:

a) spherical in order to have a maximum surface/volume ratio per amount of injected carriers the spheres could be coated or filled with growth factor or drug carrier on the outer convex and inner concave surfaces.

b) large opening of the spherical wall (0.5–1 mm) in order to assure angiopolarity,

c) wall porosity (pore size smaller than 45 μm) in order to allow hepatocytes to settle and to assure nutrition through capillaries and diffusion,

d) external shape diameter less than 3 mm in order to allow injection through a 9-gauge needle or smaller.

Materials and methods

0.2 g polymer [7] in 1.5 ml methylene chloride is mixed with 1 g NaCl crystals of less than 45 micrometers and injected through an 18-gauge hypodermic needle onto a heated layer of NaCl crystals (60°C). As evaporation of the solvent occurs spherical structures of approximately 2.5 mm external diameter with a large opening (ca. 0.5–1 mm) are formed. Following dissolving the salt in water and vacuum drying of the polymer, porous hollow microspheres result. Liver cells were isolated from syngeneic male Fisher rats using a modification of the two-step collagenase procedure of Seglen [4]: retrograde perfusion with HEPES-buffer followed by the same buffer perfusion containing 5 mM $CaCl_2$ and 0.5 mg/ml collagenase (Worthington, class II). Cells were dispersed in serum-free Williams E medium with 10 ng/ml EGF, 20 mU/ml insulin, 5 nM Dexamethasone, 20 mM pyruvate and 100 U/ml penicillin/streptomycin [5]. The viability after centrifugation to remove debris and washing with medium was 88–98% as determined by trypan blue exclusion. Cells were then allowed to settle by gravity to form a dense pellet. Aliquots of the resulting cell pellet were diluted 1:1 with cold basement membrane matrigel or cold vitrogen gel. For implantation polymer spheres were dipped into the cell suspension and implanted into abdominal wall muscle. Postoperative analysis of explanted specimens was performed by light microscopy, Scanning Electron Microscopy (SEM), histology and a few specimens were analyzed by an environmental scanning electron microscope (ESEM).

Results and conclusion

Processing options

A higher salt/polymer ratio than 1:0.2 resulted in formation of a smaller central opening of the microsphere and in a higher wall porosity. A more diluted salt/polymer mixture (salt/polymer ratio 1:0.2 diluted with more than 1.5 ml solvent) leads to a larger opening and lower wall porosity and to additional porous septal structures within the sphere. Sphere formation could not be achieved when more than 2 ml solvent was used with 1 g salt and 0.2 g polymer. When the optimal salt/polymer ratio was used (1:0.2) a temperature of the salt bed above 60°C leads to rapid sphere formation with subsequent collapse of the polymer structure and a pan-like end structure with higher wall thickness than in ordinary spheres. A temperature below 60°C resulted in a sphere formation without a central opening.

Light microscopy, histologic sections, scanning electron microscopy

In light microscopy of unfixed specimens, in histologic sections and under scanning electron microscopy (SEM and ESEM), spheres revealed formation of small arteries leading to the large opening and branching into smaller vessels growing towards the periphery of the spheres. This formation was found in all specimens examined from the second week on after implantation. Matrigel treatment caused a pronounced fibrosis within the implant and an increased number of small blood vessels. Collagen coating had no significant histologic effect compared to untreated specimens. The distribution pattern of hepatocytes was irregular with implanted cells usually found on the outside of the polymer spheres between muscle bundles or between muscle and subcutaneous tissue. 56 days postoperatively (p.o.) surviving hepatocytes located peripheral to a matrigel treated sphere could be detected. The typical blood vessel distribution to and around porous spheres is shown in Fig. 1. A detailed description of results is given in [6].

In conclusion. Differentiated angiogenesis occurs on porous hollow microspheres with a convex outer and a concave inner surface and a gradual surface transition in between. This finding is coined "angiopolarity" as a small blood vessel pole and a capillary pole could be distinguished. Intra- and perimuscular hepatocytes survived up to 56

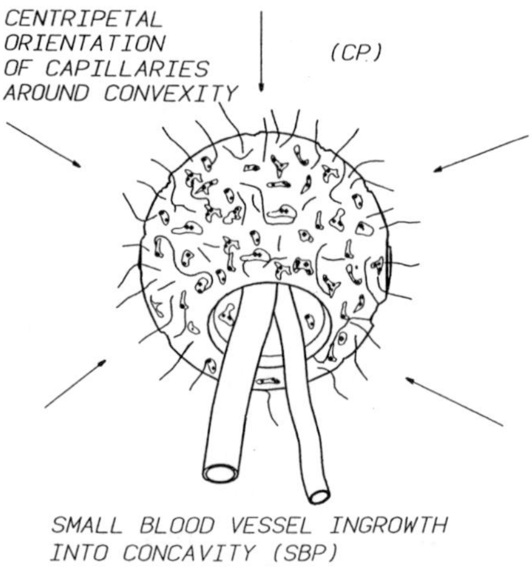

Figure 1.

days (p.o.). Polymer processing options allow the production of spheres with larger or smaller central openings than used for this animal study or of scaffolds with aspherical shapes. The angiogenic potential of these modified scaffolds will be studied in future experiments in order to find out whether angiopolarity can only be achieved with spherical implants.

References

1 Vacanti JP, Morse MA, Saltzmann WM, Domb AJ, Perez-Atayde A, Langer R: Selective Cell transplantation using bioabsorbable artificial polymers as matrices. J Pedriatic Surg. 1988; 23: 3–9.
2 Demetriou AA, Whiting JF, Feldman D, et al., Replacement of liver function in rats by transplantation of microcarrier-attached hepatocytes. Science 1986; 233: 1190–1192.
3 Arias, IM et al. (eds.) The liver. Biology and pathobiology. New York, Raven Press, 1988, p. 932.
4 Seglen PO: Preparation of isolated rat liver cells. Meth Cell Biol 1976; 13: 29–83.
5 McGowan JA, Bucher NLR: Pyruvate promotion of DNA synthesis in serum-free cultures of rat hepatocytes. In vitro 1983; 19(3): 159–166.
6 Wintermantel E, Cima L, Schloo B, Langer R: Angiopolar degradable polymer for cell delivery through injection. Biomaterials, submitted.
7 Medisorb 8515 by Dupont, Wilmington, Delaware, USA

Please address correspondence to: PD Dr. E. Wintermantel, Swiss Federal Institute of Technology, ETH-Zentrum, HG F 48.2, Rämistr. 101, CH-8092 Zürich Switzerland

Angiogenesis: Key Principles - Science - Technology - Medicine
ed. by R. Steiner, P. B. Weisz & R. Langer

Angiogenic vessels as a potential for targeted therapy in human liver metastases

A. I. Skene, R. J. L. Williams, R. D. Goldin and R. D. Rosin

Department of Surgery and Pathology, Clinical Research Centre, Harrow and St. Marys Hospital, London, England

Introduction

Primary and secondary liver tumours are a major cause of cancer related deaths worldwide with hepatic metastases outnumbering primary liver tumours by 50:1. Hepatic metastases are particularly common in association with gastro-intestinal tumours. At present surgery offers the only means of cure in these patients. However only a small minority of cases have isolated or small numbers of liver secondaries which are suitable for resection.

Loco-regional therapies for liver metastases have not shown consistent benefits over systemic treatments and are associated with specific complications. A variety of methods of enhancing relative uptake of cytotoxic agents within hepatic metastases have previously been attempted including monoclonal antibodies, liposomes and microspheres. However few of these methods have conferred a clinical benefit.

We have assessed the potential of therapy targeted against angiogenic vessels in human liver metastases.

Angiogenic vessels

The liver parenchyma is supplied by a system of specialised sinusoids. The sinusoidal endothelial cells are different from normal endothelial cells and from the angiogenic vessels at the tumour host interface. The sinusoidal lining cells have specialised receptor sites for certain glycoproteins. In a previous study [1], the affinity of the endothelial cell marker Ulex Europaeus Agglutinin 1 (UEA1) for the vasculature of hepatic metastases was compared with that of normal liver sinusoids. The normal liver sinusoids were uniformly negative for this marker whereas the tumour angiogenic vasculature were positive in all 27 tumours tested. The aim of the current study was to compare the uptake of UEA1 in hepatic colorectal metastases with that in the normal adjacent liver, when administered to isolated perfused human liver resections. This was performed in order to assess the potential of treatment of liver metastases using this method of targeting.

Materials and methods

Hepatic lobectomy specimens were obtained at operation from three patients undergoing surgery for solitary liver metastases. All major branches of the hepatic artery, vein and portal vein were cannulated and the specimen flushed with cold heparinised Hartmann's Solution. Specimens were then established onto an isolated perfusion circuit and perfused at 37°C (pressure arterial 95 mmHg, portal 30 mmHg). The circuit consisted of a perfusate reservoir, bubble oxygenator, heat exchanger, perfusion pump and manometer.

The UEA1 lectin was radiolabelled with ^{131}I using the Iodogen method and unbound isotope separated using a sephadex column. Protein labelling was confirmed using TCA precipitation and was over 95%. ^{125}I-labelled human albumin served as a control protein. 100 μCi ^{131}I-UEA1 and 50 μCi ^{125}I-albumin were administered by hepatic intra-arterial injection in three alliquots at 0, 5 and 10 min and perfusion continued for a further 15 min.

After perfusion the lobectomy specimen was bisected and five 5-mm^3 samples were biopsied and weighed from the tumour centre (TC), tumour periphery (TP), liver near tumour (LN) and liver remote from tumour (LR). Lectin localisation was assessed by subsequent gamma scintigraphy.

Results

Figures I and II show the results of the uptake of the labelled markers in the three perfusion specimens. Liver 1 demonstrated the pattern of uptake anticipated from the immunocytochemical series, with the highest ratio of UEA1 within the tumour. This pattern was not repeated in the livers 2 and 3.

Conclusion

These results suggest that UEA1 has an affinity for living liver tissue not seen in the over-layed immunocytochemical fixed tissues. These results would suggest that the inconsistency in uptake would not be useful as a clinical method of targeting using this particular lectin. The

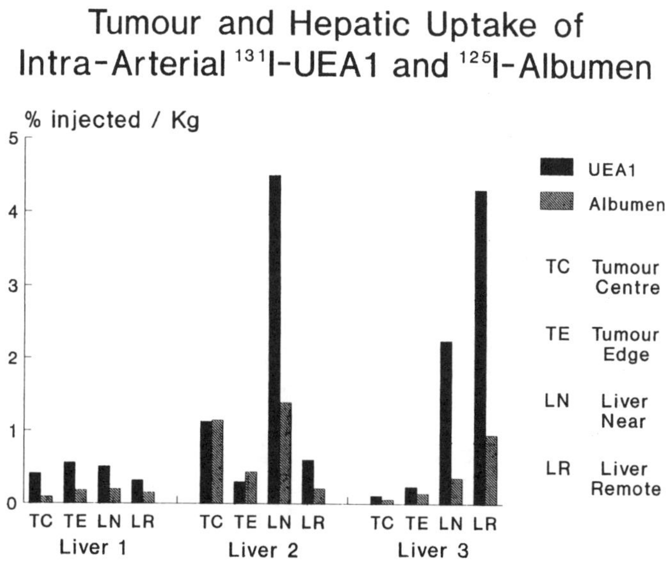

Figure I

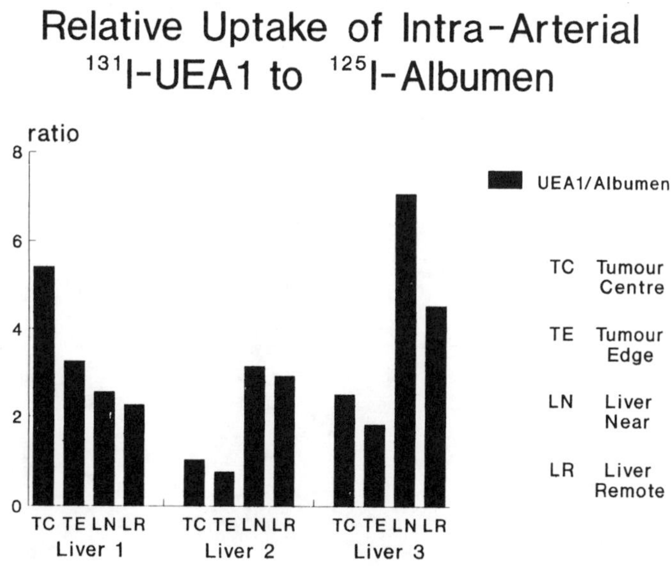

Figure II

phagocytotic activity of the living Kupffer cells may be important in the uptake of the UEA1 within the normal liver parenchyma.

Reference

1 Williams RJL, Dourakis SP, Goldin RD, Guillou PJ: Vascular endothelial markers in human liver metastases from gastrointestinal cancer. Br J Cancer 1990; 62 (Suppl XI): 8.

Angiogenesis: Key Principles - Science - Technology - Medicine
ed. by R. Steiner, P. B. Weisz & R. Langer

The effect of angiotensin II on tumor blood flow and the delivery of microparticulate cytotoxic drugs

D. J. Kerr, J. A. Goldberg[b], J. R. Anderson[b], N. Wilmott[d], A. T. Whatelely[d], C. S. McArdle[b] and J Mckillop[c]

Depts. of Medical Oncology[a], Surgery[b] and Medicine[c], University of Glasgow and Dept. of Pharmacy[d], University of Strathclyde, Glasgow, Scotland

Abstract. Colorectal hepatic metastases have a notoriously poor response to conventional systemic chemotherapy. We have synthesised cytotoxic drug (doxorubicin and mitomycin C) containing spheres 40 μm in diameter, using human albumin and ethyl cellulose as matrices. Introduction of these cytotoxic microspheres into the hepatic artery should embolise to the tumor and provide a controlled release depot for the anticancer agent. The vasoconstrictor, angiotensin II (AII) has been shown to increase tumor blood flow relative to normal tissue when administered via the hepatic artery, therefore we have investigated the effect of AII on targeting of cytotoxic microspheres to hepatic metastases. Patients with hepatic metastatic colorectal carcinoma had hepatic arterial catheters inserted at laparotomy and connected to subcutaneous injection ports. Peroperatively, ^{99m}Tc-labelled albumin microspheres were administered via the arterial catheter. Fifteen minutes later, AII was infused (10 μg per minuter for 4 min) via the catheter and ^{131}I-labelled albumin microspheres were adminstered as a bolus at the midpoint of the AII infusion. Multiple biopsies were taken of normal liver and tumor metastasis and the tissue radioactivity counted for ^{99m}Tc and ^{131}I. Further studies were performed postoperatively in which ^{99m}Tc-labelled microspheres were administered via the hepatic artery catheter and their distribution was followed using tomographic SPECT scanning. Combined results of this study suggested that AII can increase tumor SPECT scanning. Combined results of this study suggested that AII can increase tumor blood flow rates relative to normal hepatic tissue by approximately 3-fold.

Introduction

The development of hepatic metastases following apparently curative surgery for colorectal cancer is a relatively common phenomenon and it has been estimated that as many as 60% of patients have liver metastases at the time of death [1]. Systemic chemotherapy with conventional cytotoxic drugs, such as 5-fluorouracil, is relatively ineffective due to its narrow therapeutic ratio and potentially severe systemic side effects.

It has been previously demonstrated that liver metastases derive their blood supply predominantly from the hepatic artery [2] and there is increasing interest in hepatic arterial administration of cytotoxic drugs. Surgical techniques, involved in insertion of indwelling intrahepatic arterial catheters, have improved greatly, therefore repeated vascular access to the liver is now possible.

Drug targeting can occur at different levels: first order targeting is increased drug delivery to the organ or body compartment harbouring tumours; second order targeting depends on enhanced drug delivery to tumour tissue; third order targeting implies selective delivery of therapy to the cancer cell relative to surrounding host cells [3]. Regional chemotherapy allows first order targeting by enhancing drug delivery to the organ or tissue compartment harbouring tumour.

Put simply, the aim of intra-arterial chemotherapy, is to generate high cytotoxic drug concentrations within the tumour and low drug concentrations in the vascular compartment. Two assumptions are implicit in this argument, stemming from the widely held view that antineoplastic agents have steep dose-response curves:

a) the greater the exposure of tumour to the cytotoxic agent the greater the likelihood of an antitumour response. Drug exposure must take account of both peak drug concentration achieved and the duration of exposure.
b) The lower the concentration of drug in the systemic, vascular compartment, the less likely the occurence of side effects.

Using a physiologically based pharmacokinetic model, Chen and Gross [4] showed that the regional drug exposure advantage (i.e. the ratio of drug exposure after intra-arterial infusion to drug exposure following intravenous administration of the same dose) is related to the rate of drug clearance by the rest of the body and the rate of blood flow into the target organ.

One means of retaining anti-cancer agents more effectively within the liver might be to load the active agent into embolising particles. There have been reports of chemotherapeutic agents such as adriamycin and mitomycin C being carried in biodegradable particles which act as slow-release systems when administered via the hepatic artery [5, 6]. Unfortunately the distribution of arterially administered particles reflects the pattern of arterial blood-flow within the liver, and the proportion of drug reaching hypovascular tumours will be low.

Previous studies in animal models have suggested that vasoactive agents such as noradrenaline and angiotensin II modify the pattern of arterial blood-flow by causing temporary arteriolar constriction in normal blood-vessels [7]. Hepatic tumour vasculature is immature, possessing neither smooth muscle nor an adrenergic nerve supply, and is therefore unable to respond to arterial-constrictors in the same way as normal vasculature [8, 9]. The use of these agents might therefore enhance the delivery of arterially administered substances to tumours.

We described our experience with angiotensin II as a means of targeting arterially administered particles in patients with advanced hepatic metastases.

Patients and methods

The relative microsphere delivery to tumour and normal liver before and after angiotensin II was measured using a double isotope technique. ^{131}Iodine (^{131}I) and technetium (^{99m}Tc)-labelled microspheres, were prepared as described below. The particles ranged in diameter from 20–40 μm, a size which becomes trapped in the first capillary bed encountered when injected intra-arterially.

Albumin microspheres were prepared by adding an aqueous solution of human serum albumin to a constantly stirred oil phase to produce a water-in-oil emulsion. The microspheres thus formed were stabilised with glutaraldehyde. After mesh filtration and differential centrifugation, microspheres of the appropriate size were obtained [10]. The particles were then labelled with ^{131}I using the chloramine T method [11].

^{131}I microspheres were stable in phosphate-buffered saline. Less than 2.5% of radiolabel was found within the supernatant after 3 months storage. The preparation was also remarkably stable in serum with less than 2% of the radiolabel being released after 9 days incubation.

^{99m}Tc-labelled microspheres were obtained from commerical sources (TDK-5; sorin Biomedica SpA, 13040 saluggia (Verchelli, Italy). This preparation is routinely used for imaging during hepatic arterial perfusion scintigraphy; free activity does not exceed 5%.

Tracer doses of ^{131}I- and ^{99m}Tc-labelled albumin microspheres were freshly prepared under sterile conditions before each laparotomy. There were approximately 4×10^5 particles per dose.

Nine patients (mean age 56 years; range 41–70 years) with liver metastases (eight colorectal; one unknown primary) undergoing placement of a hepatic arterial catheter for regional chemotherapy were studied.

At laparotomy, a hepatic artery catheter was inserted into the gastro-duodenal artery, so that the tip of the catheter lay at the orifice of the gastro-duodenal artery without impeding hepatic arterial blood-flow. The adequacy of perfusion was checked using dilute methylene blue. A tracer dose of ^{131}I microspheres (2.5 ml) was then injected into the catheter over 5 s and flushed with 5 ml heparinised saline over 10 s.

Angiotensin II was infused into the catheter at a rate of 10 μg in 2 ml normal saline/min for 100 s. Immediately after the infusion, the tracer dose of ^{99m}Tc microspheres in 2.5 ml was injected over a 5 s period and flushed with 5 ml heparinised saline over 10 s.

Biopsies of tumour and adjacent normal liver were obtained and weighted (mean biopsy weight 0.8 g; range 0.3–1.7 g). The uptake of ^{131}I and ^{99m}Tc in the biopsies was measured using a gamma scintillation counter (Packard Instruments Auto-gamma 5000). The tissue was distributed between vials and counted with respective windows (120–

160 keV and 300–400 keV) to differentiate the activity due to the two radio-isotopes. Standards of the ^{131}I and ^{99m}Tc were also counted to estimate the overlap between spectra, and the activity in each tissue sample resulting from the individual isotopes calculated using an in-house computer programme. The relative microsphere content of the tissue samples was expressed as counts per gram of tissue and the tumour: normal ratios of activity before and after the administration of angiotensin II were assessed in this way.

The tumour: normal ratios before and after angiotensin II infusion respectively were compared using the Wilcoxon paired test.

The methodology had been approved by the local Ethical Committee and the necessary Administration of Radioactive Substances Advisory Committee certification obtained. Informed consent was obtained from all patients.

Results

The hepatic arterial infusion of angiotensin II induced a modest rise in systemic blood-pressure (an increase in the systolic pressure of up to 40 mmHg) which reached peak at the end of the 100 s infusion, then gradually declined. No rebound hypotension was observed. The results are summarised in Fig. 1.

Following angiotensin II infusion, the uptake of microspheres in tumour was greater than that in normal liver. The median improvement in tumour: normal ratio was 2.8 (range 0.8–11.7; $p < 0.05$).

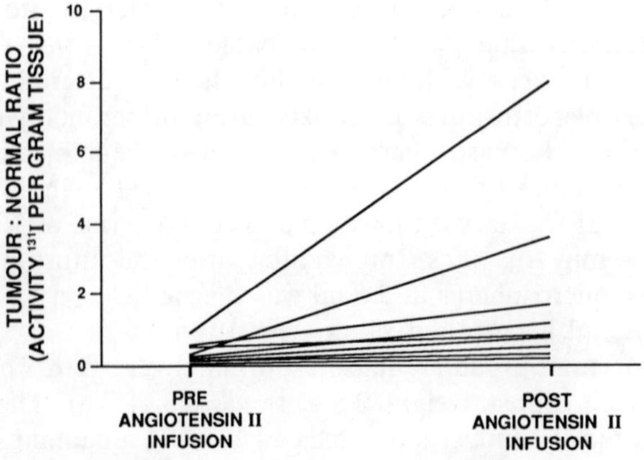

Figure 1. The ratio of microsphere delivery to tumour and normal hepatic tissue before and after angiotensin II.

Discussion

The notion of altering tumour flow with vasoactive substances is not new. Radionuclide angiography with technetium-labelled macro-aggregated albumin has shown that hepatic metastases (< 8 cm diameter) tend to be hypervascular relative to surrounding hepatic parenchyma, whereas larger metastases have a hypovascular core and hypervascular rim [12]. The tumour arterioles tend to be less reponsive to the vasoconstrictive effects of biogenic amines such as adrenaline, however this effect is dependant on blood vessel size and the type of experimental model studied as noradrenaline has been shown to increase [13], decrease [14] or have no effect [15] on tumour blood flow in different studies. The vasoconstrictor, angiotensin II, increases blood flow to hepatic metastases relative to normal tissue in patients. Sasaki et al. [16] infused angiotensin II (10 μg per minute for 4 min) through hepatic arterial catheters and followed blood flow in hepatic tumour and normal hepatic tissue by co-fusing radioactive krypton gas. Tumour blood flow increased by approximately 300% relative to normal liver during the infusion of angiotensin II and quickly returned to baseline values upon cessation of treatment. Angiotensin II can only be given by a brief infusion as it causes significant rises in systemic blood pressure, although this was not a problem in the present study.

Vasopressin has been used extensively in pharmacoangiography [17] to improve delineation of tumour blood vessels and has been shown to increase tumour blood flow in vitro in murine subcutaneous carcinomas. Similarly, the anti-hypertensive vasodilator verapamil [18] increases tumour blood flow in hind-limb transplanted carcinomas in mice.

In this study, we have used a radioactive microsphere technique to assess the targeting capactity of angiotensin II. Tracer doses of microspheres were used to minimise the possibility of the first dose of particles significantly altering the haemodynamics of liver blood-flow and hence the distribution of the second dose of microspheres.

Despite the small number of patients studied, a significant improvement in microsphere delivery to tumours is seen with angiotensin II. It was interesting to note that, despite the errors due to sampling inherent in a study of this kind, the mean improvement in microsphere delivery to tumour after angiotensin II was of a similar magnitude to the peak increase in arterial perfusion of hepatic tumours with angiotensin II described by Sasaki and his colleagues.

Unfortunately, the relative increase in arterial blood flow to hepatic tumours following angiontensin II infusion was shown by Sasaki and his co-workers to be a relatively short-lived effect. This mechanism would not therefore lend itself to the targeting of subtances administered by prolonged infusion. Nevertheless, tumour targeting by angiotensin II might be used successfully with anticancer agents given by bolus injection particularly if this took the form of a particle-bound preparation. This clearly represents a significant effecting terms of first order targeting, i.e., delivering the chemotherapeutic agent to the tumour's vascular compartment.

It might be possible to further increase second order targeting, facilitation of drug diffusion into the tumour interstitium by biodadhesion targeting. This refers to the combination of embolism with particles coated in a binding substance for tumour vascular endothelial cells. One could postulate that particle surface coating with the laminin receptor or other basement membrane components like fibronectin could conceivably enhance transvascular migration of embolic particles which were at the lower limit of size.

With the introduction of new microsphere matrices it could prove possible to encapsulate a wider range of cytotoxic agents which could be more appropriate for certian tumour types. 5-fluourouracil would be a good example of a small, water-soluble drug which, because of its saturable cellular uptake, would be a good candidate for a prolonged slow release preparation.

In summary, we would predict that manipulation of tumour bloodflow will continue to be an important aspect of chemo-embolism as a cancer therapy.

References

1 Welch JP, Donaldson GA: The clinical correlation of an autopsy study of recurrent colorectal cancer. Ann Surg 1979; 189: 496.

2 Ridge JA, Bading JR, Gelbard AS, Benua RS, Daly JM: Perfusion of colorectal hepatic metastases: relative distribution of flow from the hepatic artery and portal vein. Cancer 1987; 59: 1547.

3 Widder KJ, Senyei AE, Ramsay DF: Magnetically responsive microspheres and other carriers for the biophysical targeting of antitumour agents. Adv Pharmacol Chemother 1979; 16: 213–71.

4 Chen H-SG, Gross JF: Intra-arterial infusion of anticancer drugs: theoretic aspects of drug delivery and review of responses. Cancer Treat Rep 1980; 64: 31.

5 McArdle CS, Lewi H, Hansell D, Kerr DJ, McKillop J, Willmott N: Cytotoxic loaded albumin microspheres: a novel approach to regional chemotherapy. Br J Surg 1988; 75: 132.

6 Fujimoto S, Miyazaki M, Endoh F, et al.: Effects of intra-arterially infused biodegradable microspheres containing mitomycin C. Cancer 1985; 55: 522.

7 Burton MA, Gray BN, Self GW, Heggie JC, Townsend PS: Manipulation of experimental rat and rabbit tumour blood flow with angiotensin II. Cancer Res 1985; 45: 5390.

8 Hafstrom L, Nobin A, Persson B, Sundqvist K: Effects of catecholamines on cardiovascular response and blood flow distribution of normal tissue and liver tumours in rats. Cancer Res 1980; 40: 481.

9 Mattson J, Applegren L, Hamberger B, Peterson HI: Adrergic innervation of tumour blood-vessels. Cancer Letters 1977; 3: 347.

10 Lee TK, Sokoloski JD, Royer GP: Serum albumin beads: an injectible, biodegradable system for the sustained release of drugs. Science 1981; 213: 233.

11 Hunter WN, Greenwood FC: Preparation of Iodine (^{131}I) labelled human growth hormone of high specific activity. Nature 1962; 194: 495.

12 Gyves J, Ensminger W, Thrall J, et al.: Dependence of hepatic tumour vascularity on tumour size. Clin Res 1982; 30: 747A.
13 Iwaki A, Nagasue N, Kobayashi M, Inokuchi K: Intra-arterial chemoatherapy with concomitant use of vasoconstrictors for liver cancer. Cancer Treat Rep 1978; 62: 145.
14 Hafstrom L, Perrson B, Sundquist K: Blood flow in experimental liver tumours: effects of vasoactive drugs. Acta Chir Scand 1980; 146: 149.
15 Miyaziki M: Intra-arterial infusion treatment for patients with malignant hepatic tumour: with special reference to hepatic pharmacokinetics and to combined used of vasoactive and anti-tumour drugs. Jpn J Surg 1981; 83: 1295.
16 Sasaki Y, Shingi I, Yoshihisa H, Shunnichi N, Ishikawa O, Ohigashi H, Taniguchi K, Koyana H, Iwanaga E, Terasawa T: Changes in distribution of hepatic blood flow induced by intra-arterial infusion of angiotensin II in human hepatic cancer. Cancer 1985; 55: 311.
17 Jenkins SA, Day DW, Mooney B, Devitt P, Taylor I, Shields R: The effect of vasopressin and hepatic artery ligation on the blood supply to normal and metastatic liver tissue. Br J Cancer 1984; 50: 785.
18 Kaelin WG (Jr), Shrivastav S, Shand JG, Jirtle RL: Effect of verapamil on malignant tissue blood flow in SMT-2A tumour bearing rats. Cancer Res 1982; 42; 3942.

Angiogenesis: Key Principles - Science - Technology - Medicine
ed. by R. Steiner, P. B. Weisz & R. Langer

Effects of hypercholesterolemia on monokine-induced smooth muscle cell proliferation

Tina M. Lam, Nancy E. Whereat, Scott C. Henderson, Wilson H. Burgess[a], Ann Shaheen[a] and Howard P. Greisler

Loyola University Medical Center, Maywood, Illinois, USA, and [a]American Red Cross, Rockville, Maryland, USA

Abstract. Macrophage/smooth muscle cell interactions play a role in atherogenesis and foreign body reactions to biomaterials. This study investigates the effect of a hypercholesterolemic diet on the ability of smooth muscle cells (SMCs) to respond to monokines which are produced in response to hypercholesterolemia, biomaterials or both. Peritoneal macrophages were harvested from rabbits fed either a normal ($Mø_{NL}$) or a 2% cholesterol/ 6% peanut oil diet ($Mø_{ATH}$) (plasma cholesterol 2840 vs 42.3 [$p < 0.005$]). The macrophages were then cultured in the presence of either 1) polyglactin 910 (PG910), 2) Dacron, or 3) no biomaterial (control), and the media collected and pooled by week for the smooth muscle cell mitogenesis assays. Rabbit aortic smooth muscle cells were harvested and cultured from the same two groups of rabbits (SMC_{NL} or SMC_{ATH}), quiesced in serum free media (48 h) followed by addition of the test media and ^{3}H-TdR. The addition of either biomaterial to $Mø_{NL}$-conditioned media increased ^{3}H-TdR incorporation in both smooth muscle lines as compared to controls. PG910 resulted in significantly higher ^{3}H-TdR incorporation than Dacron (weeks 3–5, $p < 0.005$). The addition of either biomaterial to $Mø_{ATH}$ also increased ^{3}H-TdR incorporation in both smooth muscle cell lines, however, the magnitude of the response was decreased as compared to the $Mø_{NL}$-conditioned media in both cell lines ($p < 0.001$ for either SMC line). In contrast to the $Mø_{NL}$-conditioned media, the addition of Dacron to $Mø_{ATH}$ resulted in the highest level of ^{3}H-TdR incorporation in both cell lines as compared to the media without biomaterial. The SMC_{NL} had a higher response to both the monokines in conditioned media (2-fold) and to fetal bovine serum (3-fold) than the SMC_{ATH}($p < 0.001$). Although there is a generalized decrease in release of mitogens active on SMCs from $Mø_{ATH}$, the $Mø_{ATH}$ exposed to Dacron release increased amounts of mitogenic factors, most active on the SMC_{ATH} cell line. A common mode of failure of small diameter Dacron grafts in man is pseudointimal hyperplasia, and it is inviting to postulate that the Dacron/macrophage/smooth muscle cell interactions in this atherosclerotic group of patients plays a role in the pathogenesis of this lesion.

Introduction

Vascular smooth muscle cell proliferation is an important component of atherogenesis and a major factor in the failure of a significant number of vascular reconstructions [1–3]. Smooth muscle cell proliferation in vitro can be stimulated by a variety of growth factors [4–9] and inhibited by heparin-like glycosaminoglycans [10, 11] or by transforming growth factor-beta, a multifunctional peptide present in significant amounts in platelets and synthesized by a variety of cells in culture

[12–17]. Atherogenesis and the foreign body reactions to vascular grafts likely involve interactions between smooth muscle cells and macrophages, both of which can esterify and accumulate cholesterol when exposed to hypercholesterolemic serum [18–20]. Macrophages have been shown capable of secreting factors both stimulatory and inhibitory to smooth muscle cell proliferation. The present study was undertaken to examine the effect of a hypercholesterolemic diet on the ability of smooth muscle cells to respond to monokines which are produced in response to hypercholesterolemia, biomaterials, or both.

Methods

Cell culture

Rabbit aortic structure muscle cells were harvested from two groups of rabbits, one fed a normal rabbit chow diet and the other fed a diet supplemented with 2% cholesterol and 6% peanut oil for 2 months (serum cholesterol 2130 vs 110; $p < 0.005$). The cells were harvested by resecting the aorta into Hanks Balanced Salt Solution (HBSS), sharply dissecting free the adventitia and intima and mincing the remaining media into 1-mm^3 explants. The explants were cultured in Dulbecco's Modified Eagle's Medium (DMEM; glucose 4.5 g/l) containing 10% fetal bovine serum (FBS), 100 U/ml penicillin and 0.1 mg/ml streptomycin. Cells were maintained in a humidified 37°C, 5% CO_2 atmosphere and were utilized after their migration from the explants but prior to any further subculturing. Cell identification was confirmed by immunoperoxidase staining with a mouse monoclonal antibody to smooth muscle α-actin (Sigma, St. Louis, MO).

Unstimulated macrophages were harvested by peritoneal lavage with HBSS containing 10 U/ml of heparin, from the same two groups of rabbits (serum cholesterol 2840 vs 42.3, $p < 0.005$ for macrophages harvested from atherosclerotic rabbits). After a 2-h adherence separation the cells were cultured in Minimum Essential Medium (MEM) containing 10% equine platelet-poor plasma, 100 U/ml penicillin, 0.1 mg/ml streptomycin, L-glutamine 200 mM, Hepes Buffer, sodium pyruvate 100 mM, and non-essential amino acids 10 mM. Macrophage purity was determined by phase contrast microscopy, non-specific esterase staining, and identification of Fc receptors on the cell membranes. The cells were then cultured with finely shredded polyglactin 910 (PG910, 1.6 mg/cm^2), Dacron (1.6 mg/cm^2) or no biomaterial (control) and maintained at 37°C in a humidified 5% CO_2 incubator. The media was collected at the times of refeeding and frozen to −70°C until use in the bioassays.

SMC mitogenesis assays

The two different SMC lines were grown to confluence and aliquoted into 96-well microtiter plates at a concentration of 5000 cells/well. The cells were then incubated until confluence was obtained, upon which, the cells were washed with phosphate buffered saline (PBS) and refed with serum free media. Serum free medium was prepared as a 1:1 mixture of DMEM and Ham's F-12 media supplemented with transferrin 5 μg/ml, penicillin 100 U/ml, streptomycin 0.1 mg/ml, insulin 1 μM, and ascorbate 0.2 mM. Forty-eight hours after the change into serum free media, quiescence was documented by scintillation counting of tritiated thymidine uptake. Thawed conditioned media and ^{3}H-TdR (1 μCi/well) were then added to test wells of quiesced SMC in replicates of four and allowed in incubate for 48 h. The wells were then processed for scintillation counting of ^{3}H-TdR incorporation into newly synthesized DNA. Cells were washed once with normal saline and then fixed twice with 200 μl of absolute methanol at 4°C for 5 min each. This was followed by washing with dH_2O three times. The cells were then incubated twice in 5% trichloroacetic acid (TCA) for 10 min each and again washed with dH_2O. The acid precipitable material was then solubilized by incubating each well with 150 μl of 0.3 N NaOH at room temperature for 20 min. Each 150 μl sample was then transferred to 10 ml of Ready Protein scintillation cocktail and counted in a Beckman LS6800 scintillation counter. Results are expressed as percent of positive control (FBS) which gave the maximal response. For neutralizing anti-bFGF antibody studies, the anti-bFGF antibody was purchased from Upstate Biotechnology Inc. (Lake Placid, NY). The neutralizing anti-bFGF antibody (10 μg/ml final concentration) was pre-incubated with the macrophage conditioned media for 2 h prior to addition in the SMC bioassays, as described above. The concentration of antibody was determined by dose response studies in the same cell lines.

CCL64 inhibitory assay

CCL64 mink lung epithelial cells (ATCC, Rockville, MD) were trypsinized 3 days after their last passage and seeded into 24-well plates at a concentration of 5000 cells/well and incubated at 37°C in a 5% humidified CO_2 incubator for 1 h to allow attachment. The cells were then plated in a MEM medium supplemented with 10% FBS, penicillin 100 U/ml, and streptomycin 0.1 mg/ml. After attachment, 100 μl of the thawed conditioned media or pure TGF-β (R&D Systems, Minneapolis, MN) was added to the cells and allowed to incubate for another 48 h, after which the cells received a 16-h pulse of ^{3}H-TdR. ^{3}H-TdR incorporation was then measured for each week as described above for the

SMC assays, with results expressed as percent of maximal TGF-β inhibitory activity.

Western blots

Conditioned media were subjected to polyacrylamide gel electrophoresis (PAGE) (15% w/v) in the presence of Sodium Dodecyl Sulfate (SDS) using the Mighty Small gel apparatus (Hoeffer Scientific Instruments, San Francisco, CA). After electrophoresis the gels were stained with Coumassie Brilliant Blue R-250, dried, and exposed to X-ray film by our published techniques [21, 22]. Gels were analyzed by western blotting to determine immunoreactivity to polypeptides of appropriate molecular weight using anti-PDGF B chain, anti-ECGF, and anti-bFGF antibodies.

Data analysis

Tritiated thymidine incorporation into newly synthesized DNA was calculated as percent of maximum response and expressed as mean $\pm$ standard deviation and the groups statistically analyzed by the Student's t-test.

Results

Normal SMC assay

Quiescence of the SMC was documented by the PBS negative controls, 130 ± 24 CPM, and the capacity of quiescent cells to be stimulated to synthesize DNA documented by the FBS positive controls, $51{,}131 \pm 2{,}885$ CPM.

Fig. 1 shows the increased ^{3}H-TdR uptake with the addition of biomaterials to either macrophage culture. The addition of PG910 to macrophages harvested from normal diet rabbits (normal macrophages) (Fig. 1a) resulted in the highest level of ^{3}H-TdR uptake which peaked during prosthetic resorption and returned to control levels once resorption was complete. Fig. 1b reveals that the macrophages harvested from rabbits fed an atherogenic diet (atherosclerotic macrophages) with or without biomaterial result in significantly less mitogenic activity as compared to the normal macrophages. In contrast to the normal macrophages, Dacron and not PG910 results in the highest level of ^{3}H-TdR incorporation.

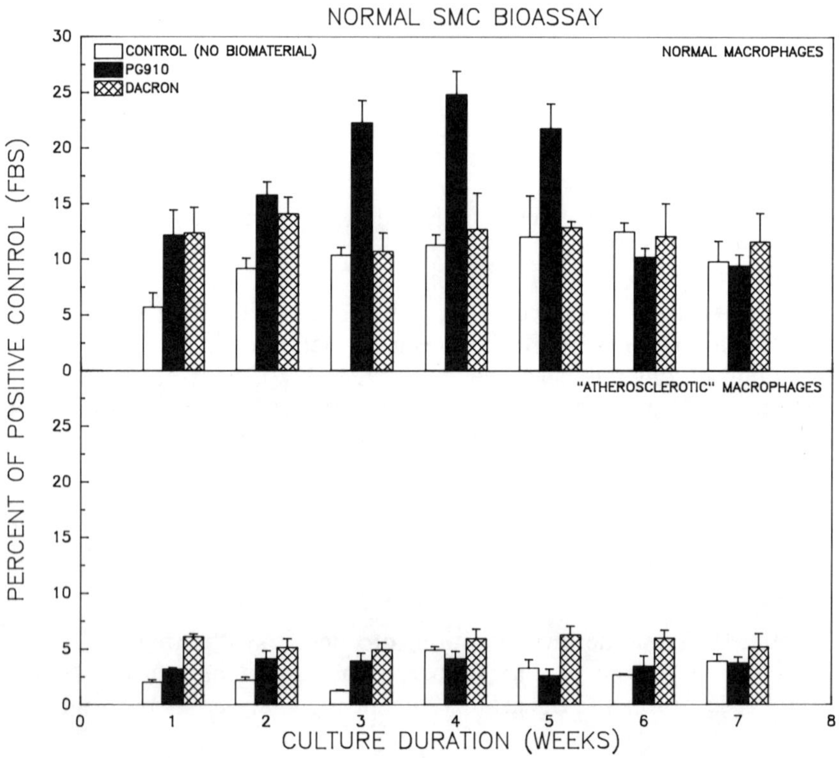

Figure 1. a) Normal rabbit aortic smooth muscle results. Studies using conditioned media from macrophages harvested from normal NZW rabbits, the results are expressed as mean CPM samples ÷ mean CPM FBS positive control × 100. (Mean ± SD). b) Normal rabbit aortic smooth muscle results. Studies using conditioned media from macrophages harvested from atherosclerotic NZW rabbits, the results are expressed as mean CPM samples ÷ mean CPM FBS positive control × 100. (Mean ± SD).

Atherosclerotic SMC assay

Quiescence was documented by the PBS negative controls, 165 ± 20 CPM and the capacity of quiescent cells to be stimulated to synthesize DNA documented by the FBS positive control, 26,940 ± 4,186 CPM.

Fig. 2 again shows the increased ^{3}H-TdR uptake with the addition of biomaterials to the macrophage cultures (normal and atherosclerotic). As can be seen, the patterns of response are similar to that seen with the normal SMC but the magnitude of response is decreased as compared with the atherosclerotic SMC.

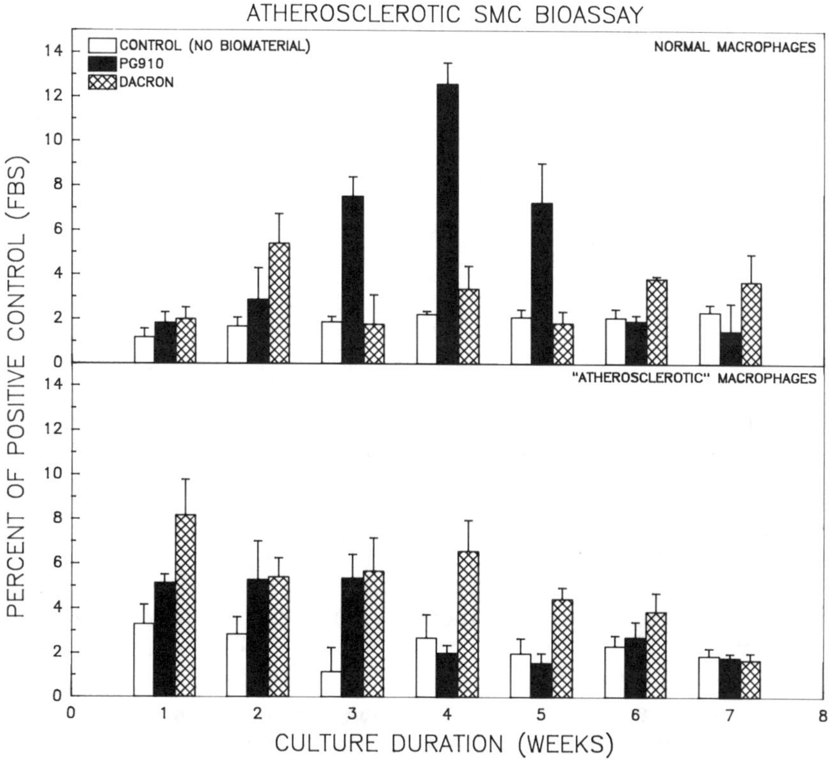

Figure 2. a) Atherosclerotic rabbit aortic smooth muscle results. Studies using conditioned media from macrophages harvested from normal NZW rabbits, the results are expressed as mean CPM samples ÷ mean CPM FBS positive control × 100. (Mean ± SD). b) Atherosclerotic rabbit aortic smooth muscle results. Studies using conditioned media from macrophages harvested from atherosclerotic NZW rabbits, the results are expressed as mean CPM sample ÷ mean CPM FBS positive control × 100. (Mean ± SD).

Growth factor identification

Preliminary immunoprecipitation studies utilizing neutralizing antibodies in the normal SMC bioassay and Western blot analyses showed no immunoreactivity with antibodies against PDGF B chain or ECGF. However, the anti-bFGF antibody did immunoreact with a band of apparent molecular weight 17,000 on the SDS PAGE produced by media from normal macrophages grown in culture with both PG910 and Dacron.

The radioimmunoassay for PDGF (Amersham International, Arlington, IL) failed to identify PDGF in any of the conditioned media.

Preliminary studies using the CCL64 assay revealed that the atherosclerotic macrophage-conditioned media contained greater amounts of TGF-β inhibitory activity than did the normal macrophage-conditioned media. The addition of biomaterials to both macrophage groups resulted in decreased inhibitory activity than without biomaterials and the addition of PG910 to normal macrophage cultures resulted in a greater decrease of TGF-β activity than in the atherosclerotic macrophages. Immunoprecipitation studies utilizing neutralizing anti-TGF-β antibody are underway.

Discussion

The regulation of arterial wall cell proliferation is central to the problems of atherogenesis and the tissue reactions to implanted blood-contacting biomaterials. All of the cell types present in the arterial wall and in the microenvironment of the implanted vascular prosthesis are capable of producing growth factors regulating the proliferation of endothelial cells, smooth muscle cells, and fibroblasts [4, 23–27].

Arterial smooth muscle cells, in contrast to their quiescent state, proliferate in response to injury and certain disease states. The smooth muscle cell is the only cell type normally present in the media of mammalian arteries and must therefore be involved not only in maintaining arterial wall tension via contraction-relaxation but also in vascular remodeling, repair, and growth. These latter functions require the retention in the adult of certain basic mesenchymal properties, namely the abilities to synthesize extracellular matrix components and to divide. This multiplicity of functions requires a whole spectrum of variation in morphology or phenotype. At one end of the spectrum is the quiescent smooth muscle cell whose major function is contraction. The cytoplasm of this cell is largely filled with thick and thin myofilaments and it has been described as being in the "contractile" state. The vast majority of cells in the media of the adult aorta are in this phenotype [29, 30]. At the other end of the spectrum is the smooth muscle cell engaged in the synthesis of extracellular matrix and/or division. The cytoplasm of this cell contains few myofilaments but large amounts of rough endoplasmic reticulum, free ribosomes, and Golgi complex. These cells have been described as being in the "synthetic" state or phenotype [29, 30]. This phenotype is present during wound repair and atherogenesis. The cells of the arterial wall, by their production of growth stimulatory and inhibitory factors affect the phenotype of the SMC and thus its capacity to respond to mitogens.

Macrophages are present early in atherogenesis, wound healing, and in the reactions to implanted vascular prostheses. Each of these processes involves angiogenesis, fibroplasia, and smooth muscle cell proliferation. Consequently the activated macrophage may be an early mediator of these events. A number of in vitro studies and in vivo angiogenesis assays have demonstrated that macrophages, when appropriately activated by a variety of test substances produce mitogens which stimulate the proliferation of fibroblasts, arterial smooth muscle cells, and several endothelial cell lines. Macrophage-derived stimulators of mesenchymal cell growth include bFGF, PDGF, IL-1 and TGF-β. Macrophage-derived inhibitors of mesenchymal cell growth include TGF-β, TNF-α, and IL-1 [23, 24, 29–32].

Past work from our laboratory has demonstrated that bioresorbable vascular prostheses produced from lactide/glycolide copolymers (e.g. PG910) implanted into the infrarenal aortas of female NZW rabbits elicit an extensive transinterstitial ingrowth of capillaries, endothelial cells, and myofibroblasts to a significantly greater extent than does similarly woven Dacron. However similarly woven PG910, when implanted into rabbits made atherosclerotic by the feeding of the 2% cholesterol and 6% peanut oil diet, resulted in much thinner arterial walls composed predominantly of lipid laden macrophages with minimal myofibroblast or endothelial cell ingrowth. The current study evaluated the effect of an in vivo atherogenic diet on the ability of SMC to respond to monokines which are produced in response to hypercholesterolemia, biomaterals, or both.

Many of the cells of diffuse intimal thickenings, and the cells of the atherosclerotic plaques that develop within them are morphologically in a synthetic phenotype and are thus different from the more quiescent contractile state cells within the media. If smooth muscle cells in the synthetic phenotype in vivo are functionally equivalent to morphologically similar synthetic state cells in culture, this may be due in part to the large number of divisions that they have likely undergone to form the thickening or plaque. Certainly senescent cells (that is, those that have exhausted their proliferative capacity) have been demonstrated in human plaques and may explain the decreased response of the SMC cultured from rabbits fed the atherogenic diet to FBS and the conditioned media.

The addition of PG910 to normal macrophages resulted in a significant increase in SMC proliferation, while Dacron led to a smaller response. These data suggest that the in vivo response seen in our previously published animal studies [33] with the PG910 prostheses may in fact be, at least in part, attributable to macrophage activation by the PG910 with subsequent ingrowth of myofibroblasts. The Dacron implanted into the same animal model shows a markedly attenuated myofibroblast ingrowth which is also consistent with the in vitro data presented here.

The presence of bFGF, a potent inducer of DNA synthesis in mesenchymal cells, may reflect either growth factor production or release of stored bFGF from injured cells. Morphologically, cell death in these groups was not observed. However, additional studies are underway to identify increased bFGF message at the transcriptional level. The apparent inhibition of cell proliferation produced by the media from atherosclerotic macrophages suggests the possible release of γ-interferon, TNF-α, as well as TGF-β, a regulatory peptide, may regulate a number of smooth muscle cell functions including proliferation, behavioral organization, and biosynthetic phenotype [34].

An interesting result was the effect of Dacron on macrophages harvested from atherosclerotic rabbits. These macrophages interacting with Dacron led to increased SMC proliferation active on both SMC lines. A common mode of failure of small diameter Dacron grafts in man is anastomotic pseudointimal hyperplasia. It is tempting to speculate that Dacron/Macrophage interactions in this atherosclerotic group of patients may play a role in the pathogenesis of this lesion.

Campbell et al.[19] have shown that there is a marked heterogeneity among the SMC phenotypes with respect to lipoprotein metabolism. Arterial SMC in the contractile phenotype show a very low capacity to bind and degrade VLDL, LDL, and β-VLDL, this was associated with minimal accumulation of cellular cholesterol, low esterification of cholesterol, and morphologically, no evidence of increased intracytoplasmic lipid. Smooth muscle cells in the synthetic phenotype, which are capable of extensive proliferation, have a higher capacity to metabolize lipoproteins, specifically β-VLDL and VLDL. The increased lipid accumulation within the synthetic state SMC in hypercholesterolemia may lead to altered cytokine release as well as altered capacity to respond to stimulatory or inhibitory agents.

Our study demonstrated that the in vivo lipoprotein challenge presented to the SMC altered the ability of the SMC to respond to the monokines released in response to hypercholesterolemia and/or biomaterials. The role of lipoproteins and macrophage/biomaterial interactions in differentially modulating smooth muscle cells and their capacity to respond to cytokines deserves further exploration.

References

1 Ross R, Glomset JA: The pathogenesis of atherosclerosis. I and II. N Engl J Med 1976; 295: 369–377, 420–425.

2 Clowes AW, Gown AM, Hanson SR, Reidy M: Mechanisms of arterial graft failure. Am J Pathol 1985; 119: 43–54.

3 Zierler RE, Bandyk DF, Thiele BL, Strandness DE: Carotid stenosis following endarterectomy. Arch Surg 1982; 117: 1408–1415.

4 Ross R, Glomset J, Kariya B, Harker L: A platelet-derived serum factor that stimulates the proliferation of arterial smooth muscle cells in vitro. Proc Natl Acad Sci USA 1974; 71: 1207.

5 Ross RE, Raines EW, Bowen-Pope DR: The biology of platelet-derived growth factor. Cell 1986; 46: 155–169.

6 Gospadarowitz D, Vlodavsky I, Savion N: The extracellular matrix and the growth of vascular endothelial and vascular smooth muscle cells. J Supramol Struct 1980; 13: 339–372.

7 Greenberg G, Hunt T: The proliferative response in vitro of vascular endothelial and smooth muscle cells exposed to wound fluids and macrophages. J Cell Physiol 1978; 97: 353–360.
8 DiCorleto PE: Cultured endothelial cells produce multiple growth factors for connective tissue cells. Exp Cell Res 1984; 153: 167.
9 Schreiber BM, Martin BM, Hollander W, Franzblau C: β-VLDL induced alterations in growth potentiating activity produced by mononuclear phagocytes. Atherosclerosis 1988; 69: 69–79.
10 Castellot JJ, Addonizio ML, Rosenberg RD: Cultured endothelial cells produce a heparin-like inhibitor of smooth muscle cell growth. J Cell Biol 1981; 90: 372–379.
11 Clowes AW, Karnovsky MJ: Suppression by heparin of smooth muscle cell proliferation in injured arteries. Nature (Lond) 1977; 265: 625–626.
12 Assoian RK, Komoriya A, Meyers CA, Miller DM, Sporn MB: Transforming growth factor-beta in human platelets. J Biol Chem 1983; 258: 7155–7160.
13 Assoian RK, Sporn MB: Type β transforming growth factor in human platelets: release during platelet degranulation and action on vascular smooth muscle cells. J Cell Biol 1986; 102: 1217–1223.
14 Child CB, Proper JA, Tucker RF, Moses HL: Serum contains a platelet-derived transforming growth factor. Proc Natl Acad Sci USA 1982; 79: 5312–5316.
15 Pircher R, Jullien P, Lawrence DA: Beta-transforming growth factor is stored in human blood platelets as a latent high molecular weight complex. Biochem Biophys Res Commun 1986; 136: 30–37.
16 Roberts AB, Sporn MB, Assoian RK, Smith JM, Roche NS, Wakefield LM, Heine UI, Liotta L, Falanga V, Kehrl JH, Fauci AS: Transforming growth factor type β: rapid induction of fibrosis and angiogenesis in vivo and stimulation of collagen formation in vitro. Proc Natl Acad Sci USA 1986; 83: 4167–4171.
17 Sporn MB, Roberts AB, Wakefield LM, Assoian RK: Transforming growth factor-β: biological function and chemical structure. Science 1986; 233: 532–534.
18 Stein O, Halperin G. Stein Y: Interaction between macrophages and aortic smooth muscle cells. Enhancement of cholesterol esterification in smooth muscle cells by media of macrophages incubated with acetylated LDL. Biochim Biophys Acta 1981; 665: 477.
19 Campbell JH, Reardon MF, Campbell GF, Nestel PJ: Metabolism of atherogenic lipoproteins by smooth muscle cells of different phenotype in culture. Arterosclerosis 1985; 5: 318–328.
20 Campbell JH, Popadynec L, Nestel PJ, Campbell GR: Lipid accumulation in arterial smooth muscle cells: influence of phenotype. Atherosclerosis 1983; 47: 279–295.
21 Greisler HP, Klosak JJ, Dennis JW, Ellinger J, Kim DU, Burgess WH, Maciag T: Endothelial cell growth factor (ECGF) attachment to biomaterials. ASAIO Transactions 1986; 32: 346–349.
22 Greisler HP, Klosak JJ, Dennis JW, Karesh SM, Ellinger J, Kim DU: Biomaterial pretreatment with ECGF to augment endothelial cell proliferation. J Vasc Surg 1987; 2: 393–399.
23 Baird A, Mormede P, Bohlen P: Immunoreactive fibroblast growth factor in cells of peritoneal exudate suggests its identity with macrophage-derived growth factor. Biochem Biophys Res Commun 1985; 126: 358.
24 Shimokado K, Raines EW, Madres DK, Barrett TB, Benditt EP, and Ross R: A significant part of macrophage-derived growth factor consists of at least two forms of PDGF. Cell 1985; 43: 277.
25 Nilsson J, Sjolund M, Palmberg L, Thyberg J, Heldin C-H: Arterial smooth muscle cells in primary culture produce a platelet-derived growth factor-like protein. Proc Natl Acad Sci USA 1985; 82: 4418.
26 DiCorleto PE, Bowen-Pope DF: Cultured endothelial cells produce a platelet-derived growth factor-like protein. Proc Natl Acad Sci USA 1983; 80: 1919.
27 Hannan RL, Kourembanas S, Flanders KC, Rogelj SJ, Roberts AB, Faller DV, Klagsbrun M: Endothelial cells synthesize basic fibroblast growth factor and transforming growth factor-beta. Growth Factors 1988; 1: 7–17.
28 Campbell JH, Campbell GR: What controls smooth muscle phenotype? Atherosclerosis 1981; 40: 347–357.

29 Chamley-Campbell JG, Campbell GR, Ross R: The smooth muscle cell in culture. Physiol Rev 1979; 59: 1–61.
30 Schmidt JA, Mizel SB, Cohen D, Green I: Interleukin 1, a potential regulator of fibroblast proliferation. J Immunol 1982; 128: 2177.
31 Tosato G, Seamon KB, Goldman ND, Sehgal PB, May LT, Washington GC, Jones KD, Pike SA: Monocyte-derived human B-cell growth factor identified as interferon β_2 (BSF-2, IL-6) Science 1988; 239: 502.
32 Old LJ: Tumor necrosis factor (TNF). Science 1985; 230: 630–632.
33 Greisler HP, Dennis JW, Endean ED, Dim DU: Derivation of neointima of vascular grafts. Circulation suppl I 1988; 78: 16–112.
34 Majack RA: Beta-type transforming growth factor specifies organizational behavior in vascular smooth muscle cell cultures. J Cell Biol 1987; 105: 456–471.

Angiogenesis: Key Principles - Science - Technology - Medicine
ed. by R. Steiner, P. B. Weisz & R. Langer

Deuterium nuclear magnetic resonance imaging of the developmental pattern of tumour blood flow

Ikram A. Burney, Ross J. Maxwell, John R. Griffiths and Stanley B. Field

MRC Cyclotron Unit, Hammersmith Hospital, London W12 OHS & CRC Biomedical Magnetic Resonance Research Group, St. George's Hospital London, SW17 0RE, England

Introduction

^{31}P Magnetic Resonance Spectroscopy has been used by various workers to study tumour bioenergetic status and also to monitor tumour response to various therapeutic modalities such as chemotherapy, radiotherapy and hyperthermia. Many of the observations reported may be explained by changes in tumour perfusion and oxygenation [1–3]. Recently non-invasive ^{2}H NMR spectroscopy techniques, measuring uptake or clearance of D_2O, have been introduced, thus enabling repetitive estimation of tumour blood flow [4, 5]. Also, imaging the distribution of uptake of this stable, flow-limited tracer provides an indication of vascular competency across a tumour [6]. In the present study we have combined ^{2}H imaging of D_2O uptake with localized ^{31}P spectroscopy to make serial measurements of tumour blood flow and bioenergetic status during the growth of a transplantable rat fibrosarcoma in order to examine whether there are significant changes in metabolism and perfusion as a tumour enlarges.

Methods

Fibrosarcomas designated $LBDS_1$ were grown subcutaneously in the flanks of 8–10-week-old BD_9 rats. ^{2}H imaging and ^{31}P spectroscopy were performed on a 4.7 T, 30-cm bore, SISCO spectrometer. The rats were anaesthetized with a combination of hypnorm and midazolam in order to restrain them whilst in the bore of the magnet. A 23G butterfly needle was inserted in one of the tail veins for administration of the tracer, D_2O.

The rat was placed on a flask containing recirculating warm water to maintain the core temperature at 37°C and positioned so that its tumour hung vertically downwards over the ^{31}P surface coil. The homogeneity of the static magnetic field was optimized by shimming at the H_2O resonance. ^{1}H images were acquired to determine the position of the tumour in the magnet. Localized ^{31}P spectra were obtained from cuboid volumes of 1–1.3 cm a side, depending on the tumour size, using an ISIS pulse sequence with a 2 cm diameter surface coil. The 90° read pulse was a 500 μs sincos adiabatic pulse. Other experimental parameters included: gradient strength of 7.5 gauss cm^{-1}, pulse repetition time of 3 s, and 160 free induction decays averaged over a period of 8 min.

Deuterium images were obtained using a 4-turn solenoid coil, tunable to both ^{1}H and ^{2}H. The rat was positioned as for ^{31}P spectroscopy but the tumour hung down into the solenoid coil. After shiming on the ^{1}H signal from water, proton images were acquired to select an appropriate slice for deuterium imaging. Deuterium images were obtained using a steady state free precession (rapid imaging) sequence. Experimental parameters included: 1 ms gaussian shaped 90° pulse; 60 ms repetition time; 10 ms echo time; 32 phase-encoding steps; 64 data points; 64 scans the resolution was 1.2 mm × 0.6 mm. The field of view was adjusted to a size of 4 × 4 cm. Following a background deuterium image, 500 μl D_2O in isotonic saline was injected intravenously. Four deuterium images were then acquired starting at 30, 270, 510 and 750 s after the injection, each image being obtained in 2 min. A deuterium uptake image was then calculated based on the rate of increase of signal in each pixel.

Results

Four tumours were studied from 30 days after transplantation at which time they were approximately 0.5 cm^3. ^{31}P spectra were obtained on 30th, 37th, 43rd and 46th day after transplantation. The inorganic phosphate (Pi) peak increased in amplitude as the tumours grew in size, with a corresponding reduction in the high energy phosphate peaks, particularly βNTP, reflecting a declining tumour bioenergetic status.

Deuterium uptake images were obtained from tumours immediately following ^{31}P spectroscopy. Each image was calculated on the basis of the rate of uptake of D_2O for each pixel. The pixels with highest uptake rate of the tracer were assigned as the high intensity pixels, the ones with half the rate of D_2O uptake were assigned as intermediate intensity pixels and so on. Pixels clearly outside the tumour area were used to set a critical intensity, below which the uptake was deemed negligible and was not assigned any value. As the tumour grew in size the number of high intensity pixels decreased. Areas of low or no flow developed in the centre, and gradually spread outwards. As the tumours grew from 0.5 to approximately 8 cm^3, the number of high intensity pixels became less

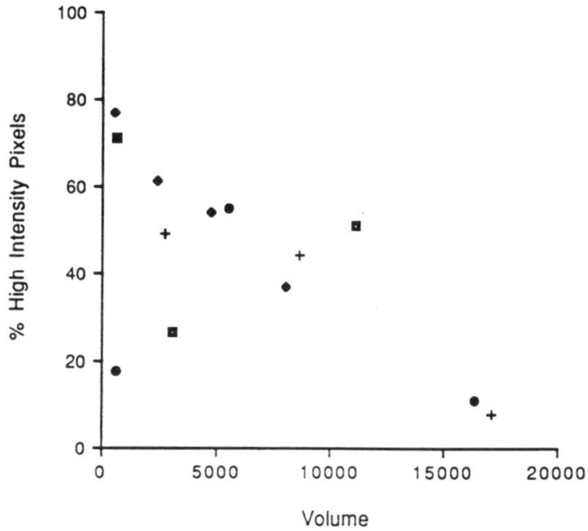

Figure 1. The percentage area of high intensity pixels as a function of tumour volume. See text for the definition of high intensity pixels. Data is from 4 rats studied either 3 or 4 times during the growth of tumours. ■, Tumour 1; ◆, Tumour 2; +, Tumour 3; ●, Tumour 4.

and less, especially at the tumour periphery. Figure 1 shows the percentage of high intensity pixels from the four tumours as a function of tumour volume.

The proportional area of tumour with high blood flow, expressed as the ratio of the area of high intensity pixels to the total cross-sectional area of the tumour was studied as a function of metabolic status of the tumour. An inverse relationship was found between the inorganic phosphate/total phosphate ($Pi/\Sigma P$) ratio, and the percentage area of high flow. High energy phosphates, expressed as βNTP/Pi, decreased in parallel with increasing tumour volume and diminishing area of high flow.

Although there was a definite trend towards decreasing area of high blood flow with corresponding deterioration in bioenergetic status of tumours with increasing size, there are also marked variations as seen in Figs. 2a and b. For example, tumour 4 shows a low number of high intensity pixels on day 30 but a high βNTP/Pi ratio, similarly tumour 1 shows a relatively low number of high intensity pixels on day 37 and high βNTP/Pi.

A deuterium uptake image was also obtained from an NMU induced mammary adenocarcinoma. The blood flow was fairly homogenous across the tumour with very few areas of low or no flow, the pattern was found to be different from the transplantable fibrosarcomas of similar size.

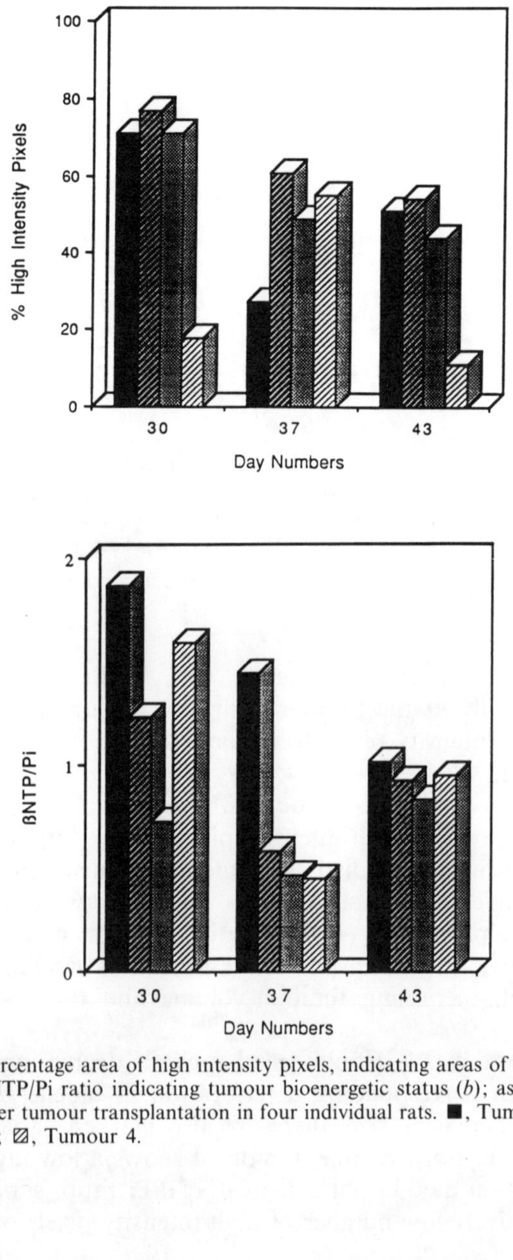

Figure 2. The percentage area of high intensity pixels, indicating areas of high tumour blood flow (*a*); and βNTP/Pi ratio indicating tumour bioenergetic status (*b*); as observed on three different days after tumour transplantation in four individual rats. ■, Tumour 1; ▨, Tumour 2; ▩, Tumour 3; ▧, Tumour 4.

Discussion

There are very few reports in the literature regarding the change in tumour blood flow and perfusion status with increasing tumour volume in different tumour models (Evelhoch et al., 1986). Our studies of direct visualization of the regional distribution pattern of blood flow by ^{2}H NMR imaging suggests a decreasing proportion of regions of high flow, particularly towards the tumour centre, with increasing tumour size in transplantable tumours, the results are consistent with the earlier observations. The changes in ^{31}P MRS parameters, i.e. decreasing Pi/Σ P and increasing βNTP/Pi, are consistent with decreasing tumour blood flow with increasing tumour size. However our preliminary results suggests a different pattern of vascular development occurs in a chemically induced mammary adenocarcinoma. This observation is consistent with the reports that there may be important differences in the vasculature which develops in primary and transplanted tumours [7]. This technique of imaging the distribution pattern of a flow-limited tracer could be usefully employed to compare the pattern of development of blood flow in other tumour models.

Repeat measurements on the same tumour revealed that the tumour bioenergetic status did not always correspond with the blood flow for all the tumours at all times, although there is a trend for both tumour metabolic status and blood flow to decline with tumour growth. This discrepancy could be explained by a number of factors, e.g. the effect of anaesthesia, maintainance of temperature within the bore of the magnet, or since ^{31}P and ^{2}H observations were not made simultaneously, to transient ischaemia of the tumour [8].

Acknowledgements. This work was supported by a grant from Medical Research Council and a grant from Cancer Research Campaign.

References

1 Evelhoch JL, Sapareto SA, Nussbaum GH, Ackerman JJH: Correlations between ^{31}P NMR spectroscopy and ^{15}O Perfusion measurements in RIF-1 murine tumour in vivo. Radiat Res 1986; 106: 122–131.

2 Vaupel P, Okunieff P, Kallinowski F, Neuringer LJ: Correlations between ^{31}P NMR spectroscopy and tissue O_2 tension measurements in a murine fibrosarcoma. Radiat Res 1989; 120: 477–493.

3 Tozer GM, Bhujwalla ZM, Griffiths JR, Maxwell RJ: Phosphorous-31 magnetic resonance spectroscopy and blood perfusion of the RIF-1 tumour following X-irradiation. Int J Radiat Oncol Biol Phys 1989; 16: 155–164.

4 Karczmar GS, Meyerhoff DJ, Speder A, Valone F, Wilkinson M, Shine MD, Weiner MW: Response of tumours to therapy studied by ^{31}P magnetic resonance spectroscopy. Invest Radiol 1989; 24: 1020–1023.

5 Kim S-G, Ackerman JJH: Quantitative determination of tumour blood flow and perfusion via deuterium nuclear magnetic resonance spectroscopy in mice. Cancer Res 1988; 48: 3449–3453.

6 Evelhoch JL, Larcombe McDouall JB, Mattielo J, Simpson NE, Seyedsadr M, McCoy CI: Deuterium NMR imaging of regional tumour blood flow in mice. 16th LH Gray Conference. 1990.

7 Field SB, Needham S, Burney IA, Maxwell RJ, Coggle JE, Griffiths JR: Differences in vascular response between primary and transplanted tumours. Br J Cancer. 1991; 63: 723–726.

8 Chaplin DJ, Olive P, Durand RE: Intermittent blood flow in a murine tumour; radiobiological effects. Cancer Res 1987; 13: 579–586.

Angiogenesis: Key Principles - Science - Technology - Medicine
ed. by R. Steiner, P. B. Weisz & R. Langer

The perfluorocarbon emulsion PFOB: potential indicator of change in blood volume using fluorine-19 NMR spectroscopy

Charles Thomas, Christopher Counsell, Janet Sansom, Pauline Wood, Ian Stratford and Gerald Adams

MRC Radiobiology Unit, NMR Laboratory, Chilton, Didcot OX11 ORD, England

Introduction

The biological response of some anti-cancer therapeutic agents is mediated via the tumour vasculature. These agents included photodynamic therapy (PDT), endocrine therapy, immunotherapy, hyperthermia, radio- and chemotherapies [1].

An important aspect is the in vivo measurement of change in blood perfusion induced by these therapeutic agents. Nuclear Magnetic Resonance (NMR) spectroscopy has been used to measure in vivo change in blood flow rate using freely diffusible tracer molecules such as D_2O [2]. Measurement of change in blood flow provides no direct estimate however of any associated change in blood volume.

This paper reports a novel approach using Fluorine-19 NMR spectroscopy in vivo to measure directly changes in the vascular perfusion volume in experimental tumours [3]. The method employs the 100% w/v emulsion of perfluorooctylbromide (PFOB) as a freely circulating blood volume indicator and the vasodilator hydralazine as a physiological modifier of blood flow to induce change in blood volume.

Material and methods

Mice and tumours

Murine KHT sarcoma and murine SCCVII squamous cell carcinoma cells (2×10^5) were implanted intradermally into the mid-dorsal pelvic region of 8–12-week-old C3H/He mice. Volume doubling time were about 2 days for the two tumours. On the day of the experiment, the volume of the tumours obtained from linear measurements of 3 diameters ($\pi/6 \times d1 \times d2 \times d3$), varied between 73 and 214 mm^3. Mice (weight = 27 ± 3 g) were anaesthetized i.p. during all experiments with a 1:1:2 mixture of hypnorm-hypnovel at a dose of 0.2 ml per 25 g body weight. Mice were kept warm inside the magnet using heating pads.

NMR spectroscopy

Mice were placed in a 4.7 Tesla, 30-cm horizontal bore super-conducting magnet (Oxford Instruments) interfaced to a SISCO 2000 spectrometer. A 7-mm external surface coil was placed over the tumour which was not less than 4 mm deep. Proton imaging was used to verify that the sensitivity of the coil did not extend below this depth. Trifluoroacetamide (CF_3CONH_2) was used as external reference for normalizing fluorine spectra. Spectra were obtained in most experiments from 256 pulses of 7 μs width at 1-s repetition time, i.e. signal was acquired in about 4 min. Signal intensity was calculated from the peak area after base line correction. Results were expressed in terms of the ratio of the intensity of the CF_3 peak of the PFOB spectrum to that of the reference peak.

Local tumour blood flow rate

The method of Xenon-133 clearance rate [4] was used for measuring changes in local tumour blood flow after hydralazine administration. Groups of eight mice bearing either KHT or SCCVII tumours were treated under anaesthesia with Xe-133 (Activity 740 MBq/ml of a saline solution). After direct tumour injection of 10 μl volumes of Xe-133 at 3 different sites, the fall in activity of the radiotracer was monitored with a whole-body γ-ray counter mounted externally over the mouse. The fall in scintillation counts was followed for 10 min with a count taken every 10 s.

Compounds

The 100% w/v emulsion of perfluorooctylbromide ($C_8F_{17}Br$) contains 52% fluorocarbon by volume [5] and was kindly supplied through Drs M. Guichard and J. G. Riess by Alliance Pharm. Corp., San Diego, California. The emulsion was given i.v. via a tail vein catheter with tubing extending to a syringe manipulated from outside the magnet. Hydralazine was injected i.v. at a dose of 5 mg/kg body weight in PBS using a second tail vein catheter. This avoided moving the mouse in the magnet during experiments.

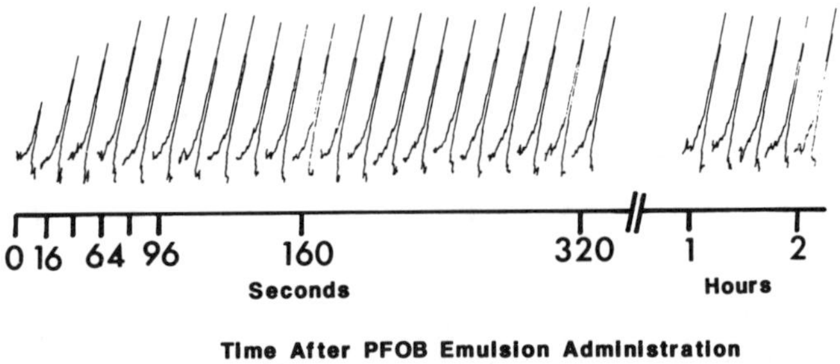

Figure 1. Illustration of the stability in time of the fluorine signal intensity from a KHT tumour after administration of the 100% w/v PFOB emulsion. A single dose of the emulsion was given (8 ml/kg) and at time 0, a second dose was given (8 ml/kg) and signal intensity monitored.

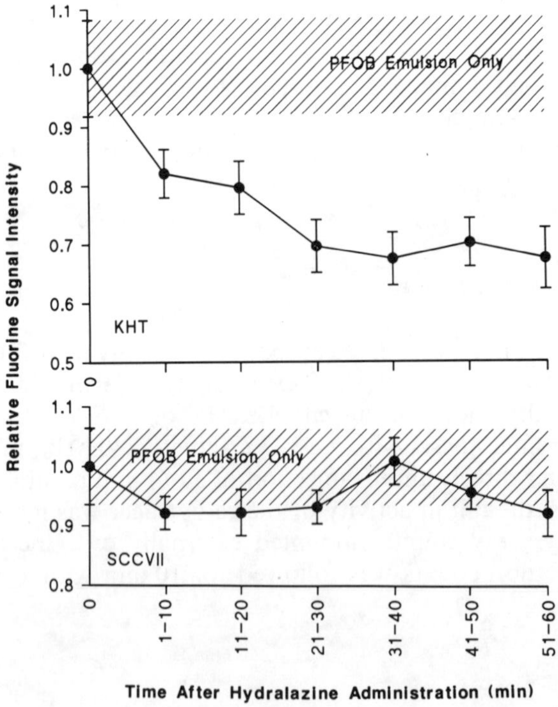

Figure 2. Time course for the response of the fluorine signal intensity from the KHT and SCCVII tumours after hydralazine administration. Mean and standard error of the mean are shown (3–6 samples per point). The hatched area shows the constancy with time of the control spectrum after PFOB emulsion administration. Reproduced from [3].

Results

Figure 1 illustrates the stability in time of the fluorine signal strength from a KHT tumour after administration of two single doses of the

Table 1. Effect of hydralazine (5 mg/kg) on Xe-133 clearance rate after *direct intratumour* injection of Xe-133 in mouse SCCVII and KHT tumours. Mean and SEM (n = 8) are shown. Reproduced from [3].

Tumour type	No hydralazine	After hydralazine
KHT	$8.03 \pm 0.73 \times 10^{-4}\,s^{-1}$	$2.93 \pm 0.48 \times 10^{-4}\,s^{-1}$*
SCCVII	$11.2 \pm 1.2 \times 10^{-4}\,s^{-1}$	$8.59 \pm 0.81 \times 10^{-4}\,s^{-1}$

*$p < 0.05$

100% w/v PFOB emulsion (2 × 8 ml/kg). It was found using 16 pulses at a repetition rate of 1 s that signal intensity equilibrated about 60 s after administration and then remained unchanged for at least 2 h.

Figure 2 shows that the signal intensity from PFOB in the KTH tumours is significantly reduced by 30% 30 min after hydralazine injection ($p = 10^{-3}$). In contrast, hydralazine caused no change in the PFOB signal from SCCVII tumours ($p = 0.4$).

Comparative studies of the effect of hydralazine on local tumour blood flow are shown in Table 1. For the KHT tumour, a 65% reduction of the mean clearance rate of Xe-133 was found after hydralazine administration ($p < 10^{-3}$) while no significant change was observed in the SCCVII tumour ($p = 0.12$). This is as expected based on the effect of hydralazine on vascular perfusion volume in these two tumours.

Discussion

A novel approach to measure direct change in blood volume was developed in this study using an emulsion of the fluorocarbon PFOB as a fluorine NMR spectroscopic tracer. The PFOB emulsion is of interest as blood volume indicator because the droplets of fluorocarbon are very small (average diameter of about 200 nm) and therefore accessible to the microcirculation [6]. After circulation of the droplets of fluorocarbon in the blood, they are distributed in the reticulo endothelial system (mainly liver and spleen) and adipose tissues. The principal route of elimination of PFOB occurred through the lungs and skin by expiration; the retention time of PFOB is dose dependent: the half life is about 4 days at a dose of of 4 g/kg [6]. The 100% w/v PFOB emulsion has also the advantage to be less toxic than other perfluorocarbon emulsions [7, 8]. Finally, PFOB gives strong NMR signal at low administered dose because of the high fluorocarbon content of the emulsion.

The speed with which the signal intensity reaches an equilibrium value (<60 s) and the stability in time of the signal intensity indicates that the PFOB emulsion is circulating freely under stationary state conditions within the vascular volume explored by the external surface coil. The fluorine NMR signal is proportional to the PFOB concentration in the vascular volume, and if during the time of observation, the signal is constant, the blood volume should be constant. Any change in signal intensity should reflect a reduction in the vascular volume. This hypothesis was tested using the vasoactive agent hydralazine which given at 5 mg/kg is known to reduce systemic blood pressure [9]. Administration of hydralazine reduced both local blood flow and blood volume in KHT tumours while no change was observed in SCCVII tumours. The results obtained with KHT tumours support the concept that hydralazine decreases blood perfusion in some experimental tumours by the steal phenomenon [10]. The presence of smooth muscle in SCCVII tumours [11] could probably explain why hydralazine has only small vascular effect in this tumour. Indeed, any reduction in blood supply due to change in blood pressure within the vessels may be compensated by the vasodilatory effects of the smooth muscle.

Thus, it is proposed that the emulsion of perfluorocarbon PFOB could be used as a fluorine-19 NMR spectroscopic indicator of the vasculature to follow quantitatively *real time* changes induced by therapeutic agents or angiogenic factors on blood volume in accessible tumours and normal tissues. This method may also be useful to detect early changes in blood volume induced by proliferation of endothelial cells in solid tumours. This spectroscopic approach should complement other vascular studies which have used fluorocarbon emulsion in NMR imaging.

Acknowledgements. The authors wish to thank John Nolan and Jill Pearson for technical help, David Papworth for statistical analysis, Paul Bates and his team for animal care and Debbie Badger for typing the manuscript. This work was partly supported by the Commission of the European Community DGXII (C. Thomas) and by the Imperial Cancer Research Fund (C. Counsell and P. Wood).

References

1 Proceedings of the XVI L. H. Gray Conference: Vasculature as a Target for Anti-Cancer Therapy, Manchester, 17–20 September 1990. Int J Radiat Biol 1991, in press.
2 Mattiello J, Evelhoch JL, Brown E, Schaap AP, Hetzel FW: Effect of photodynamic therapy on RIF-1 tumour metabolism and blood flow examined by ^{31}P and ^{2}H NMR spectroscopy. NMR Biomed 1990; 3: 64–70.
3 Thomas C, Counsell C, Wood P, Adams GE: Fluorine-19 NMR spectroscopy for measuring dynamic changes in blood perfusion volume in experimental tumours in vivo. JNCI 1991, submitted.
4 Kallman RF, de Nardo GL, Stasch MJ: Blood flow in irradiated mouse sarcomas as determined by the clearance of Xe-133. Cancer Res 1972; 32: 483–490.
5 Long DC, Riess JG, Follana R, Burgan AR, Mattrey RF, Long DM: Preparation and applications of highly concentrated PFOB emulsions. Biomat Artif Cells Artif Organs 1988; 16: 441–442.

6 Riess JG, Le Blanc M: Preparation of perfluorochemical emulsion for biomedical use: principles, materials and methods. In Lowe KC (ed): Blood substitute, Preparation, Physiology and Medical Applications. Ellis Horwood, Chichester, 1988, pp 94–125.
7 Burgan AR, Herrick WC, Long DM, Long DC: Acute and subacute toxicity of 100% PFOB emulsion. Biomat Artif Cells Artif Organs 1988; 16: 681–682.
8 Long DM, Long DC, Mattrey RF, Long RA, Burgan AR, Herrich WC, Shellhamer DF: An overview of perfluorooctyl-bromide – Application as a synthetic oxygen carrier and imaging for X-ray, ultrasound and nuclear magnetic resonance. Biomat Artif Cells Artif Organs 1988; 16: 411–420.
9 Okunieff P, Walsh CS, Vaupel P, Kallinowski F, Hitzig BM, Neuringer LJ, Suit HD: Effects of hydralazine on in vitro tumor energy metabolism, hematopoietic radiation sensitivity, and cardiovascular parameters. Int J Rad Biol Oncol Phys 1989; 16: 1145–1148.
10 Jirtle RL: Chemical modification of tumour blood flow. Int J Hyperth 1988; 4: 355–371.
11 Trotter MJ, Chaplin DJ, Olive PL: Mechanisms for intermittent blood flow in experimental tumours. In Proceedings of the XVI L. H. Gray Conference: Vasculature as a Target for Anti-Cancer Therapy, Manchester, 17–20 September 1990. Int J Radiat Biol 1991; 60 (No. 1/2).

Angiogenesis: Key Principles - Science - Technology - Medicine
ed. by R. Steiner, P. B. Weisz & R. Langer

Clinical investigations on blood perfusion in human malignancies of the pelvis and abdomen: significance for tumor therapy

Michael Molls[1], Horst Jürgen Feldmann[1] and Klaus Sievers[2]

Department of Radiotherapy[1] and Radiodiagnostics[2], West German Tumor Center, University Hospital, 4300 Essen, FRG

Abstract. Blood flow of deep pelvic and abdominal tumors was investigated with the thermal clearance method, dynamic CT and dynamic MRI. There are good correlations between the measurement values obtained by these methods. A low flow was observed in rectal cancer and soft tissue sarcoma in contrast to pancreatic cancer and hypernephroma. The temperature increase induced by regional hyperthermia was dependent on the individual tumor blood flow. Dynamic CT can be used pretherapeutically and predict the quality of a heat treatment, which is important with regard to concepts consisting in radiotherapy or chemotherapy plus hyperthermia.

Introduction

Experimental studies and first clinical observations indicate that biological properties of a tumor such as blood flow, oxygen concentration, nutrient supply, bioenergetic status and pH, influence the tumor response after irradiation, hyperthermia and radiotherapy [1–4]. Hypoxia correlates with resistance towards irradiation but sensitivity towards heat [5]. At least, the bioavailability of cytostatic drugs depends on the blood perfusion of a tumor. We have started to investigate blood flow related parameters in deep seated malignancies with the aim to characterize physiological properties of human tumors in situ. Furthermore, we wanted to find out whether the quality of a clinical heat treatment depends on the blood perfusion of a tumor.

Material and methods

37 patients with deep seated tumors received a palliative treatment consisting in the combination of radiotherapy and hyperthermia. Details of the treatment procedure, especially of deep regional hyperthermia and of intratumoral thermometry during a heat treatment are described elsewhere [6–9]. The patients suffered from pelvic recurrences of rectal cancer (n = 19), pelvic soft tissue sarcomas (n = 6), pelvic recurrences of cervical cancer (n = 6), large abdominal pancreatic cancers (n = 2), pelvic metastases of hypernephroma (n = 2), a locally

advanced bladder cancer (n = 1) and a large lymphoma of the lower leg (n = 1).

Blood flow related data were obtained by using three different methods.

Thermal clearance: After termination of heat supply at the end of a clinical hyperthermia treatment, blood flow was determined from the thermal clearance (washout of heat) measurements in tumor centers and peripheries by use of the bioheat transfer equation [3, 10].

Dynamic CT: The investigation was performed pretherapeutically. The enhancement of contrast-material after bolus injection was plotted over time and registered for the same tumor locations for which the washout rate was determined. The regions of interest were again the periphery and the center of a tumor. The enhancement of contrast-material (ΔHU) over the baseline in the early phase (30 s) after bolus injection is a reliable parameter for blood perfusion [11].

Dynamic MRI: Patients were investigated pretherapeutically by contrast-enhanced magnetic resonance imaging (1,5 T superconducting magnet). Dynamic studies were performed using a FLASH (fast low-angle shot) gradient echo sequence. After the first FLASH sequence, an intravenous Gadolinium-DTPA bolus injection was rapidly administered, and further FLASH sequences were acquired. The Gadolinium-induced enhancement of signal intensity versus noise (ΔS/N) in the early phase after bolus injection is a reliable parameter for the perfusion in the regions of interest. In the MRI investigations the regions of interest were comparable with those of the dynamic CT studies.

Results

Table 1 shows that there exist relevant differences in the flow rates of different tumor entities (in particular a low flow in rectal cancers and

Table 1. Blood flow related data for tumor centers and peripheries (all tumor entities) and different tumor entities. The flow is given in ml/100 g-min (thermal clearance method) and ΔHU (dynamic CT). The temperature increase obtained in the clinical hyperthermia treatments is given in ΔT [°C]. In general, at the treatment start the intratumoral temperature was 37°C. N = number of investigations.

	Flow rate [ml/100 g-min]		ΔT [°C]	Flow rate [ΔHU]	
Tissue category	Mean ± SD	N	Mean ± SD	Mean ± SD	N
Tumor center	6.7 ± 3.5	58	5.3 ± 2.3	6.9 ± 4.7	27
Tumor periphery	12.2 ± 4.9	110	3.5 ± 1.3	14.5 ± 11.7	42
Rectal cancer	8.9 ± 4.0	75	4.5 ± 1.6	7.9 ± 3.9	24
Soft t. sarcoma	8.8 ± 3.9	45	4.8 ± 2.0	8.3 ± 4.2	15
Cercival cancer	11.3 ± 2.9	21	3.5 ± 0.8	10.9 ± 6.9	17
Hypernephroma/ Pancreatic cancer	19.0 ± 6.3	17	2.1 ± 0.6	36.9 ± 9.0	7
Lymphoma	15.8 ± 1.6	6	3.6 ± 1.0	10.0	2
Bladder Cancer	4.4 ± 0.6	4	5.7 ± 0.6	13.0 ± 4.4	4

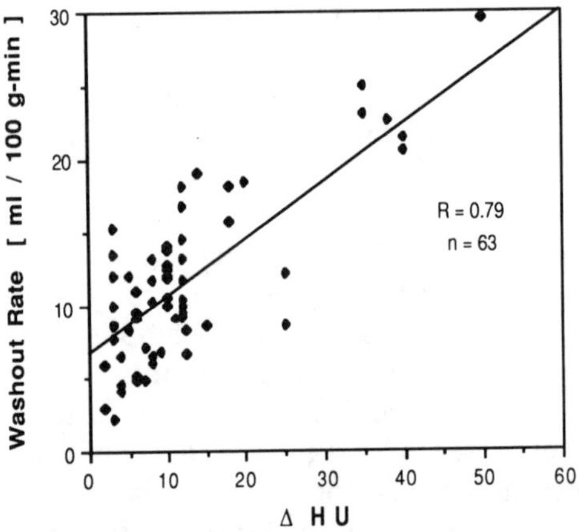

Figure 1. Correlation between thermal washout rates and ΔHU (increase of Houndsfield units) values obtained from 63 measurements in 22 tumors

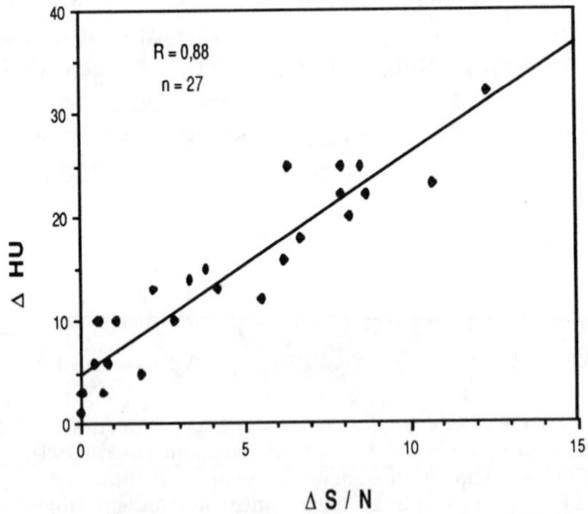

Figure 2. Correlation between ΔHU values (dynamic CT) and ΔS/N values (dynamic MRI) obtained from 27 measurements in 10 tumors. ΔS/N = increase of signal/noise ratio.

soft tissue sarcomas). In addition, when compared with tumor peripheries the lower flow occurs in tumor centers. Obviously, in those tumors (entities) or tumor regions (center vs periphery) with relatively low perfusion, higher temperatures can be induced by hyperthermia treatment (Table 1). Finally, there was a good correlation between dynamic CT data and washout rates as well as between dynamic CT data and results obtained by MRI (Figs. 1 and 2).

Discussion

Blood flow related data in deep seated tumors of the pelvis or lower abdomen were obtained by using the isotope clearance technique or the thermal washout procedure [12, 13]. However, to the best of our knowledge, our own studies provide, for the first time, results which characterize blood perfusion of different entities of deep seated tumors in more detail. The different flow rates explain the differences in temperature increase obtained for different tumor entities, tumor centers and tumor peripheries. With regard to treatment outcome after thermoradiotherapy, it is important to induce comparatively high intratumoral temperatures as there exist correlations between clinical tumor regressions and quality of heat treatments [14–16]. Concluding from our own results we would recommend to perform dynamic CT before starting with an additive hyperthermia treatment (thermoradiotherapy, thermochemotherapy). Although tumors with a high perfusion such as hypernephromas or pancreatic cancers are obviously not very suitable for hyperthermic treatment, it is appropriate to perform dynamic CT with a predictive intention also in patients suffering from these malignancies. There are indications that significant differences of blood perfusion exist not only between entities but also within entities. Thus, the pretherapeutic investigation of blood flow (dynamic CT, MRI, Positron emission tomography) can contribute to a more individual planning of treatment (decision for or against hyperthermia in addition to radiotherapy or chemotherapy). Up until now it is unknown whether such investigations will have a predictive value also for radiotherapy or chemotherapy alone. However, it is anticipated that in future, predictive assays and investigations will generally play a major role in clinical oncology [2, 3, 4].

Finally, we would like to point out that investigations on tumor blood flow in individual patients are a necessary prerequisite for therapy strategies which include drugs that influence tumor perfusion.

References

1 Vaupel P, Kallinowski F: Blood flow, oxygen and nutrient supply, and microenvironment of human tumors: A review. Cancer Res 1989; 49: 6449–6465.

2 Molls M, Feldmann HJ, Sack H: The biological basis of thermoradiotherapy: The clinician's view. J Cancer Res Clin Oncol 1990; 116: 954.

3 Molls M, Feldmann HJ: Clinical investigation on blood flow in malignant tumors of the pelvis and abdomen, in Vaupel P and Jain RK (eds.): Tumor Blood Supply and Metabolic Microenvironment: Characterization and Implications for Therapy, Stuggart, Fischer Verlag, 1991, in press.

4 Molls M, Feldmann HJ, Streffer C: Physiological, glycolytic and cytogenetic parameters as predictors for heat treatment, in Chapman JD, Dewey WC, Whitmore GF (eds.): Radiation Research, A Twentieth-Century Perspective, Vol. II, San Diego, New York, Boston, London, Sydney, Tokyo, Toronto, Academic Press, 1991, in press.

5 Overgaard J: The current and potential role of hyperthermia in radiotherapy. Int J Radiat Oncol Biol Phys 1989; 16: 535–549.
6 Müller R-D, Feldmann HJ, Molls M, John V, Stuschke M: Computertomographisch gesteuerte Positionierung von Kathetern zur Temperaturmessung bei der Hyperthermie maligner Tumoren. Strahlenther Onkol 1988; 164: 593–601.
7 Molls M: Meeting report: Deep local and regional hyperthermia. J Cancer Res Clin Oncol 1989; 115: 207–210.
8 Molls M, Feldmann HJ, Adler S, Sack H: Regional hyperthermia – a feasibility study. Strahlenther Onkol 1989; 165: 717–720.
9 Feldmann HJ, Molls M, Adler S, Meyer-Schwickerath M, Sack H: Hyperthermia in eccentrically located pelvic tumors: excessive heating of the perineal fat and normal tissue temperatures. Int J Radiat Oncol Biol Phys 1991; 20: 1017–1022.
10 Lagendijk JJW, Hofman P, Schipper J: Perfusion analyses in advanced breast carcinoma during hyperthermia. Int J Hyperther 1988; 5: 479–495.
11 Claussen C, Lochner B (eds.): Dynamische Computertomographie. Berlin, Heidelberg, New York, Springer Verlag, 1983, pp. 16–24.
12 Nyström C, Forssman L, Roos B: Myometrial blood flow studies in carcinoma of the corpus uteri. Acta Radiol Ther 1969; 8: 193–198.
13 Samulski TV, Fessenden P, Valdagni R, Kapp DS: Correlations of thermal washout rate, steady state temperatures, and tissue type in deep seated recurrent or metastatic tumors. Int J Radiat Oncol Biol Phys 1987; 13: 907–916.
14 Oleson JR, Dewhirst MW, Harrelson JM, Leopold KA, Samulski TV, Tso CY: Tumor temperature distributions predict hyperthermia effect. Int J Radiat Oncol Biol Phys 1989; 16: 559–570.
15 Issels RD, Prenninger SW, Nagele A, Boehm E, Sauer H, Jaunch K-W, Denecke H, Berger H, Peter K, Wilmanns W: Ifosfamide plus etoposide combined with regional hyperthermia in patients with locally advanced sarcomas: A phase II study. J Clin Oncol 1990; 8: 1818–1829.
16 Feldmann HJ, Molls M, Heinemann H-G, Romanowski R, Krümpelmann S, Budach V, Sack H: Thermoradiotherapy of pelvic and abdominal tumors: toxicity, thermal parameters and treatment results. Strahlenther Onkol 167; 1991: 347–348.

Angiogenesis: Key Principles - Science - Technology - Medicine
ed. by R. Steiner, P. B. Weisz & R. Langer

A bone tissue integrated single fibre laser Doppler flowmeter probe

E. Göran Salerud[a,]* and Sölve Hellem[b]

[a]*Department of Biomedical Engineering, and* [b]*Department of Oral and Maxillo-facial Surgery, University Hospital, Linköping University, S-581 85 Linköping, Sweden*

Introduction

Laser Doppler Flowmetry (LDF) is a technique in which physical properties of monochromatic light is used to estimate the flux of particles in biological tissues. The Doppler effect, absorption and scattering of light waves are the most useful physical phenomenons. The outline of the technique was first described by Stern [1] for skin tissue and today numerous microvascular beds and almost all clinical disciplines are represented. New and enhanced waveguides together with a more intelligent signal processing [2, 3] have greatly improved LDF technology. Introduction of near infrared laser diodes points out a new generation of LDF [4], portable and inexpensive. The technique is still in its infancy but the exponentially growing number of scientific reports shows an exciting future. One particular tissue of interest to investigate is bone. However, preference has been given to microvascular studies in other tissues and until recent years bone was a minor topic in the literature on microcirculation.

The theory of Havers [5] suggested vascular canals in the bone tissue and has been confirmed as fundamental in vascular topography. The physiological aspects of blood flow in bone where the live performance of flow modalities is most essential are however a challenge to microvascular research. Many techniques have been proposed to measure blood perfusion in bone tissue but none except for Laser Doppler Flowmetry fulfil most of the requirements for experimental research and clinical use [6, 7]. In previous studies with LDF large-sized probes creating trauma to the tissue and waveguides with axial asymmetry were used. The development of needle-probes and single fibre probes [3] has helped in overcoming many of these problems. To solve the special problems concerning a less traumatic measurement of perfusion in deep tissue

*To whom all correspondence should be addressed.

layers an integrated Single Fibre Laser Doppler Flowmeter (SFLDF) technique was developed.

Material and methods

Theory of LDF operation

Absorption and scattering are the two most common processes when electromagnetic radiation interacts with biological tissue. In absorption the photon ceases to exist and the energy is converted to heat. The photon scattering is a directional change of the light wave propagation. The change in amplitude of the electromagnetic wave is related to the number of scatters, while the shift in frequency is due to movement of the scatters. This shift in wavelength is known as the Doppler effect and proportional to the velocity of the scatter. Most photons undergo several collisions within the tissue matrix before they are absorbed or scattered away. These multiple interactions give the photons random direction, and therefore scatters are hit at all angles and no predominant direction is found. The number of collisions depend on tissue heterogeneity.

The flux or perfusion in the volume under study can be described by equation (1) [8].

$$\text{Flux} = \int_0^\infty \bar{v} N(v)\, dv \qquad (1)$$

where $\bar{v}$ is the mean velocity and N(v) the velocity distribution of the scatters. The implementation of the theory is shown in Figure 1.

Laser light is injected into an efferent optical fibre that transmits light to the tissue under study. In the tissue matrix the photons interact with moving scatters as well as with static structures.

A portion of the photons undergo Doppler broadening. A fraction of the frequency shifted photons are backscattered and picked up and transmitted through one or more afferent optical waveguides. They

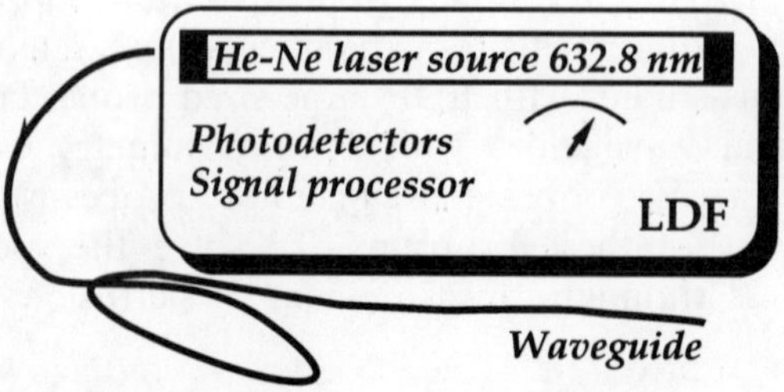

Figure 1. Schematic diagram of LDF.

finally impinge on the photodetectors and are converted to electrical currents. A signal processing unit extracts the portion of the photocurrent related to blood flow and finally generates an output signal related to the flux. The output signal from the laser Doppler flowmeter equipment is described in equation 2.

$$\text{LDF output signal} = f(c) \int_{\omega_1}^{\omega_2} \omega P(\omega)\, d\omega \qquad (2)$$

where ω is the weighting function and $P(\omega)$ the power spectral density of the photocurrent. At high fractional hematocrit values multiple scattering occurs and the output signal becomes nonlinear. Therefore the output flux has to be adjusted with a correction factor [2], f(c), dependent on the fractional hematocrit value. Theoretically it can be shown that

$$\int_0^{\infty} \bar{v} N(v)\, dv = f(c) \int_{\omega_1}^{\omega_2} \omega P(\omega)\, d\omega \qquad (3)$$

Tissue-integrated single fibre laser Doppler flowmeter probe

The technique rests on the standard signal processing algorithms found in most LDF's. The most obvious difference is the waveguide probe, containing only one fibre-optic waveguide. This waveguide transmits light bidirectionally without mixing its inherent frequencies. The tissue-integrated probe consists of a short waveguide, shielded by a hollow titanium screw. The optical waveguide connected to the equipment can be attached end-face to end-face to the tissue-integrated probe leaving the tissue to be investigated without any physical contact from the outside.

The tibias of adult rabbits were selected as the model for the experimental set-up. In both tibias four pure titanium probes were implanted and left for osseointegration. When measurements were performed a small incision was made in the skin covering the osseointegrated titanium probes and the waveguide coming from the equipment could easily be snapped on to the head of the probe. Of the four titanium probes implanted, two were perforating the first cortical layer and brought in contact with the bone marrow. The other two probes penetrate through the bone marrow and attach to the inner surface of the opposite cortical layer. Two of the probes were hollow titanium probes with no optical lens and an external waveguide could be inserted and brought into contact with the tissue to be investigated and was used as reference. Figure 2 shows the schematic *longitudinal section of the tibia.*

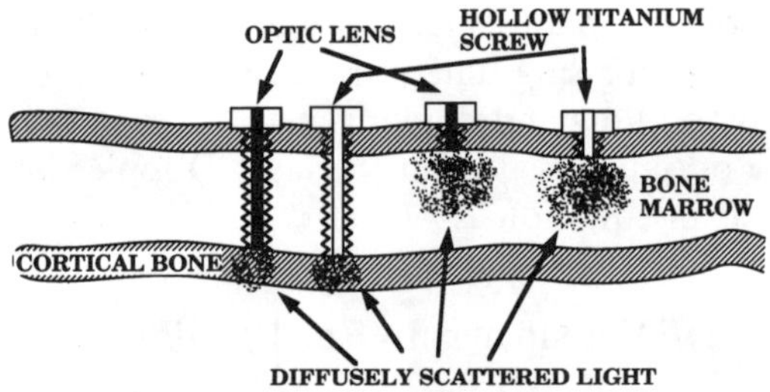

Figure 2. From left to right number one and three contains an integrated optical lens, number two and four are hollow probes. Short probes for bone marrow perfusion, long probes for cortical perfusion.

Results

The perfusion in cortical bone and bone marrow was measured at four different occasions. The first measurement was performed 3 weeks after the implantation and a second recording two weeks later. This was to ensure reliable data collection. The last two measurements were carried out 4 months after the probes were implanted. At every measurement at least one animal was sacrificed for histological documentation of the tissue integration.

The tissue volume under study in Figure 3 shows the bone marrow four months after implantation. Fat cells and sparse vessels can be seen in close contact with the probe without any signs of inflammation.

Figure 4 shows the probe reaching the second cortical layer. The distance between probe and cortical layer is less than 100 μm. This small gap does not interfere with the measurement volume of the cortical layer. An optimum tissue integration can be seen.

The response from the tissue integrated probe was instantaneous and the signal-to-noise ratio was higher compared to in earlier studies. Inserting the reference probe, the perfusion level decreases almost to zero in both cortical bone and bone marrow. The initial perfusion level was reestablished 2–3 minutes later.

The "normal" or resting value was measured with the osseointegrated probe at the beginning and the end of each experiment session which lasted for about four hours. A slow lowering of the value was seen but not found statistically significant and is probably a secondary effect of the pharmacological stress. A large difference in perfusion value was observed between cortical bone and bone marrow. A calculated mean ratio of 16.2 was found and the corresponding confidence interval was set to 10.9–21.4.

Time course for resting blood perfusion in one rabbit is shown in Figure 5. Bone marrow perfusion shows a high initial value but de-

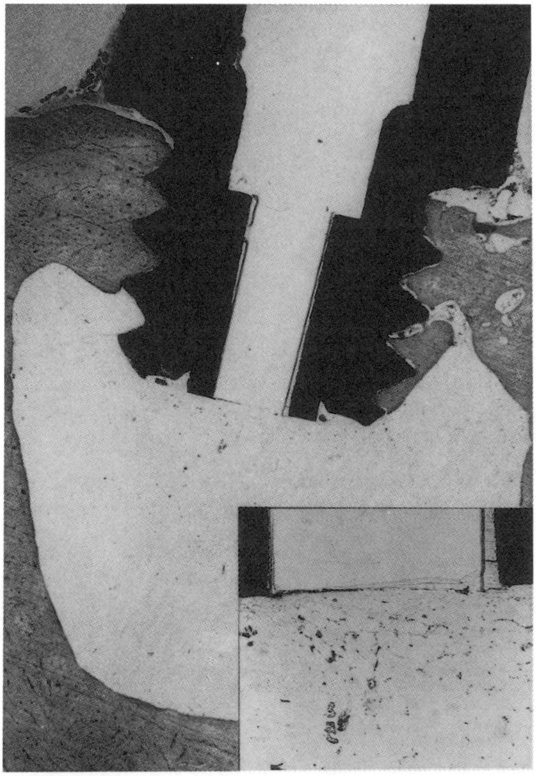

Figure 3. Cross-section of tibia and the short osseointegrated optical probe protruding the first cortical layer [H & E × 10]. Inset – interface between optical lens (∅ = 500 μm) and bone marrow [H & E × 25].

creases to a steady state level at the end of the experiment (4 months). No dramatic changes are seen in the cortical perfusion levels. Making an estimate, the ratio between cortical perfusion versus bone marrow perfusion is approximately fivefold.

Summary and conclusions

Laser Doppler Flowmetry was developed and evaluated mainly for measuring blood flow in exposed or superficial tissue layers like skin. Results showed large variation coefficients between juxtaposed skin sites with apparently homogeneous blood perfusion [9]. Spatial differences in blood flow as well as anatomical differences probably explain the obtained results. Vasomotion [10], hereto only discovered with a microscopic technique, was recorded and evaluated for the first time. Similar results had been obtained in previous bone perfusion studies [6]. The

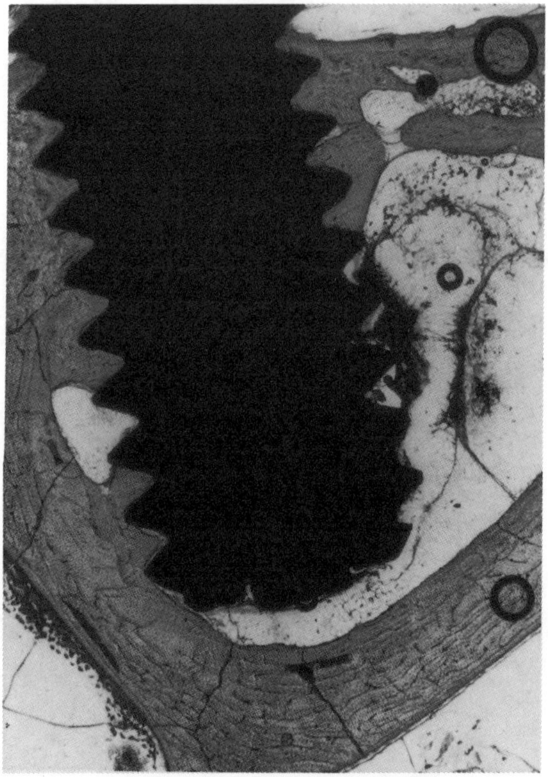

Figure 4. Cross-section of tibia and the long osseointegrated optical probe protruding through the bone marrow and reaching the cortical bone layer [H & E × 10].

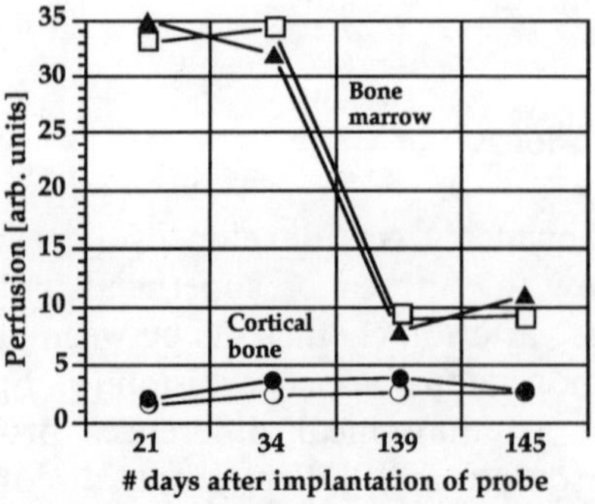

Figure 5. The bone blood perfusion recorded in one rabbit, 3–6 weeks and 4 months after implantation. Filled marks represent start of experiment and open marks four hours later.

LDF technique has also been used extensively for skin irritancy tests [11] and microvascular effects of vasoactive substances and drugs [12]. An *in vivo* model in calibrating LDF quantitatively [13] showed a linear relationship between LDF output signal and total blood flow in the range 0–300 ml min^{-1} 100 g^{-1}. This quantitative calibration is however tissue specific.

Most tissues of interest are not superficial. The single fibre LDF provides additional power in studying deep tissue physiology and trauma-induced changes in regulation of microvascular perfusion [14, 15]. Furthermore long-term studies require implantable and biocompatible probes and this could be fulfilled using the SFLDF technique in combination with a tissue-integrated titanium probe.

Implantation of the titanium probes initially creates some trauma to the vascular bed in the bone marrow but almost none in the opposite cortical bone layer. The healing process enhances the blood perfusion during the first weeks. The increased perfusion can be due to different parameters such as angiogenesis. The histological findings support this hypothesis. This gives preference to the tissue-integrated probe where interference with the tissue should be avoided. The integrated probe technique has only been evaluated in bone tissue. Modification of the probe shape and dimension, however, would make it suitable for other tissues as well.

References

1 Stern MD: In vivo evaluation of microcirculation by coherent light scattering. Nature 1975; 254: 56–58.
2 Nilsson GE: Signal processor for laser Doppler tissue flowmeters. Med & Biol Eng & Comput 1981; 22: 343–348.
3 Salerud EG, Öberg PÅ: Single-fibre laser Doppler flowmetry. A method for deep tissue perfusion measurements. Med & Biol Eng & Comput 1987; 25: 329–334.
4 Shepherd AP, Riedel GL, Kiel JW, Haumschild DJ, Maxwell LC: Evaluation of an infrared laser Doppler blood flowmeter. Am J Physiol 1987; 252: G832–G839.
5 Havers C: Citation from: Brookes M (ed): The blood supply of bone. Butterworths, London 1971.
6 Hellem S, Jakobsson LS, Nilsson GE, Lewis DH: Measurement of microvascular blood flow in cancellous bone using laser Doppler flowmetry and ^{133}Xe-clearance. Int J Oral Surg 1983; 12: 165–177.
7 Hellem S, Jakobsson LS, Nilsson GE: Microvascular reactions in cancellous bone to halothane-induced hypotension in pigs. Int J Oral Surg 1983; 12: 178–185.
8 Bonner R, Nossal R: Model for laser Doppler measurements of blood flow in tissue. Appl Optics 1981; 20: 2097–2107.
9 Tenland T, Salerud EG, Nilsson GE, Öberg P-Å: Spatial and temporal variations in human skin blood flow. Int J Microcirc: Clin Exp 1983; 2: 81–90.
10 Salerud EG, Tenland T, Nilsson GE, Öberg P-Å: Rhythmical variations in human muscle blood flow. Int J Microcirc: Clin Exp 1983; 2: 91–102.
11 Nilsson GE, Otto U, Wahlberg JE: Assessment of skin irritancy in man by laser Doppler flowmetry. Contact Dermatitis 1982; 8: 401–406.
12 Fagrell B: Microcirculatory Methods for Evaluating the Effect of Vasoactive Drugs in Clinical Practice. Acta Pharmacol Toxicol 1986; 59: 103–107.

13 Ahn H, Lindhagen J, Nilsson GE, Salerud EG, Jodal M, Lundgren O: Evaluation of Laser Doppler Flowmetry in the assessment of intestinal blood flow in cat. Gastroenterology 1985; 88: 951–957.
14 Kvernebo K, Salerud EG: Single fibre laser Doppler flowmetry in the evaluation of human muscle blood flow. In Microcirculation – an update, vol 1 (Tsuchiya et al., Eds), Amsterdam; Elsevier Science Publishers. 1987; 335–338.
15 Kvernebo K, Staxrud LE, Salerud EG: Assessment of Human Muscle Blood Perfusion with Single-Fiber Laser Doppler Flowmetry. Microvasc Res 1990; 39: 376–385.

Angiogenesis: Key Principles - Science - Technology - Medicine
ed. by R. Steiner, P. B. Weisz & R. Langer

Induction of angiogenesis by growth factors: relevance to pancreatic islet transplantation

John I. Stagner and Ellis Samols

Veterans Administration Medical Center and University of Louisville, Louisville, KY, USA

Abstract. Biodegradable pellets releasing 20 ng/day of endothelial cell growth factor alpha (αECGF) or a- or b-fibroblast growth factor (FGF) and 90 μg/day of heparin were implanted beneath the renal capsule in rats and dogs and the muscularis/serosal border of the pyloric stomach in dogs to test for angiogenesis in a potential pancreatic islet transplant site. These factors were also tested in vitro to determine whether the capillary bed of the isolated islet could be preserved. αECGF was superior to a- or bFGF in promoting endothelial cell growth and capillary formation in isolated islets. Both a- or bFGF and αECGF induced the development of a dense capillary bed in the dog stomach, whereas in the kidney site αECGF was more effective in the rat than was a- or bFGF. Priming the isolated islet as well as the transplant site prior to islet transplantation resulted in islet blood flow being established within 3 days in contrast to 7–14 days in controls.

Introduction

The transplantation of pancreatic islets as a means of treating both insulin- and noninsulin-dependent diabetes has met with limited success [1, 2]. It has been suggested that a primary reason for nonimmmune graft failure may be the lack of survival of an adequate mass of insulin producing islet β cells. Because the revascularization of transplanted islets is dependent upon vessels derived from the host [3–5], a process requiring from 7 to in excess of 30 days, it is our hypothesis that islet and β cell ischemia may be responsible for the early failure of islet transplants. This study was undertaken to determine whether capillary growth could be induced at a potential islet transplant site or within the isolated islet before islet transplantation in an effort to promote enhanced and rapid revascularization and thereby decrease the period of islet ischemia after transplantation.

Material and methods

In vivo studies

Biodegradable pellets (Innovative Research of America) releasing 20 ng/day of αECGF or a- or bFGF and 90 μg/day of heparin were implanted beneath the kidney capsule of Lewis rats and dogs and beneath the muscularis/serosal border of the dog pyloric stomach. Placebo pellets releasing 90 μg/day of heparin were implanted in the opposite kidney or opposite side of the stomach. In a separate study, pellets were prepared to release 20 μg/day of dipyridamole in addition to heparin and the growth factors to increase local levels of adenosine [6]. The respectve tissues were harvested and processed for histological evaluation 14 days after pellet implantation. All surgical procedures were performed under surgical levels of anesthesia induced by sodium pentobarbital.

In vitro studies

Rat islets were isolated by collagenase digestion followed by a dextran gradient centrifugation step [7]. Islets from 2–4 rats were pooled and were divided among culture dishes. Islets were cultured for 7 days at 37°C in RPMI-1640 medium containing 10% fetal calf serum and 90 μg/ml heparin in the presence or absence of 20 ng/ml of αECGF or a- or bFGF. The respective media were changed every other day.

Results

In vivo studies

No evidence of capillary growth was observed in the presence of placebo pellets in the dog stomach nor in the renal subcapsular space in the dog or rat. In the rat and dog, αECGF promoted the production of a dense capillary bed surrounding the pellet which extended into and beyond the serosa of the kideny and stomach. In the rat or dog kidney a- or bFGF induced significant neovascularization surrounding the pellet compared with placebo pellets. However the capillary bed was much less developed than in the presence of αECGF. In the dog stomach a- and bFGF induced the development of a dense capillary bed equivalent to that obtained by αECGF. The presence of dipyridamole appeared to enhance the angiogenic potential of both ECGF and the FGFs. In each site and species, capillary formation was noted after 7 days. Angiogenesis and continued maturation of the vessels continued throughout the 14 days of study.

In vitro studies

Isolated rat islets cultured for 7 days without ECGF did not show evidence of capillary growth nor any visible signs of the presence of intra-islet capillaries in whole islets of histological sections. These control islets maintained their spheroidal shape and smooth outline during the 7 days of study. In contrast, islets incubated in the presence of ECGF demonstrated a time-dependent proliferation of intra-islet capillaries and endothelial cells. Endothelial budding at the islet surface was noted by day 3 in culture, with maturation into endothelial tubes by days 5 and 7. The capillary bed was modestly preserved in the presence of a- and bFGF as observed in histological sections, although no new antgiogenesis was obtained.

Discussion

The islet vasculature is central to islet regulation and function as well as transplanted islet survival. Immediately after islet isolation and until the establishment of an afferent blood supply, the islets must be considered to be avascular and to be dependent upon diffusion for oxygen and nutrient supply. The native capillary bed within the islet collapses, dies and is expelled from the islet within 24–48 h after islet isolation. As shown by electron microscopy, few if any endothelial cells are present within the islet after 5–7 days in tissue culture [8]. Because the diffusion of oxygen in tissue is limited by the number of cell layers [9], the central core of insulin-producing β cells remain ischemic until blood flow is reestablished. It is these cells which are responsible for the majority of insulin secretion and are at the most risk from anoxia [10, 11]. The continued loss of interior β cells before and after transplantation may be in part responsible for the progressive diminution of β cell function reported after islet transplantation [12–15] and for the large numbers of islets required for transplantation [15] in order to compensate for what may be β cell deficient islets.

The importance of the rapid establishment of a blood supply to transplanted islets in an effort to preserve islet tissue is immediately appreciated. However the path of the blood flow within the transplanted islet has not been previously considered. Studies in the perfused rat, dog, monkey and human pancreas [16–18] have shown that the vascular sequence of intra-islet cellular perfusion and interaction is important in normal islet function. Despite the superficial differences in islet morphology, the above resports have shown that the sequence of interaction and perfusion is from the islet β cell to the α cell and thence to the δ cell. The microvascular system of the islet is lost after islet isolation and is replaced by vessels derived from the host. Whether the original sequence of intra-islet cellular perfusion is restored after revascularization is not known. However a recent study suggests that the native pattern is not reestablished as the host vessels do not penetrate the islet but form a glomerular network surrounding the islets [19]. A derangement of the path of blood flow, known to be responsible for normal islet hormone secretion, in a transplanted islet may significantly affect islet function and may be partially responsible for the reported delayed and diminished β cell response to glucose [12–15]. Whether the normal vascular sequence is maintained in islets in which the capillary bed has been preserved by treatment with angiogenic growth factors is not known. Initial short-term studies have suggested that the pattern is maintained [20]. However a long-term study must be conducted to determine whether the islet capillaries persist or are replaced by the host, possibly serving as a framework for the new vessels.

The presence of islet capillaries may reduce the time required for revascularization and may preserve islet function. However the effects of the islet capillary or endothelium upon immune rejection are controversial. It has not been clearly demonstrated that the endothelial cells are normally antigenic or Ia positive [21–24], although Ia antigens are inducible in the presence of pharmacological concentrations of interleukin-I [24]. The endothelium is important in the inflammatory response and in immune rejection by allowing or promoting the passage of monocytes into the transplanted tissue [24]. Recent studies have shown that the host-derived microvascular system of xenogenic transplanted islets as well as the vessels within the acinar tissue of transplanted pancreata are attacked and destroyed in rejection episodes prior to the destruction of islet tissue [25, 26]. The question of the possible contribution of the islet capillary endothelium to islet rejection may now be addressed as we have shown that it is possible to preserve the native islet microvascular system by treatment with angiogenic growth factors.

The use of growth factors in the induction of targeted angiogenesis may have relevance in many areas of biology and medicine, including the transplantation of tissues such as the pancreatic islet which are dependent upon the microvasculature for normal function as well as survival. The present study has shown that it is possible to preserve and to promote the growth of the islet capillary bed and to reduce the time required for islet revascularization by priming the islet and tissue site prior to islet transplantation. These studies will be of use in the examination of immune rejection and the contribution of the endothelial cell to the process of immune rejection.

Acknowledgements. Research supported through the Veterans Administration and The Diabetes Research and Education Foundation.

References

1 Sutherland DER, Moudry KC, Fryd DS: Pancreas transplant registry report. Diabetes 1989; 38: 46–54.

2 University of Michigan Pancreas Transplant Evaluation Committee: Pancreatic transplantation as a treatment of insulin-dependent diabetes mellitus. Diabetes Care 1988; 11: 669–675.
3 Jansson L, Sandler S: Influence of hyperglycemia on blood perfusion of autotransplanted pancreatic islets in diabetic rats. Diabetes 1989; 38: 196–198.
4 Menger MD, Jäger S, Walter P, Feifel G, Hammersen F, Messmer K: Angiogenesis and hemodynamics of the microvasculature of transplanted islets. Diabetes 1989; 38: 199–201.
5 Rooth P, Dawidson I, Lafferty K, Diller K, Armstrong J, Pratt P, Simonsen R, Taljedal IB: Prevention of detrimental effect of cyclosporin A on vascular ingrowth of transplanted pancreatic islets with verapamil. Diabetes 1989; 38: 202–205.
6 Adair TH, Gay WJ, Montani J-P: Growth regulation of the vascular system: evidence for a metabolic hypothesis. Am J Physiol 1990; 259: R393–R404.
7 van der Vliet JA, Meloche RM, Field MJ, Chen D: Pancreatic islet isolation in rats with ductal collagenase distention, stationary digestion and dextran separation. Transplantation 1988; 45: 493–495.
8 Parr EL, Bowen KM, Lafferty KJ: Cellular changes in cultured mouse thyroid glands and islets of Langerhans. Transplantation 1980; 30: 135–141.
9 Sutherland RM: Cell and environmental interactions in tumor microregions: the multicell spheroid model. Science 1988; 240: 177–184.
10 Selawry H, Mui MM, Paul RD, Distasio JA: Allograft survival using highly enriched populations of rat islets. Transplantation 1984; 37: 205–207.
11 Stefan Y, Meda P, Neufeld M, Orci L: Stimulation of insulin secretion reveals heterogeneity of pancreatic β cells in vivo. J Clin Invest 1987; 80: 175–183.
12 Alejandro R, Cutfield RG, Shienvold FL, Polonsky KS, Noel J, Olsen L, Dilleberger J, Miller J, Mintz DH: Natural history of intrahepatic islet cell autografts. J. Clin Invest 1986; 78: 1339–1347.
13 Mintz DH, Alejandro R, Bloom AD, Zucker D, Polonsky KS: The fate of intrahepatic islet allografts in dogs with spontaneous diabetes. Diabetes Res and Clin Prac 1988; 5: 350.
14 Rajotte RV, Warnock GL, Procyshyn AW, Wieczorek K: Intrasplenic isografts of canine pancreatic islets: metabolic study. Transplant Proc 1984; 16: 834–837.
15 Warnock GL, Rajotte RV: Critical mass of purified islets that induce normoglycemia after implantation into dogs. Diabetes 1988; 37: 467–470.
16 Samols E, Stagner JI: Intra-islet regulation. Am J Med 1988; 85: 31–35.
17 Stagner JI, Samols E, Koerker DJ, Goodner CJ: Perfusion with anti-insulin gamma globulin indicates a B to A to D cellular perfusion sequence in the pancreas of the rhesus monkey, *Macaca mulatta*. Pancreas 1991; in press.
18 Samols E, Stagner J: Islet perfusion is B-A-D in the human pancreas. Diabetes 1991; 40: 73A.
19 Menger MD, Jäger S, Walter P, Hammersen F, Messmer K: A novel technique for studies on the microvasculature of transplanted islets of Langerhans in vitro. Int J Microcirc: Clin Exp 1990; 9: 103–117.
20 Stagner JI, Samols E: The induction of capillary bed development by endothelial cell growth factor before islet transplantation may prevent islet ischemia. Transplant Proc 1990; 22: 824–828.
21 Haisch CE, Lodge PA, Huber SA: The vascular endothelial cell is central to xenogeneic immune reactivity. Surgery 1990; 108: 306–311.
22 Wood NL, Schook LB, Struder EJ, Mohanakumar T: Biochemical characterization of human vascular endothelial cell-monocyte antigen defined by monoclonal antibodies. Transplantation 1988; 45: 787–892.
23 Shienvold FL, Alejandro R, Mintz DH: Identification of Ia-bearing cells in the rat, dog, pig and human islets of Langerhans. Transplantation 1986; 41: 364–372.
24 Leszczynaski D, Schellekens H, Hayry P: Vascular endothelium in allograft rejection. Transplant Proc 1988; 20: 262–263.
25 Nakai I, Aikawa I, Oka T: Quantitative analysis of microvasculature of rat pancreas transplants in acute rejection. Diabetes 1989; 38: 111–113.
26 Menger MD, Jäger S, Walter P, Hammersen F, Messmer K: The microvasculature of xenogeneic transplanted islets of Langerhans. Transplant Proc 1990; 22: 802–803.

Angiogenesis: Key Principles - Science - Technology - Medicine
ed. by R. Steiner, P. B. Weisz & R. Langer

Role of basic fibroblast growth factor in proliferation of endothelium and smooth muscle after denuding injury in vivo

Volkhard Lindner and Michael A. Reidy

University of Washington, Department of Pathology SJ-60, Seattle, WA 98195, USA

Endothelial cell growth after injury

Balloon catheter injury of the rat common carotid artery causes complete denudation of endothelium. Using this model, endothelial regrowth from the aortic arch and the carotid bifurcation is initially rapid but cessation of endothelial outgrowth occurs after approximately 6 weeks [1]. This leaves a central portion of the common carotid artery permanently denuded of endothelium. However, if a less traumatic technique (filament loop denudation) is used to remove the endothelium, proliferation of endothelium proceeds at a continuously high rate until complete coverage of the vessel is achieved within approximately 12 weeks. The difference in endothelial regrowth in these two models allowed us to investigate which factors might be involved in controlling the growth of these cells. With an antibody raised against human recombinant basic fibroblast growth factor (bFGF) immunocytochemistry was performed on en face (Häutchen) preparations of the leading edge of the regenerating endothelium. Strong staining with this antibody was found only in those endothelial cells that continued to proliferate, i.e. at early times after balloon catheter denudation and at all times after filament denudation [1]. The staining appeared to be predominantly nuclear. Endothelial cells that had stopped proliferating, i.e. at later times after balloon catheter injury or normal endothelium in uninjured vessels showed markedly less staining. Thus a correlation between the presence of bFGF and endothelial cell proliferation seemed to exist. We then examined whether the cessation of endothelial regrowth that occurred after balloon catheter denudation could be overcome by administration of exogenous bFGF. In these experiments, systemic infusion of bFGF six weeks after balloon catheter denudation caused a dramatic increase in proliferation of the endothelial cells near the still denuded area of the vessel (^{3}H-thymidine index 38% vs 2% in controls) [2]. In subsequent experiments we investigated whether in this model endothelial outgrowth onto the denuded vessel could be increased by prolonged treatment with bFGF. Indeed, injecting bFGF twice per week (12 μg) for a period of 8 weeks caused a marked increase in reendothelialization and in some cases even complete endothelial coverage was achieved [2]. Since the experiments mentioned above still leave the question unanswered whether bFGF is required for endothelial cell growth, the following experiment was carried out. An antibody was raised against human recombinant bFGF, that specifically neutralized the mitogenic effect bFGF in an in vitro cell proliferation assay on 3T3 cells. The anti-proliferative action of this antibody was then tested on endothelium in vivo that had a high proliferation rate. Segments of the rat aorta were deendothelialized with a balloon catheter and the IgG fraction of this antibody (10 mg i.v.) was injected. Proliferation of endothelium at the wound edge was determined by autoradiography following injection of ^{3}H-thymidine (24, 32, and 40 h after denudation). Compared to controls that received nonimmune IgG no significant effect of the anti-bFGF IgG on endothelial cell proliferation was found (^{3}H-thymidine index 40% vs 33% in controls, unpublished data). This result does not rule out that bFGF is essential for endothelial cell proliferation since bFGF is not a secreted protein and therefore there may be no extracellular bFGF. Taken together with the predominant nuclear staining for bFGF in endothelial cells the possibility exists that bFGF is acting via an intracrine pathway.

Smooth muscle cell proliferation after vascular injury

Smooth muscle cell (SMC) proliferation with subsequent intimal thickening is an important event in the development of an atherosclerotic lesion. Although growth factors are thought to be involved, there is currently no direct evidence for a role of any mitogen in this process. Following balloon catheter injury of the rat carotid artery, substantial death of medial SMCs occurs which is followed by SMC proliferation and intimal thickening. One hypothesis to explain the rapid SMC proliferation in this model is that mitogens are released from these dead cells. One such mitogen could be basic fibroblast growth factor (bFGF) since it was found to be present in the vessel wall by immunoblot analysis and we have previously shown that bFGF is a potent mitogen for SMCs in vivo [3]. Immunocytochemistry showed bFGF to be localized in the nuclei of SMCs and endothelial cells with no apparent staining in the extracellular matrix of the vessel wall [4]. In this study we examined the role of endogeneous bFGF and arterial injury and found that the proliferation of SMCs after balloon catheter denudation of the rat carotid artery was significantly reduced when a neutralizing antibody raised against human recombinant bFGF (see above) was injected prior to injury [4]. Groups of rats received an intravenous injection of either nonimmune IgG or anti-bFGF IgG (Protein-G purified from rabbit serum, 10 μg/rat) prior to surgery and 5 min later the artery was denuded of endothelium by balloon catheterization. SMC proliferation in these arteries was determined by autoradiography following injection of ^{3}H-thymidine (24, 32, and 40 h after denudation). The ^{3}H-thymidine index of medial SMCs was significantly reduced after antibody injection as compared to animals receiving the nonimmune IgG (1.4% vs 16.8%). However, the intimal thickening that developed within 8 days after balloon catheterization was not significantly decreased after injection of the antibody. These data support the hypothesis that SMC proliferation after arterial injury is largely due to release of bFGF from damaged cells in the vessel wall.

Supported by NIH grants HL 03174-35 and 1P50HL 42270-01 and a grant from the Council for Tobacco Research.

References

1 Lindner V, Reidy MA, Fingerie J: Regrowth of arterial endothelium: Denudation with minimal trauma leads to complete endothelial cell regrowth. Lab Invest 1989; 61: 556–563.
2 Lindner V, Majack RA, Reidy MA: Basic FGF stimulates endothelial regrowth and proliferation in denuded arteries. J Clin Invest 1990; 85: 2004–2008.
3 Lindner V, Lappi DA, Baird A, Majack RA, Reidy MA: Role of basic fibroblast growth factor in vascular lesion formation. Circ Res 1990; 68: 106–113.
4 Lindner V, Reidy MA: Proliferation of smooth muscle cells after vascular injury is inhibited by an antibody againt basic fibroblast growth factor. Proc Natl Acad Sci USA 1991; in press.

Angiogenesis: Key Principles - Science - Technology - Medicine
ed. by R. Steiner, P. B. Weisz & R. Langer

Capillary Growth in the Heart

Margaret D. Brown and Olga Hudlicka

Department of Physiology, University of Birmingham, Birmingham B15 2TT, England

Abstract. Experiments were performed to test the hypothesis that increased stretch and/or tension of myocytes in the absence of changes in blood flow could induce capillary growth in the heart. Chronic treatment with either dobutamine (rabbits) or alinidine (rats) which increased force of contraction and/or stroke volume respectively without increases in coronary blood flow led to enlargement of the anatomical size of the capillary bed, with no change in cardiac weight, thus supporting the role of external mechanical factors in angiogenesis.

Introduction

In the normal adult heart, capillary growth has been reported to occur in response to interventions which increase cardiac work and thus enhance oxygen requirements, e.g., high altitude hypoxia [1, 2], hyperthyroid status [3, 4] and exercise training [5, 6]. In all of the above, coronary blood flow would be increased. Increased flow per se resulting from chronic treatment with coronary vasodilators also leads to myocardial capillary neoformation [7–9]. Growth of capillaries in the heart may therefore result from factors associated with increased blood flow, either with or without increased demand for oxygen.

The most significant increase in capillary supply (up to 70% greater capillary density) was reported in rabbit hearts after 4 weeks chronic reduction in heart rate to 50% of resting by electrical bradycardial pacing [10]. With this intervention, total coronary blood flow was unchanged but flow per beat was significantly greater [11] and it was suggested [12] that mechanical factors associated with an increase in stretch and/or tension of the capillary wall occurring during a prolonged diastole, when flow was greater and capillary diameters would be wider [13], could represent a signal for angiogenesis.

However, it is also possible that during bradycardial pacing when stroke volume was enlarged by 72 + 7% with a concomitant increase in force of contraction, stretch and/or tension of the myocytes themselves could exert mechanical influences upon the vessel wall from outside and could act as stimuli for angiogenesis. To distinguish between these possible factors, experiments were performed in which stroke volume and/or force of myocyte contraction were increased in the absence of increases in coronary blood flow, and the chronic effects on myocardial capillary supply studied.

Materials and methods

1) The specific bradycardic drug *alinidine* (Boehringer Ingelheim) was used to decrease heart rate and increase stroke volume. The acute effects of alinidine were assesed in 9 pentobarbitone-anaesthetized Sprague Dawley rats. Left ventricular coronary flow was estimated using microspheres together with indices of cardiac performance (dP/dt_{max} from a catheter introduced into the left ventricle, minute and stroke work calculated from cardiac output/stroke volume from thermodilution × mean blood pressure) at the peak effect on heart rate of 3 mg/kg alinidine i.p (approximately 20 min post administration). This dose was subsequently administered twice a day i.p. for a period of 5 weeks to 6 animals, controls ($n = 4$) receiving vehicle (saline).

2) The beta-adrenergic inotropic agent *dobutamine* was used at a dose which increased myocardial force of contraction and stroke volume without significantly increasing coronary blood flow. The acute effects of dobutamine were assessed in 6 pentobarbitone-anaesthetized New Zealand Red rabbits 20 min into a 30 min i.v. infusion (20 μm/kg/min). Coronary blood flow (microspheres) and cardiac performance (cardiac work, dP/dt_{max}) were measured as described above. This dose of dobutamine (Dobutrex) was then administered as a continuous i.v. infusion for a total of 14 days to 5 animals using portable infusion pumps [9]. Control animals ($n = 4$) received saline.

3) *Capillary growth* was assessed in the hearts of rats and rabbits after the treatment period by quantification of all anatomically-present capillaries stained histochemically for alkaline phosphatase in 12-μm frozen sections of left ventricular free wall. In rat hearts, numbers of capillaries and fibres (outlines enhanced by background eosin staining) were counted using unbiased sampling in 8 square fields (each 0.0169 mm^2) per heart to yield estimates of capillary-per-fibre ratio. In rabbit hearts, capillary density (numbers of capillaries/mm^2) was assessed from counts of vessels in 10 square fields (each 0.0676 mm^2) per heart.

Results

Table 1 summarises the acute effects of alinidine in rats and dobutamine in rabbits on coronary blood flow and cardiac performance at the doses subsequently administered on a chronic basis. Alinidine reduced heart rate by $28 + 2\%$ ($p < 0.001$) with a concomitant $34 + 11\%$ increase in stroke volume index ($p < 0.05$). As a result of the decreased heart rate and a significant negative inotropic effect (dP/dt_{max} was $38 + 5\%$ lower, $p < 0.01$), coronary blood flow was halved, but remained unchanged when calculated on a per beat basis.

The dose of dobutamine employed produced a significant sustained increase in force of contraction (dP/dt_{max} increased by $24 + 6\%$, $p < 0.01$) without any effects on heart rate, blood pressure or coronary blood flow, either total or per beat. The increase in stroke volume index ($30 + 13\%$) was of borderline significance ($p < 0.1 > 0.05$), likewise that in stroke work index (increased by $18 + 10\%$).

Neither alinidine nor doubtamine administered chronically had any effect on heart/body weight ratios ($0.320 + 0.010$ vs $0.321 + 0.009$ for alinidine-treated and control rats, $0.209 + 0.010$ vs $0.197 + 0.016$ for dobutamine-treated and control rabbits). In rats, capillary supply was evaluated as capillary-per-fibre ratio since use of background eosin staining made it possible to delineate myocyte outlines. Capillary-per-fibre ratio was significantly increased after 5 weeks alinidine treatment (1.54 ± 0.02 vs 1.31 ± 0.04 for controls, $p < 0.001$). In rabbits, capillary density, which was estimated because it was not possible to detect fibre outlines, showed a significant increase compared with controls after two weeks dobutamine treatment ($2698 \pm 64/mm^2$ vs $2089 \pm 210/mm^2$, $p < 0.05$).

Discussion

In these experiments, an increase in the anatomical size of the coronary capillary bed has been demonstrated in the absence of any cardiac hypertrophy following chronic treatment with either dobutamine or alinidine. It is unlikely that either drug could have induced capillary growth as a result of increased demand for oxygen or increased blood flow. The dose of dobutamine used was chosen specifically to have a modest inotropic action, the low oxygen cost of which is shown by the insignificant effects on coronary blood flow during infusion. Ziada et al. [9] used two drugs, adenosine and a xanthine derivative, which increased coronary conductance in rabbit hearts by 65% and 33% respectively upon acute infusion and only led to increases in capillary density of 29% and 17% after 3–5 weeks continuous infusion. The 29% increase in capillary density seen after only two weeks infusion of dobutamine, which increased coronary conductance by $31 \pm 17\%$ (n.s.) is not therefore likely to be the result of mechanical factors associated with blood flow.

Other determinants of myocardial oxygen consumption and hence coronary flow – heart rate and cardiac stroke work [14] – were not significantly changed by dobutamine and thus would not have contributed any stimulus to capillary growth through increases in flow,

Table 1. Haemodynamics, coronary blood flow and cardiac performance measured in pentobarbitone-anaesthetised animals during peak effects of alinidine (3 mg/kg i.p.) in rats or dobutamine (20 μg/kg/min i.v.) in rabbits compared with control animals

	Heart rate b.p.m	Blood pressure mm Hg	Coronary blood flow mls/g/min	Flow/beat mls/g	Stroke volume index mls/beat/kg	Stroke work J/100 g/beat	dP/dt_{max} mm Hg/s
Rats							
Control (n = 7)	395 ± 24	115 ± 9	5.22 ± 0.98	1.29 ± 0.22	0.84 ± 0.09	40.5 ± 7.6	3460 ± 300
Alinidine (n = 9)	286 ± 8***	97 ± 5	2.64 ± 0.36*	0.99 ± 0.09	1.12 ± 0.09*	42.8 ± 3.5	2130 ± 160**
Rabbits							
Control (n = 5)	284 ± 16	107 ± 6	2.21 ± 0.26	0.97 ± 0.06	0.62 ± 0.13	84.7 ± 18.5	3280 ± 490
Dobutamine (n = 5)	290 ± 16	102 ± 5	3.19 ± 0.33	0.98 ± 0.06	0.71 ± 0.13	100.2 ± 20.8	4030 ± 580*

*$p < 0.05$, **$p < 0.01$, ***$p < 0.01$ vs control animals

either overall or per beat. The fact that the pressure-rate product (systolic blood pressure × heart rate) was not altered during acute dobutamine infusion ($+6 \pm 8\%$) indicates further that myocardial oxygen demands were changed little.

The reduction of total coronary blood flow and absence of changes per beat on acute injection of alinidine, a result of the decreased heart rate and negative inotropic effect, eliminates the possibility of any mechanical flow-associated factors acting as stimuli to angiogenesis. Stroke volume was increased because of the longer filling time and would presumably mean a larger end-diastolic volume and hence stretch of myocytes. The higher oxygen costs of a more dilated heart [15] would be more than offset by the lower heart rate and contractility, and stroke work was not altered by the drug.

In conclusion, capillary growth has been shown in the heart following chronic interventions which altered the mechanical external forces acting on coronary vasculature, either by stretch of myoctyes (with increased stroke volume during bradycardia) or by increased force of contraction (with increased dP/dt_{max}).

References

1 Turek Z, Grandtner M, Kreuzer F: Cardiac hypertrophy, capillary and muscle fibre density, muscle fibre diameter, capillary radius and diffusion distance in the myocardium of growing rats adapted to a simulated altitude of 3500 m. Pflügers Arch 1972; 335: 19–28.
2 Kayar S, Banchero N: Volume overload hypertrophy elicited by cold and its effect on myocardial capillarity. Resp Physiol 1985; 59: 1–14.
3 Chilian WM, Wangler RD, Peters KC, Marcus ML, Tomanek RJ: Thyroxine-induced left ventricular hypertrophy in the rat. Anatomical and physiological evidence for angiogenesis. Circ Res 1985; 57: 591–597.
4 Breisch EA, White FC, Hammond HK, Flynn S, Bloor CM: Myocardial characteristics of thyroxine stimulated hypertrophy. A structural and functional study. Basic Res Cardiol 1989; 84: 345–358.
5 Mattfeldt T, Kramer KL, Zeitz R, Mall G: Stereology of myocardial hypertrophy induced by physical exercise. Virchows Arch Abt A Path 1986; 409: 473–484.
6 Anversa P, Ricci R, Olivetti G: Effects of exercise on the capilalry vasculature of the rat heart. Circulation 1987; Suppl 75: 1–12.
7 Ljungqvist A, Tornling G, Unge G, Jurdahl EB, Larsson B: Capillary growth in the heart and skeletal muscle during dipyridamole treatment and exercise. Progr Appl Microcirc 1984; 4: 9–15.
8 Mall G, Mattfeldt T, Rieger P, Frolov VA: Ultrastructual morphometric study on the rat heart after chronic ethanol feeding. Virchows Arch A Path Anat Histol 1980; 389: 59–77.
9 Ziada AMAR, Hudlicka O, Tyler KR, Wright AJA: The effect of long-term vasodilatation on capillary growth and performance in rabbit heart and skeletal muscle. Cardiovasc Res 1984; 18: 724–732.
10 Wright AJA, Hudlicka O: Capillary growth and changes in heart performance induced by chronic bradycardial pacing in the rabbit. Circ Res 1981; 49: 469–478.
11 Hudlicka O, West D, Kumar S, El Khelly F, Wright AJA: Can growth of capillaries in the heart and skeletal muscle be explained by the presence of an angiogenic factor? Br J Exp Path 1989; 70: 237–246.

12 Hudlicka O: Capillary growth: role of mechanical factors. NIPS 1988; 3: 117–120.
13 Tillmans TH, Ikeda S, Hansen H, Sarma JS, Fauvel JH, Bing RJ: Microcirculation in the ventricle of the dog and turtle. Circ Res 1974; 34: 561–569.
14 Rooke GA, Feigl EO: Work as a correlate of canine left ventricular oxygen consumption and the problem of catecholamine oxygen wasting. Circ Res 1982; 50: 273–286.
15 Sarnoff SJ, Braunwald E, Welch GH Jr, Case RB, Stainsby WN, Macruz R: Hemodynamic determinants of oxygen consumption of the heart with special reference to the tension time index. Am J Physiol 1958; 192: 148–156.

Angiogenesis: Key Principles - Science - Technology - Medicine
ed. by R. Steiner, P. B. Weisz & R. Langer

Angiogenic role of lactic acid in the mechanism of neovascularization

György Imre and Éva Csonka

2nd Dept. of Ophthalmology and 2nd Inst. of Pathology, Semmelweis University of Medicine, Budapest, Hungary

Ocular neovascularizations are severe complications of different eye diseases. For this reason we began to work on the mechanism of neovascularization 28 years ago. Michaelson [1] was the first who supposed that vasoformative factors may be released in hypoxic tissues, and since then we have found out that in the pathogenesis of a proliferative retinopathy retinal areas with bad perfusion are essential. Neovascularization begins on the border of these areas. So the starting point of our investigations was hypoxia. In hypoxic tissues anaerobic glycolysis prevails, and if the venous circulation deteriorates which is also a very important condition of neovascularization, lactic acid, the end-product of the glycolysis, will accumulate. The possibility that an increased lactic acid concentration may have an important role in the mechanism of neovascularization has been confirmed by the following data:

1. Lactic acid concentration is increased in vascularizing tissues. To prove this there are several indirect data and, in connection with corneal vascularization direct data as well. In our experiment [2] on rabbit corneas a severe alkali burn was inflicted. After vascularization had begun, both the avascular part of the burnt corneas and an identical part of control corneas were excised, dried and the lactic acid concentration was determined by Hohorst's enzymatic method. The lactic acid concentration of the avascular part of alkali burnt vascularizing corneas was significantly increased (Tab. 1). In cases of rubeosis iridis we found [3] a significant increase in lactic acid concentration in the aqueous humour (Tab. 2). But the concentration of lactic acid in the aqueous humour can definitely increase only if the rate of aqueous flow is decreased as we found in cases of rubeosis iridis (Tab. 3).

2. Increasing the lactic acid concentration of avascular tissues leads to vascularization. The vascularization of the cornea began earlier and was significantly more intensive after intracorneal injections of lactic acid than after injections of other solutions (Tab. 4) which caused a

Table 1. Lactate concentration on the 8th day; μg/mg dry weight, mean ± SD (no. of corneas)

Alkali burned	6.49 ± 2.44 (7)
Controls	2.76 ± 2.43 (7)

$p < 0.01$

Table 2. Lactate concentration of the aqueous humour; mean ± SD

Rubeosis iridis (11 eyes)	Control eyes (10)
10.68 ± 2.47 mM/l	7.68 ± 2.08 mM/l

$p < 0.01$

Table 3. Tonographic values of 28 eyes with rubeosis iridis; mean ± SD

C: 0.09 ± 0.09	F: 0.61 ± 0.72 μl/min
in 20 eyes < 0.15	in 21 eyes < 1.0
in 8 eyes > 0.15	in 7 eyes > 1.0

Table 4. Length of vascularization on the 6th day

		Length (mm), mean ± SD (No. of corneas)
2 Intracorneal inj., 0.05 ml, 0.1%	Lactic acid	1.34 ± 0.35 (5)
	Sodium hydroxide	1.00 ± 0.10 (5)*
	Acetic acid	0.58 ± 0.23 (5)**
	Hydrochloric acid	0.46 ± 0.23 (5)**

*p < 0.05; **p < 0.01.

Table 5. The lactic acid concentration of the vitreous, mM/l; mean ± SD (No. of eyes)

Controls	After posterior lens puncture
7.65 ± 2.41 (9)	13.61 ± 6.51 (9)

$p < 0.05$

more serious corneal damage [4]. Intravitreal injections of lactic acid in 0.1% concentration caused in 8 out of 16 kitten eyes a preretinal neovascularization. Originally this was our first experiment [5]. Later Cunha-Vaz [6] proved that one single intravitreal injection of lactic acid may cause venular endothelial proliferation, the first step of retinal neovascularization. This was observed also after puncture of the lens [7] and we found [8] a significantly increased lactic acid concentration in the vitreous after posterior lens puncture (Tab. 5).

Table 6. Length of vascularization on the 6th day

		Length (mm), mean ± SD (No. of corneas)
2 Intra-corneal inj., 0.05 ml, 0.1%	D-Lactate	0.23 ± 0.40 (11)
	L-Lactate	0.46 ± 0.38 (11)

$p < 0.02$

Table 7. Lactate concentration on the 14th day; μg/mg dry weight, mean ± SD (No. of corneas)

Avascular swelling	1.78 ± 0.15 (6)
Controls	2.97 ± 0.90 (6)

$p < 0.02$

3. L-lactate produced by the organism induces more intensive corneal vascularization than D-lactate does (Tab. 6), a compound foreign to the organism [9].

4. Prolonged swelling of the cornea is an important condition for its vascularization. The compactness of the cornea prevents its vascularization. Swelling of the cornea, however, is not always accompanied by vascularization. In cases of experimentally produced avascular swelling the lactic acid concentration of the cornea was significantly decreased (Tab. 7). Avascular swelling was induced by filling the anterior chamber of rabbit eyes with silicon oil [10].

5. ^{14}C-lactic acid in 10^{-4} M concentration is taken up in vitro by the endothelial cells from calf aorta after 20 and 60 min treatment (Fig. 1). Lactic acid in 10^{-3} and 10^{-4} M concentration essentially increases the ^{3}H-thymidine incorporation into the endothelial cells from calf aorta after 4 and 6 h treatment in tissue culture. After 48 h the control cells were in a stationary phase but thymidine uptake of the lactic-acid-treated endothelial cells lagged (Fig. 2). Accordingly on the 4th day the cell number significantly increased on the effect of lactic acid in 10^{-3} M concentration (Tab. 8) [11].

Our experiments confirm the important role of lactic acid in the mechanism of neovascularization and make it probable that the vasoformative endothelial cells are stimulated by the lactic acid itself. But how does lactic acid stimulate the endothelial cells? Lactic acid like other organic anions also diffuses through the endothelial cells into the blood. If in a tissue there is an increased lactic acid concentration, its effect will depend on the venous circulation. If the venous circulation is intact, the blood leads the lactic acid away, and there is no neovascularization as there is no retinal neovascularization in cases of central retinal vein occlusion. Chan and Little [12] proved that in such cases there is a widespread endothelial cell death, no endothelial cells will

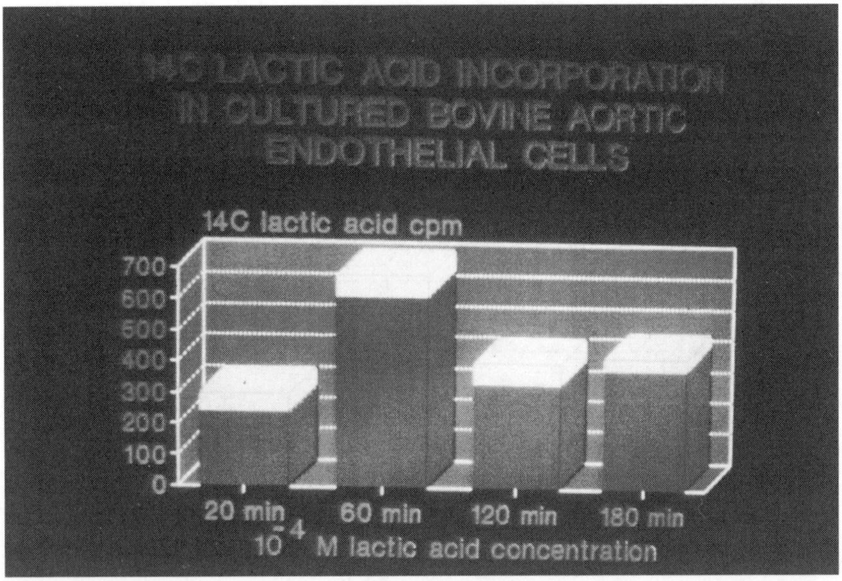

Figure 1. ^{14}C Lactic acid incorporation in cultured bovine aortic endothelial cells.

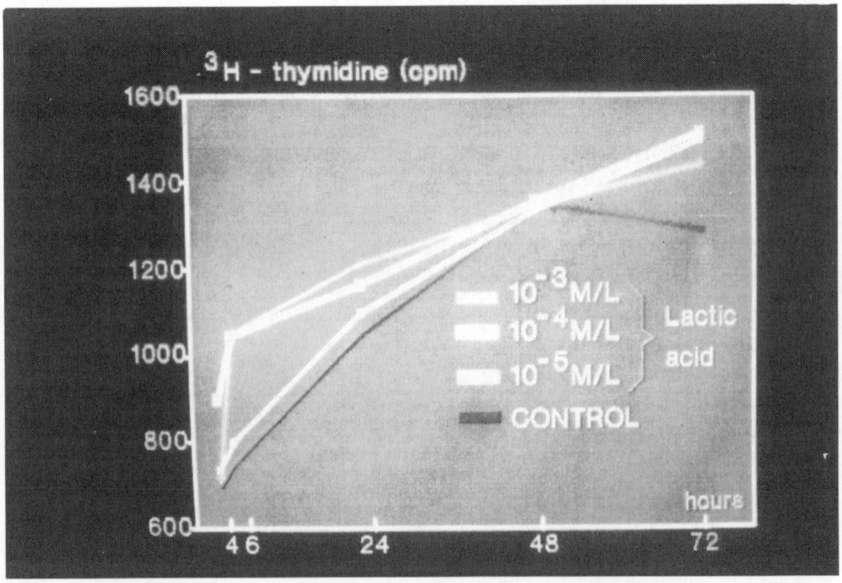

Figure 2. ^{3}H-thymidine incorporation on the effect of lactic acid in cultured endothelial cells.

Table 8. Endothelial cell number in thousands on 4th day of cultivation; mean $\pm$ SD

Control	101.5 ± 24.3	
10^{-2} M Lactic acid	toxic	
10^{-3} M Lactic acid	144.4 ± 35.9	$p < 0.01$
10^{-4} M Lactic acid	124.3 ± 24.5	n.s.

remain to build up new vessels. But if there is only a deteriorated venous flow and the intracellular lactate concentration slowly increases, the endothelial cells are compelled, as a defence, to use up the lactic acid and hereby be activated and proliferate the forming of new vessels.

It is probable that other factors, e.g. prostaglandins, different mitogens and growth factors, can influence the neovascularization, e.g. by increasing the lactic acid concentration. If there are fibroblasts or macrophages growing into a hypoxic tissue the metabolism of these cells may increase the lactic acid concentration further. With the increased lactic acid concentration due to increased glycolysis and deteriorated drainage of metabolites, the situation in which the endothelial cells are compelled to use up the lactic acid and are activated in this way, explains vascular growth.

References

1 Michaelson IC: The mode of development of the vascular system of the retina, with some observations on its significance for certain retinal diseases. Trans. Ophthal. Soc. U.K. 1948; 68: 137–180.
2 Imre G, Bögi J: Corneal vascularization, in Trevor-Roper P (ed): The cornea in health and disease. Int Congr Symp Ser London, Royal Soc Med. 1981, vol 40, pp 233–235.
3 Imre G: The role of lactic acid in neovascularizations. Graefes Arch klin exp Ophthal 1977; 201: 263–266.
4 Imre G: The mechanism of corneal vascularization. Acta Morphol Acad Sci Hung 1966; 14: 99–104.
5 Imre G: Studies on the Mechanism of Retinal Neovascularization. Role of Lactic Acid. Br J Ophthal 1964; 48: 75–82.
6 Cunha-Vaz J: Physiopathogenesis of retinitis proliferans and new vessel formation, in François J (ed): Blutzirkulation in der Uvea, in der Netzhaut und im Sehnerven. Stuttgart, Enke, 1978, pp 247–253.
7 Deem CW, Futterman S, Kalina RE: Induction of endothelial cell proliferation in rat retinal venules by chemical and indirect physical trauma. Invest Ophthal Visual Sci 1974; 13: 580–585.
8 Imre G, Pálfalvi M: Lactate concentration of the vitreous after lens puncture. Data relating to the mechanism of neovascularization. IRCS Medical Sci, The Eye 1977; 5: 231.
9 Imre G: Corneal Vascularization. Role of Lactic Acid. Acta Morphol Acad Sci Hung 1969; 17: 171–173.
10 Imre G, Pál M: Avascular swelling of the cornea. Acta Medica Acad Sci Hung 1968; 25: 97–100.
11 Imre G: Pathogenesis of proliferative retinopathy and rubeosis iridis of diabetics, in Henkes H (ed): Proc of the VIIth Congress of Europ Soc of Ophthal Helsinki, Yliopistopaino, 1985, pp 510–511.
12 Chan CC, Little HL: Infrequency of retinal neovascularizations following central retinal vein occlusion. Ophthalmology 1979; 96: 256–262.

Angiogenesis: Key Principles - Science - Technology - Medicine
ed. by R. Steiner, P. B. Weisz & R. Langer

Hyperoxia (145 mmHg pO_2) and tissue normoxia (20–40 mmHg pO_2) modulate human vascular cell functions

D. G. S. Thilo-Körner[a], U. Marre[a], N. Savion[b], F. W. Hehrlein[c] and R. H. Bödeker[d]

[a]*Dept. Internal Medicine*, [c]*Dept. Cardiac Surgery*, [d]*Institute Med. Documentation and Statistics, University of Giessen, Klinikstr. 36, D-6300 Giessen (FRG), and*
[b]*Goldschleger Eye Res. Institute, University of Tel Aviv, Israel*

Introduction

Between aggression and defence, the endothelial cell, as the pluripotent control cell of the vessel wall, fulfills its functions in regulated interactions between the blood, the smooth muscle cells (SMC) and the fibroblasts of the vessel wall [1–5]. Oxygen influences various cellular functions like the cell proliferation, the angiogenesis, the lipid metabolism and others [2–11]. The oxygen partial pressure (pO_2) in the arterial blood is between 70–100 mmHg, in the venous blood around 40 mmHg, in the aortic media between 30–40 mmHg and in the bovine pulmonary artery between 9–22 mmHg pO_2 [7, 11]. Therefore, many years ago, we postulated that 145 mmHg pO_2 (21 vol% O_2) applied in vessel wall cell cultivation is hyperoxic, does not resemble in vivo tissue normoxia and induces an unphysiologically high oxygen stress to the vascular cells in vitro, thereby, artificially modulate their functional behavior [2, 4, 5, 12].

Materials and methods

Human umbilical cord vein endothelial cells (HUVEC) were isolated according to the method of Thilo-Körner with the Thilo-Körner MBT-cannula and alpha-chymotrypsin [3, 12, 13]. For SMC-cultivation we obtained vessel material from patients recieving an aorto-coronary bypass (n = 120; 36–71 y; vena saphena magna [VsSMC] arteria mammaria [AmSMC]) and from the Aorta abdominalis (Aa), Vena cava (Vc), art. lumbalis (AL) (n = 15; 18–86 y). SMC were cultivated with the explant-technique in RPMI 1640 medium under a pO_2 of 20–

145 mmHg in our glass incubator system [5, 12]. During proliferation, the tPA-antigen (ELISA; American Diagnostic, New York, N.Y.; USA) and the PAI (Imunlyse™ PAI-1-ELISA-kit; Biopool AB, Umea, Sweden) were determined in the culture medium. Compounds: unfractionated heparin (Liquemin^R, 10 μg/ml = 1.4 IE WHO-standard, Hoffmann-La Roche, Grenzach, FRG) and low molecular weight heparin (LMW-heparin, Sandoz, Nürnberg, FRG; MW 7600, aPTT/anti-Xa-activity-ratio 4.2; 1 μg/ml = 0.04 U aPTT or 0.173 U anti-Xa-activity); acetylsalicyclic acid (ASS) 10^{-5} M, dipyridamole (DIP) 2×10^{-6} M.

Results

HUVEC, bovine aortic endothelial cells (BAEC) and VsSMC differ in their cell size- and volume-distribution. Human arterial and venous SMC are pO_2-sensitive in this respect [2, 4, 5]. HUVEC show no morphological cell alterations, however, a 30% higher cell density under 40 mmHg in comparison to 145 mmHg pO_2 after five days in culture. No data were available on the influence of pO_2 on the SMC-outgrowth from the explants. This outgrowth of venous and arterial SMC is pO_2-sensitive (Fig. 1). After 3 weeks there is a 2-fold higher outgrowth under 40 in comparison to 145 mmHg pO_2 with no difference between female and male patients. The pO_2 modulates the proliferation of all human SMC tested in our system (Fig. 2). Like in VsSMC we can measure a 8.8- (20 mmHg), 12.8- (40 mmHg), 12.5- (85 mmHg) and a 5.6-fold (145 mmHg pO_2) and in AaSMC a 25.1- (20 mmHg), 17.2- (40 mmHg) and 18.2-fold (85 mmHg pO_2) proliferation increase after seven days in culture. The sensitivity of Aa-, AL-, Vc and VsSMC against the antiproliferative effect of unfractionated or LMW-heparin, changes with decreasing pO_2 [2, 4, 5]. In contrast to heparin, ASS in therapeutic concentration retards VsSMC-proliferation by 22% (40 mmHg) and 13% (145 mmHg pO_2) and in 78% of our proliferation experiments. DIP does this in 70%.

There are only a very few publications suggesting that SMC participate in the fibrinolytic activity of the vessel wall [4, 5, 15, 16]. Therefore, we examined if human SMC release tPA and PAI-1 during their proliferation [4, 5]. Human arterial and venous SMC stain positively for tPA (immunperoxidase method). Their proliferation increase is combined with a decrease in tPA-release and a gradual increase in PAI-1 (Fig. 3). There is a 32.2 ± 7.6-fold higher tPA-release than PAI-1 in the first five days. The average tPA-release is 2.3-fold higher in VsSMC in comparison to AmSMC. ASS induces nearly a 2-fold increase of tPA-Ag under 40 mmHg pO_2 in AmSMC. ASS significantly increases the tPA-release from VsSMC under 40 mmHg pO_2 ($n = 9$, $p = 0.006$); for

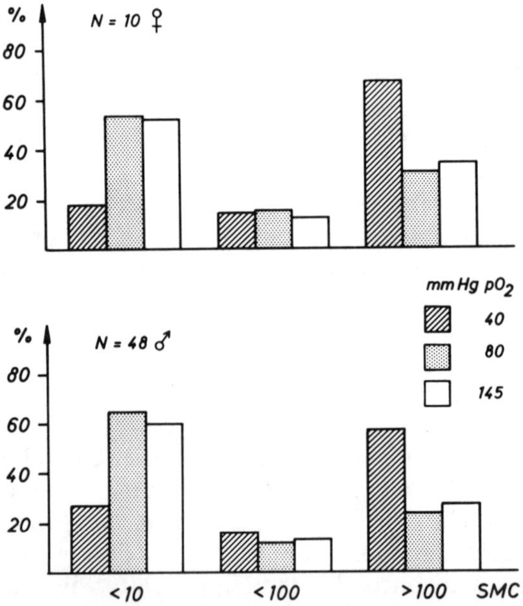

Figure 1. Outgrowth of VsSMC from 1895 explants under 40 or 145 mmHg pO_2. Aorta-coronary-bypass-patients: n = 10 females, n = 48 males, 41–69 y, judged by light microscopy into three catagories (under 10, 100, over 100 SMC per explant) day 7.

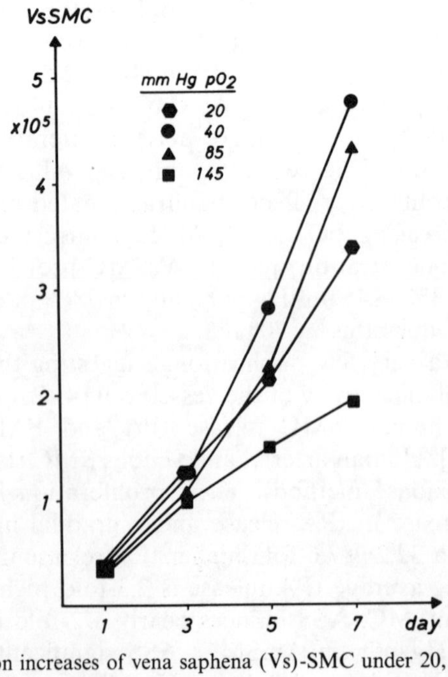

Figure 2. Proliferation increases of vena saphena (Vs)-SMC under 20, 40, 85 and 145 mmHg pO_2; each point n = 6.

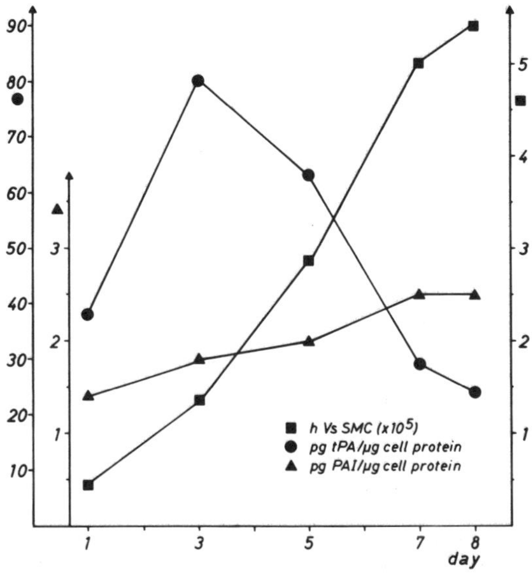

Figure 3. Relation between the VsSMC-proliferation and the release of tissue plasminogen activator-antigen (tPA) and plasminogen activator inhibitor (PAI) (per μg cell protein) into the culture medium during 8 days under 40 mmHg pO_2; $n = 6$.

DIP the p-value is 0.064. The tPA-Ag-release is not dependent on the patient's age ($p = 0.87$).

We were able to demonstrate that there is a similar cholesterol-uptake in VsSMC after 3 days under 145 and 40 mmHg pO_2 [5, 17]. Our results show a significant alteration of the phospholipid-metabolism of human arterial and venous SMC. Unexpected were the results of the conversion of cholesterol to cholesterol esters in BAEC, BASMC, HUVEC, VsSMC as well as AmSMC [5, 17]. AmSMC demonstrate a 2-fold higher conversion rate than VsSMC which is 15- and 7-fold in relation to BASMC.

Discussion

145 mmHg pO_2 (21 vol% O_2) still used in cell cultivation does not represent the in vivo pO_2. Our results clearly show that this hyperoxic pO_2 modulates the proliferation, cellular drug senstivity for heparin and aspirin, the cholesterol and especially the phospholipid metabolism of human vascular wall cells. Human arterial and venous SMC secrete tPA-antigen and PAI-1 during proliferation modulated by platelet function modulators and the pO_2. Our and other results underline the importance of pO_2 in the modulation of cellular functions and

strengthen our hypothesis that vascular cells should be examined under tissue normoxia (e.g. 20–40 mmHg pO_2) in order to obtain data with physiological and clinical relevance. This is also important for testing compounds with possible antiproliferative, fibrinolysis- and lipid-modulating properties and the extrapolation of the results obtained to the in vivo situation in humans.

References

1 Thilo-Körner DGS: Aspekte der Gefäßwandzellforschung. Relevanz für die Klinik? Hämostaseologie 1986; 7: 1–13.

2 Thilo-Körner DGS: Einfluß von Heparin auf die glatten Muskelzellen der Gefäßwand. In Bruhn HD et al. (ed): Aktuelle Probleme der Antikoagulation in Klinik und Praxis, Stuttgart, Schattauer, 1986, pp. 201–218.

3 Thilo-Körner DGS: Die Endothelzelle-ein vitro Modell der Gefäßwandforschung. Bio-Enginieering 1989; 6: 32–42 (Thilo-Körner-cannula, 1988; patent # P 38 14 805.6, DGM 88 05 792.5).

4 Thilo-Körner DGS, Marre U, Savion N, Moosdorf R, Bödeker RH: Modulation of proliferation, fibrinolysis, lipid metabolism on human A/VSMC by pO_2, aspirin and dipyridamole, in Descovich GS et al. (eds.): Atherosclerosis and Cardiovascular disease, Bologna, Editrice Compositori, 1989, vol. 4, pp. 105–112.

5 Thilo-Körner DGS, Marre U, Savion N, Moosdorf R, Bödeker RH: Human endothelial- and smooth muscle cells: partial oxygen pressure modulates their proliferation, fibrinolysis and lipid metabolism, in Delius W, Lippert P, Wieland H: Preventive Cardiology, Munich, Urban & Schwarzenberg, 1989, pp. 79–100.

6 Balin AK, Fisher AJ, Carter DM: Oxygen modulates growth of human cells at physiologic partial pressures. J exp Med 1984; 160: 152–166.

7 Benitz WE, Coulson JD, Lessler DS, Bernfield M: Hypoxia inhibits proliferation of fetal pulmonary arterial smooth muscle cells in vitro. Pediatr Res 1986; 20: 966–972.

8 Broxmeyer HE, Cooper S, Gabig T: The effects of oxidizing species derived from molecular oxygen on the proliferation in vitro of human granulocyte-macrophage progenitor cells. Ann NY Acad Sci 1989; 554: 177–185.

9 Kan M, Kato M, Yamane I: Long-term serial cultivation and growth requirements from human umbilical vein endothelial cells. In vitro Cell Devl Biol 1985; 21: 181–188.

10 Knighton DR, Hunt TH, Scheuenstuhl H, Halliday BJ, Werb Z, Banda MJ: Oxygen tension regulates the expression of angiogensis factor by macrophages. Science 1983; 221: 1283–1285.

11 Niinikoski J, Heughan C, Hunt TK: Oxygen tension in the aortic wall of normal rabbits. Atherosclerosis 1973; 17: 353–359.

12 Thilo-Körner DGS, Heinrich D, Temme H: Endothelial cells in culture. A literature survey on isolation, harvesting, cultivation, medium and serum composition, cell counting, gas atmosphere and confluency rates, in Thilo-Körner DGS, Freshney RI (eds.): The endothelial cell – a pluripotent control cell of the vessel wall, Karger, Basel, 1983, pp. 158–202.

13 Thilo DGS, Müller-Küsel S, Heinrich D, Käufer I, Weiss E: Isolation of human venous endothelial cells by different proteases. Artery 1980; 8: 259–266.

14 Björkerud S: Impaired fibrinolysis-inducing capacity for postinjury phenotype of cultivated human arterial and human atherosclerotic intimal smooth muscle cells. Circ Res 1988; 62: 1011–1018.

15 Booyse FM, Scheinbuks J, Radek J et al: Immunological identification and comparison of plasminogen activator forms in cultured human endothelial cells and smooth muscle cells. Thromb Res 1981; 24: 495–504.

16 Goldsmith GH, Ziats NP, Robertson AL: Studies on plasminogen activator and other proteases in subcultured human vascular cells. Exp Mol Pathol 1981; 35: 257–264.

17 Savion N, Thilo-Körner DGS: Lipoproteins, coagulation factors and vascular cell function, in Delius W et al. (eds.): Preventive Cardiology, Munich, Urban & Schwarzenberg, 1989, pp. 45–78.

Angiogenesis: Key Principles – Science – Technology – Medicine
ed. by R. Steiner, P. B. Weisz & R. Langer

Impact of basic fibroblast growth factor (bFGF) on wound healing in chronically ischemic tissue

Markus Kamler, Hans-Anton Lehr[a], John H. Barker[b], Rainer K. Saetzler, Eberhard Uhl[a] and Konrad Messmer[a]

Department of Experimental Surgery, University of Heidelberg, FRG; [a]Institute for Surgical Research, Ludwig-Maximillians University, Munich, FRG; [b]Division of Plastic and Reconstructive Surgery, University of Louisville, USA

Introduction

Fibroblast growth factor has been implied as a central mediator in wound healing, including epithelialization and ingrowth of capillaries [1]. It was the aim of our animal study to investigate the effects of recombinant human basic FGF (rhbFGF) on wound healing in chronically ischemic tissue.

Methods

Our experiments were performed on 6–12-week-old hairless mice (hr/hr) [2]. Chronic ischemia was induced to the ears by ligating two of the three main vessel bundles entering the ear base and verified by measuring tissue pO_2 with a membrane-covered multiwire PO_2 electrode [3].

Circular full-thickness wounds (∅2.5 mm) were created on the ear tissue by sharp excision of skin and subcutaneous tissue. bFGF (0.5 g/50 1; n = 10 animals) or vehicle (vaseline; n = 10 animals) were administered locally in 3-day intervals onto the wounds which were protected by a semi-permeable transparent dressing.

Epithelialization and capillary ingrowth were assessed at three day intervals by intravital fluorescence microscopy in awake animals, immobilized in a plexiglass restrainer [4]. In order to assess epithelialization, we measured the wound surface area by digital planimetry. Using fluorescein-isothiocyanate-labelled dextran (FITC-dextran, M_r 150,000; Sigma Chemical Co., St. Louis/Mo/USA) as plasmamarker we assessed vessel diameters and red cell velocities in the tissue surrounding the wound area, as well as the non-capillarized area in the wound center.

Results

Verification of ischemia

$TcpO_2$ decreased from 24 mmHg (Med Q3/Q1: 28/14) to 6 mmHg (8/3) after ligation of the vascular bundles and only slowly recovered to values of 12 mmHg (16/8) at day 12 after ligation. Ischemia resulted in a delay of wound closure time by three days as compared to wound healing in normal tissue [5].

Effect of bFGF treatment

In bFGF-treated animals the wound surface area was significantly reduced from day 6 on after wound excision ($p < 0.01$). While in control animals epithelialization was not completed prior to day 12, the wounds were closed at day 9 in all bFGF-treated animals ($p < 0.01$). Despite these pronounced effects of bFGF on wound epithelialization, capillary ingrowth was found unaffected by bFGF-treatment.

Discussion

Using the hairless mouse model for the intravital fluorescence microscopic study of wound healing, we have demonstrated that bFGF accelerates the epithelialization of standardized wounds in chronically ischemic tissue, and that this effect is not associated with enhanced ingrowth of capillaries. In other experimental models of wound healing, which do not allow differential assessment of wound epithelialization and neovascularization, caution should therefore be exercised when extrapolating from wound closure data to the process of wound vascularization.

References

1 Gospodarowicz D, Ferrara N, Schweigerer L, Neufeld G: Structural characterization and biological functions of fibroblast growth factor. Endocrinology 1987; 8: 95–114.

2 Barker JH, Hammersen F, Bondar I, Uhl E, Galla TJ, Menger MD, Messmer K: The hairless mouse ear for in vivo studies of skin microcirculation. Plast Reconstr Surg 1989; 83: 948–959.

3 Kessler M, Höper J, Krumme BA: Monitoring of tissue perfusion and cellular function. Anesthesiology 1976; 45: 186–199.

4 Kamler M, Lehr HA, Saetzler RK, Galla TJ, Messmer K: Impact of ischemia on tissue oxygenation and wound healing: Improvement by vasoactive medication. Advances in Medicine and Biology, Plenum Press, NY, 1991 (in press).

5 Bondar I, Uhl E, Barker JH, Galla TJ, Hammersen F, Messmer K: A new model for studying microcirculatory changes during dermal wound healing. Res Exp Med 1991 (in press).

Angiogenesis: Key Principles - Science - Technology - Medicine
ed. by R. Steiner, P. B. Weisz & R. Langer

Endothelial cell proliferation is induced by radiation in cultured explants of human urothelium and oesophageal mucosa

Carmel Mothersill, C. B. Seymour, D. Mulvin and T. P. Hennessy

Radiobiology Group, Nuclear Energy Board, Clonskeagh Square, Dublin 14, Department of Urology, Beaumont Hospital, Dublin 9, and Department of Surgery, St James's Hospital, Dublin 8, Ireland

Abstract. Normal oesophageal mucosa obtained during upper abdominal surgery or urothelium obtained from kidney transplants was placed in explant culture and exposed to ^{60}Co gamma radiation after 48 h. Cultures were maintained for two to six weeks after exposure and were monitored at various intervals for the development of features associated with malignant transformation. Endpoints examined included proliferation rate, frequency of proliferating cells, cell type distribution and degree of differentiation of the different cell types. The results indicate that following exposure to gamma rays (2.5–10 Gy) an increased overall growth rate of the surviving cells can be observed 2–4 weeks later. Analysis of the results using autoradiography confirms that a higher level of cell proliferation occurs in treated cultures than in the control untreated cultures.

When the distribution of different cell types in the culture is examined, the increase in growth can be seen to be due to greatly increased numbers of endothelial cells. These proliferated over the surface of the epithelial cells and are more strongly positive for endothelial cells markers than endothelial cells occurring in control cultures. The degree of differentiation of endothelial cells into capillary like structures is also more apparent in carcinogen treated cultures.

Foci expressing both epithelial and endothelial markers also occur. The results suggest that exposure of tissue fragments to radiation stimulates the growth and development of endothelial cells in resulting cell cultures. The effect may be due to a direct action of the treatment on the endothelial cells but it is more likely that it results from a secondary effect mediated by traumatic response of damaged epithelial cells.

Introduction

Angiogenesis is possibly the key step in the progression from an initiated mass of cells to a malignant and probably lethal neoplasm. Experimental investigations, particularly those involving in vitro systems, seldom address the problem. This is partly because of the lack of multicell or organ models suitable for studying the whole process of malignant transformation, but much of the reason for the lack of development of useful systems must be due to conceptual constraints which view cancer as a disease originating in a single cell and which regard events happening after the initiation of the neoplastic cell as responses by the tissue/organism to the signals from the transformed cell. This conceptual framework makes processes such as angiogenesis relevant in the physiological and therapeutic sense but irrelevant in the more basic areas where quantification in terms of dose response is the key issue of interest.

Because of this, most systems available for the study of transformation use criteria such as focus formation or soft agar colony formation to determine the frequency of transformation events [1, 2]. If a 'disease process' approach was taken and full expression of malignancy rather than initiation of a single cell was regarded as more important, in vitro studies would probably relate more closely and in a more meaningful way to the vivo situation.

As an approach to developing such a 'bridging' system, explants of human tissues were established and treated with carcinogens. The effects were monitored in cells growing out from the explant over the life span of the culture.

Materials and methods

Human tissues were obtained from transplant donors or from non malignant surgical specimens. Tissue was minced to pieces 2×2 mm^2 and placed in a solution of 0.1% w/v trypsin petite type containing 1 mg/ml collagenese type 4 for 30–60 min at 37°C. The tissue fragments were than plated singly in 25-cm^2 tissue culture flasks (NUNC) containing 2 ml RPM1 1640 medin with 20% v/v Foetal Calf Serum 1 mIU/ml insulin, 1 μg/ml hydrocortisone and necessary antibiotics at non-cytotoxic concentrations [3]. Cultures were incubated at 37°C in 5% CO_2 in air until required. Tissue explants took about 48 h to attach firmly to the culture flask. They were then treated with ^{60}CO gamma irradiation and returned to the incubator. At various times after irradiation, cultures were removed and stained for the presence of various epithelial endothelial or oncogene markers, using standard immunocytochemical techniques.

Results

Fig. 1 shows the survival curves obtained for oesophageal and urothelial cells exposed to increasing doses of irradiation. These show a normal acute dose response after three days but repopulation by surviving cells has obliterated the dose response by 28 days post treatment and cell numbers are similar for irradiated and control cultures. In Fig. 2 the cell types composing these cellular outgrowths are analyzed. It is apparent that in both tissues, endothelial cell proliferation occurs and that after high doses of radiation, endothelial cells (or cells bearing endothelial antigens factor VIII and EN4) form the greater part ot the culture.

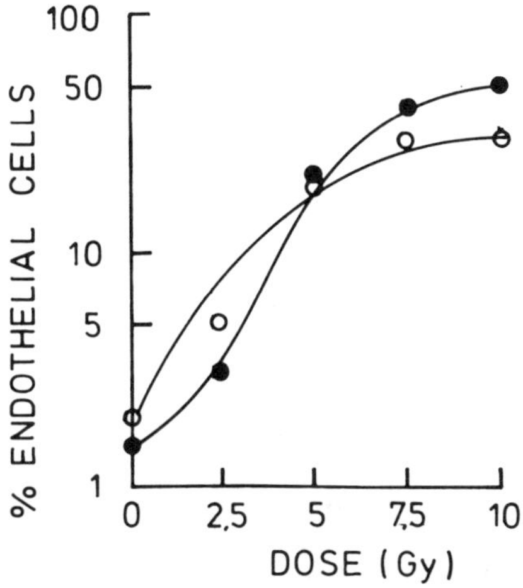

Figure 1. The mean percentage of endothelial cells detected in 5 random grid squares of oesophageal (○) or urothelial (●) explant outgrowths 25 days post irradiation.

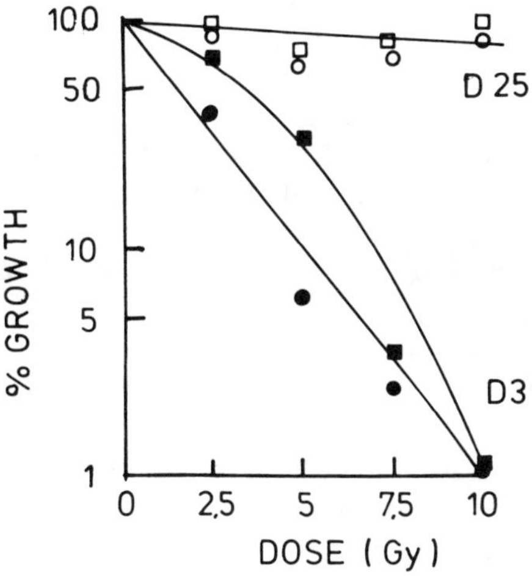

Figure 2. The percentage growth of oesophageal (●, ○) and urothelial (■, □) explant outgrowths detected 3 (●, ■) and 25 (○, □) days post irradiation.

Morphological examination of the cultures reveals that the endothelial cells mainly proliferate over the surface of non-proliferating epithelial cells. In certain areas of irradiated cultures, characteristic foci can be seen. These contain morphologically distinct proliferating cells which bear endothelial epithelial antigens and also express cMyc oncoprotein. The frequency of these foci increases with time post irradiation (Tab. 1).

Table 1. Numbers of foci bearing epithelial and endothelial cell markers detected 25 days post irradiation in outgrowths from urothelial and osophageal tissue explants exposed to Cobalt 60 rays

Dose (Gy)	Oesophageal mucosa	Urothelium
0	1.6 ± 0.7	1.0 ± 0.4
5.0	11.5 ± 3.3	6.3 ± 1.5
10.0	24.4 ± 11.6	33.6 ± 9.2

Discussion

The survival curve data presented in Fig. 1 shows that, after acute doses of radiation, growth of cells is inhibited/cells are killed in a dose dependent manner. It is interesting to note that twenty five days after the irradiation, the dose response effect has disappeared and survival is virtually the same at each dose. This indicates that cells which survive the radiation exposure are dividing rapidly and are capable of achieving the same plateau phase numbers eventually.

Analysis of the composition of the explant outgrowths at twenty five days post irradiation shows that the main components are endothelial cells. Their proliferation is clearly the dominant factor leading to the repopulation of the explant culture. This suggests that endothelial cell proliferation is actively stimulated by exposure of tissue fragments to radiation.

The foci which develop on the monolayer are of considerable interest as possible neoplastic cell precursors. The fact that epithelial and endothelial cell markers are both expressed by these cells suggests that the cells have features common to both basic cell types making them very interesting candidates for a pre- or early neoplastic unit.

Further work is obviously required to see if these cells can be isolated and cultured. The technique may offer a way of studying the involvement of basic angiogenic mechanisms at a very early stage.

Acknowledgements. We wish to acknowledge support from the Radiation Protection Programme, Contract No. B170023 and the Irish Cancer Research Advancement Board.

References

1 Franks LM, Teich N: Introduction to the Cellular and Molecular Biology of Cancer. Oxford Science Publications, Oxford, New York, Tokyo 1988.

2 Freshney RI: Animal Cell Culture. A Manual of Basic Techniques. 2nd Edn. Alan R. Liss, New York 1988.

3 Mothersill C, Cusack A, McDonnell M, Hennessy TP and Seymour CB: Differential Response of Normal and Tumour Oesophageal Explant Cultures to Radiation. Acta Oncologica 1989; 27 Fasc. 3: 275–280.

Angiogenesis: Key Principles - Science - Technology - Medicine
ed. by R. Steiner, P. B. Weisz & R. Langer

The direct effect of light therapy on endothelial cell proliferation in vitro

Lucy Ghali and Mary Dyson

United Medical and Dental Schools, London, England

Abstract. During repair, new blood vessels are formed by the process of angiogenesis. The prerequisite to blood vessel formation is the proliferation of endothelial cells. The purpose of this study was to find out if light therapy is capable of affecting endothelial cell proliferation in vitro. The direct effect of light at wavelengths of either 660 or 820 nm was studied on primary cultures of bovine aortic endothelial cells using varying energy densities, of either 1, 2, 4, or 8 J/cm^2. The proliferation of the endothelial cells was assessed over a period of five days after a single irradiation.

Introduction

Many investigators have found that low power lasers can be used effectively to treat skin disorders, such as ulcers or wounds, in order to accelerate healing processes [1–3]. Low power lasers have also been used in other fields of medicine including rheumatology [4] and surgery [5]. However, clinical cases are not always well documented, and experimental results are sometimes contradictory [6]. Different investigators have used different lasers each with its characteristic wavelength; exposure times and frequency of treatments have varied; irradiance has rarely been the same for two experiments. There is considerable data but comparisons of the results are difficult.

Materials and methods

Primary cultures of endothelial cells were grown from bovine aortic tissue. Cells were maintained in a 5% CO_2 humidified incubator using Minimum Essential Medium (MEM) with 10% Fetal Calf Serum (FCS), 1% Pen/Strep, and 1% L-Glutamine (Life Technologies Ltd, Paisley). When confluent, the endothelial cells were trypsinized and resuspended at a cell density of 1×10^5 cells/ml; 1 ml of this cell suspension was plated into each well of a 24-well tissue culture plate. The cells were then irradiated with one of two different wavelengths: 660 nm (noncoherent) and 820 nm (coherent). Energy densities of either 1, 2, 4, or 8 J/cm^2, with irradiation times of 113, 226, 452, and 904 s respectively were used. The light was produced by a Biotherapy 3ML device (Omega Universal Technologies). There were eight experimental groups, each with a paired control which received a sham irradiation. The physical characteristics of the light source were the same for both the coherent and noncoherent sources and were as follows: 1) Active medium (Gallium Aluminium Arsenide); 2) Average power output: 15 mw; 3) Spot size at 4 cm distance: 1.5 cm diameter, area = 1.76 cm^2; 4) Power density: 15 mw/1.76 cm^2 = 0.0088 w/cm^2; 5) Source spectral band width at 100% intensity = 0.01 nm; 6) Total angle of divergence = 23° parallel to junction plane; 7) Pulsing frequency = 18.24 Hz; 8) Pulse duration = 43.75 m s. After irradiation the cells were resuspended in fresh medium at a density of $1 \times 10^3/100\ \mu l$. Twelve 100 μl volumes of each cell suspension were pipetted into each well of a 96-well culture plate for each time period. Cell growth was arrested by removing culture medium and fixing the cells, either at 6 h or at daily intervals for 5 days. Cell proliferation in the 96-well plates was assessed by methylene blue dye uptake [7]. Relative endothelial cell proliferation in experimental groups and their corresponding sham irradiated controls was compared, the results being analysed statistically using the Mann Whitney-U test.

Results and discussion

In this study the proliferation rate of endothelial cells in vitro after light therapy has been evaluated. Figure 1 shows that, after direct irradiation with light of a wavelength of 660 nm, there was a statistically significant increase in cell number starting from day 2 ($p < 0.05$) after

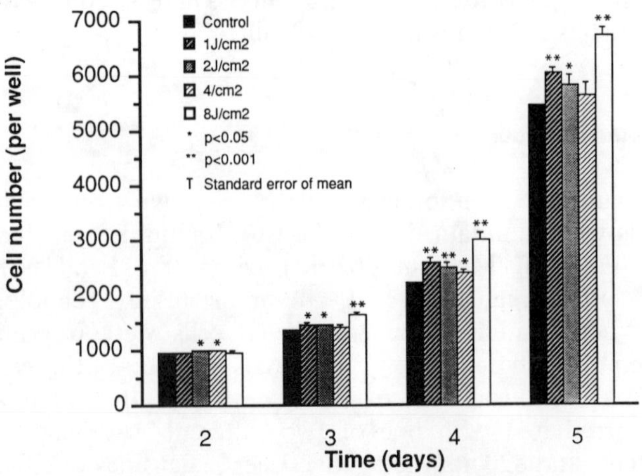

Figure 1. Direct effect of 660 nm light on endothelial cell proliferation.

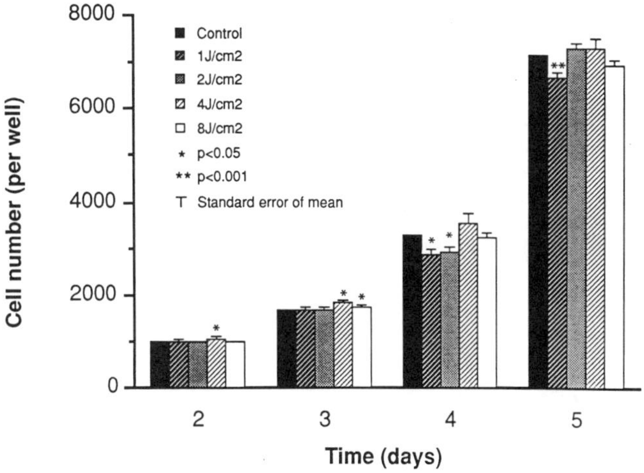

Figure 2. Direct effect of 820 nm light on endothelial cell proliferation.

irradiation, reaching its maximum at 5 days ($p < 0.001$). When irradiated with 820 nm, the cells showed less response and, at 1 J/cm^2, the effect was inhibitory ($p < 0.001$) (Fig. 2). It is of considerable interest that an inhibitory as well as a stimulatory effect was observed, since there are occasions where these different effects can be exploited; for example in a situation where overgrowth of blood vessels is a dominant pathology (in a number of disease states such as tumour growth, rheumatoid arthritis, haemangioma, and several common causes of blindness [8–10]), inhibition would be desirable. Whereas, in a situation where vessel growth is relatively slow, stimulation would be advantageous, e.g. in wound healing, tissue implantation and where the circulation is deficient. From the results obtained, we conclude that appropriate selection of low level laser wavelength may allow fine control of angiogenesis.

References

1 Mester E, Ludany G, Selyei M, Szende B, Tota GJ: The stimulating effect of low power laser rays on biological systems. Laser Rev 1948; 1: 3.
2 Mester E, Mester AF, Mester AM: The biomedical effects of laser application. Lasers Surg Med 1985; 5: 31.
3 Longo L, Evangelista S, Tinacci G, Sesti AG: Effect of diodes-laser silver arsenide aluminium (Ga-Al-As) 904 nm on healing of experimental wounds. Lasers Surg Med 1987; 7: 444–447.
4 Goldman JA, Chiapella J, Casey H, Bass N, Graham J, McClatchey W, Dronavalli RV, Brown R, Bennett WJ, Miller SB, Wilson CH, Pearson B, Haun C, Persinski L, Huey H, Muckerheide M: Laser therapy of rheumatoid arthritis. Lasers Surg Med 1980; 1: 93–101.

5 Trelles MA, Mayayo E: Bone fracture consolidates faster with low-power laser. Lasers Surg Med 1987; 7: 36–45.
6 McCaughan JS Jr, Bethel BH, Johnston T, Jaussen W: Effect of low dose argon irradiation on rate of wound closure. Lasers Surg Med 1985; 5: 607–614.
7 Oliver MH, Harrison NK, Bishop JE, Cole PJ, Laurent GJ: A rapid and, convenient assay for counting cells cultured in microwell plates: Application for assessment of growth factors. Journal of Cell Science 1989; 92: 513–518.
8 Folkman J: Tumor angiogenesis. Adv Cancer Res 1985; 43: 175.
9 Folkman J, Klagsbrun M: Angiogenic factors. Science 1987; 235: 442.
10 Hudlicka O, Tyler KR: Angiogenesis: The growth of the vascular system. London, Academic Press, 1986, pp 115–147.

Angiogenesis: Key Principles - Science - Technology - Medicine
ed. by R. Steiner, P. B. Weisz & R. Langer

Angiogenesis during tumor progression in human malignant melanoma

Domenico Ribatti[a], Angelo Vacca[b], Wanda Palma[c], Mario Lospalluti[c] and Franco Dammacco[b]

[a]*Institute of Human Anatomy, Histology and Embryology*, [b]*Department of Biomedical Sciences and Human Oncology*, [c]*Institute of Dermatology, University of Bari Medical School, Policlinic, I-70124 Bari, Italy*

Introduction

Human and animal cancers can progress in a series of multiple steps toward an increasing aggressiveness [1]. This process is characterized by a number of genetic and epigenetic changes which result in a repeated rise and a prevalence of new subpopulations with enhanced and competitive growth [2].

Human malignant melanoma is one of the tumors in which a succession of stages of progression is well recognized both histopathologically and clinically. Specifically, nevocytic and dysplastic nevi are supposed to be precursor entities of malignant tumors [3]. In addition, progression stages in primary melanomas can be identified by the increasing Breslow thickness, as suggested by the linear correlation between thickness and occurrence of metastases [4]. Clinical [5] and experimental [6] studies have also shown that melanoma progression parallels a higher growth rate and metastatic potential by tumor cells.

Folkman and his coworkers clearly demonstrated that tumor begins its growth in an avascular phase followed by a vascular one in which further tumor growth is angiogenesis-dependent [7]. Also, there is a close relation between growth rate of the tumor cells and magnitude of associated neovascularization, at least in experimental tumors [8].

On the basis of these studies, it was of particular interest to find out whether distinct biologic steps of melanoma progression could be associated with different degrees of related angiogenesis.

Materials and methods

Tissue specimens

Twenty-six nevi (15 nevocytic, 11 dysplastic), 44 stage I primary melanomas and 14 melanoma metastases were studied. Representative parts of tissues were surgically removed, snap frozen with liquid nitrogen and stored at $-80°C$ until stained by immunohistochemistry. Routine histologic procedures served for diagnosis and classification. Dysplastic nevi were diagnosed according to Clark et al. [3] and Elder et al. [9]. Primary melanomas were grouped according to the Breslow thickness [4].

Immunohistochemistry

The anti-laminin monoclonal antibody (MAb) L-8271 (Sigma Chemical Co., St Louis., MO) was used. This MAb localizes the basal membrane of endothelia and served to highlight all the blood vessels. Acetone-fixed 8-μm cryostat sections were stained by an indirect immunoperoxidase method previously described [10]. In short, sections were incubated with MAb L-8271 or a control IgG1 from mouse P3X63/Ag8 myeloma [11] followed by biotin-labeled horse anti-mouse Ig (Vector Inc., Burlingame, CA) and avidin-horseradish-peroxidase conjugate (Vector Inc.). The sections were red-stained with 3-amino-9-ethylcarbazole (Sigma) as chromogen, counterstained with hematoxylin and mounted in buffered glycerin.

Morphometric evaluation of vascular density

By using a planimetric point-count method [10], the number and the diameter of blood vessels were evaluated a) within the tumor, b) at the base of the tumor, and c) within adjacent not involved tissues. Six randomly chosen microscope fields were analysed at 160× magnification by a 144-point mesh inserted in the eyepiece of a Leitz-Dialux 20 photomicroscope. The total number of blood vessels transversally cut located at the mesh intersection points was evaluated. The vessel diameter was calculated as the mean value of the two orthogonal diameters in the vessels transversally cut only. Mean value and SD for both parameters were determined in each sample. The statistical significance of the differences between the mean values was determined by Student's t-test for unpaired data. $p < 0.01$ was considered significant.

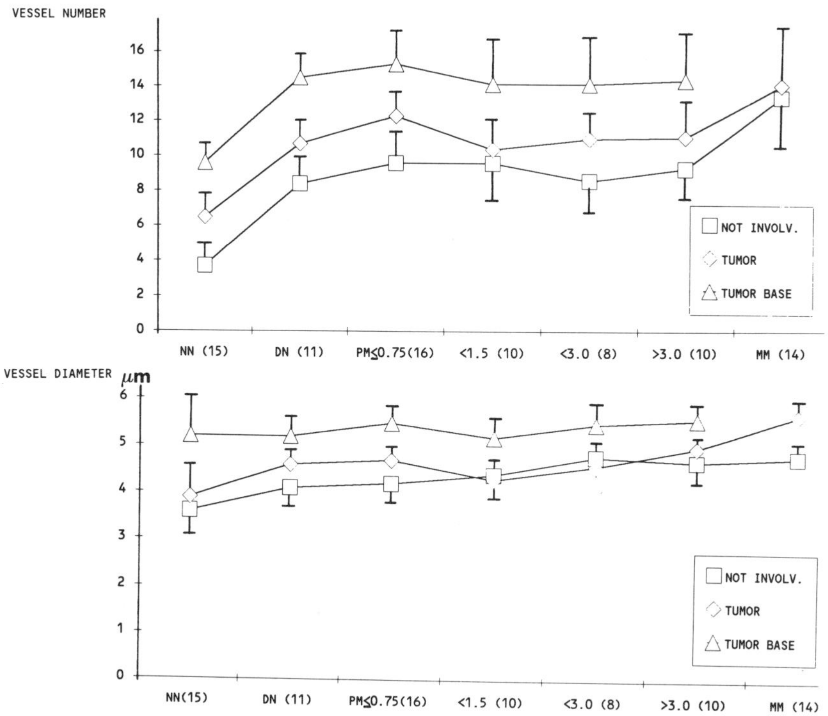

Figure 1. Vessel number and diameter in various melanocytic lesions, as assessed by immunohistochemical staining followed by a planimetric point count method (see Materials and Methods). Number of tissues tested is shown in parentheses. Abbreviations: NN = nevocytic nevi; DN = dysplastic nevi; PM = primary melanomas, grouped according to the Breslow thickness (mm); MM = melanoma metastases.

Results

As shown in Fig. 1, the vessel number was significantly increased in the three areas of dysplastic nevi analyzed, as compared with the nevocytic nevi. A further increase was observed in early primary tumors (≤ 0.75 mm Breslow thickness). However, this parameter was substantially unchanged in the evolutionary steps of primary tumors. In addition, a significant increase was observed in the metastases, within the tumor tissue and the surrounding not involved tissue, as compared with the corresponding areas of advanced (> 3.00 mm) primary tumors.

Fig. 1 also shows that the vessel diameter was not significantly different between nevi and primary tumors. A significant difference was only observed between metastatic tissues and nevocytic nevi. In all cases the values of the vessel number and diameter were higher at the base of lesions.

The vessel density in nevocytic and dysplastic nevi, as well as in a metastatic tissue, is outlined in Fig. 2.

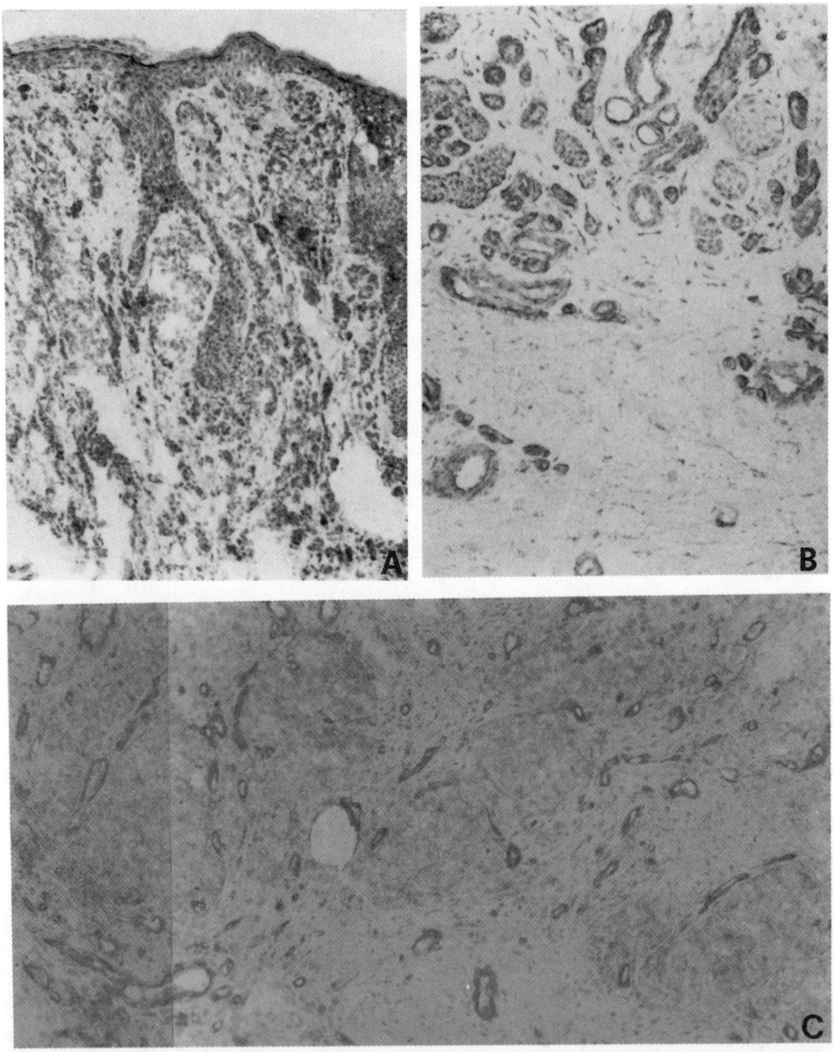

Figure 2. Immunoperoxidase staining of vessels in cryostat sections from a nevocytic nevus (*A*), a dysplastic nevus (*B*) and a subcutaneous metastatic melanoma (*C*).

Discussion

This study investigated the relationship between different grades of biologic malignancy and vascularization in melanocytic lesions. Results showed that transition from nevocytic to dysplastic nevi and, to a lesser extent, to early primary tumors, is associated with a more pronounced vascularization, as compared with the other steps of melanoma progression. Thus, malignant transformation of melanocytes, and its putative precursor lesion such as the dysplastic nevus [3], imply a remarkable vessel density.

A positive relation between melanoma thickness and vascularization was demonstrated by Doppler ultrasound technique [12], but not by a standardized immunohistochemical procedure, like that used in this study. Dissimilarity in the systems to measure vascularization is likely to account for this difference.

Our data agree with other reports on tumor progression in breast cancer [13] and in experimental models of β-cells of the pancreatic islets [14] in which angiogenic capacity appears before the cells become tumorigenic. In addition, our data provide further evidence that the angiogenic response may be an independent event in the course of tumor progression and that it does not necessarily correlate with the malignancy grade of a given neoplasm [10].

Therefore, it is tempting to hypothesize that components other than tumor cells, i.e. macrophages, lymphocytes, mastcells and angiogenic factors stored in the basement membrane and in the extracellular matrix, might regulate vasoproliferation associated to the tumor progression.

Acknowledgements. The authors thank Professor L. Roncali for critical reading of the manuscript. This work was supported in part by the 'Minstero dell'Università e della Ricerca Scientifica e Tecnologica', the Finalized Project 'Clinical Applications of Oncology Research' of the Italian Research Council, Rome, and 'Consorzio Interuniversitario Nazionale per la Bio-Oncologia' (C.I.N.B.O), Italy.

References

1 Dulbecco R: Cancer progression: the ultimate challenge. Int J Cancer 1989; S4: 6–9.
2 Frost P, Fidler IJ: Biology of metastasis. Cancer 1986; 58: 550–553.
3 Clark WH Jr, Elder DE, Guerry D IV, Epstein MN, Greece MH, Van Horn M: A study of tumor progression. The precursor lesions of superficial spreading and nodular melanoma. Hum Pathol 1984; 15: 1147–1165.
4 Breslow A: Thickness, cross-sectional areas and depth of invasion in the prognosis of cutaneous melanoma. Ann Surg 1970; 72: 902–908.
5 Kopf AW, Gross DF, Rogers GS, Rigel DS, Hellman LJ, Levenstein M, Welkovich B, Friedman RJ, Roses DF, Bart RS, Mintzis MM, Gumport SL: Prognostic index for malignant melanoma. Cancer 1987; 59: 1236–1241.
6 Herlyn D, Elder DE, Bondi E, Atkinson B, Guerry D IV, Koprowski H, Clark WC Jr: Human cutaneous nevi transplanted onto nude mice: Lesional steps in tumor progression. Cancer Res 1986; 46: 1339–1343.
7 Folkman J: Tumor angiogenesis. Adv Cancer Res 1985; 43: 175–203.
8 Rofstad EK: Growth and vascular structure of human melanoma xenografts. Cell Tissue Kinet 1984; 17: 91–101.
9 Elder DE, Greene MH, Bondi EE, Clark WH Jr: Acquired melanocytic nevi and melanoma: the dysplastic nevus syndrome, in Ackerman AB (ed): Pathology of malignant melanoma, New York, Masson, 1981, pp 185–215.
10 Ribatti D, Vacca A, Bertossi M, De Benedictis G, Roncali L, Dammacco F: Angiogenesis induced by B-cell non-Hodgkin's lymphomas. Lack of correlation with tumor malignancy and immunologic phenotype. Anticancer Res 1990; 10: 401–406.

11 Vacca A, Buchegger F, Carrel S, Schreyer M, Mach JP: Imaging of human T cell xenografts in nude mice by radiolabeled monoclonal antibodies and $F(ab')_2$ fragments. Cancer 1988; 61: 58–67.
12 Srivastava A, Laidler P, Highes LE, Woodcock J, Shedden EJ: Neovascularization in human cutaneous melanoma: a quantitative morphological and Doppler ultrasound study. Eur J Cancer Clin Oncol 1986; 22: 1205–1209.
13 Brem S, Jensen H, Gullino P: Angiogenesis as a marker of preneoplastic lesions of the human breast. Cancer 1978; 41: 239–244.
14 Folkman J, Watson K, Ingber D, Hanahan D: Induction of angiogensis during the transition from hyperplasia to neoplasia. Nature 1989; 339: 58–61.

Angiogenesis: Key Principles - Science - Technology - Medicine
ed. by R. Steiner, P. B. Weisz & R. Langer

Tumor growth regulation by modulation of basic fibroblast growth factor

Janet L. Gross, William F. Herblin, Kristi Eidsvoog, Robert Horlick and Steven S. Brem[a]

The Du Pont Merck Pharmaceutical Co., P.O. Box 880400, Wilmington, DE 19880-0400, USA, and [a]Northwestern Memorial Hospital Chicago, IL 60611-2906, USA

Introduction

Fibroblast growth factors (FGF) were initially characterized for their mitogenic properties [1]. Since then, members of this family of growth factors have been shown to be growth stimulatory for a variety of cell types in vitro, including glial cells [2–4]. The importance of FGF in tumor biology is supported by the demonstration that FGFs stimulate tumor angiogenesis [5], a process critical for the sustained growth of solid tumors [6]. In addition, cells transfected with bFGF exhibit amplified autocrine growth in monolayer culture and in soft agar, suggesting a role in cell transformation [7]. Finally, several recently described oncogenes, int-2, hst/KS3, and FGF-5, discovered in human tumors encode proteins which share 40–45% homology with bFGF [8–11].

There is now evidence to support the roles for both acidic [12] and basic FGFs [3, 4, 13, 14] in the development and progression of malignant neoplasms. Glioma cells can be distinguished from several other tumor types on the basis of their high level of expression of multiple immunoreactive forms of bFGF, high-affinity receptors for bFGF, and their mitogenic response to bFGF [3, 4]. In this report we have examined both rat and human glioma lines for the presence of message for bFGF, other FGF-related oncogenic proteins, and high-affinity FGF receptors [15]. In addition, we demonstrate that two of several monoclonal antibodies [16] generated against human recombinant bFGF (hr-bFGF) inhibited hr-bFGF stimulated proliferation of cultured human glioma cells. One of these antibodies antagonized angiogenesis in a rat kidney capsule model, and retarded rat C6 glioma growth in nude mice.

Materials and methods

Rat C6 glioma, and human SNB19 and SNB56 cells were maintained as previously described [17].

Poly A+ mRNA was isolated and reverse transcribed using the Fast Track and Red Module kits and protocols from Invitrogen. Oligonucleotide primers were used to amplify target cDNA using the GeneAmp kit and protocol from Cetus. PCR reactions were carried out at 94°C for 1 min, primer annealing temperature for 1 min, 72°C for 2 min for 30 cycles followed by a 7-min chase at 72°C. Primer annealing temperature was calculated as described by Sambrook et al. [18]. Southern analysis was performed using ^{32}P endlabeled oligonucleotide probes to verify results.

For [^{3}H]thymidine incorporation studies, SNB-56 cells were plated in duplicate at 1×10^5 cells per 35 mm plate. After plating, cells were washed with phosphate-buffered saline (PBS) and incubated for 24 h at 37°C in serum free media, followed by 24 h in chemically defined media (CDM, [3]). After a PBS wash, cultures were incubated at 37°C for 1 h with 1 μCi/ml [^{3}H]thymidine in the presence or absence of indicated concentrations of purified IgG's, and 1 ng/ml hr-bFGF was added. Eighteen hours later, cultures were washed 3 × with ice-cold PBS, and solubilized with 1 ml of 0.2 N NaOH. DNA was precipitated with 20% trichloracetic acid onto GF/B glass fiber filters, and radioactivity measured in a Packard Tricarb scintillation counter.

The rat kidney capsule assay was modified from the procedure of Hayek et al. [19]. Male "C-D" rats (150 gm) were used, and the anesthesia was Ketamine (80 mg/kg, i.p.) and Acetylpromazine (2.5 mg/kg, i.p.).

Nude (nu/nu) mice on an outbred Swiss background were bred and maintained as previously described [20]. Tumors were established by s.c. injection of 10^7 C6 cells per site. Mice were pair matched for body weight and sex when tumors reached approximately 50 mg. Injections (i.v.) of 100–150 μg MOPC21 (control) or anti-hr-bFGF DG2 IgG were adjusted per mouse weight and were given once daily for 15 days. Sodium Brequinar was given i.p., Q1D × 9 at 12 mg/kg [20]. On day 16, mice were sacrificed and tumor weights were measured. Growth inhibition was calculated after subtracting the calculated tumor weight at the start of treatment [20]. Data were analyzed for statistical significance using the pooled variance t-test.

Results

In support of roles for bFGF in glioma cells, message levels for bFGF and related family members were examined. Both rat C6 and SNB-19 cells contained message for bFGF as well as for high affinity bFGF receptors, FGFR1 and FGFR2 (Tab. 1). In addition to containing message for FGFR1 and bFGF, SNB-19 cells also contained message for the following bFGF-related proteins: FGF-1, FGF-5,

Table 1. Characteristics related to bFGF and receptors in glioma lines

Property	C6	SNB-19	SNB-56
Immunoreactive bFGF (Intracellular, ng/10 6 cells)	89	142	84
Message			
FGF-1	–	+	
FGF-2	+	+	NT
FGF-3	–	–	NT
FGF-4	–	–	NT
FGF-5	–	+	NT
FGF-6	–	–	NT
FGF-7	–	+	NT
High affinity receptor			
Sites/Cell ($\times 10^3$)	7.8	15	2.5
K_d (pM)	46	18	30
Low affinity receptor			
Sites/Cell ($\times 10^6$)	9	6	10
K_d (nM)	42	16	40
Message receptor			
FGFR1*	+	+	NT
FGFR2	–	+	NT
FGFR3	–	–	NT
FGFR4*	+	–	NT

The bFGF quantitation and receptor binding studies were analyzed in rat C6 glioma and human SNB-19 and SNB-56 glioma lines, maintained at confluence. mRNA was isolated from 1×10^8 cells in log phase growth. *2IgG subtype. NT = not tested. FGF1–7 are: aFGF, bFGF, *int-2*, *hst*, FGF-5, FGF-6, and KGF [21] FGF-R 1–4 are: *flg*, *bek* cek-2, and FGFR-4.

and FGF-7 (Tab. 1, [21]). The inability to detect similar messages in the rat C6 glioma cells may be due to species specificity of the primers.

Having demonstrated a role for bFGF and the possibility of a role for FGF related proteins in gliomas, we examined the effects of several monoclonal antibodies to hr-bFGF [16] on DNA synthesis in SNB-56 human glioma cells. Distinct from the other anti-bFGF antibodies, DE6 and DG2 IgGs inhibited control [^{3}H]thymidine incorporation as well as hr-bFGF stimulated [^{3}H]thymidine incorporation in cultured human glioma SNB-56 cells (Tab. 1). This inhibition was dose-dependent, with DG2 being more potent than DE6; similar results were obtained with rat C6 cells (J.L. Gross, unpublished data).

The DG2 anti-bFGF IgG was shown to inhibit hr-bFGF induced angiogenesis in the rat kidney capsule model [16]. In order to demonstrate the uniqueness of this antibody, DG2 as well as the other anti-bFGF antibodies, were examined for their anti-angiogenic effects in this model. As shown in Table 3, hr-bFGF strongly stimulated angiogenesis, resulting in an average score of 3.2. Among the anti-bFGF

Table 2. Inhibition of hr-FGF induced [^{3}H]thymidine incorporation by monoclonal antibodies to bFGF

Antibody addition	[^{3}H]thymidine incorporation (Avg. DPM) −hr-bFGF	+hr-bFGF
None	7520	14327
MOPC21 (8.9 nM)	6723	11521
AE4A5 (11.0 nM)	7350	14478
AE8A4 (9.1 nM)	11515	11251
AF11D1 (7.3 nM)	6721	11298
CE5D4 (7.8 nM)	8350	12013
DE6E7(10.6 nM)	4173	7517
DG2D7 (16.0 nM)	4098	8666
FE8D1 (10.0 nM)	10087	15405

[^{3}H]thymidine incorporation was performed on duplicate 35-mm cultures of SNB-56 human glioma cells, as described in Materials and Methods. The numbers in parentheses indicate the concentration of each antibody per dish.

Table 3. Effect of anti-bFGF monoclonal antibodies on hr-bFGF-induced angiogenesis in the rat kidney capsule angiogenesis model in vivo

Pretreatment hr-bFGF (μg)	Monoclonal antibody	Angiogenic score	n
0	None	0.37 ± 0.1	45
1	None	3.2 ± 0.8	58
1	MOPC21	2.8 ± 0.9	15
1	AE4A5	2.9 ± 0.4	15
1	AE8A4	3.0 ± 0.8	15
1	AF11D1	3.1 ± 0.8	15
1	CE5D4	2.7 ± 0.7	15
1	DE6E7	2.8 ± 0.6	15
1	DG2D7	1.3 ± 0.9	15
1	FE8D1	3.1 ± 0.7	15

Antibodies to hr-bFGF were mixed with hr-bFGF so that the final concentration of hr-bFGF and antibody were 1 μg and 40 μg, respectively, unless otherwise stated. The mixtures were incubated with Affi-Gel Blue Beads, placed on Gelfoam, and surgically implanted under the left kidney capsule of anesthetized rats, as previously described [16, 18]. After 7 days, the angiogenic score was evaluated: 0 = no response; 4 = strong angiogenic response. The data are represented as the mean +/− standard deviation; "n" represents the number of rats analyzed in each treatment group.

antibodies tested only DG2 significantly reduced, but did not eliminate, angiogenesis induced by hr-bFGF (Tab. 3). DE6 IgG was inactive in this model (J.L. Gross, unpublished data).

Since the DG2 IgG was distinct from other monoclonal antibodies to hr-bFGF, this antibody was tested for its anti-tumor activity against rat C6 gliomas xenografted in nude mice. As shown in Table 4, i.v. administration of this antibody significantly reduced the tumor burden

Table 4. Effect of DG2 anti-hr-bFGF IgG on the growth of C6 rat glioma xenografts in vivo

Group	Net tumor weight (mg/mouse) (±SEM)	Tumor growth inhibition (%)	P value
Experiment 1			
Control	4814 ± 565	NA	NA
Brequinar Na	4193 ± 459	13	0.405
DG2 IgG	3224 ± 289	33	0.026
MOPC21 IgG	5604 ± 746	0	0.410
Experiment 2			
Control	5544 ± 607	NA	NA
Brequinar Na[1]	3608 ± 391	35	0.013
DG2 IgG[2]	3804 ± 462	31	0.031
MOPC21 IgG[2]	4744 ± 511	14	0.323

NA, not applicable; [1]12 mg/kg, Q1D × 9, i.p.; [2]100–150 μg, Q1D × 15, i.v.
Nude mice bearing rat C6 gliomas were injected as indicated and tumor weights were measured on day 16 as described in Materials and Methods. The number of mice evaluated per group were 10 in experiment 1, and 15 in experiment 2.

(33% and 31% tumor growth inhibition) in two independent experiments. C6 tumor growth was marginally affected by Brequinar sodium, an anti-metabolite that reduced to a greater degree the growth of other solid tumors, such as breast and colon carcinomas [20]. Administration of DG2 i.p. or DE6 i.v. were not effective in this tumor model. In addition, i.v. administration of DG2 IgG to nude mice bearing human colon carcinoma DLD-2 tumors had no effects on tumor growth (J.L. Gross, unpublished data).

Discussion

We have previously demonstrated that human and rat glioma cells exhibit high levels of immunoreactive bFGF and high-affinity receptors for hr-bFGF relative to other solid tumor lines examined [4]. In addition, human and rat glioma lines contain bFGF as shown by immunofluorescence (S. Brem, unpublished data), and respond to exogenous hr-bFGF in culture with increased growth [3, 4] and enhanced activity of the protease plasminogen activator [J. Gross, unpublished data]. Consistent with the expression of bFGF and bFGF receptors, both human and rat glioma cells contained message for both bFGF and FGFR1 (*flg*). This confirms our previous observations, that among several different kinds of cultured tumor lines, high-affinity binding of [125]hr-bFGF correlated with message levels for *flg* [22]. The findings of message for other FGFs and other FGF receptors suggests that expression of members of the FGF family, besides acidic and bFGF, may be significant in brain neoplasms.

We have confirmed and extended the results of Reilly et al. [16]. Among a panel of seven monoclonal antibodies against hr-bFGF which inhibited high affinity hr-bFGF binding to cultured BHK cells [16], only two (DE6 and DG2) antibodies reduced hr-bFGF stimulated [^{3}H]thymidine incorporation in Swiss 3T3 fibroblasts [16] and in human and rat glioma cells and only DG2 IgG inhibited angiogenesis in a rat kidney capsule model in vivo [16]. The findings that DG2 and DE6 IgGs reduced [^{3}H]thymidine levels below those exhibited by control cells suggest a role of endogenous bFGF in autocrine growth stimulation. Experiments are underway to determine the epitopes unique to these two antibodies. The DG2 antibody also significantly retarded the growth of rat C6 glioma tumors in vivo, but not the growth of human DLD-2 colon carcinoma tumors in vivo. These results are of interest because rat C6 cells respond to hr-bFGF in vitro, whereas cultured DLD-2 cells lack high affinity hr-bFGF receptors and are unresponsive to hr-bFGF in culture [4]. Also, as shown by the lack of significant efficacy by Brequinar sodium against C6 glioma tumor growth in vivo, growth of this tumor type may be particulary difficult to control by standard anti-cancer agents in preclinical models in vivo.

Members of the FGF family may play a significant role in the growth and propagation of human gliomas. Since this tumor kills patients by invasion and local dissemination rather than by metastasis, peptides which stimulate growth, invasion and vascularization of this tumor in the cranium are most likely to be significant. There is now ample evidence to support roles for members of the FGF family in cultured glioma cells, as well as in brain neoplasms from patients [12–14 and S. Brem, unpublished observations]. These results suggest that interference with expression [23] or signal transduction in the FGF family may represent a form of therapy for malignant neoplasms of the central nervous system.

References

1 Rifkin DB, Moscatelli D: Recent developments in the cell biology of basic fibroblast growth factor. J Cell Biol 1990; 109: 1–6.
2 Morrison RS, de Vellis J: Growth of purified astrocytes in a chemically defined medium. Proc Natl Acad Sci USA 1981; 78: 7205–7209.
3 Morrison RL, Gross JL, Herblin WF, Reilly TM, LaSala PA, Alterman RL, Moskal JR, Kornblith PL, Dexter DL: Basic fibroblast growth factor-like activity and receptors are expressed in a human glioma cell line. Cancer Res 1990; 50: 2524–2529.
4 Gross JL, Morrison RS, Eidsvoog K, Herblin WF, Kornblith PL, Dexter, DL: Basic fibroblast growth factor: a potential autocrine regulator of human glioma cell growth. J of Neuroscience Res 1990; 27: 689–696.
5 Shing YJ, Folkman J., Haudenschild C, Lund D, Crum R, Klagsburn M: Angiogenesis is stimulated by a tumor-derived endothelial cell growth factor. J Cell Biochem 1985; 29: 275–287.
6 Folkman J: What is the evidence that tumors are angiogenesis dependent? J Natl Cancer Inst 1990; 82: 4–6.

7 Neufeld G, Mitchell R, Ponte P, Gospodarowicz D: Expression of human basic fibroblast growth factor cDNA in baby hamster kidney-derived cells results in autonomous cell growth. J Cell Biol 1988; 106: 1385–1394.

8 Smith R, Peters G, Dickson C: Multiple RNAs expressed from the *int-2* gene in mouse embyronal carcinoma cell lines encode a protein with homology to fibroblast growth factors. The EMBO J 1988; 7: 1013–1022.

9 Yoshida T, Miyagawa K, Odagiri H, Sakamoto H, Little PFR, Terada M, Sugimura T: Genomic sequence of hst, a transforming gene encoding a protein homologous to fibroblast growth factors and the *int-2*-encoded protein. Proc Natl Acad Sci USA 1987; 84: 7305–7309.

10 Delli Bovi P, Curatola AM, Kern FG, Greco A, Ittmann M, Basilico C: An oncogene isolated by transfection of Kaposi's sarcoma DNA encodes a growth factor that is a member of the FGF family. Cell 1987; 50: 729–737.

11 Zhan X, Mates B, Hu X, Goldfarb M: The human *FGF-5* oncogene encodes a novel protein related to fibroblast growth factors. Mol Cell Biol 1988; 8: 3487–3495.

12 Libermann TA, Friesel R, Jaye M, Lyall RM, Westermark B, Drohan W, Schmidt A, Maciag, T, Schlessinger J: An angiogenic growth factor is expressed in human glioma cells. The EMBO J 1987; 6: 1627–1632.

13 Murphy PR, Myal Y, Sato Y, Sato R, West M, Friesen HG: Elevated expression of basic growth factor messenger ribonucleic acid in acoustic neuromas. Mol Endocrinol 1989; 3: 225–231.

14 Takahasi J, Mori H, Fukumoto M, Igarashi K, Jaye M, Oda Y, Kikuchi H, Hatanaka M: Gene expression of fibroblast growth factors in human gliomas and meningiomas: Demonstration of cellular source of basic fibroblast growth factor mRNA and peptide in tumor tissues. Proc Natl Acad Sci USA 1990; 87: 5710–5714.

15 Partanen J, Makela P, Eerola E, Korhonen J, Hirvonen H, Lasesson-Welsh L, Alitalo, K: FGFR-4, a novel acidic fibroblast growth factor receptor with a distinct expression pattern. The EMBO J 1991; 10: 1347–1354.

16 Reilly TM, Taylor DS, Herblin WF, Thoolen MJ, Chiu AT, Watson DM, Timmermans PBMWM: Monoclonal antibodies directed against basic fibroblast growth factor which inhibit its biological activity in vitro and in vivo. BBRC 1989; 164: 736–743.

17 Gross JL, Behrens DL, Mullins DE, Kornblith PL, Dexter DL: Plasminogen activator and inhibitor activity in human glioma cells and modulation by sodium butyrate. Cancer Res 1988; 48: 291–296.

18 Sambrook J, Fritsch EF, Maniatis T: Molecular Cloning: A laboratory Manual (Cold Spring Harbor Lab., Cold Spring Harbor, NY) 2nd Edn, 1989.

19 Hayek A, Culler FL, Beattie GM, Lopez AD, Cuevas P, Baird A: An in vivo model for the study of the angiogenic effects of basic fibroblast growth factor. BBRC 1987; 147: 876–880.

20 Dexter DL, Hesson DP, Ardecky RJ, Rao GV, Tippett DL, Dusak BA, Paull KD, Plowman J, DeLarco BM, Narayanan VL, Forbes M: Activity of a novel 4-quinolinecarboxylic acid, NSC368390 [6-fluoro-2-(2′-fluoro-1, 1′-biphenyl-4-yl)-3-methyl-4-quinolinecarboxylic acid sodium salt], against experimental tumors. Cancer Res 1985; 45: 5563–5568.

21 Yayon A, Klagsburn, M: Autocrine regulation of cell growth and transformation by basic fibroblast growth factor. Cancer Metast Rev 1990; 9: 191–201.

22 Eidsvoog, KE, Horlick RA, Herblin WF, Gross JL: Evidence for two putative basic fibroblast growth factor receptors in normal and transformed human cell lines. J Cell Biochem 1991; Suppl. 15F: 228.

23 Morrison RS: Suppression of basic fibroblast growth factor expression by antisense oligodeoxynucleotides inhibits the growth of transformed astrocytes. J Biol Chem 1991; 226: 728–734.

Angiogenesis: Key Principles - Science - Technology - Medicine
ed. by R. Steiner, P. B. Weisz & R. Langer

A protein RNase inhibitor (RNasin) expresses anti-angiogenic properties in mice

Igor J. Polakowski[a], Kenneth Lewis[b] and Robert Auerbach[a]

Laboratory of Developmental Biology, University of Wisconsin[a] and Promega Corporation[b], Madison, Wisconsin 53706, USA

Angiogenin is a protein which in addition to its ability to induce angiogenesis can function as a ribonuclease [1].[1] First described and isolated by Vallee and his collaborators, it has since been fully sequenced and subjected to extensive biochemical analysis. Angiogenin is a potent angiogenesis-inducing molecule. Angiogenin shows selective ribonuclease activity, readily breaking down tRNA but yielding larger intermediate fragments than pancreatic ribonulease A when acting on 18S or 28S RNA. Angiogenin has been synthesized by recombinant technology and site-specific mutagenesis studies have been carried out to help identify those moieties significant for ribonuclease and angiogenic activity of the molecule. Interestingly, while in most instances the efficacy of angiogenin as an angiogenesis inducer and as a ribonuclease were correlated, mutants have now been identified which show an increase in ribonuclease activity with a concomitant decrease in angiogenic activity. Recently, Vallee and his colleagues have suggested that the Rnase active site is required but not sufficient for angiogenic activity and that a second site on the molecule is required in addition [2]. What mechanism allows angiogenin to act as an inducer of new blood vessel formation is not known, although binding studies suggest that angiogenin can act directly on endothelial cell targets.

Placental ribonuclease inhibitor (PRI) binds to angiogenin with 100 times more efficiency than it does to other ribonucleases [cf. 1]. PRI abrogates the ability of angiogenin to induce neovascularization in the chick embryo chorioallantoic membrane assay. Since angiogenin is present in high amounts in serum, it is likely that its activity is inhibited unless special local conditions call for its action. Such inhibition may well be mediated by ribonuclease inhibitors such as PRI.

[1]This recent review provides a thorough overview and excellent bibliography of the many studies of B. L. Vallee and his group.

The purpose of our experiments was to determine whether PRI can inhibit angiogenesis in a mouse model. Using RNasin^{T}, a recombinant protein based on placental ribonuclease inhibitor nucleotide sequence, we have measured the induction of endothlial cell migration/angiogenesis in a sponge implant, the disc angiogenesis assay developed by Fajardo et al. [3].

Materials and methods

Mice

All experiments were carried out with C57BL/6 mice raised in our own facility.

Angiogenesis assay

The disc angiogenesis system we used was modelled after the previously published protocols of Fajardo et al. [3]. In brief, polyvinyl sponge discs were prepared as described previously. A central core was removed and permitted to absorb known amounts of test preparations. The loaded sponge was then coated with ethylene vinyl copolymer (ELVAX) to achieve a slow-release formulation. The core was reinserted into the sponge disc and the whole disc was then covered on its broad surfaces by filter paper (Whatman # 1) made impermeable by filling the pores with plastic filler (plexiglas/ethylene dichloride formulation). Assembled sponges were then inserted subcutaneously into the flank of adult mice and recovered after 12–14 days in vivo.

Analysis

Discs were embedded in paraffin, sectioned tangentially (7–10 μm) and stained with hematoxylin/eosin to permit visualization of the area of the disc penetrated by migrating cells. The total area of the disc in each section was calculated as was the area containing stained cells. Initially, measurements were made on projected images, while in later experiments digitization and subsequent calculations were made using a computerized image analysis system (IAN+) developed in our laboratory [4].

RNasin

Special procedures[2] were used to reduce the amount of endotoxin in the recombinant protein samples. RNase inhibition studies were carried

[2]The procedures used to achieve purification include proprietary information.

out in vitro to establish specific activity of each preparation. RNasin inactivation was accomplished by heating of the RNasin preparation for 10 min at 58°C.

Results

A total of 60 discs containing RNasin (2300 units/disc) and 60 control discs (heat-inactivated RNasin or PBS) were implanted, each animal receiving one active RNasin and one inactive control disc. 50+ discs were recovered for each of the two groups. The extent of preparation was measured by calculating the disc area and the penetrated area for each implant. While the mean penetration of control discs was 24.9%, penetration in the RNasin-containing discs was 10.3. The difference was highly significant ($p < 0.001$).

Preliminary experiments were carried out using angiogenesis-inducing factors such as basic fibroblast growth factor (bFGF), the results confirming the earlier reports of Fajardo et al. Interestingly, addition of RNasin to FGF-containing cores reduced the angiogenic reaction although the resultant core penetration was greater than that seen with RNasin alone.

Discussion

There are many possible explanations for our demonstration that RNasin can inhibit the angiogenic response in mice as measured in the disc angiogenesis assay. Although RNasin has as its only known function the ability to inhibit ribonucleases the simplest assumption is that its effect on angiogenesis is associated with ribonuclease inhibition. If so, then it becomes signficant that ribonucleases other than angiogenin appear not to induce neovascularization. Moreover, the binding affinity of RNasin to angiogenin far exeeds its affinity to other ribonucleases. This then leads to the proposition that RNasin's inhibition of angiogenesis is effected by its action on angiogenin.

One must overlook the possibility that the activity of RNasin in our test system is not based on inhibition of ribonuclease activity at all. There may well be additional domains in the RNasin molecule that do not affect the active site of ribonucleases. Indeed one such domain may bind with a second site on the angiogenin molecule essential for its angiogenic activity.

However, one must still wonder why RNasin can inhibit the vascular reaction induced by exogenous inducers of angiogenesis such as FGF. One possibility would be that angiogenin (or other ribonucleases) participate at some point in the angiogenic cascade distal to that of the initiators of the reaction. But the distinct possibility remains that RNasin exerts its effects on angiogenesis by an as yet undiscovered pathway.

References

1 Fox EA, Riordan JF: Molecular Biology of Angiogenin. In Chien S (ed); Molecular Biology of the Cardiovascular System, Chapter 8, Philadelphia, Lea & Febiger, 1990, pp. 139–154.
2 Hallahan TW, Shapiro R, Vallee BL: Dual site model for the organogenic activity of angiogenin. Proc Natl Acad Sci USA 1991; 99: 2222–2226.
3 Fajardo LF, Kowalski J, Khwan HH, Prionas SD, Allison AC: The disc angiogenesis system. Lab Invest 1988; 58: 718–724.
4 Obeso J, Weber J, Auerbach R: A hemangioendothelioma-derived cell line: its use as a model for the study of endothelial cell biology. Lab Invest 1990; 63: 259–269.

Angiogenesis: Key Principles - Science - Technology - Medicine
ed. by R. Steiner, P. B. Weisz & R. Langer

Reactions of microvascular endothelial cells to tumor necrosis factor in vivo

Yury Persidsky and Arkady Frolov

Kiev Research Institute of Epidemiology and Infectious Disease, 252038 Kiev, USSR

Introduction

Bacterial and some viral infections may lead to a state of shock complicated by multiple system organ failure, adult respiratory distress syndrome, and disseminated coagulation defects. Catastrophic sequences of infections may be attributed to the effects of the cytokine tumor necrosis factor (TNF-α) [1]. Most of the above-mentioned deleterious consequences of infections are connected with the damage of microvessels in internal organs [2]. Microvascular endothelial cells (MEC) may be considered major cellular targets for TNF-α actions [3]. The main aim of our study was to investigate by means of transmission (TEM) and scanning (SEM) electron microscopy the reactions of MEC to relatively high doses of rhTNF-α.

Materials and methods

Recombinant human TNF-α at a dose of 10^6 U was intravenously injected to S57BL/6 mice. Mice were sacrificed 1, 2, 4, 8 and 24 h after injection. Small pieces of lung, liver, kidney, brain were prepared for SEM and TEM investigation and examined in Superprobe-733 (Jeol) and JEM 100 CX (Jeol).

Results

The study of MEC in various organs revealed common structural features depending on the time following TNF injection. During the first 1–2 h the so-called "stress-reaction" appeared in the MECs. It consisted in bulging of nuclear zone and increased numbers of filaments. The changes mentioned above were accompanied by signs of fluidity of cell membranes such as formation of blebs and projections (Fig. 1).

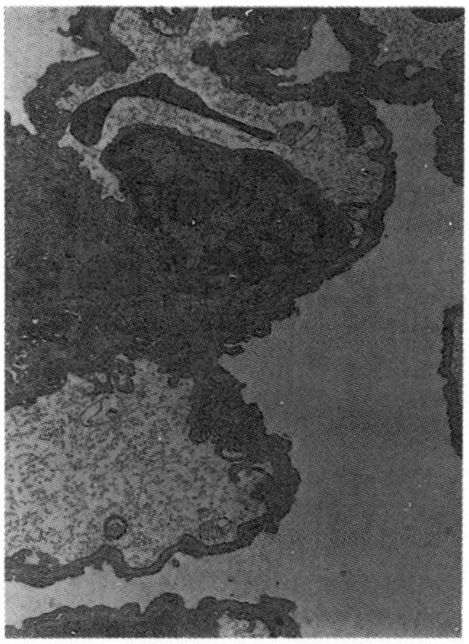

Figure 1. Vacuolation and partial detachment of endothelial cell to interalveolar septal capillary in lung. TEM × 5800.

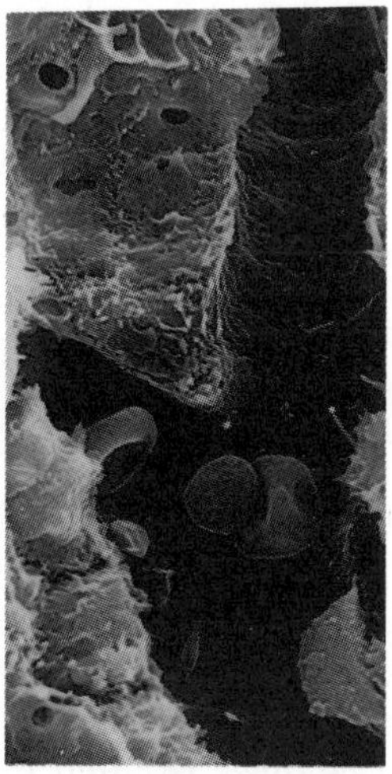

Figure 2. Decrease of the number of pores in liver sinusoidal endothelial cells. SEM × 3600.

Progressive decrease of the number of pores and fenestrae in the cytoplasm of liver sinusoid MEC appeared to be connected with membrane instability (Fig. 2). There were two possibilities concerning further development of MEC morphology alterations 4 h after TNF injection. Part of the MECs demonstrated necrobiotic injury. Other MECs showed the growth of a number of biosynthetic organelles. Such reorganization of MEC structure was the most pronounced in the venular part of the microvessels, and the same transformation was noted in capillaries. The outline of these MECs was very irregular, marginating and spreading

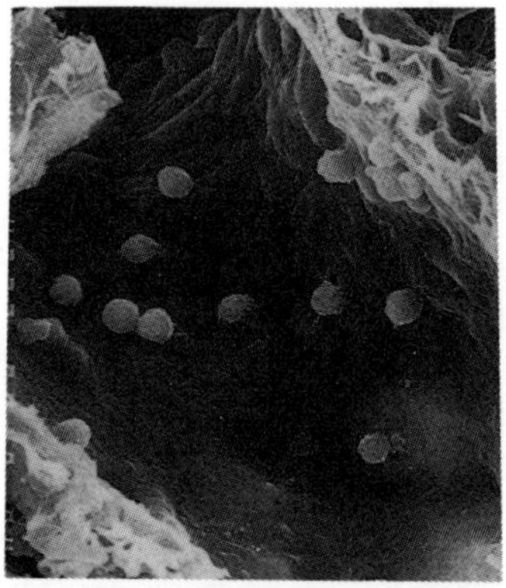

Figure 3. Uneveness of endothelial covering in liver central vein. Adhesion of polymorphonuclear leucocytes to luminal surface of endothelium. SEM × 1200.

leucocytes were revealed on their luminal cytolemmas (Fig. 3). Structural changes in MECs resulted in specific damage in organs. Interstitial and alveolar edema developed in the lung. Kupffer cell activation and acute lipid dystrophia of hepatocytes were revealed in the liver. Damage of peritubular and glomerular capillaries, necrobiotic changes of tubular epithelial cells were found in the kidney.

Discussion

As recently reported, the above mentioned ultrastructural features accompanied by complex functional changes were related to the process of activation of MEC [2]. This special state included the expression of adhesive molecules and chemotactic factors for leucocytes, increased procoagulant activity, synthesis of endothelium derived relaxing factor [2, 4, 5]. These functional changes resulted in a permeability increase, adhesion and activation of leucocytes, and fibrin deposition in lumen of microvessels. This damage of microvessels led to multiple organ failure (respiratory distress syndrome in lung, tubular necrosis in kidney, acute fatty liver, brain edema). The results of our investigation proved that MEC of internal organs were one of the major cellular targets for TNF action in vivo.

References

1 Tracey K, Cerami A: Cachectin/tumor necrosis factor and other cytokines in infectious diseases. Curr Opin Immunol 1989; 1: 454–461.

2 Pober J: Cytokine-mediated activation of vascular endothelium. Am J Pathol 1988; 133: 426–433.

3 Beutler B, Cerami A: The biology of cachectin/TNF-α-primary mediator of the host response. Ann Rev Immunol 1989; 7: 625–655.

4 Cavender D, Edelbaum D, Ziff M: Endothelial cell activation induced by tumor necrosis factor and lymphotoxin. Am J Pathol 1989; 134: 551–560.

5 Schleef R, Bevilaqua M, Sawdey M, Gimbrone M, Loskutoff D: Cytokine activation of vascular endothelium. Effects on tissue-type plasminogen activator inhibitor. J Biol Chem 1988; 263: 5797–5803.

Angiogenesis: Key Principles - Science - Technology - Medicine
ed. by R. Steiner, P. B. Weisz & R. Langer

Studies on tumor induced angiogenesis

J. L. Ambrus, C. M. Ambrus, P. Forgach, S. Stadler, J. Halpern, S. Sayyid, P. Niswander and C. Toumbis

The Roswell Park Cancer Institute, State University of New York at Buffalo, Departments of Medicine, Ophthalmology, and Experimental Pathology, Buffalo, New York 14263, USA

Abstract. Methods were developed to test angiogenic response to human tumor implants and various biologic agents in the cornea of rabbits and non-human primates (*Macaca arctoides*). Crude PDGF preparations were found to have significant angiogenic effect. Purified, recombinant PDGF preparations were also effective inhibitors (e.g. pentoxifylline (Px) (which also were found to release PgI2 and t-PA) inhibited human tumor implant induced angiogenesis and reduced spontaneous metastases in 3 transplantable murine tumors (Furth-Columbia Wilm's tumor in Furth-Wistar rats, C-1300 neuroblastoma in A/J mice and HM-Kim mammary carcinoma in Wistar rats) but not in the NIH adenocarcinoma in Balb/c mice. Sodium diethyldithiocarbamate (DDTC), a metal complexing agent with special affinity to copper and anti-thyroid as well as, immune stimulating activity was shown to be anti-angiogenic and to potentiate the effect of Px. The anti-fibrinolytic agents epsilon amino caproic acid (EACA) and tranaxamic acid (t-AMCHA) were anti-angiogenic. DDTC and Px were synergistic from this point of view.

Introduction

Malignant tumor growth appears to be angiogenesis dependent and angiogenesis is directly or indirectly induced by tumors. Angiogenesis is an important stage of carcinogenesis, particularly in the development of newly established metastases. A large number of angiogenic substances have been described [reviews: 1–5] and anti-angiogenic factors identified [6–11].

In the present study we wanted to explore factors which may be involved in tumor induced angiogenesis and possible antagonists. In earlier studies [12–16] we found that in the tortuous A–V anastomosis rich tumor microvasculature there is sluggish flow and continuous aggregation of platelets which release various platelet factors including the platelet derived growth factors (PDGF). We have explored the angiogenic effect of crude platelet extracts and certain recombinant PDGF's.

We also wanted to explore possible inhibitors of tumor induced angiogenesis. McAuslan and others [17, 18] reported that copper containing protein factions and copper ions themselves had angiogenic activity. It was thought that copper prosthetic groups may be involved in angiogenesis. In earlier studies we have found [19] that the copper complexing agent, sodium diethyldithiocarbamate (NaDDTC) is relatively non-toxic in animals except for a mild antithyroid activity. Renoux et al. [20–22] reported that this agent has immune modulatory activity. It increases cell mediated immunity, recruits and activates T-cells. Others have found that this compound inhibits cis-platinum induced nephrotoxicity, acts as a radiation sensitizing agent and inhibits superoxide dismutase [23–26]. For these reasons we wanted to explore this compound alone and in combination with other potential anti-angiogenic agents.

In earlier studies we found that the methylxanthine derivative pentoxifylline increases red cell membrane fluidity, inhibits platelet aggregation and white cell adhesion, thus promoting the microcirculation [27–29]. We also found that this agent can prevent metastases of certain experimental tumors [30–32].

In earlier studies we investigated plasmin inhibitors [33–36]. We hypothesized that the fibrinolysin system plays a role in keeping the abnormal microvasculature of tumors open. For this reason, we thought that fibrinolysin inhibitors such as epsilon aminocaproic acid (EACA) or tranaxamic acid (t-AMCHA) may reduce vascularization of neoplastic tissue. We also wanted to explore combinations of some of the above materials which presumably would work through different mechanisms producing an anti-angiogenic effect.

Materials and methods

In preliminary experiments we explored angiogenic and anti-angiogenic effects in the dorsal air sac technique of Folkman et al. [37] modified by Phillips and Kumar [38] using ICR/Ha male mice and in the chorioallantoic membrane assay of Gullino [39] and Jacob et al. [40, 41]. On the basis of these results experiments were planned using the corneal angiogenesis method of Folkman et al. [5, 6] in New Zealand white rabbits and *Macaca arctoides* monkeys. The materials to be studied were incorporated into purified Elvax-40 membranes inserted into the cornea. We found that Elvax-40 (ethylene-vinyl acetate copolymer, Dupont Co.) contained various interfering substances and for this reason it was necessary to purify it. It was washed with 2.5% ethanol at 37°C for 8 h. At least 50 washing cycles were employed. The material was then dissolved in methylene chloride at 37°C to give a 10% solution (w/v). Material to be tested was added to this solution, thoroughly mixed and then poured into a glass mold resting on a block of dry ice. After 10 min, the solidified membrane was transferred to a Nitex screen in a glass petri dish resting on a block of dry ice. After 30 min the membrane was dried in a dessicator under vacuum at room temperature. These membranes were found not to be angiogenic by themselves. These membranes were inserted into the cornea. The operated area was inspected daily with a slit-lamp microscope using a green filter at 40 × 's magnification. Final scoring was done 7 days after implantation by counting the vessels which appear to penetrate the operated area per high powered field. The enclosed figures show the actions scored: 1+ (5–20 ingrowing vessels), 2+ (21–50 ingrowing vessels), 3+ (> 50 ingrowing vessels).

Materials studied included crude platelet derived growth factor prepared from normal human platelets kindly supplied by Dr Allan Lipton, University of Pennsylvania, Hershey Medical Center, Hershey, PA. Extracts of 10^9 platelets/ml were incorporated into the purified Elvax-40/ml. The platelet extract was prepared by the method of Witkowski et al. [42]. Recombinant B/B homodimer of PDGF was kindly supplied by Drs. O. Friedman and P. Kelly of the Collaborative Research Laboratories, Boston, MA. Isolation, characterization and assay of platelet derived growth factor was reported by Ross et al. [43]; Antoniades et al. [44, 45]; Westermark et al. [46] and Heldin et al. [47].

Human malignant melanoma cell culture was established from biopsies, normal keratinocyte cultures were established from the surrounding normal skin area of the same patients.

Sodium diethyldithiocarbamate (NaDDTC) was obtained from the Sigma Chemical Co., (St. Louis, MO). Pentoxifylline (Trental) was kindly supplied by Drs. Philip Savitsky and Barton Cobert from the Hoechst Roussel Pharmaceutical Co., (Somerville, NJ). Epsilon aminocaproic acid (Amicar) was kindly supplied by the Lederle Co. Aminocaproic acid (Cyclokapron, Kabi Vitrum) was kindly supplied by the Kabi Pharmaceutical Co., (Stockholm, Sweden).

Results and discussion

Table 1 summarizes the results in both rabbit and monkey corneas. Purified Elvax-40 had no angiogenic effect but when crude PDGF was incorporated significant angiogenesis developed more so in rabbits than in monkeys. Recombinant *E. coli* derived PDGF, BB homodimer was angiogenic in both species. On the other hand, Maione et al. [48] reported that platelet factor 4 is anti-angiogenic. It is possible that PDGF and platelet factor 4 are both released in the vascular bed of tumors, but in most instances PDGF predominates. Human malignant melanoma cells were highly angiogenic but keratinocytes cultured from surrounding normal skin had no effect. When malanoma cells were combined with NaDDTC, angiogenesis was inhibited more in monkeys than in rabbits. The hemorheologic agent, pentoxifylline also had inhibitory activity. When pentoxifylline and NaDDTC were combined

Table 1. Effect of human melanoma-induced angiogenesis

Corneal implant	Mean degree of angiogenesis* in rabbits	in *Macaca arctoides* monkeys (n = 4)
Purified Elvax-40 (E)	0	0
E + Crude PDGF** from 10^9 human platelets/ml	3	2
E + rPDGF B/B homodimer 50 half max. u/ml	2	2
E + 10^9 human malignant melanoma cells (M)/ml	3	3
E + 10^9 human keratinocytes/ml	0	0
E + M + NaDDTC*** 25 mg/ml	2	1
E + M + Pentoxifylline 10 mg/ml	2	2
E + M + Px + NaDDTC	1	0
E + M + EACA**** 20 mg/ml	2	2
E + t-AMCHA***** 20 mg/ml	2	2

*No. of Capillaries HPF (40 ×)	Designation
0–4	0+
5–20	1+
21–50	2+
>50	3+

**PDGF = Platelet derived growth factor. ***NaDDTC = Sodium diethyldithiocarbamate; ****EACA = Epsilon aminocaproic acid (Amicar); *****t-AMCHA = Transaxamic acid (Cyclokapron).

there was significant potentiation of anti-angiogenic activity by these two agents, which presumably work through different mechanisms. The two anti-fibrinolytic agents epsilon aminocaproic acid and tranaxamic acid were both effective in slightly reducing tumor induced angiogenesis.

Figure 1 shows a rabbit with an implanted Elvax-40 membrane. Figure 2 shows high powered fields of the operating microscope through which the degree of angiogenesis was scored. It shows 1+ (Figure 2), 2+ (Figure 3) and 3+ (Figure 4) reactions. In an earlier paper [49] we pointed out that there is host against tumor defense but also tumor against host defense. The latter includes the production of prostaglandins (particularly PgE2) which is angiogenic [4] and inhibits antitumor immunity including interferon production [50, 51].

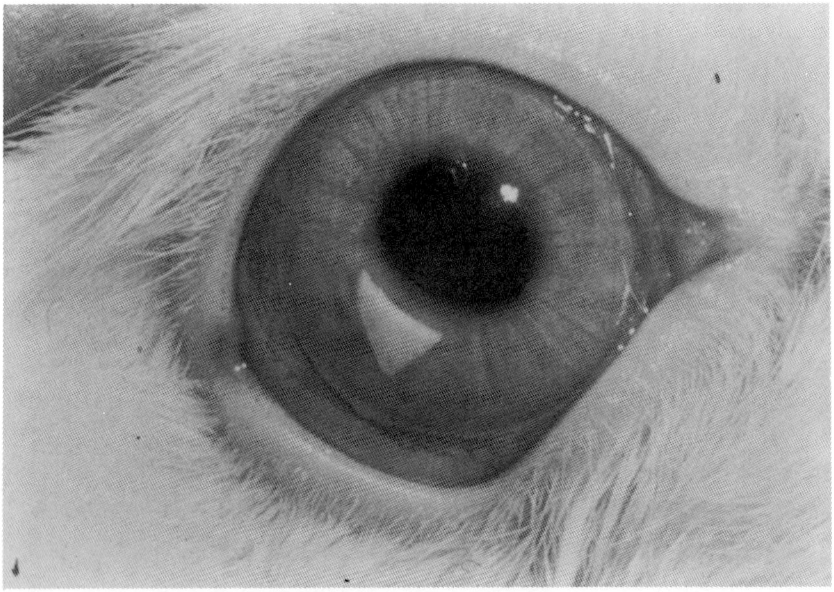

Figure 1. Elvax-40 membrane in rabbit cornea.

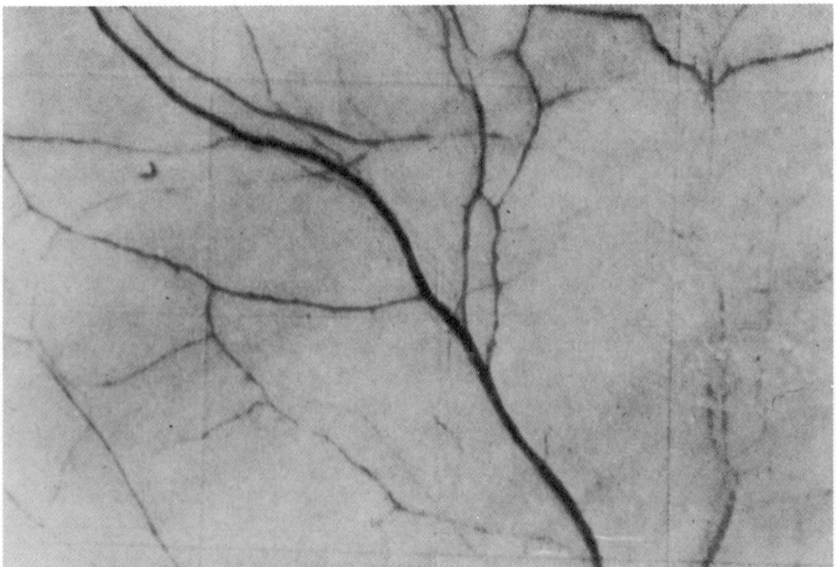

Figure 2. Grade I angiogenic reaction.

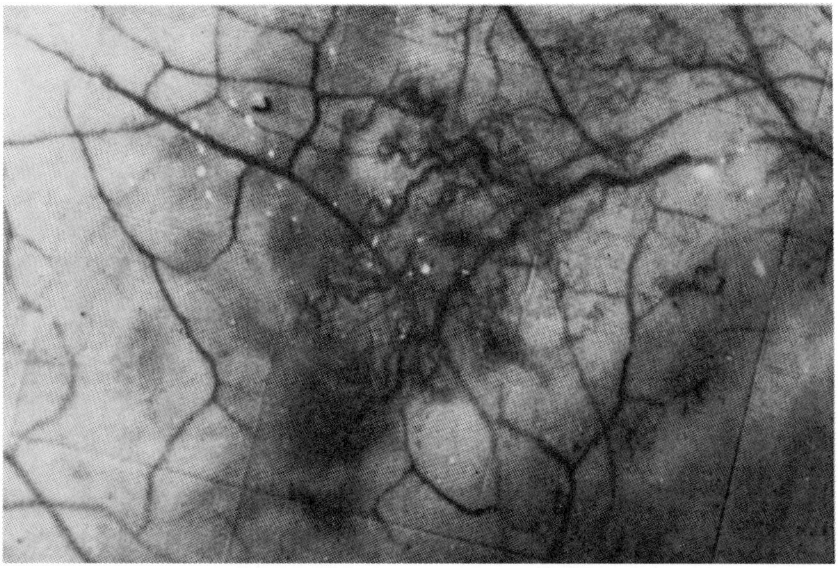

Figure 3. Grade II angiogenic reaction.

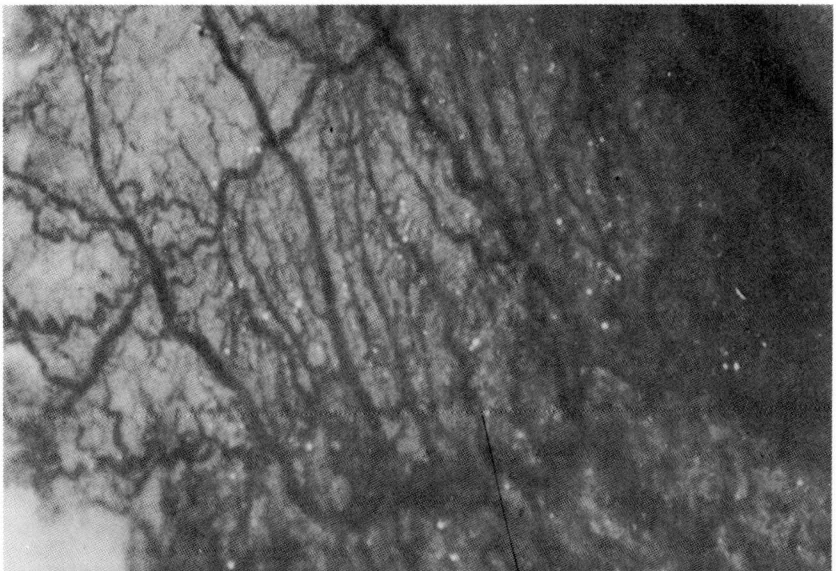

Figure 4. Grade III angiogenic reaction.

Conclusions

It appears from these data that platelet aggregation and release derived PDGF may be a contributory factor in angiogenesis developing in abnormal vascular beds. Human malignant cells as opposed to normal cells appear to be angiogenic in both rabbits and monkeys. This effect is inhibited by the copper complexing agent NaDDTC and the hemorheologic agent pentoxifylline. The two work presumably through different mechanisms and synergize with each other. Two anti-fibrinolytic agents epsilon aminocaproic acid and tranaxamic acid were also shown to have some anti-angiogenic effect. This suggests that the fibrinolysin system may play a role in keeping newly formed capillary networks open. Anti-angiogenic agents may have therapeutic potential in preventing growth of newly established metastatic tumors. New studies with systemic therapy are being planned.

References

1 Klinworth GK: Corneal Angiogenesis: A Comprehensive Critical Review. Springer Verlag, NY, 1991.
2 Rifkin DB, Klagsbrun M: Angiogenesis: Mechanisms and Pathobiology. Cold Spring Harbor Laboratory Press, 1987.
3 Folkman J: Tumor angiogenesis. In Becker FF (ed): Biology of Tumors, Plenum Press, NY, 1975, pp 355–388.
4 Gullino PM: Angiogenesis factor(s). In Baserga R (ed.): Tissue Growth Factors. Springer-Verlag, 1981, pp 428–449.
5 Folkman J: Tumor angiogenesis. Adv Cancer Res 1974; 19: 331–358.
6 Langer R, Folkman J: Angiogenesis inhibitors, in Sartorelli, AC, Lazo, JS, Bertino J (eds): Molecular Actions and Targets for Cancer Chemotherapy Agents. Academic Press, NY, 1981, pp 511–525.
7 Brem H, Folkman J: Inhibition of tumor angiogenesis mediated by cartilage. J Exp Med 1975; 141: 427–439.
8 Cariou R, Harrousseau JL, Tobelem G: Inhibition of human endothelial cell proliferation by heparin and steroids. Cell Biol Int Rep 1988; 12: 1037–1047.
9 Moses MA, Sudhalter J, Langer R: Identification of an inhibitor of neovascularization from cartilage. Science 1990; 248: 1408–1409.
10 Yang EY, Moses HL: Transforming growth factor beta-1 induced changes in cell migration; proliferation and angiogenesis in the chicken choriallantoic membrane. J Cell Biol 1990; 111: 731–741.
11 Pepper MS, Belin D, Montesano R, Orci L, Vassali JD: Transforming growth factor-beta 1 modulates basic fibroblast growth factor-induced proteolytic and angiogenic properties of endothelial cells in vitro. J Cell Biol 1990; 111: 743–755.
12 Ambrus JL, Ambrus CM: Blood coagulation in neoplastic disease. In Gastpar H (ed): Onkohamostaseologie (Hematologic Problems in Cancer). F. K. Schattauer Verlag, Stuttgart-New York, 1976, pp 167–193.
13 Ambrus JL, Ambrus CM, Gastpar H: Studies on platelet aggregation and platelet interaction with tumor cells. In de Gaetano G, Garattini S (eds): Platelets: A Multidisciplinary Approach. Raven Press, NY, 1978, pp 467–480.
14 Ambrus JL, Ambrus CM: Anticoagulants, fibrinolytic enzymes and platelet interaction with tumor cells. In Brodsky I, Kahn SB, Conroy JF (eds): Cancer Chemotherapy III (The 46th Hahnemann Symposium). Grune and Stratton, Inc., NY, 1978a, pp 97–109.

15 Ambrus JL, Ambrus CM, Gastpar H: The role of platelet aggregation in metastatic dissemination and thromboembolic complications in cancer patients. Prevention with pentoxifylline. In Manrique, RV, Muller R (eds): Disorders of Blood Flow: New Therapeutic Aspects. Excerpta Medica, Amsterdam-Oxford-Princeton, 1981, pp 29–42.
16 Ambrus JL, Gastpar H: Plättchenaggregationshemmer in der Krebsforschung (Platelet aggregation inhibitors in oncology). In Kostering H (ed): Onkologie und Blutgerinnung (Oncology and Blood Coagulation). Roche Editions, Basel, Switzerland, 1983, pp 201–214.
17 McAuslan BR, Hoffman H: Endothelium stimulating factor from Walker carcinoma cells. Relation to tumor angiogenic factor. Exp Cell Res 1979; 119: 181–190.
18 McAuslan, BR: A new theory of neovascularization based on identification of an angiogenic factor and its effect on cultured endothelial cells. In Jimenez de Asua L, Shields R, Levi-Montalcini R, Iacobelli S (eds): Control Mechanisms in Animal Cells. Raven Press, NY, 1980, pp 285–292.
19 Ambrus CM, Ambrus JL, Harrisson JWE: Effect of sodium diethyl-dithio-carbamate on thyroid activity. Am J Pharm 1951; 123: 129–301.
20 Renoux G, Renoux M: Thymus-like activities of sulfur derivatives on T-cell differentiation. J Exp Med 1977; 145: 466–471.
21 Renoux G, Renoux M: The effects of sodium diethyl-dithiocarbamate, azathioprine, cyclophosphamide, or hydrocortisone acetate administered alone or in association for 4 weeks on the immune responses of BALB/c mice. Clin Immunol Immunopathol 1980; 15: 23–32.
22 Renoux G: Levamisole and sodium diethyldithiocarbamate. Trends Pharmacol Sci 1981; 2: 248–249.
23 Borch RF, Plasants ME: Inhibition of cis-platinum nephrotoxicity by diethyldithiocarbamate rescue in a rat model. Proc Natl Acad Sci 1979; 76: 6611–6614.
24 Risas DA, Esinitis-Risas C, Bisl RH, Stankova L, Head C: Biphasic radio-sensitive of human lymphocytes by diethyldithiocarbamate: possible involvement of superoxide dismutase. Int J Radiat Biol 1980; 38: 257–266.
25 Masos L, Jarvis JA: The paradox effect of carbon disulfide and diethyldithiocarbamate of adrenal dopamine-beta-hydroxylase. G Ital Med Lav 1981; 3: 107–108.
26 Guarnieri C, Flamisni F, Ventura C, Rossoni-Caldarera C: Inhibition of rat heart superoxidase dismutase activity by diethyldithiocarbamate and its effect on mitochondrial function. Biochem Pharmacol 1981; 30: 2174–2176.
27 Ambrus JL: Microvascular changes in diabetes, in Kerstein MD (ed): Diabetes and Vascular Disease. J. B. Lippincott Co., Phil. Pa., 1990, pp 9–39.
28 Gastpar H, Ambrus JL, Ambrus CM, Spavento P, Weber FJ, Thurber LE: Study of platelet aggregation in vivo III. Effect of pentoxifylline. J Med 1977; 8: 191–197.
29 Gastpar H, Ambrus JL, Ambrus CM: Studies on platelet aggregation in vivo VII. Effect of pentoxifylline on circulating tumor cells. J Med 1978; 9: 265–268.
30 Gastpar H, Ambrus JL, Ambrus CM: Platelet aggregation inhibitors in the prevention of metastasis in neoplastic disease. A review of experimental and clinical investigations. J Med 1981; 12: 15–33.
31 Ambrus JL, Ambrus CM, Gastpar H: Effect of phosphodiesterase inhibitors on platelet aggregation and tumor metastasis. Proc. Conference on Blood Rheology and Microcirculation, Capri, Italy. La Ricerca XI (suppl 1) 1981; 197–207.
32 Gordon S, Witul M, Cohen H, Sciandra J, Williams P, Gastpar H, Murphy GP, Ambrus JL: Studies on platelet aggregation inhibitors in vivo. VIII. Effect of pentoxifylline on spontaneous tumor metastasis. J Med 1979; 10: 435–443.
33 Ambrus CM, Ambrus JL, Lassman HB, Mink IB: Studies on the mechanism of action of inhibitors of the fibrinolysin system. Ann NY Acad Sci 1968; 146: 430–447.
34 Ambrus JL, Ambrus CM, Stutzman L, Schimert G, Niswander KR, Woodruff MW, Magoss IV: Treatment of fibrinolytic hemorrhage with proteinae inhibitors: A preliminary report. Ann NY Acad Sci 1968a; 146: 625–641.
35 Ambrus CM, Ambrus JL, Lassman HB, Mink IB: On the heterogeneity of activity by various types of plasmins and the spectra of inhibition of some plasmin inhibitors. Res Comm Chem Pathol Pharmacol 1970; 1: 67–85.
36 Ambrus JL, Schimert G, Lajos TZ, Ambrus CM, Mink IB, Lassman HB, Moore RH, Melzer J: Effect of anti-fibrinolytic agents and estrogens on blood loss and blood coagulation factors during open heart surgery. J Med 1971; 2: 65–81.

37 Folkman J, Merler E, Abernathy C, Williams G: Isolation of a tumor factor responsible for angiogenesis. J Exp Med 1971; 133: 275–288.
38 Phillips P, Kumar S: Tumour angiogenesis factor (TAF) and its neutralization by a xenogeneic antiserum. Int J Cancer 1979; 23: 82–88.
39 Gullino PM: Angiogenesis and oncogenesis. J Natl Cancer Inst 1978; 61: 639–643.
40 Jakob W, Jentzsch KD, Mauersberger B, Heder G: The chick embryo chorioallantoic membrane as a bioassay for angiogenesis factors: Reactions induced by carried materials. Exp Pathol 1978; 15: 241–249.
41 Jakob W, Jentzsch KD, Mauersberger B, Oehme P: Demonstration of angiogenesis activity in the corpus luteum of cattle. Exp Pathol 1977; 13: 231–236.
42 Witkowski E, Kepner N, Leitzel K, Rogers C, Jefferson LS, Lipton A: Enhanced tumor growth in vivo by a factor in human platelets and rat liver. Cancer Res 1982; 42: 2350–2352.
43 Ross R, Glomset J, Kariya B, Harker L: A platelet dependent serum factor that stimulates the proliferation of arterial smooth muscle cells in vitro. Proc Natl Acad Sci 1974; 71: 1207–1210.
44 Antoniades HN, Scher CD: Radioimmunoassay of a human serum growth factor for Balb/c-3 T3 cells: Derivation from platelets. Proc Natl Acad Sci 1977; 74: 1973–1977.
45 Antoniades HN: Human platelet-derived growth factor (PDGF): Purification of PDGF-II and separation of their subunits. Proc Natl Acad Sci 1981; 78: 7314–7317.
46 Westermark B, Wasteson A: A platelet factor stimulating human normal glial cells. Exp Cell Res 1976; 98: 170–174.
47 Heldin CH, Westermark B, Wasteson A: Purification platelet derived growth factor and partial characterization. Proc Natl Acad Sci 1979; 76: 3722–3726.
48 Maione TE, Gray GS, Petro J, Hunt AJ, Donner Al, Bauer SI, Carson HF, Sharpe RJ: Inhibition of angiogenesis by recombinant human platelet factor 4 and related peptides. Nature 1990; 247: 77–79.
49 Attallah A, Lee JB, Ambrus JL, Ambrus CM, Karakousis C, Takita H: Prostaglandin E2 production by human tumor. Defense mechanism against the host? A preliminary study. Res Comm Chem Pathol Pharmacol 1984; 43: 195.
50 O'Connor MGI, Ambrus JL, Chadha KC: Regulation of in vitro human leukocyte interferon production: Effect of prostaglandin synthetase and phosphodiesterase inhibition. Res Comm Chem Pathol Pharmacol 1986; 54: 379.
51 Ambrus JL, Poiesz BJ, Lillie MA, Stadler S, Di Berardino L, Chadha KC: Interferon and interferon inhibitor levels in patients infected with varicella-zoster virus, AIDS, ARC, Kaposi's sarcoma and in normal individuals. Am J Med 1989; 87: 405–407.

Angiogenesis: Key Principles - Science - Technology - Medicine
ed. by R. Steiner, P. B. Weisz & R. Langer

The mode of action of anti-angiogenic steroid and heparin

Ikuo Harada[a], Toshiyuki Kikuchi[a], Yutaka Shimomura[a], Mikio Yamamoto[b], Hideki Ohno[c] and Noboru Sato[b]

Departments of [a]Orthopaedic Surgery, [b]Biochemistry and [c]Hygiene, National Defense Medical College, 3–2 Namiki, Tokorozawa, Saitama 359, Japan

Introduction

Reports on the effect of dexamethasone (DM) on angiogenesis are conflicting. Gross et al. [1] reported that DM, applied locally as a sustained-release polymer, suppressed angiogenesis and tumor growth in a rabbit V2 carcinoma implanted in the rabbit cornea. However, Crum et al. [2] showed little or no anti-angiogenic activity DM even in the presence of heparin.

Data on the effect of heparin on angiogenesis are also conflicting. Although heparin itself does not initiate angiogenesis and has been shown to inhibit the growth of capillary endothelial cells (ECs) [3], it appears to gain stimulatory activity in neoplastic or non-neoplastic cell-induced angiogenesis when bound to copper ions [4] or to some growth factors probably like acidic fibroblast growth factors [5]. However, heparin becomes inhibitory in the presence of glucocorticoids; a combined heparin-steroid treatment inhibited neovascularization and induced tumor regression [6].

Recently, the anti-angiogenic action of heparin-steroid combinations have been reported to induce basement membrane breakdown, capillary retraction and endothelial rounding [7]. However, the precise mechanism by which the heparin-steroid combination induces regression of growing vessels remains uncertain. The purpose of the present study was to examine the effects of heparin and a steroid (DM) on angiogenesis and to elucidate their mode of action.

Materials and methods

We employed an in vitro angiogenesis model system [8] in this study using microvessel fragments and mesenchymal myofibroblastic cells

(Mfc), both from rat adipose tissues. In this system, feeder-layered myofibroblastic cells induce capillary formation from microvascular fragments seeded on Mfc by producing an endothelial cell mitogen (Mf-ECGF) and an extracellular matrix that includes type I collagen [8, 9]. The effect of heparin (Fluka AG, Switzerland) or DM (Banyu Co., Tokyo) on in vitro angiogenesis was examined by the addition of both drugs to the system. Six days later, capillary growth among the several cultures was compared according to the method of Sato et al. To explore the mode of action of heparin and DM, we examined their effects on capillary ECs and Mfc by determining proliferation, collagen [11] and collagenase synthesis [12], or Mf-ECGF production [10]. Capillary ECs were harvested from bovine adrenal cortex following the method of Folkman et al. [13].

Results and conclusion

Heparin (10–100 μg/ml) and DM (10^{-8}–10^{-6} M) each clearly inhibited capillary growth in vitro at all concentrations tested in a dose-dependent manner. Representative samples of the results are shown in Figure 1: Fig. 1A, control cells cultured with no additives; Fig. 1B, cells cultured with heparin (50 μg/ml); Fig. 1C, cells cultured with DM (10^{-8}).

Heparin inhibited EC proliferation as well as its collagen and collagenase synthesis (Table 1A). With respect to Mfs, heparin had no effect on proliferation and on Mf-ECGF production, but it inhibited collagen synthesis; its effect on collagenase synthesis was not determined (Table 1A). DM, on the other hand, had no effect on EC proliferation and on collagen synthesis. However, it did inhibit Mfc proliferation and collagen synthesis (Table 1B).

Table 1. Action of heparin (10 to 100 μg/ml) or dexamethasone (10^{-8} to 10^{-10} M) on endothelial cells (ECs) and myofibroblastic cells (Mfc).

A. Heparin, at all concentration levels.

	ECs	Mfc
Proliferation	↓	→
Collagen synthesis	↓	↓
Collagenase synthesis	↓	ND
Mf-ECGF production	(—)	→

B. Dexamethasone, at all concentration levels.

	ECs	Mfc
Proliferation	→	↓
Collagen synthesis	→	↓
Mf-ECGF production	(—)	→

↓: inhibition, →: no effect, ND: not determined.

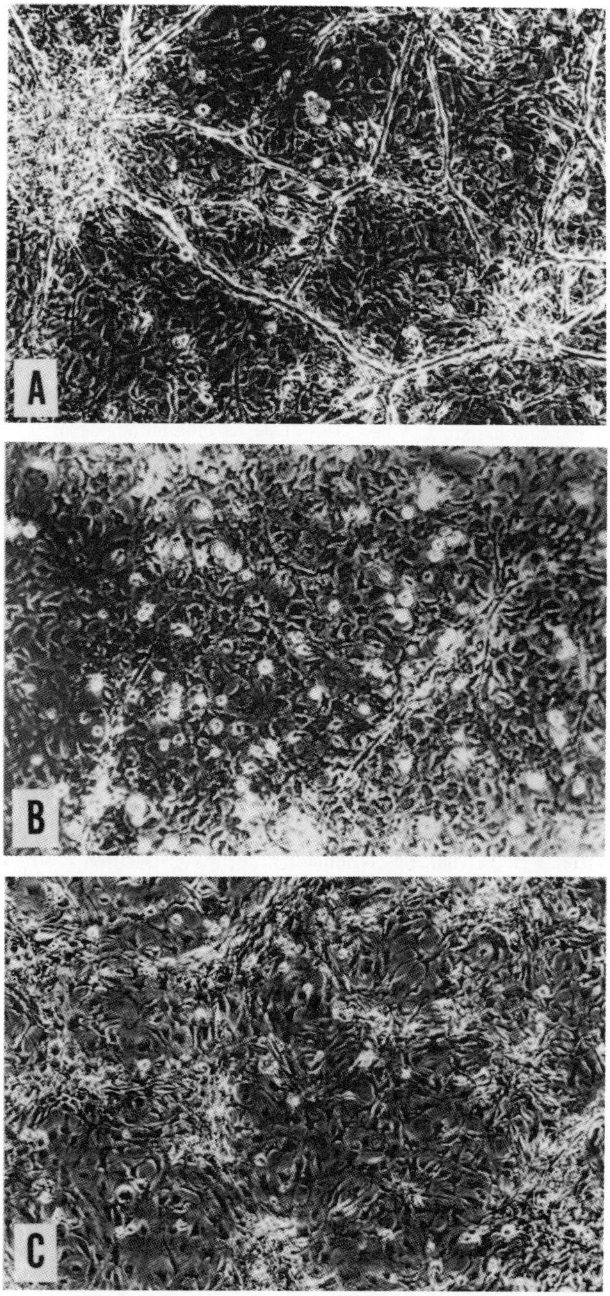

Figure 1. Representative samples illustrating inhibition of in vitro capillary growth by heparin and dexamethasone. Photographs show growing capillaries from microvascular fragments seeded on the layer of myofibroblastic cells (Mfc) 6 days after introduction of the additives. *A* Control cells with no additive; *B* cells with heparin (50 μg/ml); *C* cells with dexamethasone (10^{-8} M).

Thus, our data reveal that the angiogenic inhibitory mechanisms of heparin and DM are distinct from each other, but together the one supplements the other; in combination, they have a potent anti-angiogenic effect. Thus, acting in concert, anti-angiogenic steroids and heparin inhibit EC growth and migration (collagenase synthesis), and prevent extracellular matrix formation in basement membrane and stroma.

References

1 Gross J, Azizkhan RG, Biswas C, Bruns RR, Hsieh DST, Folkman J: Inhibition of tumor growth, vascularization, and collagenolysis in the rabbit cornea by medroxyprogesteron. Proc Natl Acad Sci USA 1981; 78: 1176.

2 Crum R, Szabo S, Folkman J: A new class of steroids inhibit angiogenesis in the presence of heparin or a heparin fragment. Science 1985; 230: 1375.

3 Gospodarowicz D, Massoglia S, Cheng J, Fujii DK: Effect of fibroblast growth factor and lipoproteins on the proliferation of endothelial cells derived from bovine adrenal cortex, brain cortex, and corpus luteum capillaries. J Cell Physiol 1986; 127: 121.

4 Raju KS, Alesandri G, Gullino PM: Characterization of a chemoattractant for endothelium induced by angiogenesis effectors. Cancer Res 1984; 44: 1579.

5 Thornton SC, Mueller SN, Levine EM: Human endothelial cells: use of heparin in cloning and long-term serial cultivation. Science 1983; 222: 623.

6 Folkman J, Langer R, Linhardt RJ, Haudenschild CC, Taylor S: Angiogenesis inhibition and tumor regression caused by heparin or heparin fragment in the presence of cortisone. Science 1983; 221: 719.

7 Ingber DE, Madri JA, Folkman J: A possible mechanism for inhibition of angiogenesis by angiostatic steroids; induction of capillary basement membrane dissolution. Endocrinology 1986; 119: 1768.

8 Sato N, Sawasaki Y, Senoo A, Fuse Y, Hirano Y, Goto T: Development of capillary networks from rat microvascular fragments in vitro: the role of myofibroblastic cells. Microvasc Res 1987; 33: 194.

9 Sato N, Nariuchi H, Tsuruoka N, Nishihara T, Beitz JG, Calabresi P, Frackelton Jr AR: Actions of TNF and IFN-γ on angiogenesis in vitro. J Invest Dermatol 1990; 95: 85S.

10 Sato N, Fukuda K, Nariuchi H, Sagara N: Tumor necrosis factor inhibiting angiogenesis in vitro. J Natl Cancer Inst 1987; 79: 1383.

11 Tsuruoka N, Sugiyama M, Tawaragi Y, Tsugimoto M, Nishihara T, Goto T, Sato N: Inhibition of in vitro angiogenesis by lymphotoxin and interferon-γ. Biochem Biophys Res Commun 1988; 155: 429.

12 Hori H, Nagai Y: Purification of tadpole collagenase and characterization using collagen and synthetic substrates. Biochem Biophys Acta 1979; 566: 211.

13 Folkman J, Haudenschild CC, Zetter BR: Long-term culture of capillary endothelial cells. Proc Natl Acad Sci USA 1979; 76: 5217.

Angiogenesis: Key Principles - Science - Technology - Medicine
ed. by R. Steiner, P. B. Weisz & R. Langer

Angiostatic activity of anticancer agents in the chick embryo chorioallantoic membrane (CHE-CAM) assay

Rudolf Steiner

Research Division, Department of Surgery, University Hospital, CH-8091 Zürich, Switzerland

Abstract. Most anticancer agents are screened for antiproliferative and differentiation-inducing activity on tumor cells, but not for differential effects on vascular endothelium. Clinically active cytotoxic or cytostatic agents and selected analogs were tested in a closed, ethically accepted and inexpensive in vivo fertilized egg system [CHE-CAM]. The results revealed characteristic effects on vessel numbers between day 7 and 11 and on embryo weight at day 12. Hitherto unknown angiostatic activities were discovered with antitumor antibiotics, plant alkaloids, antimetabolites and other classes of compounds. The information provided by the CHE-CAM assay could represent a valuable complementation to the existing drug-screening systems for solid tumors and suggests a selective angiostatic role of some antitumor agents currently used in combination treatments.

Introduction

Solid tumors of more than 1–2 mm diameter in size are for their further survival and growth dependent on a functional capillary system which is induced by the tumors themselves and is supplied by the normal microenvironment [1]. Endothelial cells in small growing solid tumors and in inflammation tend to have a much higher mitotic rate compared to normal resting tissues [2, 3]. The relationship between vascular development and tumor requirements for nutrients is likely to be fragile and is governed by tumor-induced angiogenesis factors which in turn act directly or indirectly on the peritumoral vascular system.

The possiblity therefore exists that the destruction of solid tumors by antitumor agents is mediated in part by local loss of this essential vasculature. J. Denekamp suggested in 1982 that some antitumor drugs may damage tumor-induced endothelial cells and thereby indirectly slow down or arrest tumor growth [4]. To our knowledge no systematic study on angiostatic effects of cytotoxic agents has so far been published [5].

The aims of the study were: 1) to utilize the chicken-embryo chorioallantoic membrane (CHE-CAM) assay as a model to investigate the effects of currently used anticancer agents on growing CAM vessels and on embryo weight, 2) to develop a quantitative angiogenesis assay for pharmacological dose-response and time-course experiments, and 3) to qualitatively assess concomitant drug effects in other CAM-tissues (epithelium, stroma, large pre-existing CAM vessels).

Material and methods

Preparation of eggs

Fertilized chicken eggs (American Hy-Line, white) were purchased from a local hatchery. To start incubation the eggs were removed from the 4°C storage area, transferred to an egg incubator (Covatutto 120) set at 37°C with 80% humidity for optimal growth conditions and manually turned over twice daily. On incubation day 3 (ID 3) the eggs were disinfected with peracetic acid (100 ppm) and an air-sac at the narrow pole was made by gently aspirating 2 ml albumen with a 30-gauge needle through a punched hole. On ID 4 a cut square of the shell of about 2 × 2 cm above the air-sac was removed with a pair of fine tip forceps by slowly lifting it off exposing the white inner shell membrane below. The shell membrane was subsequently carefully peeled off avoiding shell particles to drop onto the membranes of the developing embryo. The window was resealed with adhesive tape and the eggs returned to the incubator.

The in-ovo assay was preferred to the in vitro shell-less system as the latter proved more reliable for time-course studies as well as for quantitation of embryo weight changes.

Drug preparation

The drugs were dissolved or suspended in appropriate vehicles (normal saline, distilled water or other solvents as indicated). 4 to 5 μl of drug containing vehicle or vehicle only as controls were added under sterile conditions onto glassfiber disks of 3 mm diameter (prepared from Whatman GF filters) or applied to dried methylcellulose disks.

The chicken embryo chorioallantoic membrane [CHE-CAM] assay

The microcirculation within the chorioallantoic membrane is well suited for in vivo observations and can be used to monitor angiogenic and angiostatic activity of agents. Since the second and third order capillary vessels progressively increase in numbers from day 6 to day 14 we choose to quantitatively analyse the drug effects from ID 7 to ID 11 On ID 7 (day 0) the dried drug or vehicle (control) containing disks were aseptically placed under stereoscopic guidance onto an appropriate well vascularized site onto the chorioallantoic membrane which at this time point is growing exponentially. Each drug concentration was tested in a series of 5 eggs. The second and third order vessels were counted with the help of a stereomicroscope at 10 × magnification in a 0.5-mm wide CAM-zone ('annulus') around the circumference of the disks.

The average counts of second and third order CAM vessels on day 0 [ID 7] radiating from the circumference of the disks was 10 + 2 vessels/annulus. The vessel counts were repeated at 24 h (day 1/ID 8), 48 h (day 2/ID 9), 72 h (day 3/ID 10) and 96 h (day 4/ID 11) respectively. Representative drug effects were recorded photographically. After completion of each count the shell-window was resealed and the egg immediately returned to the incubator. In general no significant infections were encountered during the five days test period.

On ID 12 the embryos were weighed and gross malformations were recorded. A selection of specimens of the CAM and other embryonic tissues were obtained at different time points, fixed in 10% phosphate-buffered formalin (pH 7.0), embedded in paraffin, sectioned at 2–5 μm and stained with hematoxylin and eosin (H&E).

Quantitation

Vessel counts (ID 7/8/9/10/11) and embryo weights (ID 12) of treated and control groups were compared with the Student's t-test. The vessel (v) count or embryo weight (ew) reduction rate RRv or RRew were calculated with the formula:

$$\text{RRv} = \text{vc} - \text{vt: vc} \quad \text{or}$$

$$\text{RRew} = \text{ewc} - \text{ewt: ewc}$$

(vc/ewc: vessel count/embryo weight in the controls; vt/ewt: vessel counts/embryo weight in the treated groups)

Qualitative evaluation of drug effects

Three distinct degrees of drug-induced effects on the CAM could be recorded by in vivo stereomicroscopy at 10 × magnification and classified as follows:

1) Selective angiostatic i.e. inhibitory effects on growing small size CAM vessels (2nd/3rd order vessels) with or without angiogenic rebound effects. Preexisting CAM vessels not affected.
2) Selective angiotoxic effects on preexisting larger CAM vessels not involving other CAM tissues.

3) CAM epithelium and/or CAM stroma involved. Swelling of the epithelial membrane and reactive fibrosis with more or less severe angiotoxic effects.

Representative observations of these qualitative drug effects were recorded photographically. In selective cases histological sections were performed.

Results

The maximal reduction rates of 2nd and 3rd order vessels at 48 h after drug exposure with the most widely used cytotoxic agents are summarised in Table 1. The angiostatic effects were compared with appropriate vehicle controls and with the angiostatic heparin [50 μg]/hydrocortisone [60 μg] combination.

Angiostatic effects tended to become clearly visible at 24 or 48 h after drug exposure. Very rarely did they occur after 96 h.

Interestingly anthracyclines such as adriamycin, epirubicin, or esorubicin and the podophyllin-derivative etoposide (VP-16) exerted a selective strong angiostatic activity on small growing CAM vessel (2nd and 3rd order vessels) without affecting the larger pre-existing vessels. The vessel reduction rates of these angiostatic drugs were in the range of 85 to 95%. Etoposide was angiostatic at very low drug concentrations with only minor concomitant irritations on the CAM. Mitoxantrone inhibited

Table 1. Reduction of CAM vessels at 48 h (%) and reduction of embryo weight (%).

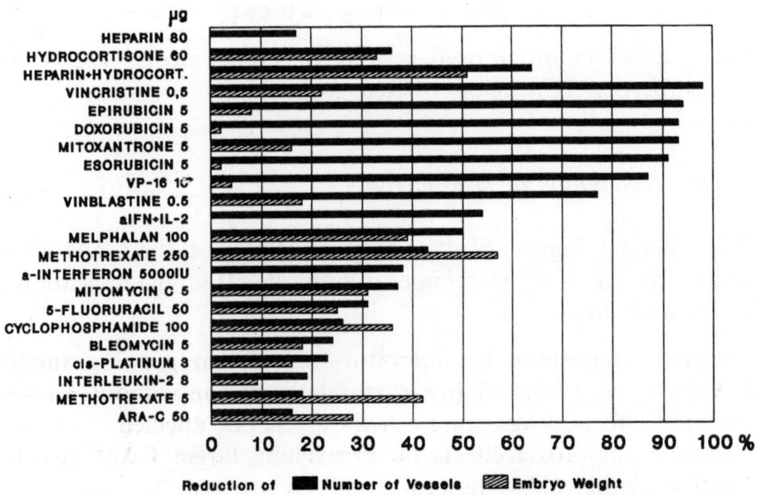

the small vessels in the low dose range (1 μg and 5 μg/disk) but was toxic to pre-existing vessels at higher doses (10 μg/disk) [data not shown].

Vincristine and vinblastine both damaged vessels of all calibers nearly completely, but reversibly, even far beyond the region of the 'annulus'. In addition vincristine induced a stromal reaction in the CAM at 1 μg/disk which was irreversible at higher doses. Bleomycin provoked more toxic effects to the epithelium than to the CAM vessels at 50 μg/disk.

Cyclophosphamide, melphalan, 5-fluorouracil, methotrexate and cis-platinum were only weak angiostats but exerted with the exception of cis-platinum profound inhibitory effects on the weight of the growing embryo at all concentrations tested.

The specific angiostatic compounds such as adriamycin and its analogues or etoposide did not reduce the weight of the growing embryo significantly at the optimal concentration for capillary inhibition. In sharp contrast to vincristine and vinblastin who were found to be toxic to vessels of all calibers and to the embryo as well.

Interferon alpha had a maximal capillary reduction rate of 42% at 5000 IU/disk [6, 7]. Interleukin-2 was similarly active as an angiostat at the dose level of 8 μg/disk. The drug combination interferon alpha plus interleukin-2 at intermediate dosages (2500 IU alpha interferon + 8 μg interleukin-2) had a higher inhibition rate (54%) as compared to both drugs alone. The embryo weights were slightly increased under alpha interferon but decreased in the interleukin-2 group.

Discussion

The progressive growth of solid tumors is angiogenesis-dependent [1]. The establishment of an adequate blood supply requires migration and proliferation of endothelial cells originating from the host. Irrespective of whether angiogenesis is triggered by a simple mechanism or by multiple ones the in vivo environment provides many factors which contribute to the control of this process in an active or permissive fashion. The search of antitumor agents which can influence the 'final common pathway' of the angiogenic reaction would therefore seem to be a major task towards improving the existing therapies for solid tumors.

Our results confirmed and expanded Denekamps hypothesis that certain anticancer drugs are selectively damaging growing endothelial cells. Anthracyclines and VP-16 were shown to be selective angiostats at low dosages whereas cyclophosphamide and cis-platinum were not.

It is interesting to note that clinical oncologists tend to combine 'strong angiostatic' drugs with 'weak angiostats' when treating tumor patients.

These combinations may be effective by influencing secondary drug resistance phenomena and by improving drug delivery to the tumor- or metastatic sites [8].

The information provided by the CHE-CAM assay could represent a valuable complement to the existing drug screening systems and thereby help to select better angiostatic compounds or more effective drug combinations for solid tumors [9].

A more detailed description of our results will be published elsewhere [10].

Acknowledgments. I thank Mr. Bao-Xi Qu [WHO-Research Fellow] for technical assistance. The expert help of Elfriede Gschwend in histology is gratefully acknowledged. This work was supported in part by the regional Cancer Leagues St. Gallen/Appenzell/Thurgau and by the Eugen & Elisabeth Schellenberg Foundation.

References

1 Folkman J: What is the evidence that tumors are angiogenesis-dependent? JNCI 1990; 82: 4–6.
2 Denekamp J: Vasculature as a target for tumour therapy. In: Messmer K, Hammersen F, eds. Progress in applied microcirculation. Vol. 4. Hammersen F, Hudlicka O, eds. Basel, Karger 1984, 28–38.
3 Sholley MM, Cavallo T, Cotran RS: Endothelial proliferation in inflammation. Am J Path 1977; 89: 277–290.
4 Denekamp J: Endothelial cell proliferation as a novel approach to target tumour therapy. Br J Cancer 1982; 45: 136–139.
5 Lazo JS: Endothelial injury caused by antineoplastic agents. Biochem Pharmacol 1986; 35: 1919–1923.
6 Sidky YA, Borden EC: Inhibition of angiogenesis by interferons: effects on tumor- and lymphocyte-induced vascular responses. Cancer Res 1987; 47: 5155–5161.
7 White CW, Sondheimer HM, Crouch EC, Wilson H, Fan LL: Treatment of pulmonary hemangiomatosis with recombinant interferon alpha. N Engl J Med 1989; 320: 1197–1200.
8 Kerbel RS: Inhibition of tumor angiogenesis as a strategy to circumvent acquired resistance to anti-cancer therapeutic agents. BioEssays 1991; 13: 31–36.
9 Steiner R, Qu BX, Gschwend E: The chick embryo chorioallantoic membrane [CHE-CAM] assay: a complementary screening system for anticancer activity. AACR (Washington DC) 1990.
10 Steiner R: Angiostatic activity of anticancer agents in the chick embryo chorioallantoic membrane (CHE-CAM) assay. Submitted to JNCI.

Angiogenesis: Key Principles - Science - Technology - Medicine
ed. by R. Steiner, P. B. Weisz & R. Langer

In vitro activity of novel sulphonic derivatives of distamicyn A

M. Mariani, A. Paio, M. Ciomei, W. Pastori, C. Franzetti, G. Melegaro, M. Grandi and N. Mongelli

Farmitalia Carlo Erba, Erbamont Group, Research and Development Center, Nerviano and Milano, Italy

Abstract. The successful growth of tumors is dependent on the process of vascularization elicited by the tumor itself. As confirmed by many authors, there is a correlation between the presence of factors that stimulate tumor growth and angiogenesis. One of the approaches we have explored to control angiogenesis has been to synthesize compounds able to complex growth factors. A number of sulphonated derivatives of distamycin A were found active in inhibiting the binding of bFGF and PDGFβ on Swiss 3T3 cells with ID_{50} values ranging between 142–587 μM for bFGF and 28–79 μM for PDGFβ. The effect of these new derivatives in inhibiting angiogenesis was initially explored in the chorioallantoic membrane assay. It was observed that the selected compounds were active in this model system at the concentration of 350 nm/pellet. These new molecules present low or no cytotoxic activity on M5076 murine reticulosarcoma cells, the ID_{50} values being higher than 60 μM after 72 h continuous exposure in vitro.

Introduction

New vessels formation is believed to be a crucial event in the growth of solid tumor [1]. Several factors (eg. bFGF, PDGFβ) have been described to participate in the tumoral angiogenic response directly stimulating the migration and proliferation of the normally quiescent endothelial cells [2, 3].

In the frame of a research project aimed at developing pharmacological agents capable of modulating neovessel formation and particularly tumoral angiogenesis, a number of novel polyanionic SO_3 group bearing-derivatives of distamycin A were synthesized.

This report presents results obtained in testing these compounds for their binding affinity to bFGF and PDGFβ on Swiss 3T3 cells.

In addition, the test products were evaluated for their capacity to inhibit neovascularization in the chorioallantoic membrane assay, and for in vitro cytotoxicity on M5076, a murine reticulosarcoma cell line.

Materials and methods

Binding assay

Confluent monolayers of Swiss 3T3 cells were washed with binding medium (DMEM + 0.2% BSA + 100 μg/ml Bacitracin + 20 mM Hepes).

To each well, 0.2 nM ^{125}I-bFGF or ^{125}I-PDGFβ were added in the presence of different amounts of unlabeled growth factor or the test products.

Specifically bound radioactivity was measured after 4 h at 4°C.

CAM assay

The test was performed as described by Crum et al. [4]. Briefly, chick embyros were removed from their shells on day 3 of development, placed in plastic Petri dishes and maintained at 37°C, 3% CO_2. On day 5, embryos were treated with saline solution or the test molecules incorporated into a 0.5% methylcellulose disk placed on top of growing CAMs. A group of ten eggs was used for each dose.

After 48 h, embryos were observed with a stereomicroscope at 10 × for the presence or absence of capillaries around the disks. We considered positive embryos presenting avascular zones with at least 4 mm in diameter.

Results are reported as % of CAM positives at the tested concentration.

Cytotoxicity evaluation

M5076 murine reticulosarcoma cells (50,000 cells/ml) were treated at seeding with different concentrations of the compounds dissolved in saline solution immediately before use. After 72 h treatment, cells were counted using a Coulter Counter and viability tested by a dye exclusion technique.

Results and conclusions

The structures of the derivatives investigated are shown in Fig. 1. It can be seen that the test compounds have a common skeleton of four methyl-pyrrolic rings with naphtalen groups, but vary in the number and position of the SO_3 groups. Furthermore, FCE 26644 has the naphtalen group bound to the skeleton in position 2′ instead of 1′ as in the other molecules depicted.

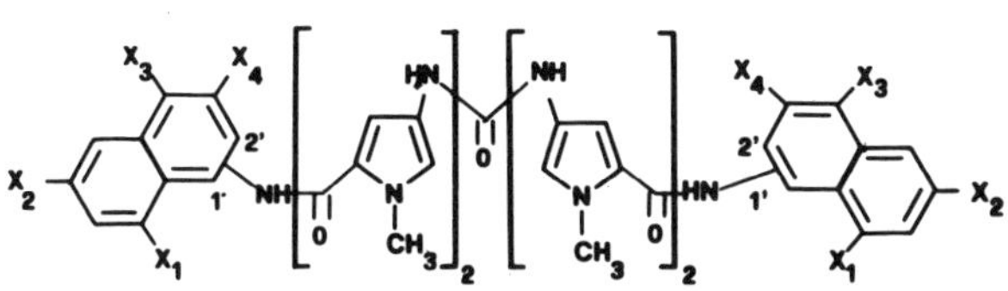

FCE 26605	$X_1 = X_2 = X_3 = SO_3Na$	$X_4 = H$	1' N
FCE 26492	$X_1 = X_3 = SO_3Na$	$X_2 = X_4 = H$	1' N
FCE 26752	$X_1 = X_2 = X_4 = SO_3Na$	$X_3 = H$	1' N
FCE 26604	$X_1 = X_2 = SO_3Na$	$X_3 = X_4 = H$	2' N

Figure 1. Structure of distamycin A derivatives.

Results presented in Table 1 indicate that all compounds are capable of complexing bFGF and PDGFβ with ID_{50} values ranging between 142–587 μM for bFGF and 28–70 μM for PDGFβ, respectively.

Whether this effect is due to a formation of a complex between the compound and the growth factor or with the cellular receptors is currently under formal investigation.

Table 1. Effect of distamycin derivatives on the binding of bFGF and PDGFβ

Compounds	ID_{50} (μM)* bFGF	ID_{50} (μM)* PDGFβ
FCE 26605	153 ± 15	46 ± 3
FCE 26492	587 ± 20	70 ± 0
FCE 26752	390 ± 102	28 ± 3
FCE 26644	142 ± 18	55 ± 3

*Dose inhibiting 50% growth factor binding.

Table 2. Activity of distamycin derivatives on CAM assay

Compounds	Dose* (nm/pellet)	Activity** (% of positive CAMS)
Saline solution	—	0
FCE 26605	350	60
FCE 26492	350	65
FCE 26752	350	71
FCE 26644	350	100

*Dose affecting angiogenesis without detectable toxicity to chick embryo.

**Evaluated as $\frac{\text{No. of CAM with inhibiting area}}{\text{No. of CAM treated}} \times 100.$

Table 3. Cytotoxic activity on M5076 cells

Compounds	ID_{50}* (μM)
Doxorubicin	0.4
FCE 26605	177
FCE 26492	138
FCE 26752	315
FCE 26644	60

*Dose inhibiting by 50% cell growth after 72 h of treatment.

Table 2 shows that all derivatives inhibit angiogenesis in the CAM assay, as judged by the ability to produce avascular zones greater than 4 mm in diameter in the test conditions employed. FCE 26644 was found to be the most active compound, being consistently active in 100% of the CAMs at the concentration of 350 nm/pellet.

In order to determine if a general cytotoxic effect played a role in the angiogenesis inhibition observed in the CAM assay the possible activity of these molecules on the proliferation of M5076 murine reticulosarcoma cells was examined.

Representative data are presented in Table 3. As indicated by the high ID_{50} values determined after 72 h exposure, it can be seen that these molecules have only marginal cytotoxic activity. By way of comparison, it should be noted that in the same test conditions doxorubicin is about 1000-fold more cytotoxic than these novel chemicals.

In conclusion, these data indicate that these novel non cytotoxic compounds are able to inhibit angiogenesis in the CAM assay and to bind bFGF and PDGFβ.

These findings and the efficacy of these molecules in reducing the in vivo growth of the M5076 tumor as reported in the accompanying paper by Sola et al. [5], also suggest that inhibition of tumor growth via the pharmacological modulation of angiogenesis is feasible.

References

1 Folkman J: What is the evidence that tumors are angiogenesis dependent? J Natl Cancer Inst 1990; 82: 4–6.
2 Folkman J, Klagsbrun M: Angiogenic factors. Science 1987; 235: 442–447.
3 Beitz JG, Kim IS, Calabresi P, Frackelton R: Human microvascular endothelial cells express receptors for platelet-derived growth factors. PNAS 1981; 88: 2021–2025.
4 Crum R, Szabo S, Folkam J: A new class of steroids inhibits angiogenesis in the presence of heparin or a heparin fragment. Science 1985; 230: 1375–1378.
5 Sola F, Biasoli G, Pesenti E, Farao M, Della Torre P, Mongelli N, Grandi M: In vivo activity of novel sulphonic derivatives of distamycin A. Proc. Int. Symposium on Angiogenesis. St. Gallen, Switzerland, March 13–15, 1991. Abst. No. 90; *in this book*.

Angiogenesis: Key Principles - Science - Technology - Medicine
ed. by R. Steiner, P. B. Weisz & R. Langer

In vivo activity of novel sulphonic derivatives of distamycin A

F. Sola, G. Biasoli, E. Pesenti, M. Farao, P. Della Torre, N. Mongelli and M. Grandi

Farmitalia Carlo Erba, Erbamont Group, Research and Development Center, Nerviano and Milano, Italy

Abstract. Solid tumor growth can be modulated through inhibition of vascularization elicited by angiogenic factors. With the objective to complex these factors, new derivatives of distamycin A were synthesized and evaluated in vitro [1] and in vivo for their ability, after i.v. administration, to inhibit bFGF-induced vascularization and the growth of M5076 murine reticulosarcoma implanted i.m. The tested compounds were able to block angiogenesis with inhibition values ranging between 70–100%. Moreover, they were found to be capable of inducing tumor inhibition with values ranging between 40% and 95% at non-toxic doses.

Introduction

Polyanionic compounds [2] are able to form a complex with proteins such as growth factors and thus block their activity. It has been reported that several factors (angiogenic factors) induce neo-vascularization, a process that is believed to play a crucial role in initiating and maintaining solid tumor growth [3]. Thus, we have investigated if sulphonated distamycin A derivatives able to complex bFGF, PDGF and to inhibit angiogenesis in the CAM (chorioallantoic membrane assay) [1], were active in vivo in inhibiting bFGF-induced neo-vascularization. We have also investigated if these novel compounds present activity in inhibiting solid tumor growth using the M5076 murine reticulosarcoma model.

Materials and methods

Pathogen-free C57 Bl/6 Female mice and C3H Female mice were obtained from Charles River CD (Calco, Italy). Mice were used when 6 to 8 weeks old.

The M5076 murine reticulosarcoma tumor was prepared as single cell suspension; 5×10^5 viable cells were implanted i.m. into C57 Bl/6 mice.

Drug treatment was performed i.v. on day 1, with 200 mg/kg of each compound. Tumor growth was monitored weekly until the time of death. Results are reported as inhibition of the area under the curve (A.U.C.) calculated from the single values.

bFGF-Gelfoam implants

Collagen sponges (Upjohn, USA) were cut into approximately 7 × 10 × 10-mm pieces and loaded with saturating amounts of a 30-μg/ml bFGF solution in PBS/BSA 0.1%. Negative control sponges were impregnated with PBS/BSA 0.1%. C3H mice were anesthetized, and a 1-cm-long dorsal midline skin incision was made 3 to 4 cm caudal to the occipital ridge.

Sponges were introduced into the subcutaneous pouch and skin was sutured with a staple gun.

After 15 days, mice were sacrificed and sponges surgically extracted and fixed in a 4% paraformaldehyde solution. Histological examination of the vasculature was evaluated by counting the number of vessels on 10 fields/sponge using 25× magnification. Results are reported as percent of vascularization using the formula reported in Table 1.

Results and conclusions

The structure of the tested compounds are presented in the accompanying report by Mariani et al. [1].

Results presented in Table 1 show that all tested molecules are able to significantly inhibit neo-vascularization induced by bFGF-Gelfoam sponges implanted subcutaneously in mice, with values of inhibition ranging between 70–100%, when evaluated 15 days after the administration of a standard dose of 200 mg/kg. Fig. 1 shows the histological appearance of representative sponges.

Table 1. Activity of novel sulphonic derivatives of distamycin on bFGF-Gelfoam-sponge

Compounds	Sponge	Dose* (mg/kg)	% vasc.**	Tox.***
—	PBS + BSA 0.1%	—	0	0/5
—	bFGF 30 μg/ml	—	80	0/5
FCE 26492	" " "	200	0	0/5
FCE 26605	" " "	200	25	0/5
FCE 26752	" " "	200	32	0/5
FCE 26644	" " "	200	25	0/5

*Treatment was given i.v. at day 1.

**% vascularization: $\frac{\text{n treated sponges} - \text{n negative sponges}}{\text{n positive control sponges}} \times 100.$

n = mean number of vessels counted in 10 fields of histological section (Fig. 1).

***No. of toxic deaths/No. of treated mice.

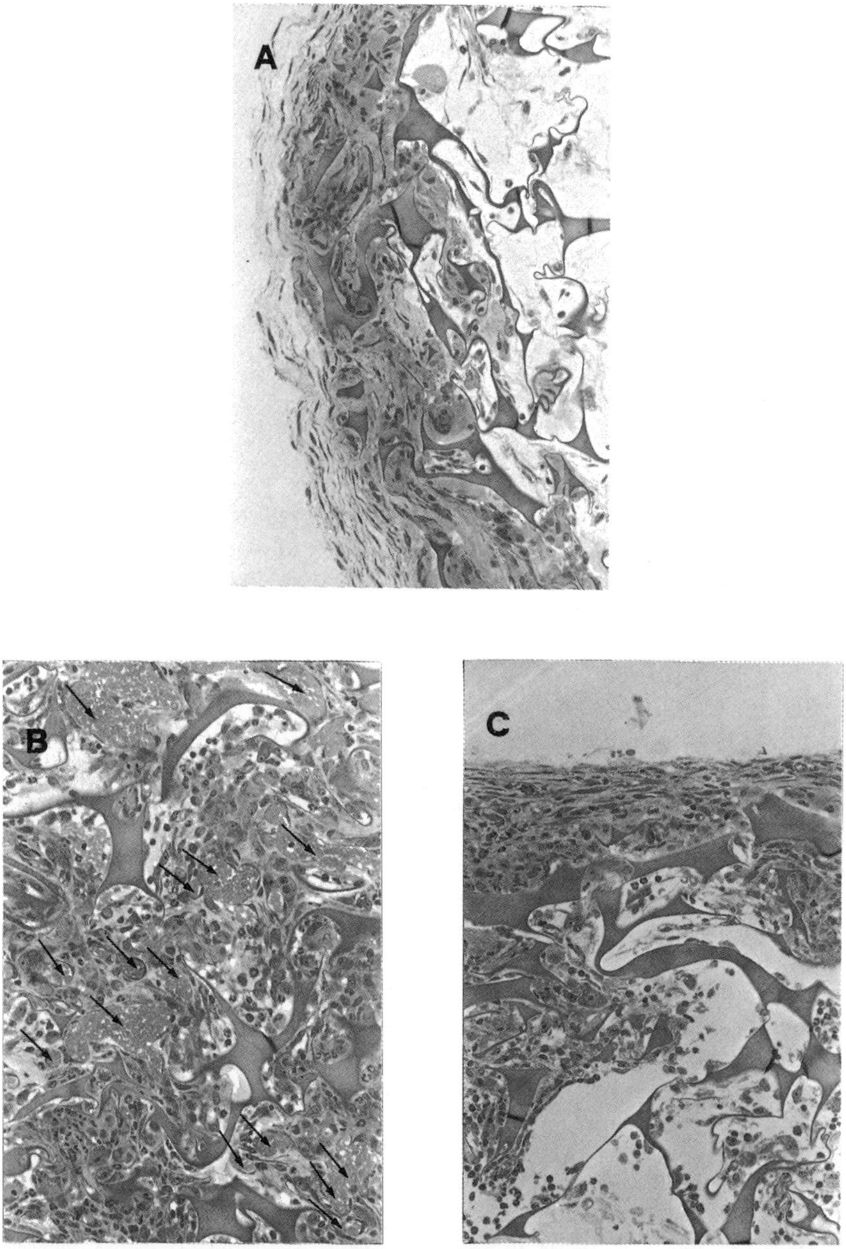

Figure 1. *Panel A*. Histological section (25× magnification) of a negative control sponge, i.e. loaded with PBS + BSA 0.1%. Parenchymal cells are present in the peripheral section of sponge. *Panel B*. Histological section of a positive control sponge, i.e. loaded with bFGF 30 μg/ml in PBS + BSA 0.1%. Arrows indicate transversal section of vessels. *Panel C*. Histological section of a sponge loaded with bFGF 30 μg/ml in PBS + BSA 0.1% and taken from a mouse treated with a single dose of 200 mg/kg FCE 26644. Only a few parenchymal cells, but no vessels, are present.

Table 2. Activity of novel sulphonic derivatives of distamycin on M5076 murine reticulosarcoma growth

Compounds	Dose (mg/kg)	% Inhib. A.U.C.*	T/C %**	Tox.***
FCE 26492	200	39	117	0/10
FCE 26605	200	57	111	0/10
FCE 26752	200	86	122	0/10
FCE 26644	200	92	150	0/10

Treatment was given i.v. at day 1.

* % inhibition area under the curve: $100 - \frac{\text{A.U.C. treated mice}}{\text{A.U.C. control}} \times 100.$

**$\text{T/C\%} = \frac{\text{median survival time of treated mice}}{\text{median survival time of controls}} \times 100.$

***No. of toxic deaths/No. of treated mice.

The values of tumor growth inhibition observed in mice implanted i.m. with M5076 murine reticulosarcoma cells are shown in Table 2. It can be seen that the tested compounds showed a clear antitumor effect; moreover, after treatment with FCE 26644, a 50% increase in survival time was also observed. At the dosage tested, none of the compounds showed gross toxicity.

In conclusion, these novel molecules are able to complex bFGF both in vitro [1] and to inhibit angiogenesis induced by FGF-loaded sponges, presumably by complexing this growth factor vivo. In addition, these compounds at well tolerated doses were found capable of significantly inhibiting the in vivo growth of the M5076 tumor. Since M5076 cells were not affected in their multiplying ability even after prolonged exposure to high concentrations of these chemicals, it is tempting to advance that tumor growth inhibition is, at least in part, attributable to interference with tumor-induced angiogenesis. Studies are in progress to better evaluate the spectrum of antitumor activity and the optimal administration schedule of these and related molecules. More in general, these preliminary results indicate that solid tumor growth can be influenced through compounds capable of inhibiting neo-vascularization.

References

1 Mariani M, Paio A, Ciomei M, Pastori W, Franzetti C, Melegaro G, Grandi M, Mongelli N: In vitro activity of novel sulphonic derivatives of distamycin A. Proc. International Symposium on Angiogenesis. St. Gallen, Switzerland, March 13–15, 1991. Abst. No. 91; *in this book.*

2 Tanaka NG, Sakamoto N, Inove K, Korenaga H, Kadoya S, Ogawa H, Osad Y: Antitumor effect of an antiangiogenic polysaccharide from an arthrobacter species with or without a steroid. Cancer Res 1989; 49: 6727–6730.

3 Folkman J, Klagsbrun M: Angiogenic factors. Science 1987; 235: 442–447.

Angiogenesis: Key Principles - Science - Technology - Medicine
ed. by R. Steiner, P. B. Weisz & R. Langer

Antiangiogenesis strategies in cancer therapy with special reference to Krestin

S. Kumar, K. Saitoh and P. Kumar

Christie Hospital (NHS) Trust, Manchester M20 9BX, England

Most cancer treatments aim to eradicate cancer cells directly, however the progressive growth of solid tumours is dependent on their ability to induce the formation of new blood vessels (angiogenesis) [1, 2]. This dependency of tumour growth on angiogenesis, makes the vasculature an important target for restraining tumour growth [3–6]. A number of key events that occur during angiogenesis include endothelial cell migration, proliferation and synthesis of basement membrane [7]. Table 1 lists some of the antiangiogenic strategies that have been used to interfere with the formation of new blood vessels, some with dramatic effects and others with marginal success. The administration of Krestin, a protein bound polysaccharide isolated from *Coriolus versicolor*, induces a marked reduction in tumour growth in animal models [8]. The mechanism of its action is not fully understood. We report here that part of its anti-tumour effect may be mediated via the inhibition of the tumour vasculature.

Materials and methods

Krestin was purified from mycelia of the CM-101 strain of *Coriolus versicolor*. The brownish looking powder was dissolved in hot phosphate buffered saline (PBS), cooled to room temperature and was given intraperitoneally to test mice. The control animals received an equal volume of PBS.

Mouse dorsal air sac assay

ICR mice were anaesthetised and a dorsal air sac was produced by injecting air. The air was evenly distributed by manual compression of the skin. An incision was made and two sealed millipore chambers containing mouse 134 hepatoma cells, as a source of angiogenesis activity, were implanted as shown in Figure 1. The incision was sealed.

Table 1. Some inhibitors of angiogenesis*

Cartilage extract	Cromolyn
Aorta extract	ICRF 159
Native hyaluronan	GPA 1734
Gangliosides	Heparin + steroid
Placental ribonuclease inhibitor	Suramin + steroid
Platelet factor 4	Minocycline
Laminin peptides	Maltose tetrapalmitate
Thrombospondin	Sulphated polysaccharides
Protamine	Krestin
Prostaglandin inhibitors	Fumagillin
D-Pencillamine	

*See references 3, 6 and 14–17.

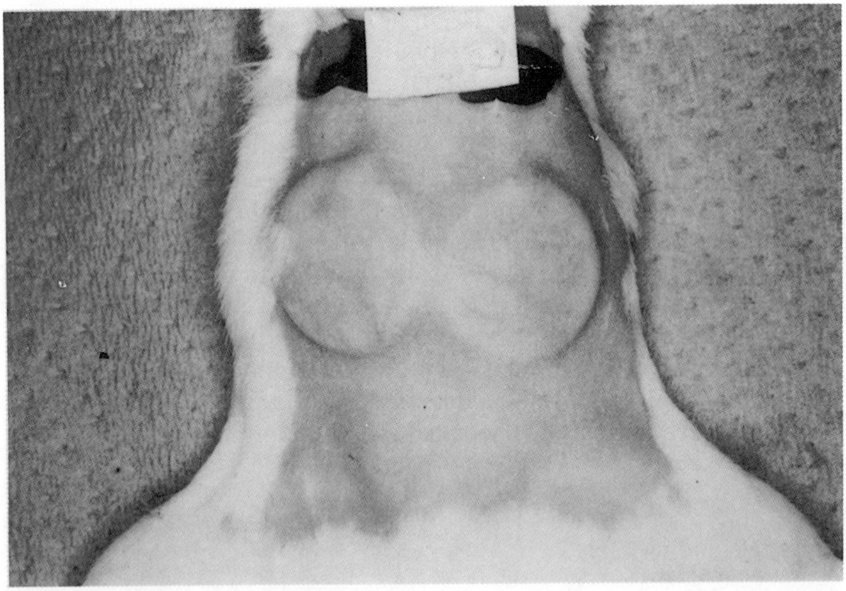

Figure 1. Mouse dorsal air sac assay: Outline of two sealed millipore chambers containing hepatoma cells can be seen lying under the skin (Day 0).

PBS or Krestin (1, 10, 50 or 100 mg/kg) was administered intraperitoneally on day 1, 3 and 5 and the experiment was terminated on day 7.

The areas immediately underneath the millipore chambers were examined for evidence of vascularisation and photographed. The test site was divided into two halves. One half was homogenised to measure alkaline phosphatase activity using a commercial kit. The other half was fixed in glutarlaldehyde in 0.07 M Sorensen's phosphate buffer pH 7.4 for 2 hours at room temperature for transmission electron microscopy. The tissue specimens were processed according to a procedure described previously [9].

Stained grids were examined systematically in the electron microscope. To ensure standardization of the sampling only cross-sections of whole blood vessels were photographed and quantified for diameter and content of Weibel-Palade bodies (WPB). The Kruskal-Wallis and Mann-Whitney U tests were used for statistical analysis.

Results

Hepatoma cells induced marked neovascularisation in PBS treated control mice (Fig. 2). In these mice the vasculature under the millipore chamber and in its immediate vicinity contained varying densities of dilated, serpentine or tortuous blood vessels some with chaotic branching. The increase in vascularity was accompanied by an infiltration of mast cells, which appeared to lie in close association with the newly formed blood vessels. There was no noticeable increase in inflammatory cells. In contrast mice that received Krestin, showed significant inhibition of angiogenesis (Fig. 3; $p < 0.05$) and maintenance of some order and normality amongst the vasculature (Figs 4 and 5). The measurement of alkaline phosphatase activity confirmed this observation (Table 2). In tissue samples from Krestin treated mice, alkaline phosphatase activity was 45.3% of that from PBS treated control mice.

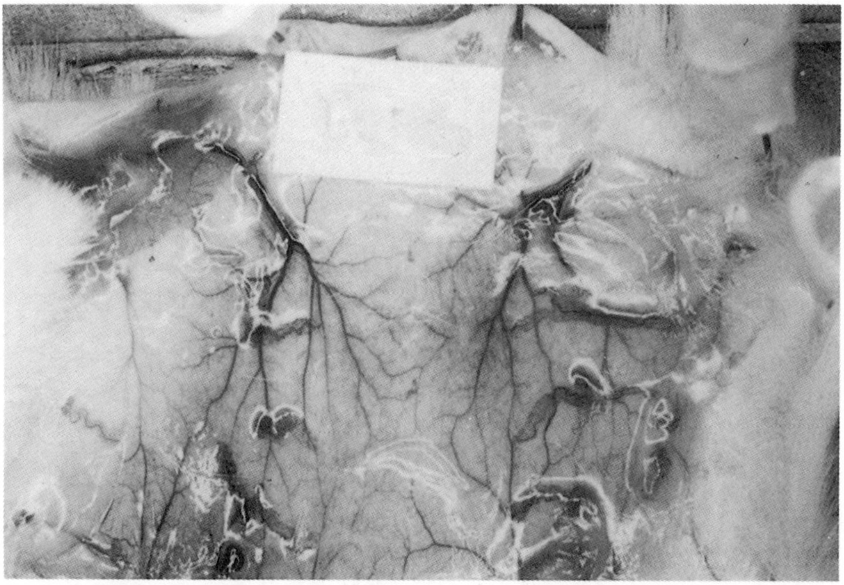

Figure 2. Mouse dorsal air sac assay: Induction of angiogenesis by hepatoma cells sealed in millipore chambers (Day 7).

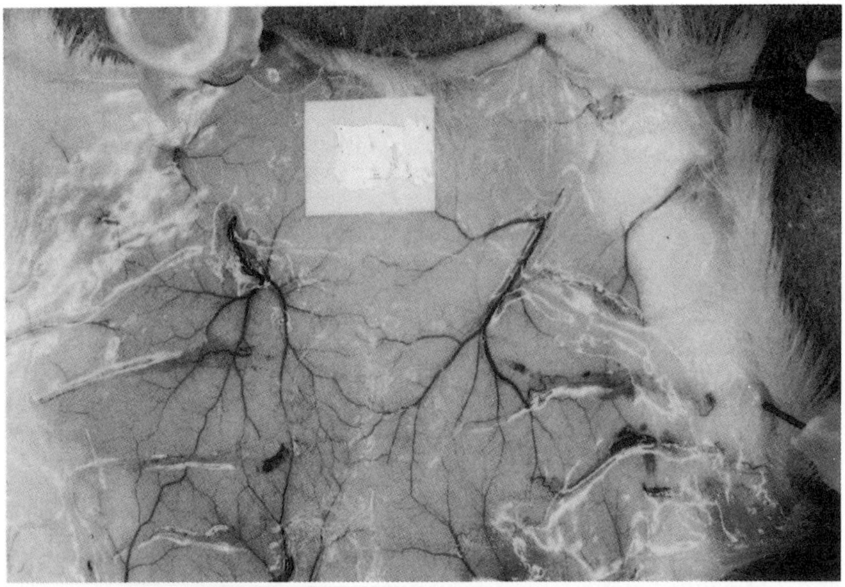

Figure 3. Mouse dorsal air sac assay: Inhibition of hepatoma induced angiogenesis following intraperitoneal administration of Krestin (Day 7).

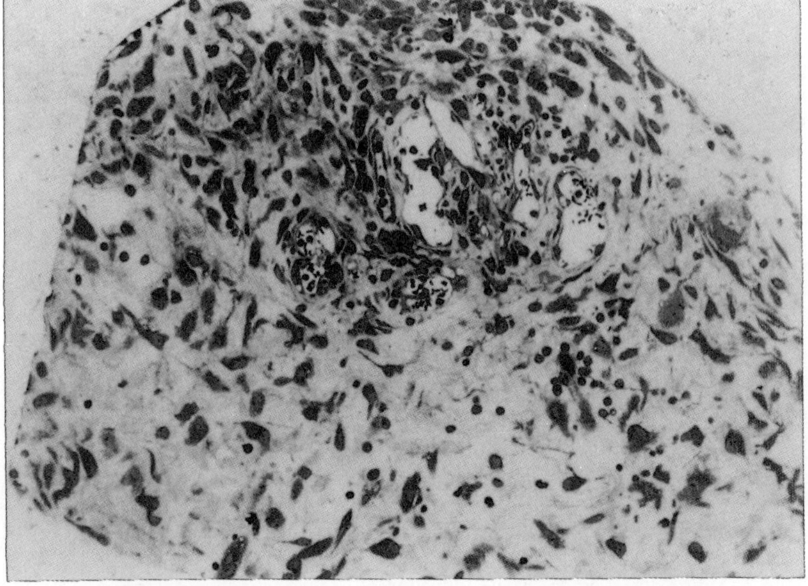

Figure 4. Legend overleaf

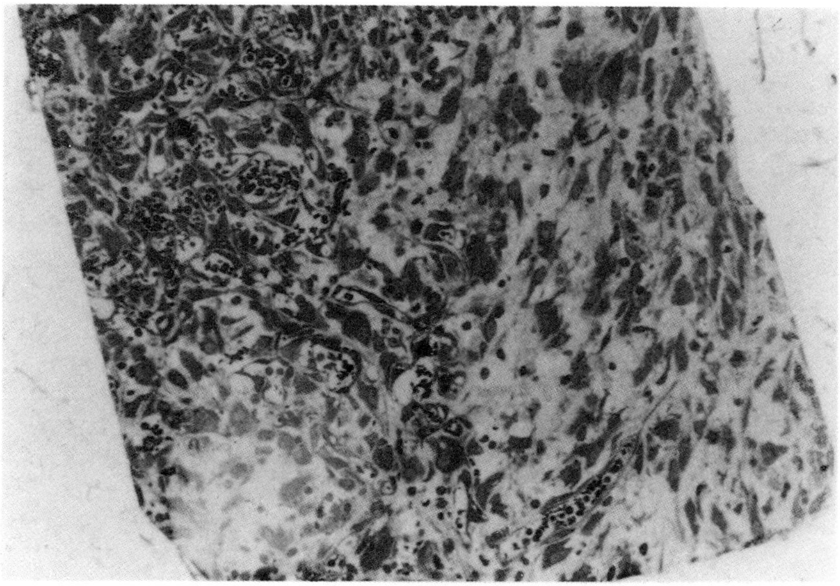

Figures 4 and 5. Mouse dorsal air sac assay: Same as Fig. 2 and 3—Toluidine blue stained semi-thin sections of a tissue from a mouse treated with Krestin (Fig. 4) shows a relative lack of blood vessels compared with PBS treated controls (Fig. 5).

Table 2. The effect of Krestin on alkaline phosphatase activity in the mouse dorsal air sac assay

Treatment	Alkaline phosphatase Units/μg mean $\pm$ S.D. (range)	% Decrease
Control: PBS treated	117 $\pm$ 35 (78–159)	0
Test: Krestin treated	64 $\pm$ 27 (38–90)	45.3

In the electron microscope, fewer blood vessels were observed in Krestin treated tissues compared with the control mice (not quantified). The median diameter of blood vessels in tissues from Krestin treated mice tended to be slightly smaller [median (range): 9.4 μm (5.0–16.6)] compared with controls [11.1 μm (5.0–22.5)]. This difference was not statistically significant.

Weibel-Palade bodies have a unique morphology and thus are fairly easy to identify (Fig. 6). They were clustered in their distributions, some endothelial cells having many WPB, while others had few (Table 3). The median number of WPB in both Krestin treated and control tissues was 2 per endothelial cell section there being no significant difference.

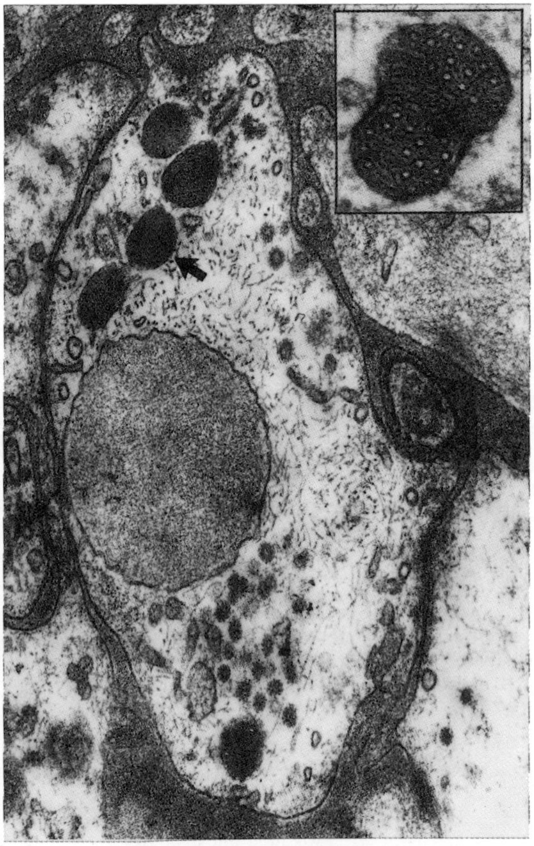

Figure 6. Electron micrograph of an endothelial cell showing typical profiles of Weibel-Palade bodies. Numerous microfilaments are present in the endothelial cell. Inset shows internal tubules in a Weibel-Palade body.

Table 3. The effect of Krestin treatment on the number of Weibel-Palade bodies in endothelial cell sections in the mouse dorsal air sac assay

Source of tissue	Endothelial cells counted	Weibel-Palade bodies/cells Median (Range)
Control: PBS treated	140	2 (1–8)
Test: Krestin treated	125	2 (1–6)

Discussion

The most significant finding of the present study is that intraperitoneal injection of Krestin markedly inhibited tumor induced neovascularisation.

In vivo, in several experimental animal tumour models, Krestin has tumoristatic effects. The mechanism of its action has not been fully elucidated [8]. Krestin biosynthetically labelled with ^{14}C reaches a maximum level in the blood between 1–2 hours after oral administration followed by a slow decrease. Its biological half-life based on the amount excreted in expired air, faeces and urine has been calculated to be approximately 4 hours. Of particular interest is the fact that Krestin accumulated in endothelial cell lysosomes. *In vitro*, Krestin, enhanced the production of interleukins (IL_1, IL_2, IL_6, IL_8), tumour necrosis factor, GM-CSF and prostaglandins by human monocytes [10]. Metastatic sublines of Dunning rat prostatic tumours express high levels of ras-oncogenes. Recently, Mickey [11] found that administration of Krestin significantly ($p < 0.001$) reduced the expression of c-HA-ras in the metastatic lines and this decrease correlated with decreases in tumour growth rates in the Krestin treated animals vs control animals hosting tumours but not receiving Krestin. Klein and Vanky [12] have reported that the presence of Krestin, in tissue culture media, enhanced the cytotoxic effect of lymphocytes against tumour cells.

Weibel-Palade bodies have been proposed to be a marker of growing capillaries in tissue angiogenesis [13]. Since Krestin caused a marked inhibition of vascularisation, a lack of demonstrable difference between Krestin treated and control mice tissues is rather surprising. It is possible that Krestin has no direct effect on endothelial cell proliferation or migration, but instead induces antiangiogenesis indirectly. Several possibilities can be suggested. For example, it may inhibit synthesis of the extracellular matrix components of blood vessels. Indeed, in a preliminary investigation, we have observed that although Krestin had no effect on tissue cultured endothelial cell migration or proliferation, it significantly reduced the synthesis of collagen by microvascular endothelial cells *in vitro*.

References

1 Folkman J: What is the role of endothelial cells in angiogenesis? Lab Invest 1984; 51: 601–604.
2 Folkman J: What is the evidence that tumours are angiogenesis dependent? JNCI 1990; 82: 4–6.
3 Ingber D, Fujita T, Kishimoto S, Folkman J: Synthetic analogues of fumigillin that inhibit angiogenesis and suppress tumour growth. Nature 1990; 348: 555–557.
4 Denekamp J: The current status of targetting tumour vasculature as a means of cancer therapy: an overview. Int J Radiat Biol 1991; 60: 401–408.
5 Auerbach R: Vascular endothelial cell differentiation: organ specificity and selective affinities as a basis for developing anti-cancer strategies. Int J Radiat Biol 1991; 60: 1–10.
6 Kumar S, Arnold F: Can metastasis be restrained? In: Breast Cancer: treatment and prognosis. (Ed.B Stoll) Blackwell Scientific Publications, Oxford, pp. 287–299, 1986.
7 Ausprunk DH and Folkman J: Migration and proliferation of endothelial cells in preformed and newly formed blood vessels during angiogenesis. Microvas Res 1977; 14: 53–65.

8 Fujii T, Sugita N, Kobayashi Y, Saito K, Iijima H, Matsunaga K, Ando T, Oguchi Y, Morita I, Yhoshikumic: Treatment with Krestin combined with mitomycin C, and effect on immune response. Oncology 1989; 46: 49–53.

9 Kumar P, Erroi A, Sattar A, Kumar S: Weibel-Palade bodies as a marker of neovascularisation induced by tumour and rheumatoid angiogenesis factors. Cancer Res 1985; 45: 4339–4348.

10 Matsushima K, Hirose K, Oppenheim J: Induction of cytokine production by a plant-polysaccharide in human PBMC. J Cancer Res Clin Oncol 1990; 116: 871.

11 Mickey DD: Alteration of gene expression in prostatic tumour. J Cancer Res Clin Oncol 1990; 116: 871.

12 Klein E, Vanky E: The effect of an immunomodulator on lymphocyte-tumour cell interactions *in vivo*. J Cancer Res Clin Oncol 1990; 116: 871.

13 Sattar A, Kumar P, Kumar S: Rheumatoid- and osteo-arthritis: quantitation of ultrastructural features of capillary endothelial cells. J Path 1986; 148: 45–53.

14 Wilks JW, Scott PS, Urba LIC, Cocuzza JM: Inhibition of angiogenesis with combination treatments of angiostatic steroids and suramin. Int J Radiat Biol 1991; 60: 73–77.

15 Buick N: Tumour angiogenesis: The role of oncogenes and tumour supressor genes. Cancer Cells 1990; 2: 179–185.

16 Blood CH, Zetter BR: Tumour interactions with the vasculature: angiogenesis and tumor metastasis. BBA 1990; 1032: 89–118.

17 Paweletz N, Kivierim M: Tumour-related angiogenesis. Clin Rev Oncol Haematol 1990; 9: 197–242.

Angiogenesis: Key Principles - Science - Technology - Medicine
ed. by R. Steiner, P. B. Weisz & R. Langer

FDA regulations for growth factors and related products

Mrunal S. Chapekar and Kurt C. Gunter

Cytokine and Cellular Biology Branch, Division of Biological Investigational New Drugs, Center for Biologics Evaluation and Research, Food and Drug Administration, Rockville, MD 20892, USA

Abstract. Biological products, including the majority of growth factors, are regulated by the Center for Biologics Evaluation and Research (CBER) in the U.S. Food and Drug Administration (U.S.FDA) under two statutes; U.S. Federal Food Drug and Cosmetics (FDC) act and the U.S. Public Health Service act. As stipulated in the U.S. Code of Federal Regulations (21 CFR) under the FDC act, the testing of new products in humans is conducted under an Investigational New Drug (IND) application. The primary objective of the FDA in all Phases of an investigation is assure safety of human subjects. During phase II and phase III studies, additional assurance regarding the scientific quality of the clinical investigation is required. A marketing approval is granted by CBER following the review of Product License Applications (PLA) and Establishment License Applications (ELA). CBER's review process provides guidance to the manufacturers of biological products towards the development of safe and effective biological products for human use. Information pertinent to preclinical issues and clinical trial design is presented here with a special emphasis on the non-hematopoietic growth factors.

Introduction

Continuing progress in the recombinant DNA technology has accelerated the development of new biological products for clinical use. Among the new class of biological products are growth factors, which, in addition to the induction of cellular proliferation, also exhibit other effects including, cell differentiation, chemotaxis, matrix formation and angiogenesis. Non-hematological growth factors such as fibroblast growth factor, transforming growth factor, and platelet-derived growth factor have been evaluated for their activity as potential wound healing agents [1], whereas, hematopoietic growth factors have been clinically investigated for their effects on bone marrow precursor cells [2]. In addition, products, such as antibodies to growth factors and other protein inhibitors capable of blocking the biological effects of growth factors, have been investigated for their potential usefulness in such diverse diseases as autoimmune diseases and cancer [3–5]. The intent of this publication is to provide information regarding the preclinical requirements for filing an investigational drug application for a new biological product with CBER in the U.S. FDA, and to discuss issues related to clinical trial design for growth factors and related products.

FDA regulations for biological products

Biological products are also considered to be drugs and are regulated by CBER under two statutes: The U.S. Food Drug and Cosmetic Act (FDC act) which defines drug as "an article intended for use in the diagnosis, cure, mitigation, treatment, or prevention of disease in man" [6] and the U.S. Public Health Service (PHS) act, section 351 of which defines the biological products and requires that a biological agent introduced into interstate or international commerce should be prepared in a licensed facility [7]. Title 21 in the Code of Federal Regulations (CFR) further expands and elaborates on these regulations [8]. 21 CFR 610, specifically, defines these regulations with respect to the requirements for potency (610.10), safety (610.11), sterility (610.12), purity (610.13) and identity (610.14) for licensed biological products. Every lot of a biological product is released by CBER prior to its use in human subjects (section 610.1). Moreover, the regulations specified in section 211 (Good manufacturing Practices or GMP), section 58 (good laboratory practices or GLP) and section 312 (IND regulations) of 21 CFR, applicable to devices and drugs, are also pertinent to biological products. CBER has, in addition, drafted Points to Consider and related documents (Tab. 1) which reflect CBER's general expectations regard-

Table 1. "Points to Consider" and related documents

Points to Consider in the Manufacture of in Vivo Monoclonal Antibody Products Subject to Licensure (1983).
Points to Consider in the Production and Testing of Interferon Intended for Investigational Use in Humans (1983).
Points to Consider in the Production and Testing of New Drugs and Biologicals Produced by Recombinant DNA Technology (1985).
Guidelines for Research Involving Recombinant DNA Molecules (1986).
Points to Consider in the Manufacture and Testing of Monoclonal Antibody Products for Human Use (1987).
Points to Consider in the Characterization of Cell lines Used to Produce Biologicals (1987).
Guidelines on Validation of the Limulus Amebocyte Lysate Test as an End Product Endotoxin Test For Human and Animal Parenteral Drugs, Biological Products and Medical Devices (1987).
Guideline for Submitting Documentation for the Stability of Human Drugs and Biologics (1987).
Points to Consider in the Collection, Processing, and Testing of Ex-Vivo-Activated Mononuclear Leucocytes for Administration to Humans (1989).
Points to Consider in the Manufacture and Clinical Evaluation of in vitro tests to Detect Antibodies to Human Immunodeficiency Virus, Type I (1989) (Draft).
Cytokine and Growth Factor Prepivotal Trial Package: With a Special Emphasis on Products Identified for Consideration Under 21 CFR 312 Subpart E, "Drugs Intended to Treat Life-Threatening and Severly Debilitating Illnesses (1990) (Draft).

ing the product manufacture and characterization, and preclinical and clinical study design. These documents may be obtained at no charge from the Congressional, Consumer, and International Affairs Staff of the U.S. FDA (HFB-142), Metro Park North Room 109, 5600 Fishers lane, Rockville, MD 20857.

There are four phases involved in the development of a biological product: Pre-Investigational New Drug Application (Pre IND) phase, IND phase, Product License Application (PLA) and Establishment License Application (ELA) review and approval, and post licensing surveillance (phase IV). The concurrent filing of an Establishment License Application (ELA) is required for a new manufacturing facility for a biological product under PLA review to assure adequate control of the manufacturing.

The Investigational New Drug (IND) application is a proposal through which sponsors seek FDA's permission to begin initial testing of a biological product in human subjects. The revision of the IND regulations in 1987 (IND rewrite) for 21 CFR 312 established a specific format for IND applications and defined the requirement for submission of protocol modifications and informational amendments. This rewrite, in addition, defined the responsibilities of sponsors and investigators, and emphasized FDA-Sponsor communication during the IND process. Since for early phase studies safety is the primary concern, CBER has 30 days to review the application to ensure that the human subjects would not be exposed to unreasonable and significant risk.

A pre-IND meetind between sponsor and CBER is useful in resolving issues related to product characterization, preclinical testing, and clinical plan for an investigational biological agent.

A typical IND application contains information on product manufacturing and characterization, preclinical activity and toxicology data, and the proposed clinical protocol(s).

Product manufacturing and characterization

Sufficient information related to the product source material, purification process, and product characterization (potency, purity, identity) should be provided in an IND application to enable CBER to assess the safety and quality of an investigational biological product.

To demonstrate potency of a biological product, development of an adequate bioassay is essential. Use of an international (World Health Organization, National Institute for Biological Standards and Control, UK) or a national (National Cancer Institute, NIH, USA) reference standard for comparison is encouraged when such a standard is available. For products where reference standards are unavailable, an in-house reference standard should be established early in the development process.

Sterility determination is mandatory for injectable products. For topical products, sterility determination is highly desirable and bioburden determination is essential for safety evaluation. Requirements for sterility testing are outlined in 21 CFR 610.12 and methodologies for general safety determination have been provided in 21 CFR 610.11.

The identity of a biological product should be established using physicochemical methods such as polyacrylamide gel electrophoresis, high pressure liquid chromatography, peptide mapping, isoelectric focusing, N- and C- terminal sequencing and/or amino acid analysis. Reducing and non-reducing polyacrylamide gel electrophoresis should be performed to detect protein breakdown or protein aggregation, especially for products containing cysteine and cystine residues.

Stability determination should be made on the final formulation of an investigational agent and both physico-chemical methods as well as bioassay should be used in these studies. Expiration dating for a biological product should preferably be based on actual experimental data.

Contaminants in an investigational agent should be identified and minimal acceptable levels for such contaminants in each lot should be established. The presence of endotoxin, chemicals used during the purification process, nucleic acids, and foreign proteins are of concern (please refer to "Points to Consider in the Production and Testing of New Drugs and Biologicals Produced by Recombinant DNA Technology," 1985).

The presence of adventitious agents in the biological products derived from continuous cell lines, hybridomas, tissues and body fluids, poses a special concern and inclusion of viral inactivation and removal steps and validation of viral inactivation and removal during purification process is required. For products derived from human tissues, adequate screening of the donors for transmission of human diseases is essential (please refer to "Points to Consider in the Characterization of Cell Lines used to Produce Biologicals," 1987). Monoclonal antibodies used in affinity columns in the purification process will raise the same safety concerns as those involved with direct administration of such antibodies (please refer to "Points to Consider for Manufacture and Testing of Monoclonal Antibody Products for Human use," 1987).

Lot-to-lot consistency for the investigational agent should be demonstrated in the IND application prior to initiating clinical trials because the presence of variable levels of contaminants may compromise product safety and can alter immunogenicity. Variations in the product potency may, moreover, change the effective dose.

Preclinical studies

Toxicology studies should be performed preferably in two suitable animal species to justify the dose, range, route and duration of intended treatment in initial clinical studies. The species chosen should possess the functional receptor for the biological agent under study. Although non-human primate studies are not mandatory for biological products, such studies are useful in determining the possible product-related toxicities for species-specific proteins in human subjects. Dosing at the levels which produce toxicity in animal studies is also important in anticipating possible clinical toxicities. Preclinical animal studies should include appropriate controls and periodic determination of laboratory parameters including neutralizing antibody levels. Additional animal toxicology studies should be performed concurrent with phase I studies if prolonged treatment is anticipated in phase II or III studies. Preclinical teratogenicity and reproductive toxicology studies are recommended prior to licensure of a biological product.

In certain cases, if there is a reason to suspect that a biologic may possess carcinogenic potential, appropriate carcinogenicity studies may be necessary for complete product characterization if a suitable animal model is available.

Information on preclinical pharmacokinetics and pharmacodynamics for a product is essential to determine the optimal biological dose. For topical products, some information on the extent of absorption from the wound surface and tissue distribution is important. If the absorption data is unavailable, toxicology data after intravenous administration of the product may be acceptable.

Some preclinical efficacy data on a biological product is desirable prior to submission of an IND application. Animal efficacy studies in a disease model are useful in establishing an optimal dose range, route, and duration of treatment for a biologic in human subjects. The rabbit ear ulcer model and diabetic animal models for impaired wound healing are examples of some of the disease models that have been employed for evaluation of non-hematological growth factors as wound healing agents.

Clinical protocols

Phase I

These studies include initial introduction of drug in clinical trials. Phase I trials are designed to elucidate side effects associated with increasing doses of a biological product. Additional data may also be obtained from these studies on pharmacokinetics, absorption, metabolism, and distribution of such a product. These studies are performed with a limited number of human subjects. The population selected is critical and children are normally excluded from phase I studies. Phase I trials may be acceptable in patients if such patients have no other alternative therapy available and risk/benefit ratio for the product is favorable. Alternatively, phase I studies for some biological products such as growth factors are permitted in normal volunteers if the preclinical data indicate minimal risk. The clinical dosing in initial studies should be based on surface area or weight of a human subject for injectable products, and on unit wound area for topical products.

Phase II

These trials are performed in the patient population for whom such treatment is intended, and the main objective of these studies is to determine minimally toxic and maximally effective dose of a product and to determine appropriate clinical efficacy endpoints. Phase I and II populations should be carefully monitored for the development of neutralizing antibody to an investigational biologic, specifically if the desired treatment needs to be given on a chronic basis.

An end-of phase II meeting between sponsor and CBER is strongly recommended prior to initiation of phase III study to discuss validity of phase II data, product related issues and proposed Phase III plan.

Phase III

Phase III trials are conducted in the patient population for whom the biological product will be approved for marketing. These trials are conducted in a large number of patients and are adequate and well-controlled studies (21 CFR 314.126).

Typically, efficacy endpoints are classified into primary and secondary endpoints. Primary endpoint should indicate clear clinical effect for patients and a single primary endpoint should be specified in these studies. For phase III trials, sample size should be adequate to detect a statistically significant difference between control and treated population. Moreover, methodologies and hypotheses to be used for efficacy analyses should be provided in these protocols. More than one phase III trial is usually needed to conclusively demonstrate product efficacy.

Phase I–III trials may be placed on clinical hold for safety reasons and Phase II and Phase III studies, in addition, may be placed on clinical hold for study design issues.

Phase IV

Phase IV studies are initiated after product licensing and are usually designed to satisfy requests from CBER made prior to product licensure. These may include: alternative dose or schedule, different patient population not covered in phase I–III trial such as elderly or children, stage of disease, or prolonged exposure to the product to address long-term safety concerns if appropriate preclinical data are unavailable.

In the "Second U.S. FDA Workshop on Cytokines and Growth Factors" held at Bethesda, MD, USA, in June 1990, the discussions related to the design of the clinical trials for wound healing agents focussed on, time to complete healing, percentage of patients completely healed, or time to complete closure, as some of the useful primary efficacy endpoints. Other secondary endpoints which indicate clinical benefit such as frequency of infections or reduced antibiotic use, were also considered as important. Additionally, the panel members suggested that demonstration of durability and quality of healing and assessment of time to relapse, may be useful in determination of product efficacy in these types of studies. Regarding the methods for documenting wound healing, periodic assessment of wound volume, wound biopsies, quantitation of granulation tissue were thought to be useful criteria.

Conclusions

The development of an investigational biologic is a multistage process. Because of the diversity of such products, CBER reviews these products based on scientific rationale, case-by-case approach and risk/benefit assessment. Evaluation of safety and effectiveness of growth factors and related products performed by CBER is based upon the same regulations as those used for other biologicals; however, the specific concerns may vary depending on the nature and clinical application of these products.

References

1 Lynch SE, Colvin RB, Antoniades HN: Growth factors in wound healing. Synergistic effects of partial thickness porcine wounds. J Clin Invest 1989; 84: 640–646.
2 Foon KA: Biological Response Modifiers – The new immunotherapy. Cancer Res 1989; 49: 1621–1639.
3 Baird A, Mormede P, Bohlen P: Immunoreactive fibroblast growth factor (FGF) in a transplantable chondrosarcoma: Inhibition of tumor growth by antibodies. J Cell Biochem 1986; 30: 79–85.

4 Miossec P, Elhamiani P, Chichehian B, D'Angeac AD, Sany J, Hirn MJ: Interleukin 2 inhibitor in Rheumatoid synovial fluid: Correlation with prognosis and soluble IL-2 receptor levels. Clin Immunol 1990; 10(2): 115–120.
5 Mendelsohn J: Antiepidermal growth factor receptor monoclonal antibodies as potential anticancer agents. J Steroid Biochem Mol Biol 1990; 37(6): 889–892.
6 Laws enforced by the U.S. Food and Drug Administration. "Public Health Service Act. Biological Products," section 262 and 263. Washington D.C.: U.S. Government Printing Office, HHS Publication No. (FDA) 89–1051.
7 Food and Drug Administration Acts. "Federal Food, Drug and Cosmetic Act, as Amended and Related Laws", Chapter V, Washington D.C.: U.S. Government Printing Office, HHS publication No. (FDA) 89–1051.
8 Code of Federal Regulation, Title 21.1990. Chapter 1 – Food and Drug Administration. Washington D.C.: Office of the Federal Register National Archives and Records Administration.

Angiogenesis: Key Principles - Science - Technology - Medicine
ed. by R. Steiner, P. B. Weisz & R. Langer

Angiogenesis in MEDLINE and beyond

F. Grossenbacher[a], R. Steiner[c], A. Schlosser[b], A. Escher[a], G. K. Uhlschmid[c]

Documentation Service of the Swiss Academy of Medical Sciences (DOKDI), P.O. Box, CH-3000 Bern 9, Switzerland[a]; University Hospital Library, CH-8091 Zürich[b]; Department of Surgery, Research Division, University Hospital, CH-8091 Zürich, Switzerland[c]

I. INTRODUCTION

The scientific literature on angiogenesis is growing fast as shown in Fig. 1 for the years 1966 to 1990.

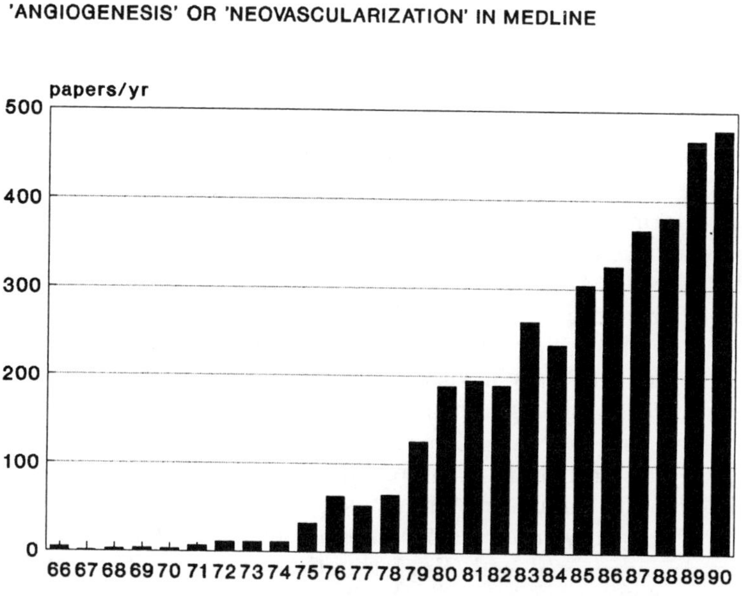

Figure 1. Papers on angiogenesis or neovascularization in MEDLINE between 1966 and 1990.

How to find the relevant references for your work in this rapidly expanding field? Traditional searching methods will probably not keep you up-to-date and you may miss important information by browsing through your favourite journals, books and annual reports. The aim of this article is to outline possible search strategies on angiogenesis in **MEDLINE** and **SciSearch.**

Computer aided literature retrieval serves as a convenient adjunctive way to rapidly access biomedical databases ONLINE via different gateway systems (hosts) or OFFLINE via compact disks (CD-ROM).

Important biomedical databases are:

	Producer:
– Medline	National Library of Medicine, Bethesda [NLM]
– Embase..............	Excerpta Medica, Amsterdam
– Biological Abstracts...	Biosis, Philadelphia
– SciSearch	Institute for Scientific Information, Philadelphia

Selected Hosts are:

- Data-Star
- Dialog
- BRS
- Dimdi
- NLM

All searches in this paper were done online in Medline or SciSearch on the computer host 'Data-Star'. It should be kept in mind, that the search language and commands are very different from host to host. Many features and capabilities of Medline in particular and of online-searching in general cannot be discussed in this short overview. For futher information do contact the professional searchers or reference librarians in your area.

MEDLINE (MEDical Literature Analysis and Retrieval System on-LINE) is produced and compiled by the National Library of Medicine in Bethesda MD, USA, and covers the whole field of medicine including dentistry and veterinary medicine, pharmacology, toxicology, biochemistry, medical psychology, microbiology, molecular biology, biomedical technology, and others.

Over 3200 journals from more than 70 countries are regularly indexed. The hole database contains over 6 million references from 1966 todate. The yearly input reaches about 300,000 articles. 70% of the documents are stored with English author abstracts. The database is updated weekly.

SAMPLE DOCUMENT:

AN 91215195.
AU Folkman-J.
IN Children's Hospital, Boston, MA 02115.
TI How the field of controlled-release technology began, and its central role in the development of angiogenesis research.

SO Biomaterials 1990 Nov, VOL: 11 (9), P: 615–8, ISSN: 0142-9612 35
Refs.
LG EN.
PT HISTORICAL-ARTICLE; JOURNAL-ARTICLE; REVIEW; REVIEW-TUTORIAL.
DE ANIMAL;
BOSTON;
CELLS-CULTURED;
DELAYED-ACTION-PREPARATIONS/*HI (history);
DRUG-IMPLANTS/HI (history);
HISTORY-OF-MEDICINE-20TH-CENT;
NEOVASCULARIZATION/*PP (physiopathology);
POLYMERS/CH (chemistry);
RESEARCH.
AB The first controlled-release methodology was developed in 1962 and was based upon diffusion of small molecules (molecular weight less than 500) through the wall of silicone rubber tubing. The problem of controlled-release of large molecules such as proteins was not solved until more than a decade later, when it was discovered that microchannels formed around lyophilized proteins embedded in polymers such as ethylene vinyl acetate copolymer (EVA) and poly (hydroxy ethylmethacrylate) (pHEMA), as water permeated through the polymer. These studies led to the development of a 5-year contraceptive implant, e.g., Norplant; rapid progress in the field of angiogenesis research; and a method for regulating cell shape and growth control in vitro. Author.
YR 90.
ZN Z1-542-363-300.
JC A4P.
IM 9108.
ED 910604.

Abbreviations

Label	*Contents*	*Example*
AN	Accession number of the NLM	91215195.
AU	Author(s)	Folkman-J.
IN	Institution (Author affiliation)	Children's Hospital, Boston, MA 02115.
TI	Title of article	How the field . . .
SO	Source	Biomaterials.
LG	Language of article	LG = EN

PT	Publication type	Historical-Article; Review.
DE	Descriptors	Neovascularization; Drug-Implants/HI (history)
AB	Abstract	The first control . . .
YR	Year of publication	90.
ZN	Country of publication of the journal	Z1-542-363-300.
JC	Journal coden	A4P.
IM	Index Medicus month	9108.IM.
ED	Entry date (NLM Computer)	910604.ED.

BOOLEAN OPERATORS:

You will always have the possibility to combine your search terms with the following operators:

ANGIOGENESIS *and* NEOVASCULARIZATION
. . . . *or* . . . .
. . . . *not* . . . .
. . . . *same* . . . .
. . . . *with* . . . .

Example:

(ANGIOGENESIS **or** NEOVASCULARIZATION) **and** (NEOPLASMS) **not** (TRANSPLANTATION)

MEDLINE THESAURUS:

The Medline thesaurus consists of over 15,500 controlled headings, which are currently used for indexing Medline references. It is recommended to carry out bibliographic searches in Medline using these controlled headings or **MeSH** terms (**Me**dical **S**ubject **H**eadings) rather than just entering free text terms or keywords.

For example, using the **MeSH** term **NEOPLASM** avoids problems with synonyms such as CANCER/TUMOR/TUMOUR. All MeSH terms are arranged in subject groups which in turn are organized hierarchically in so-called 'tree structures'. The alphabetical and hierarchical list of the MeSH is available in bookform, online or offline on CD-ROM.

1) Entry of **ANGIOGENESIS** and **NEOVASCULARIZATION** in the annotated list of MEDICAL SUBJECT HEADINGS:

ANGIOGENESIS see NEOVASCULARIZATION

ANGIOGENESIS FACTOR
D11.303.132 D24.185.348.157
D24.185.526.198
91; was see under GROWTH SUBSTANCES 1973–90

NEOVASCULARIZATION
C23.739.668
80
X ANGIOGENESIS

i.e. all articles on angiogenesis are indexed under the MeSH term **NEOVASCULARIZATION**!

2) Online MeSH terms: **ANGIOGENESIS** and **NEOVASCULARIZATION**:

D–S – SEARCH MODE – ENTER SEARCH TERMS
1_: . .tree R2 = *hierarchical listing of the term angiogenesis-factor*

[explanations (not used for searching) are italicised in the following]

			year of entry into MeSH:
T1	#	. GROWTH-SUBSTANCES-PIGMENTS-VITAMINS-NON-MESH	
T2	#	.. GROWTH-SUBSTANCES	
T3		**... ANGIOGENESIS-FACTOR**	***'91***
T4	#	. IMMUNOLOGIC-AND-BIOLOGIC-FACTORS-NON-MESH	
T5	#	.. BIOLOGICAL-FACTORS	*'89*
T6	#	... GROWTH-SUBSTANCES	
T7		.... ANGIOGENESIS-FACTOR	*'73*
T8	#	... MITOGENS	*'72*
T9	#	.... ANGIOGENESIS-FACTOR	*'73*

D–S – SEARCH MODE – ENTER SEARCH TERMS
1_: T3 *Angiogenesis-factor*
RESULT 177 *number of hits*
2_: . .d all

QN	DOCS	SEARCH TERMS	
1	177	(ANGIOGENESIS-FACTOR.DE. .W.)	*all entries on angiogenesis-factors*

END OF DISPLAY

II. SEARCH STRATEGIES IN MEDLINE

A. HOW TO KEEP A BROAD PERSPECTIVE ON ANGIOGENESIS

D–S – SEARCH MODE – ENTER SEARCH TERMS
1_: ↓. .mesh angiogenesis↓ *MeSH-term for angiogenesis*
THES ANGIOGENESIS$ *all entries on angiogenesis*
R1 ANGIOGENESIS –→ NEOVASCULARIZATION
R2 3 ANGIOGENESIS-FACTOR *all entries on angio-factors*
END OF THES

D–S – SEARCH MODE – ENTER SEARCH TERMS

1_: ↓R1 OR R2↓	*all entries on angiogenesis or neovascularization*
RESULT 1536	*number of hits*

2_: ↓. .d all↓

QN DOCS SEARCH TERMS

RESULT 1536 (*NEOVASCULARIZATION.DE. .W.*) (*ANGIOGENESIS-FACTOR.DE. .W.*)

B. HOW TO LIMIT THE PREVIOUS SEARCH TO REVIEW ARTICLES ONLY

PT REVIEW 2_: ↓1 AND REVIEWS$1.PT,DE.↓	*all review articles on angio or neovasc [1:and]*
RESULT 146	*number of hits*

Sample of references found:

AN 91105850.
AU Liotta-L-A, Steeg-P-S, Stetler-Stevenson-W-G.
IN Laboratory of Pathology, National Cancer Institute, National Institutes of Health, Bethesda, Maryland 20892.
TI Cancer metastasis and angiogenesis: an imbalance of positive and negative regulation.
SO Cell 1991 Jan 25, VOL: 64 (2), P: 327–36, ISSN: 0092–8674 113 Refs.
LG EN.
PT REVIEW.
AN 90304162.

AU Klagsbrun-M, D-Amore-P-A.
IN Laboratory for Surgical Research, Children's Hospital, Boston, Massachusetts.
TI Regulators of angiogenesis.
SO Annu-Rev-Physiol 1991, VOL: 53, P: 217–39, ISSN: 0066–4278 140 Refs.
LG EN.
PT REVIEW

C. HOW TO BROWSE THROUGH TITLES

It is possible to limit the display to fields you would like to browse by marking the title field [**ti**] only:

D–S – SEARCH MODE – ENTER SEARCH TERMS

3_: ↓. .print **ti**/1–10↓	*print first ten titles on angiogenesis-factors only*

SAMPLE OF FIRST TITLE:

1

TI Selective inhibition by magnosalin and magnoshinin, compounds from "Shin-i" (Flos magnoliae), of adjuvant-induced angiogenesis and granuloma formation in the mouse pouch.

C. HOW TO SCAN ANGIOGENESIS RESEARCH IN DIFFERENT BIOMEDICAL FIELDS

By checking relevant descriptor terms which can serve as guides for improved search strategies.

SEARCH PROFILE USED FOR AN ONLINE-SEARCH:

QN	DOCS	SEARCH TERMS	*Question No. 1*
1	1723	ANGIOGEN$.TI,DE. OR NEOVASCULARIZATION.TI,DE.	*Root angiogenesis either in title or in MeSH*
2	6703	**WOUND-HEALING #**	*all records on*
3	50	1 AND 2.MJ	*angio & woundheal.*
4	17	..LIMIT H = Y	*human only*
5	38980	**EYE-DISEASES #**	*ophthalmology*
6	556	1 AND 5.MJ.	*and eye diseases*
7	495	..LIMIT H = Y	*human only*
8	40	7 AND REVIEW$1.PT,DE................	*reviews only*
9	236662	NEOPLASMS #	*all tumors*
10	283	1 AND 9.MJ.	
11	184	..LIMIT H = Y	
12	37	11 AND REVIEW$1.PT,DE.	
13	54237	**SKIN-DISEASES #**	
14	49	1 AND 13.MJ	*angio & skin*
15	35410	**TRANSPLANTATION.DE.**	
16	29	1 AND 15.MJ	*angio & transpl.*
17	6	..LIMIT H = Y	

END OF DISPLAY

Sample references referring to question 12: Number of hits: 37 *angiogenesis and neoplasms limited to reviews only*

SAMPLE OF REFERENCES FOUND:

AU Blood-C-H, Zetter-B-R.
IN Department of Surgery, Children's Hospital, Harvard Medical School, Boston, MA 02115.
TI **Tumor** interactions with the **vasculature**: **angiogenesis** and **tumor metastasis.**

SO Biochim-Biophys-Acta 1990 Jun 1, VOL: 1032 (1), P: 89–118, ISSN: 006-3002 614 Refs.
LG EN.
PT **REVIEW**; REVIEW-ACADEMIC.
DE AMINO-ACID-SEQUENCES;
ANIMAL;
HUMAN;
MOLECULAR-SEQUENCE-DATA;
NEOPLASM-CIRCULATING-CELLS;
NEOPLASM-METASTASIS/*;
NEOPLASMS/*BS (blood supply);
NEOVASCULARIZATION/*PP (physiopathology).

AN 86079123.
AU Folkman-J.
IN Department of Surgery, Harvard Medical School, Boston, Massachusetts 02115.
TI How is **blood vessel** growth regulated in normal and **neoplastic** tissue?
G.H.A. Clowes memorial Award lecture.
SO Cancer-Res 1986 1986 Feb, VOL: 46 (2), P: 467–73, ISSN: 0008-5472 45, Refs.
LG EN.
PT **REVIEW.**
DE ANGIOGENESIS-FACTOR/*IP (isolation & purification),
PH(physiology);
ANIMAL;
CAPILLARIES/GD (growth & development);
ENDOTHELIUM/*CY (cytology);
FIBROBLAST-GROWTH-FACTOR/PH (physiology);
GROWTH-SUBSTANCES/*IP (isolation & purification);
HEPARIN/PD (pharmacology), *PH (physiology);
HUMAN;
MAST-CELLS/PH (physiology);
NEOPLASMS/*BS (blood supply);
NEOVASCULARIZATION/*;
STEROIDS/PD (pharmacology);
STRUCTURE-ACTIVITY-RELATIONSHIP;
SUPPORT-NON-U-S-GOVT;
SUPPORT-U-S-GOVT-P-H-S.

E. HOW TO ADRESS SPECIFIC QUESTIONS

Example: Which of the growthfactors are involved in angiogenesis?

A POSSIBLE SEARCH PROFILE COULD BE:

QN	DOCS	SEARCH TERMS
1	2010	ANGIOGEN$.TI,DE. OR NEOVASCULARIZATION
2	2039	TRANSFORMING WITH GROWTH WITH FACTOR$1.TI,DE.
3	35	**1 AND 2**
4	447	BASIC WITH FIBROBLAST WITH GROWTH WITH FACTOR$1.TI,DE.
5	49	**1 AND 4**
6	114	PLATELET WITH FACTOR WITH ‘4’.TI., DE.
7	3	**1 AND 6**
8	4315	EPIDERMAL WITH GROWTH WITH FACTOR$1.TI,DE.
9	26	**1 AND 8**
10	53	ANGIOGENIN.TI.
11	53	**1 AND 10**
12	729	LAMININ.TI.
13	8	**1 AND 12**
14	194	THROMBOSPONDIN.TI.
15	5	**1 AND 14**

END OF DISPLAY

E. HOW TO ASK OPEN QUESTIONS

Example: Do Retinoids play a role in angiogenesis?

SEARCH PROFILE USED:

QN DOCS SEARCH TERMS
1 1723 ↓ANGIOGEN$.TI,DE. OR NEOVASCULARIZATION.TI,DE.↓
D–S – SEARCH MODE – ENTER SEARCH TERMS
2_: ↓. .MESH RETINOIDS↓

THES RETINOIDS$
R1 # 3 RETINOIDS '84

D–S – SEARCH MODE – ENTER SEARCH TERMS
2_: ↓. .TREE R1↓ = *hierarchical listing of the category RETINOIDS in the online-MeSH*

D–S – SEARCH MODE – ENTER SEARCH TERMS

2_: ↓ R1 # ↓		*MeSH Retinoids*
RESULT	13931	*Number of hits*
3_: ↓1 AND 2↓		*Neovasc and all Retinoids*
RESULT	13	*Number of hits*

SAMPLE OF TWO REFERENCES FOUND:

AN 90314724.
AU Majewski-S, Wolska-H, Jablonska-S, Wcasik-M.
IN Department of Dermatology, School of Medicine, Warsaw.

TI Effects of systemic **etretinate** treatment on natural cytotoxicity, immune **angiogenesis** and neutrophil adherence in patients with various forms of **psoriasis**.
SO Arch-Immunol-Ther-Exp-(Warsz) 1989, VOL: 37 (3–4), P: 459–64,
ISSN: 0004-069X.
LG EN.
DE ADULT;
ETRETINATE/*TU (therapeutic use);
FEMALE;
HUMAN;
KILLER-CELLS-NATURAL/*IM (immunology);
MALE;
MIDDLE-AGE;
NEOVASCULARIZATION/*PA (pathology);
NEUTROPHILS/*IM (immunology);
PSORIASIS/*DT (drug therapy), IM (immunology);
SUPPORT-NON-U-S-GOVT.
RN 54350-48-0 (Etretinate).
AB The effects of **etretinate** treatment on natural killer cell activity, on **angiogenic capibility** of peripheral blood mononuclears and on neutrophil adherence were studied in 20 patients with various forms of **psoriasis**. . .

AN 90058469.
AU Oikawa-T, Hirotani-K, Nakamura-O, Shudo-K, Hiragun-A, Iwaguchi-T.
IN Division of Cancer Therapeutics, Tokyo Metropolitan Institute of Medical Science, Japan.
TI A highly potent **antiangiogenic** activity of **retinoids**.
SO Cancer-Lett 1989 Nov 30, VOL: 48 (2), P: 157–62, ISSN: 0304-3835.
LG EN.
DE ANIMAL;
CHALCONE/PD (pharmacology);
CHICK-EMBRYO;
DOSE-RESPONSE-RELATIONSHIP-DRUG;
IN-VITRO;
NEOVASCULARIZATION/*;
QUINONES/PD (pharmacology);
RETINOIDS/*PD (pharmacology);
SUPPORT-NON-U-S-GOVT;
TRETINOIN/PD (pharmacology);
VITAMIN-A/AA (analogs & derivatives), PD (pharmacology).

RN **11103-57-4 (Vitamin-A);**
127-47-9 (retinol-acetate);
302-79-4 (Tretinoin);
70563-58-5 (herbimycin);
94-41-7 (Chalcone);
95906-67-5 (3,5-di-tert-butylchalcone-4′-carboxylic-acid).

AB Four **retinoids, i.e. retinol (vitamin A), retinoic acid, retinyl acetate and synthetic chalcone carboxylic acid (Ch 55),** were examined for their effects on **embryonic angiogenesis** using 4.5-day **chorioallantoic membranes of chick embryo**. . . .

F: HOW TO SEARCH FOR YOUR COLLEAGUES WORK

Example: Which papers were published by Robert Langer in 1976?

1_: ↓. .ROOT LANGER-R↓

	ROOT LANGER-R$	*all Langer-R in Medline*
R1	267 DOCS	LANGER-R
R2	5 DOCS	LANGER-R-D
R3	1 DOC	LANGER-R-H
R4	5 DOCS	LANGER-R-M
R5	1 DOC	LANGER-R-R
R6	14 DOCS	LANGER-R-S
R7	1 DOC	LANGER-R-S-JR
R8	1 DOC	LANGER-RUHL-H

END OF ROOT

D–S – SEARCH MODE – ENTER SEARCH TERMS

1_: ↓R1↓

RESULT	*267*	*number of Langer-R publications*
2_: ↓1 AND YR = 76↓		*number of Langer-R publications in 1976*
RESULT	*3*	*number of hits*

Two documents displayed:

1
AN 76224366.
AU **Langer-R**, Brem-H, Falterman-K, Klein-M, Folkman-J.
TI Isolations of a cartilage factor that inhibits tumor neovascularization.
SO Science **1976 Jul 2**, VOL: 193 (4247), P: 70-2, ISSN: 0036-8075.

2
AN 77056397.
AU **Langer-R**, Folkman-J.

IN Department of Surgery, Children's Hospital Medical Center, Boston, Massachusetts.
TI Polymers for the sustained release of proteins and other macromolecules.
SO Nature **1976 Oct 28**, VOL: 263 (5580), P: 797–800, ISSN: 0028-0836.

G. HOW TO LOOK AT THE IMPACT OF A PARTICULAR PUBLICATION IN THE CITATION INDEX

On the database **SciSearch (SCIN)** the citation rate of the first author of a given paper can be established from 1980 onwards. You can start with an important paper published in the past and find more recently published papers that cite it. Thus you can study the evolution of concepts and ideas.

Example 1: How many times has the above shown 1976 Nature paper of Robert Langer been cited from 1980 up to 1991?

D–S – SEARCH MODE – ENTER SEARCH TERMS
1_: ↓. .ROOT LANGER-R-1976$
↓

	number of citations 1980 to date:	code for papers:
R1	1 DOC	LANGER-R-1976-V19-P70
R2	26 DOCS	LANGER-R-1976-V193-P70
R3	1 DOC	LANGER-R-1976-V194-P70
R4	2 DOCS	LANGER-R-1976-V201-P70
R5	**75 DOCS**	**LANGER-R-1976-V263-P797**
R6	1 DOC	LANGER-R-1976-V63-P797
R7	1 DOC	LANGER-R-1976-V93-P70

END OF ROOT

i.e. the Langer paper has been cited in 75 articles.

R5 LANGER-R-1976-V263-P797 = AN 77056397.

Example 2: Who cited Langer's 1976 Nature paper in the years 1991/1990?

SAMPLE OF 2 DOCUMENTS FOUND:

1 SCIN
AN 384020455 9100.
AU RIBATTI-D, VACCA-, RONCALLI-L, DAMMACCO-F.
TI ANGIOGENESIS UNDER NORMAL AND PATHOLOGICAL CONDITIONS.
SO HAEMATOLOGIA, 1991, V76, N4, P311-320.

LG EN.
PT R: REVIEW.
NU 120 cited references.

21 SCIN
AN 349009349 9028
AU JONES-N-S, GLENN-M-G, ORLOFF-L-A, MAYBERG-M-R.
TI PREVENTION OF MICROVASCULAR THROMBOSIS WITH CONTROLLED-RELEASE TRANSMURAL HEPARIN.
SO ARCHIVES-OF-OTOLARYNGOLOGY-HEAD-NECK-SURGERY, 1990, V116, N7, P779-785.
LG EN.
NU 27 cited references.

III. CONCLUDING REMARKS

A few rules for searching in MEDLINE and beyond:

There are many different ways of doing searches on biomedical databases whether ONLINE or OFFLINE, but a few general rules may apply to all:

- Formulate your question always as clear as possible.
- Make a search concept before you start.
- Connect the concepts with logical operators (AND, OR).
- Look up the thesaurus for Medical Subject Headings (MeSH-terms).
- Choose the appropriate database according to the questions you are asking.
- Make yourself familiar with the features and structures of the database and the gateway system (host) you will use.
- Use limit options to restrict your search (how wide and how far back should your search go?).
- Be aware of the costs of searching online!
- Save the most successful search strategy as your 'profile' (this is possible for many CD-ROM systems and as SDI system).
- Decide on how to record your search (i.e. on personal filing systems like *Reference Manager*, *ProCite*, *Endnote* etc.).
- Consult a professional searcher for advice.

The Medline database is available in many libraries for local offline searching. These CD-ROM modes reduce the cost of searching and make searches easier for novices. But the full potential of any biomedical database is usually only realized through the services of professional searchers in regional or national literature retrieval centers. Feel free to contact them *personally* for advice on how to optimize your search strategies.

Through **SDI** (**S**elected **D**issemination of **I**nformation) references for each update are searchable. It is possible to give **standing SDI orders** to the host. This service may enhance your awareness for what is happening in your specified field, it can also be useful to someone who has completed major literature searches in INDEX MEDICUS or SCIENCE CITATION INDEX before starting research and wants to keep up-to-date on relevant papers as she or he proceeds with her or his research.

Last but not least . . . do not forget to **search for books on angiogenesis**! Books are **not** listed in Medline. You can ask your reference librarian to search for you in the database **'BOOKS IN PRINT'** or **'Catline'**.

The following books on ANGIOGENESIS were found in **'Catline'**:

1. ANGIOGENESIS: mechanisms and pathobiology. Ed. by DB Rifkin & M Klagsbrun. Cold Spring Harbor Laboratory, N.Y., 1987, 161 pp.
2. OCULAR CIRCULATION AND NEOVASCULARIZATION. Ed. by D BenEzra. Nijhoff, Boston, 1987, 573 pp.
3. ANGIOGENESIS: the growth of the vascular system. O Hudlicka & KR Tyler. London, Academic Press, 1986, 221 pp.
4. ANGIOGENESIS: Proceedings of the British Microcirculation Society. Ed. by F Hammersen & O Hudlicka. Karger, Basel, 1984, 90 pp.
5. TUMOR ANGIOGENESIS. AS Fenselau. U.S. Dept. of Health & Human Services, Public Health Service, NIH, International Cancer Research Data Bank Program, Springfield, 1983, 46 pp.
6. TUMOR BLOOD CIRCULATION: angiogenesis, vascular morphology and blood flow of experimental and human tumors. Ed. by H-I Peterson, CRC Press, 1979, 229 pp.

Selected readings on literature retrieval:

- Huth JE: Searching the Literature. In: How to Write and Publish Papers in the Medical Sciences, by Edward J Huth, 2nd. ed., Williams & Wilkins, Baltimore, 1990, pp. 12–54.
- Warling B, Gilman LB: Manual versus MEDLINE searches [letter]. Am J Psychiatry 1991; 148: 686–687.
- Grefsheim SF, Shipman BL, Schwartz DG, Dow SC: Searching Medline at the University of Michigan. MD-Comput 1991; 8: 32–36.
- Mitchell JA, Johnson ED, Proud VK: New thoughts about medical students as effective searchers of MEDLINE. Acad Med 1990; 65: 434–37.
- Oncology Overview on Antiangiogenesis. National Cancer Institute [NCI], Cancer Information Dissemination and Analysis Center, in press, 1991.

Acknowledgement

We owe a special debt to all those who whole-heartedly supported this book to be published by providing a manuscript despite their many other pressures of the day.

We wish to thank Douglas A. Lauffenburger, Professor of Chemical Engineering and Cell & Structural Biology, University of Illinois at Urbana-Champaign, for his unselfish help in support of the symposium.

Special thanks are due to Dr. Martin Büechi, Attaché for Science and Technology at the Embassy of Switzerland, Washington D.C. for his public relations activities overseas; to Prof. Dr. med. Friedbert Scharfetter and to PD Dr. med. Thomas Sigrist, Kantonsspital St. Gallen, who facilitated the final preparatory phase of the meeting.

We are indebted to Pamela D. Brown (MIT, Cambridge, MA); to Marianne Mock and to Jacqueline Blind-Fahrni (St. Gallen) who assisted the operation with secretarial competence.

We thank Dr. Andreas Bally, Biology Editor of Birkhäuser Verlag, for his courage to produce this volume as an 'outstanding' item in the 'Experientia Supplementa' series.

We are grateful to open-minded trustees of the American National Science Foundation, the Swiss Society for Oncology, the Swiss Cancer League, the Cancer Leagues of Bern, Zürich and St. Gallen/Appenzell, and the Mobil Foundation (USA).

We thank the following American, European and Japanese industries who gave financial support in the initial stages of the project or helped to lessen the financial burden in the aftermath of the scientific success of the first 'International Symposium on Angiogenesis': Abbott (Schweiz) AG, Alza Corp. (USA), Bayer (Deutschland/Schweiz) GmbH, Becton Dickinson (Schweiz) AG, BMA Biomedicals (Schweiz) AG, Boehringer Ingelheim (Schweiz) AG, Du Pont De Nemours (USA), Farmitalia Carlo Erba (Italy), Gibco BRL (European Division), Glaxo (Schweiz) AG, Haemonetics SA (Switzerland), Hewlett Packard (Schweiz) AG, F. Hoffmann-La Roche AG (Basel), Immuno (Schweiz) AG, Johnson & Johnson Medical (Schweiz) AG, Lederle Cyanamid (Schweiz) AG, Kontaktgruppe für Forschungsfragen [KGF] (Ciba-Geigy/Roche/Lonza/Sandoz), Mercedes-Benz (Schweiz) AG, Merck & Co. Inc. (USA), Novo Nordisk Pharma (Schweiz) AG, Pfizer (Schweiz) AG, Roussel (Suisse) AG, Takeda Chemical Industries, LTD (Japan), Upjohn (USA).

Finally, the book would never have materialized without the selfless interest of Hans-Peter Kaeser, Typographer SWB, and his professional advice every inch of the way from planning and arranging the symposium to creating this record of a memorable scientific event which took place at the Hochschule St. Gallen on March 13 to 15, 1991. It was an instructive experience to work with him who took up and approved some of our whims (such as incorporating the schematic diagram of the human cranium taken from Paul Klee's painting "One who understands" [1934] as an overlay to angiogenesis in vitro) and just as judiciously discouraged others.

The Editors

Author Index